STATISTICAL MECHANICS

OF PHASES AND PHASE TRANSITIONS

STATISTICAL MECHANICS

OF PHASES AND PHASE TRANSITIONS

Steven A. Kivelson • Jack Mingde Jiang • Jeffrey Chang

PRINCETON UNIVERSITY PRESS

PRINCETON AND OXFORD

Published by Princeton University Press
41 William Street, Princeton, New Jersey 08540
99 Banbury Road, Oxford OX2 6JX

press.princeton.edu

All Rights Reserved

ISBN 978-0-691-24974-2
ISBN (pbk.) 978-0-691-24973-5
ISBN (e-book) 978-0-691-24972-8

British Library Cataloging-in-Publication Data is available

Editorial: Ingrid Gnerlich and Whitney Rauenhorst
Production Editorial: Jill Harris
Cover Design: Wanda España
Production: Jacqueline Poirier
Publicity: William Pagdatoon
Copyeditor: Bhisham Bherwani

Cover image: Visualization of local Ising-nematic correlations obtained from scanning tunneling microscopy on the surface of a cuprate high-temperature superconductor—a real-world manifestation of the random-field Ising model. Image courtesy of C. L. Song, E. J. Main, F. Simmons, S. Liu, B. Philabaum, K. A. Dahmen, E. W. Hudson, J. E. Hoffman, and E. W. Carlson.

This book has been composed in MinionPro and Avenir Next

Printed on acid-free paper. ∞

Printed in China.

10 9 8 7 6 5 4 3 2 1

CONTENTS

5 Broken Symmetries 106

PREFACE

This book is intended for use in a second quarter undergraduate physics course on statistical mechanics. It is assumed that the student has already encountered the foundations of thermodynamics and statistical mechanics. These prerequisites are reviewed in Chapter 2—for example, the laws of thermodynamics and the formal relation between the Helmoltz free energy and the partition function of statistical mechanics. Likewise, some computational methods from quantum mechanics and complex analysis are used, though we have included succinct reviews of these topics so as to be mostly self-contained. The goal here is to obtain an understanding of how these formal definitions can be analyzed to yield insight into the real-word physics that characterizes distinct phases of matter and the nature of the phase transitions between them.

The most satisfying perspective on these problems involves considerations that are conceptually deep and profoundly abstract. In our experience, it is easier to master these considerations by focusing on specific examples and carrying out explicit calculations. For this reason, a large portion of this text involves an analysis of a simple model problem: the Ising ferromagnet. We will analyze this model using a variety of increasingly sophisticated approximate mean-field theories, controlled asymptotic expansions, and in a few fortunate cases, exact solutions. Whenever we adopt a more general viewpoint—based on symmetry or (in the final chapter) the renormalization group—we always refer to the simpler, more explicit calculations carried out in the earlier chapters.

The first seven chapters form a coherent introduction to the basic ideas of phase transitions and critical phenomena, while Chapters 8, 9, and 10 sketch out advanced topics paving the way for further studies. The final three chapters and a few sections in earlier chapters are marked with asterisks (*) to denote that they discuss more advanced topics that can be omitted without loss of continuity.

This text is designed for an active-learning approach, where emphasis is placed on students working through the material on their own, rather than passively digesting received wisdom. Therefore, the main text at times abbreviates computational details in favor of continuity of discussions and more physical interpretations. At the end of each chapter, the omitted calculations are presented as guided steps in worksheets. The worksheets are designed to be challenging but not opaquely so, such that they may be worked through at a reasonable pace. They can be used as-is, or as templates for instructor-designed work-sheets to be worked on collaboratively by small groups of students in class. Editable LaTeX versions of worksheets, printable handouts, and associated supplementary resources are available on our accompanying website at https://press .princeton.edu/books/paperback/9780691249735/statistical-mechanics-of-phases -and-phase-transitions. All readers are strongly encouraged to work through the worksheets, as they are integral to the pedagogical vision of the authors.

We feel that one special aspect of this book stems from how it came to be written and by whom. It was based on a one quarter course in which Chang (then a junior majoring in physics) was a student, Jiang (then a graduate student in physics) was the head TA, and Kivelson (then and still a professor of physics) was the instructor. In the several years it has taken us to complete this text, we have debated over and over the relative merits of optimal concision versus expanded discussions of the intuitive underpinnings. Both Chang and Jiang were unified in opposing leaping over complex manipulations with the promise that "it is obvious that . . ." (Admittedly, sometimes it is necessary to turn to the worksheets for a complete treatment of all the intermediate steps.) We hope that the compromises we have reached will make the discussion uniquely accessible to serious students.

This book covers some of the most beautiful and most profound intellectual constructs in human history. However, the material is not simple. Thermodynamics is mathematically straightforward, but is extraordinarily abstract—it is hard, at first, to distinguish many of its most powerful principles from tautologies. Statistical mechanics, by contrast, is unimaginably computationally challenging—the difficulty here is to obtain the essential physics without getting lost in details. But in both cases, the rewards in understanding and insight are well worth the effort.

Steve A. Kivelson, Jack M. Jiang, and Jeffrey Chang

A NOTATIONAL NOTE

In this book, we use the following conventions:

- The Boltzmann constant, k_B, is set to 1, so that temperature is measured in units of energy and entropy is unitless.
- Unless otherwise stated, we work in the canonical ensemble, i.e., fixed temperature, volume, and particle number. The capital letter F represents the Helmholtz free energy and is called simply the "Free Energy."
- Wherever Fourier transforms are performed, we use the non-unitary angular frequency convention, i.e.,

$$\mathcal{F}[f(x)] = \tilde{f}(k) = \int_{-\infty}^{\infty} e^{-ikx} f(x)\,dx$$

$$\mathcal{F}^{-1}[\tilde{f}(k)] = f(x) = \frac{1}{2\pi} \int_{-\infty}^{\infty} e^{ikx} \tilde{f}(k)\,dk.$$

- Greek indices $(_{\alpha\beta\gamma...})$ are used for vector components, whereas Latin indices $(_{ijk...})$ are used to label sites on a lattice, i.e., discrete points in space.
- A spatial vector is labeled with an arrow, $\vec{r}$, whereas an order parameter or spin vector is bolded, $\mathbf{S}$.
- We use the uppercase M to represent the total magnetic dipole moment (an extensive quantity), and lowercase m to represent the magnetization, and the magnetic moment density (an intensive quantity). The integral of m over the volume of the sample gives the total magnetic moment M. This is opposite to the convention of the typical symbols used in other disciplines of physics (especially electromagnetism).
- We make use of Einstein summation notation where repeated indices in a term implies a summation unless otherwise noted, e.g.:
  A matrix multiplication $AB = C$ is represented as

$$A_{ij}B_{jk} \equiv \sum_j A_{ij}B_{jk} = C_{ik}$$

- The improper integrals involving infinities are written *sans rigueur*, e.g.:

$$\int_{-\infty}^{\infty} \text{ or } f(x)\big|_{-\infty}^{\infty} \text{ to mean } \lim_{a\to\infty} \int_{-a}^{a} \text{ or } \lim_{a\to\infty} f(x)\big|_{x=-a}^{x=a}$$

ACKNOWLEDGEMENTS

We would like to acknowledge helpful input from Ben Feldman and Sophia Kivelson. We benefited from comments on the manuscript from Margaret Kivelson, Gilles Tarjus, Stephen Shenker, Zhaoyu Han, Patricia Burchat, Anthony Zee, Katherine Xiang, Aaron Lin, Spencer Guo, Jordi Montana, and Timothy Guo. We have also benefited greatly from the participation of the many students in Physics 171 at Stanford University. We are grateful to Katherine Xiang for drafting Figure 5.3. Steven Kivelson would like to acknowledge many years of support through research grants from the Division of Materials Research of the National Science Foundation.

STATISTICAL MECHANICS

MECHANICS

OF PHASES AND PHASE TRANSITIONS

1 OVERVIEW

In our experience, the study of statistical mechanics elicits a transition from a phase of ignorance to a phase of bliss.

The existence of phase transitions is a remarkable fact of nature. Unlike most other phenomena in physics, phase transitions can involve a dramatic change in the properties of a material. Water can boil. Steam can condense back into water. A piece of iron can become magnetized. And most pure metals, when cooled sufficiently close to absolute zero, gain the ability to superconduct—to conduct electricity without any resistance at all.

Phase transitions are among the clearest and best understood examples of emergent properties—properties that involve the collective behavior of many constituent parts. When water boils, nothing intrinsic about its molecules is affected. Studying the details of the molecular interaction does not lead to insight about the nature of boiling. To properly appreciate the striking transformation from liquid to gas, a macroscopic point of view is required—a view that takes into account the statistical properties of the vast numbers of water molecules and their interactions with one another. This is the approach we will take in this book.

In this chapter, we will start with a bird's-eye overview about phases of matter, their transitions, and the way they are characterized. We begin with a qualitative description. Along the way, we will see how ideas from statistical mechanics can help us build an understanding of these phenomena. We assume that students have already been introduced to the basic ideas of statistical mechanics—the prerequisite knowledge is reviewed in Chapter 2.

1.1 Phases

In childhood we are taught of three phases of matter: solid, liquid, and gas. Each of these phases is characterized by a distinct set of physical properties: a gas is highly compressible, a liquid has fixed volume but still can flow, and a solid is rigid.[1] These physical qualities are emergent—an individual molecule cannot "flow" or "be rigid";

---

1. As we will discuss in Section 1.2.2, liquids and gases are not truly distinct phases of matter.

only a large collection of molecules can. A phase of matter is a collective effect that only makes sense if there are a lot of particles.

There are, in fact, many more phases of matter beyond these three, each with its own fascinating and mysterious properties. The superconducting phase has zero electrical resistance; a superfluid flows without any viscous damping. In ferromagnets, the spins of the electrons align preferentially in one direction, giving rise to an overall magnetic dipole moment. In a nematic liquid crystal (the material behind many modern-day digital screens), rod-like molecules are oriented preferentially in one direction, yet they freely flow past one another. All these ordered phases of matter share the same property that their individual constituents are organized with some form of collective "long-range order."

1.1.1   Phase Diagrams

The phase diagram of water is shown in Figure 1.1. A phase diagram maps out the equilibrium phases of a system as a function of a set of external control parameters—in this case, the temperature T and the pressure P. Depending on the control parameters, the properties of the material, such as its density ρ and its compressibility κ, will vary. By measuring these properties at different points on the phase diagram, we can obtain equations of state such as $\rho = \rho(T, P)$ describing how a

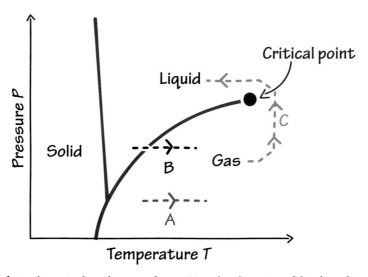

FIGURE 1.1. A schematic phase diagram of water. Note that the region of the phase diagram labeled "solid" actually comprises many structurally distinct crystalline phases. The colored dotted lines represent distinct "trajectories" through the phase diagram. Trajectory B represents the heating of water past its boiling point; a phase transition from liquid to gas occurs where it intersects the phase boundary, indicated by the magenta line. There is no phase transition encountered along trajectories A and C.

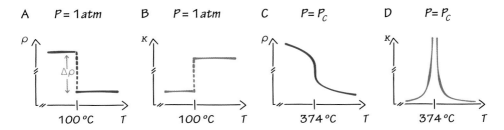

FIGURE 1.2. At a phase transition, observable properties have nonanalytic behavior. At atmospheric pressure, water boils at 100°C, hence both the density (A) and the compressibility (B) change discontinuously as a function of temperature. At the critical pressure, P_c, the density (C) varies continuously through the transition, whereas the compressibility (D) diverges.

material's observable quantities depend on the control parameters. Similarly, other thermodynamic quantities such as the free energy $F(T, P)$ may be obtained as functions of the control parameters.

1.1.2 Phase Transitions

Within a phase, all observables change smoothly as the control parameters are varied. To be precise, the equations of state are analytic functions. (An analytic function is one which is well behaved in the sense that it is continuous, has derivatives to all orders, and converges to its Taylor expansion.) This is the case, for example, along the blue path A in Figure 1.1.

In contrast, at phase boundaries, material properties can undergo discontinuous changes. One dramatic (yet familiar) example is the boiling of water, the brown path B in Figure 1.1. At atmospheric pressure, a minute change in temperature from 99.999°C to 100.001°C causes a sudden, striking transformation: the molecules fly apart, the density plummets by a thousand-fold, and the essentially incompressible water becomes highly compressible steam.

Such changes in the material properties are illustrated in Figures 1.2A and B, where we show the density ρ and the compressibility κ along a path through the phase diagram corresponding to the brown path B in Figure 1.1. The discontinuities in $\rho(P, T)$ and $\kappa(P, T)$ occur at the liquid-gas phase boundary. In contrast, along any path on the phase diagram that avoids all phase boundaries, such as the blue path A or the green path C, the evolution of physical quantities is everywhere smooth, without discontinuities in any thermodynamic observable.

1.1.3 The Critical Point

A particularly interesting feature of the phase diagram of water is the critical point at $(P_c, T_c) = (218 \text{ atm}, 374°C)$, where the liquid-gas phase boundary ends. Across any point on the phase boundary, there is a discontinuous jump in the density

$\Delta\rho \equiv \rho_{liquid} - \rho_{gas}$ between the liquid and gas phase, as illustrated in Figure 1.2A. At successively higher points along the phase boundary, the liquid and gas become more similar in density, i.e., $\Delta\rho$ becomes smaller and smaller, eventually reaching zero at the critical point (Figure 1.2C). Beyond this point, the distinction between the liquid and the gas disappears.

The critical point exhibits many unusual properties. For instance, upon approaching the critical point, the density becomes extremely susceptible to even slight variations in pressure, i.e., the compressibility diverges (Figure 1.2D). Additionally, a fluid near its critical point appears turbid, like a saturated culture of bacteria. This is visible evidence that the fluctuations in density become extremely pronounced near the critical point; when local regions of higher or lower density grow so large that they scatter visible light, this gives rise to a milky appearance.

1.1.4 Continuous and Discontinuous Transitions

The two cases we have described—the liquid-gas transition and the critical point— are examples of a discontinuous and a continuous phase transition, respectively. This classification of phase transitions, after Ehrenfest, stems from the fact that the free energy is always a continuous function of parameters, even at the point of a phase transition.[2] Transitions are classified as either discontinuous or continuous depending on whether or not any first derivatives of F (e.g., density or entropy) are discontinuous.

Continuous phase transitions share many properties with the critical point of water. There is a divergence in a susceptibility, indicating extreme sensitivity to external perturbations. Additionally, fluctuations (such as those responsible for critical opalescence) become relevant on all lengthscales.

Discontinuous phase transitions, such as a liquid-gas transition, have a host of different properties. The defining characteristic is that there is a discontinuity in a first derivative of the free energy, e.g., as water vaporizes its density plummets discontinuously (Figure 1.2A). Another characteristic of a discontinuous transition is a nonzero latent heat: upon crossing the phase transition, a particular amount of heat per unit volume is absorbed or released.[3] Furthermore, discontinuous transitions can exhibit metastability: water can be chilled below its freezing point without the formation of any ice crystals, which is known as supercooling (Section 4.4.3).

---

2. Formally, the continuity of the free energy follows from the fact that it is a bounded, convex function. Physically, since its derivatives are state functions (such as the entropy or the magnetization), these must everywhere be well-defined (although possibly themselves discontinuous) functions.

3. The latent heat of water vapor is the reason why the steam over a pot of soup feels so hot: the water vapor releases heat as it condenses. It is also why it feels so cold to get out of a swimming pool on a windy day: the water absorbs heat as it evaporates off your skin, a process accelerated by the wind!

For historical reasons, discontinuous and continuous transitions are also known as 1st-order and 2nd-order transitions, respectively. This nomenclature comes from thermodynamics: the density ρ is a first derivative of the free energy, whereas the compressibility κ is a second derivative. At a 1st-order (discontinuous) transition such as the boiling of water, the first derivative ρ jumps discontinuously, whereas at a 2nd-order (continuous) transition such as the critical point, the second derivative κ diverges.

1.1.5   Distinction between Phases

In the phase diagram of water (Figure 1.1), the liquid-gas phase boundary does not extend indefinitely: there is no phase boundary past the critical point. This means the liquid and gas phases are not truly distinct. By choosing a path in the $T - P$ plane that circles beyond the critical point (e.g., the green path C in Figure 1.1), it is possible to start in the "liquid" phase and end in the "gas" phase without ever crossing a phase boundary. In this sense, the two are really part of the same fluid phase! In contrast, the solid phase is distinct from the fluid phase because the two are always separated by a phase boundary.

In order to classify the phases of matter in an unambiguous way, a clear distinguishing criterion must be established. It may be tempting to define a gas as a "highly compressible" fluid, but since even liquids are somewhat compressible, there is no clear threshold number that can be used to make this definition precise. Phases must be defined by a clear-cut property—something which is either present or absent, with no in-between. To this end, phases are often classified on the basis of symmetry.[4]

1.2   Symmetries

Symmetry can seem abstract, but we will come to appreciate it as a powerful and precise way to make sense of the phases of matter. As you may recall, the very laws of physics have a set of fundamental symmetries. For example, when light passes through empty space, its speed is the same regardless of the direction of propagation. This reflects the rotational symmetry of space. Likewise, there is translation and reflection symmetry of space, and translation and reversal symmetry of time.

Phases of matter, however, are often less symmetric than are the laws of physics. If light passes through a piece of ice, its speed depends on the direction of propagation:

---

4. Any property that is binary, in the sense that a phase either exhibits it or does not, can in principle be used to distinguish phases of matter. As an aside, we list (without explanation) some examples of other features that can characterize a phase of matter other than its symmetries: A system can be a fluid (able to flow) or a solid (able to support a shear). A collection of electrons can form a "metal" (with a finite resistance in the limit $T \to 0$), an "insulator" (with a diverging resistance in the limit $T \to 0$), or a "superconductor" (with a vanishing resistance for $T < T_c$). In gauge theories, for instance of the strong interaction, there are "confining" phases (in which quarks are confined), and "deconfined" phases, which are distinguished by a form of "topological order."

Cubic

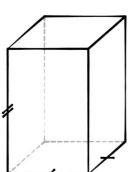

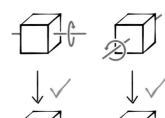

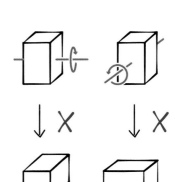

Tetragonal

FIGURE 1.3. An illustration of cubic symmetry (above) and tetragonal symmetry (below). In both cases the action of three transformations is depicted, corresponding to a 90 degree rotation about each of the three axes. A check mark or a cross indicates whether or not a transformation is a symmetry.

the index of refraction depends on how the beam of light is oriented with respect to the axes of the crystal. Crystalline solids such as ice, manifestly, do not have the same rotational symmetry as empty space.

1.2.1   The Cubic-to-Tetragonal Transition

As a first example of how symmetry can differentiate phases of matter, consider a crystal with cubic symmetry, such as common table salt. On a microscopic level, the atoms in a crystal are arranged in a regular lattice with a basic repeating unit, known as a unit cell. In a cubic lattice, the unit cell is a cube, as shown in Figure 1.3. The three axes of the crystal are equivalent and mutually perpendicular, and the lengths of the sides of the unit cell (lattice parameters) are identical.

If a crystal is heated, it undergoes thermal expansion (until at high enough temperatures it melts). On the basis of symmetry, the three axes in a cubic crystal should each lengthen in exactly the same way. If a cubic crystal is cooled, we might expect it to also contract isotropically—but this is not always the case! Rather, in certain crystals, one of the lattice parameters can become distinct from the other two, transforming the lattice from cubic symmetry to *tetragonal* symmetry. This shape-shifting behavior is illustrated in Figure 1.3. In a tetragonal crystal, two of the three axes are identical and the third one is either longer or shorter. All axes are perpendicular.

The transition from a cubic to a tetragonal phase involves a change in the symmetry of the crystal. As such, it must occur at a well-defined critical temperature, T_c, because symmetry is crisply defined: either all three axes are equivalent, or only two of them are. (You could also imagine that at an even lower temperature, T_c', the two equivalent axes in a tetragonal crystal could become different, yielding an orthorhombic crystal with three distinct axes. We will return to this point when we discuss other sorts of structural phase transitions in Section 5.2.2).

Intuitively it is pretty clear that some sort of symmetry has been lost in going from the cubic to the tetragonal phase. What this means mathematically is the following: there are certain transformations which preserve the symmetry of the cubic lattice. Such a transformation, when applied to a cubic lattice, leave the system indistinguishable from its initial state. For example, if we chose one of the three axes and rotated the crystal by 90 degrees around that axis, the lattice would look exactly the same after the rotation, as illustrated in the top right of Figure 1.3. However, this is not the case for a tetragonal lattice. As illustrated in the bottom right of Figure 1.3, only for one of the three axes is a rotation by 90 degrees a symmetry transformation. Evidently, a tetragonal lattice has a smaller set of symmetry transformations than a cubic lattice. We will come to see this as a general pattern of phases of matter: low-temperature phases tend to have reduced symmetry compared to high-temperature phases.

Finally, observe that there was nothing *intrinsic* in the cubic crystal about which of the three axes becomes unique—after all, above T_c, the three axes are equivalent in every conceivable manner. Rather, it is a random occurrence as to which axis will become distinct as the crystal is cooled below T_c. The symmetry is broken *spontaneously*. However, it is possible to force one axis to become the longer one by pulling the crystal along that axis as it is cooled. Such an external perturbation which favors one symmetry-broken state over the others is called a symmetry-breaking field.[5] Even if a symmetry-breaking field is then removed after the system is below T_c, that axis will remain elongated; the system remembers its thermal history.

1.2.2   Solids, Liquids, and Gases

From the perspective of symmetry, we can explain why the liquid and gas "phases" are actually the same phase of matter while the solid phase is definitely distinct.

---

5. Here a "field" refers to a physical quantity that acts over a region of space, such as electric/magnetic fields or more general quantities such as strain.

The question to ask is "what symmetry transformations leave these phases of matter invariant?"

The solid state is invariant under a smaller set of spatial translations than the liquid or the gas. A spatial translation is a transformation in which all the atoms in a material are uniformly shifted by a displacement, $\vec{r} \to \vec{r} + \vec{R}$. Since liquids and gases are homogeneous in space, they are invariant under translations of arbitrary magnitude and direction, i.e. $\vec{R}$ can be any three-dimensional vector. In contrast, in a crystalline solid, only certain translations are symmetries. Because the average position of atoms in a crystal is regular and periodic, a translation will leave a crystal invariant only if it places each unit cell in a new location where it overlaps with an identical copy. Consequently, in a crystal, $\vec{R}$ must be a lattice vector of the form $\vec{R} = n_a\vec{a} + n_b\vec{b} + n_c\vec{c}$, where $\vec{a}, \vec{b}, \vec{c}$ are basis vectors of the lattice, and n_a, n_b, n_c are integers representing the number of units to translate along each direction. The translation symmetry of free space is spontaneously broken in a crystal.

Stated more precisely, in a crystal, the probability of finding an atom at a given point in space is a periodic function of position, with a periodicity represented by the crystalline lattice. In contrast, in a fluid state, the probability of finding an atom at any point in space is the same as at any other (indeed, all thermodynamic quantities are independent of position). Therefore the symmetry of the crystal is lower than that of liquids and gases: a fluid is invariant with respect to any translation, while the crystal is invariant only with respect to a discrete subset of translations. There is no symmetry-based distinction between a liquid and a gas.

Beyond the translational symmetries we have discussed, a crystal also spontaneously breaks the rotational symmetry of free space. For example, if a liquid were to freeze into a tetragonal crystal, it would no longer be invariant under arbitrary rotations—it would have a reduced, 4-fold discrete rotational symmetry (rotations by $0°, 90°, 180°$, and $270°$ about the tetragonal axis). The transition from a liquid to a crystal is an example where a continuous symmetry is spontaneously broken. If, upon further cooling, the tetragonal crystal were to enter an orthorhombic phase (where all three axes are distinct), the rotational symmetry would be reduced to an even smaller subset ($0°$ and $180°$). In this example a discrete symmetry is spontaneously broken.

1.2.3 Ferromagnets

Ferromagnets are materials that spontaneously develop a net magnetization below a certain temperature (known as the Curie point). Since the north pole of a ferromagnet points in a particular direction, the ferromagnetic phase manifestly breaks the rotational symmetry of space. If ferromagnetic order develops in a liquid (a ferrofluid), it will break the continuous rotational symmetry of free space; if it develops in a crystal, it will break the discrete rotational symmetry of the crystal. As discussed

Table 1.1. Names of a few common models of ferromagnets where the magnetization is an N-dimensional vector.

$N = 1$	$N = 2$	$N = 3$
Ising	XY	Heisenberg

in Section 5.2.5, the ferromagnetic phase also spontaneously breaks time-reversal symmetry.

The microscopic origin of ferromagnetism comes from quantum mechanics. At a cartoon level, each atom can be thought of as having a miniature magnetic dipole moment, arising from the quantum mechanical spin of an unpaired electron (see Section 5.2.5). In a ferromagnetic material, the spins of neighboring atoms have a strong enough tendency to align in the same direction that, at low temperatures, the spins are aligned over macroscopic scales. At high temperatures, the thermal energy always randomizes the spin orientations so that on average there is no net magnetization.

There are a handful of common models to describe different categories of ferromagnets, summarized in Table 1.1. The simplest one is an Ising[6] ferromagnet, where the spins preferentially point along an "easy axis"; below the Curie point, the average spin is either up or down along this axis. In an XY ferromagnet, the spins are confined to lie in an "easy plane"; in this case, for $T < T_c$ the magnetization is a two-dimensional vector in the easy plane. In a Heisenberg ferromagnet, the spins are free to point in all directions equivalently, and so the magnetization is specified by a three-dimensional vector. In all these cases, the magnetization is representable by an N-dimensional vector.

In many ferromagnetic crystals, including the strong rare earth magnets, the spins are Ising-like because of the strong, anisotropic interactions between the spins and the crystal lattice (known as spin-orbit coupling). There are no real examples of XY ferromagnets—though, remarkably, a superfluid can be thought of as some kind of XY ferromagnet (see Section 5.2.6). A ferrofluid has Heisenberg symmetry. The connection between these models and real ferromagnets is discussed more in depth in Chapter 5.

1.3 Universality

The overall features of the phase diagram of water are not unique to water. They are common to many fluids. For instance, xenon also exhibits a liquid-gas phase boundary terminating at a critical point. The difference is that the critical pressure

---

6. Named after the German physicist Ernst Ising (1900-1998), who apparently did little physics after working on his eponymous model.

Table 1.2. The critical temperature and pressure of a selection of fluids.

	Ne	Ar	Kr	Xe	N_2	O_2	CO	CH_4
T_c (K)	44.8	150.7	209.4	289.8	126.0	154.3	133.0	190.3
P_c (atm)	26.9	48.0	54.1	58.2	33.5	49.7	34.5	45.7

and temperature of xenon are $P_c = 58$ atm and $T_c = 290$ K. The critical points of some common fluids are listed in Table 1.2.

In Figure 1.4, we have plotted the phase diagram of a number of fluids in the $\rho - T$ plane, with a special choice of units: for each fluid, we have divided ρ by ρ_c and T by T_c.[7] Astoundingly, if we do this, the shape of the phase boundary is almost exactly the same for all fluids! This is especially remarkable considering how these materials differ in other respects—carbon monoxide is poisonous, methane flammable, and neon inert. Despite this, they share a qualitative similarity—in terms of the existence of a critical point—and a semi-quantitative one as well, in the shape of the phase boundary. As far as phases are concerned, the only differences between these fluids seem to be their numerical values of ρ_c and T_c.

Even more remarkably, a completely different physical system, a uniaxial ferromagnet (or Ising ferromagnet), has a phase diagram similar enough to the fluid phase diagram to warrant a comparison. Its phase diagram in the $T - H$ plane is shown in the right panel of Figure 1.5. Here H is the strength of an external magnetic field applied along the easy axis of the sample. The magnetic field gives the spins an energetic bias to align in the same direction as H. At low temperature, this effect leads to a discontinuous phase boundary between the "up phase" and the "down phase." (This is analogous to the liquid-gas phase boundary.) However, the up and down phases are not truly distinct phases in the $T - H$ phase diagram because the distinction between them disappears above the Curie point. Above this temperature, the magnetization encounters no abrupt jump as the sign of the external magnetic field is changed. Comparing the schematic phase diagrams in Fig. 1.5, it is apparent that the fluid and the uniaxial ferromagnet share a topological similarity.

1.3.1 Critical Exponents

The correspondence between these two seemingly unrelated systems extends beyond qualitative similarities. Consider measuring the fluid density in the vicinity of the liquid-gas critical point. At temperatures slightly below T_c, the density of the two coexisting phases varies as

$$\rho(T) - \rho_c \sim \pm |T_c - T|^\beta, \quad \beta \simeq 1/3, \tag{1.1}$$

---

7. For any $T < T_c$, there are two "phases"—the liquid phase and the vapor phase—with densities $\rho_{\text{liquid}}(T)$ and $\rho_{\text{gas}}(T)$, respectively. The critical temperature is the point at which $\rho_{\text{liquid}}(T_c) = \rho_{\text{gas}}(T_c) \equiv \rho_c$.

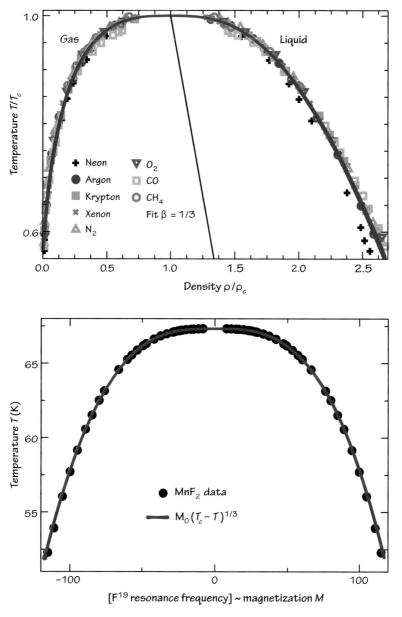

FIGURE 1.4. (top) The phase diagram of eight different fluids in the $\rho - T$ plane. The two branches of the curves indicate the densities of the two phases that coexist at a given T—the gas (small ρ) and the liquid (large ρ). The critical points of different fluids occur at different values of ρ_c and T_c. However, the values of ρ and T have here been scaled by their critical values listed in Table 1.2; remarkably this rescaling results in the phase boundaries all lying on the same curve. (bottom) An analogous phase diagram of a uniaxial ferromagnet. The curves indicate the magnetization at a given T. This figure has been reproduced with permission from James P. Sethna's textbook, *Statistical Mechanics: Entropy, Order Parameters, and Complexity* (Oxford University Press, Oxford, 2021).

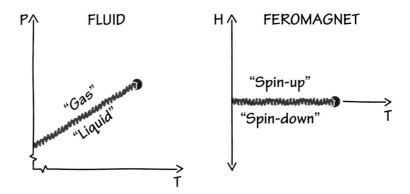

FIGURE 1.5. Schematic phase diagrams of a fluid (left) and a ferromagnet (right) near their critical points. The pressure P exerted on a fluid is analogous to the external magnetic field H applied to a ferromagnet. In both cases there is a discontinuous phase boundary which terminates at a critical point.

with a + for the liquid phase and − for the gaseous phase. The small value of β is reflected in the very flat top of the phase boundary in the top panel of Figure 1.4. As illustrated in the bottom panel of Figure 1.4, the analogous measurement for a uniaxial ferromagnet—measuring the magnetization m as a function of the temperature below the critical point—yields

$$m(T) \propto |T - T_c|^{\beta}, \quad \beta \simeq 1/3, \tag{1.2}$$

with the same critical exponent β within experimental error!

The commonality between the liquid-gas and the uniaxial ferromagnet critical points extends further. If a fluid is held at its critical pressure, the compressibility κ diverges (Figure 1.2D) as the critical temperature is approached from above, with a functional dependence of

$$\kappa \propto |T - T_c|^{-\gamma}, \quad \gamma = 1.24. \tag{1.3}$$

This is the same power law observed for the magnetic susceptibility in a uniaxial ferromagnet,

$$\chi \propto |T - T_c|^{-\gamma}, \quad \gamma = 1.24 \tag{1.4}$$

as $T \to T_c^+$!

Not all continuous phase transitions have exactly these same exponents. For instance, for an XY ferromagnet the exponent β is 0.35. Mysteriously, this value of β is identical to an analogous exponent observed in the transition to the superfluid phase of Helium-4!

In general, physical quantities in the vicinity of a critical point often take the form of power laws

$$y \propto |x|^z, \tag{1.5}$$

where x is a control parameter (like $T - T_c$) and y is a measurable thermodynamic quantity (such as a susceptibility). The critical exponent z describes the nature of the singularity.[8] In Worksheet W1.1, you get to work through some examples of power laws.

Certainly, we will want to understand the origin of these power laws at critical points, why they are so universal, and why certain transitions have the same or different critical exponents. Ideally, we would like to be able to predict the values of these critical exponents.

1.3.2 Universality Classes

In the 1970s and 1980s, many of the mysteries of continuous transitions came to be successfully understood. Today, we have a unified framework for understanding a variety of phase transitions.

Phase transitions can be classified into various universality classes. All the phase transitions within a universality class share the same values of critical exponents and share additional universal features. For instance, the uniaxial ferromagnet and the liquid-gas transition both fall into the 3D ISING universality class. Other phase transitions in this universality class, such as the order-disorder transition of β-brass (see Section 5.2.1), have the same 3D ISING critical exponents. The easy-plane ferromagnet and the superfluid, on the other hand, fall in the 3D XY class and have their own set of exponents. The transition to a ferromagnetic state in a ferrofluid falls in yet another class, the 3D HEISENBERG class.

What determines the universality class of a transition is somewhat abstract. We will return to this topic throughout the book. In the end we will find that it depends on some very general features of the broken symmetries and on the dimension of space, but on no other details of the system involved.

One of the shared properties within a universality class is the exponents of various power laws (Eq. 1.5). Conventionally, the critical exponents are represented by various Greek letters. We will encounter a whole bunch of other critical exponents throughout the text, summarized in Table 1.3. The values of these critical exponents have been determined theoretically, either by exact solution of model problems or by extensive numerical simulation of such models. They are summarized in Table 1.4. Notice that the values depend on both the symmetry index N as well as the spatial dimensionality d. In many cases these theoretical predictions agree well with experiments in a variety of systems.

One of our goals in the book, in addition to understanding the existence of distinct phases of matter and sharply defined phase transitions between them, is to understand the origin of these critical exponents and why they are universal.

---

8. A singularity is a point where a function is nonanalytic—a cusp, a discontinuity, a divergence, ...

Table 1.3. A summary of the critical exponents discussed in this text. The reduced temperature $t = (T - T_c)/T_c$ is a dimensionless measure of distance from a critical point. Exponents δ and η are defined for $t = 0$. $G(r)$ is the correlation function (Eq. 2.28) and ξ is the correlation length (Eq. 2.31).

Exponent	Definition	Name	First Seen
α	$c \sim |t|^{-\alpha}$	specific heat	Eq. 10.8
β	$m \sim |t|^{\beta}$	magnetization	Eq. 1.1
γ	$\chi \sim |t|^{-\gamma}$	susceptibility	Eq. 1.3
δ	$m \sim h^{1/\delta}$	magnetization	Eq. 4.39
η	$G(r) \sim r^{-(d-2+\eta)}$	anomalous dimension	Eq. 10.1
ν	$\xi \sim |t|^{-\nu}$	correlation length	Eq. 7.28

Table 1.4. Critical exponents of various models in different numbers of spatial dimensions. Numbers given as rational fractions are obtained from exact analytic approaches while those given as decimals are from accurate numerical simulations, rounded to two decimal places.

Universality Class	α	β	γ	δ	η	ν
2D ISING	0	1/8	7/4	15	1/4	1
3D ISING	0.11	0.32	1.24	4.79	0.04	0.63
4D ISING	0	1/2	1	3	0	1/2
3D XY	−0.01	0.35	1.32	4.78	0.04	0.67
4D XY	0	1/2	1	3	0	1/2
3D HEISENBERG	−0.12	0.36	1.39	4.91	0.04	0.71
4D HEISENBERG	0	1/2	1	3	0	1/2

1.4 Problems

1.1. *Phase diagram of simple polynomials.* The qualitative properties of a polynomial—for instance, the number of distinct real-valued roots—depend on the values of its coefficients.

(a) Draw a phase diagram that maps out the number of real-valued roots of the polynomial $f(x) = x^2 - ax + b$ as a function of a and b. How many phases are there? What is the functional form of the phase boundary?

(b) Now consider the polynomial $g(x) = x^3 - cx + d$. Draw a phase diagram in the $c - d$ plane displaying the number of real-valued roots, and determine the functional form of the phase boundary. There is a special point on the phase boundary; what is special about it? *Hint:* At the special point on the phase boundary, the extremum of the function is simultaneously a root: $g(x^*) = g'(x^*) = 0$.

(c) We can draw a different (more interesting!) phase diagram with the same polynomial, $g(x) = x^3 - cx + d$. Consider the *largest real root* of $g(x)$,

$$x_{max} = \max_x \{x | g(x) = 0\}, \tag{1.6}$$

as a function of the control parameters c and d. Draw the phase diagram of x_{max}. What is different about this phase boundary compared to parts (a) and (b)? Does the overall topology of the phase diagram resemble any of the physical systems discussed in this chapter?

Remarkably, the phase diagrams of many physical systems can be approximately determined using simple polynomials, as we have done here. The reason is that, near to a critical point, the free energy may be approximated by a Taylor expansion, i.e., a simple polynomial. We discuss this concept in Chapter 6.

1.2. *Power laws and universality.* Near critical points, thermodynamic quantities are often well described by power laws of the form $y = ax^z$. The pre-factor a depends on the details of the physical system, but the exponent z is universal.

(a) Give a few examples of power laws from prior physics classes.

(b) What is the area of a square with side length ℓ? Of a triangle? Of a circle with diameter ℓ? What is common among these formulas, and what is different?

(c) What is the volume of a cube with side length ℓ? Of a sphere with diameter ℓ? What is different from part (b)?

(d) In d dimensions, what is the formula for a shape's hypervolume V as a function of its side length? You may leave the answer in terms of an arbitrary constant a which depends on the details of the shape.

1.3. *Scale invariance and fractals.* Consider the effect of a scale transformation $\ell \mapsto b\ell$, whereby all lengths are scaled up by a factor of b, akin to the action of a magnifying glass.

(a) What happens to the hypervolume of a d-dimensional shape upon a rescaling $\ell \mapsto b\ell$?

Remarkably, a physical system at its critical point *remains unchanged* upon a change of scale: it is scale invariant. This property is reminiscent of the self-similarity of *fractals*, which are sometimes encountered in recreational mathematics. An example is the Sierpinski triangle, which is formed by starting with an equilateral triangle, dividing it into four equally sized equilateral triangles, removing the central triangle, and repeating the process for each of the remaining triangles ad infinitum.

(a) The Sierpinski triangle is not scale invariant, but it is self-similar: there are *particular* values of b which leave the shape unchanged under $\ell \mapsto b\ell$. What are they?

(b) What is the dimensionality of the Sierpinski triangle? *Hint:* Under a scaling by b, the triangle generates c copies of itself. Find b and c, and use the relation from part (a).

A system at a critical point exhibits patterns with fractal-like shapes, with structures on all lengthscales. The dimension of a fractal need not be an integer; it can even be an irrational number.

1.4. *Universality in a random walk.* Consider a particle on a 1D line which starts at $x = 0$. Suppose that at each time step, it can either step to the left, $s_i = -a$, with probability $1/2$, or it can step to the right, $s_i = +a$, with probability $1/2$. After N steps, the position of the particle is $S = \sum_{i=1}^{N} s_i$.

(a) How does the mean squared displacement, $\langle S^2 \rangle$, depend on the number of steps N?

(b) For large N, what is the probability distribution of S? *Hint:* Consider the central limit theorem.

Now consider a different random walk where each step is distributed as

$$r_i = \begin{cases} -2a & \text{with probability } 1/2; \\ a & \text{with probability } 1/4; \\ 3a & \text{with probability } 1/4. \end{cases} \tag{1.7}$$

(c) After many steps $N \gg 1$, what is the distribution of the position of the particle, $R = \sum_{i=1}^{N} r_i$?

(d) Compare and contrast the distributions of the R random walk and the S random walk after a few steps and after very many steps.

This is an example of universality: the R and S random walks have different microscopic behaviors, but after very many steps, they converge to the same asymptotic behavior.

1.5 Worksheets for Chapter 1

W1.1 Critical Exponents

If we plot certain properties of materials as a function of a tuning parameter close to a phase transition, the functional dependence can often be described by some power law. Here let us gain some familiarity with the shapes of various power laws, and identify the correct power laws from real experimental data.

1. Sketch the following power laws:

    (a) $m \propto (T_C - T)^\beta$, where $0 < \beta < 1$.

    (b) $c \propto |T - T_C|^{-\alpha}$, where $0 < \alpha < 1$.

When you have access to graphing tools, try plotting out these power laws and observe how they change as you vary α and β!

2. A phase transition is considered "continuous" if the first derivative of the free energy is continuous across the phase boundary. Are the above transitions continuous given that:

   (a) m is a first derivative of the free energy with respect to an applied field, h?

   (b) c is a second derivative of the free energy with respect to temperature T?

3. A critical exponent and the associated power law describe the behavior of various physical quantities *close to* the critical point. Let us see what this means in practice.

   Suppose we have a magnet whose normalized magnetization close to $T = T_c$ behaves as:
   $$m(T) = \sqrt{1 - (T/T_c)^2} \text{ for } T < T_c. \tag{1.8}$$

   (a) What is the meaning of the temperature scale T_c?

   (b) Close to the phase boundary ($0 < 1 - |T/T_c| \ll 1$), the normalized magnetization can be written as:
   $$m(T) = a(T_c - T)^\beta. \tag{1.9}$$

   Find the value of the constant of proportionality a and the critical exponent β such that Eq. 1.9 correctly describes the behavior of Eq. 1.8 close to the phase boundary (see Figure W1.1).

   Hint: Find the quantity that is small close to T_c, and expand Eq. 1.8 w.r.t. that quantity.

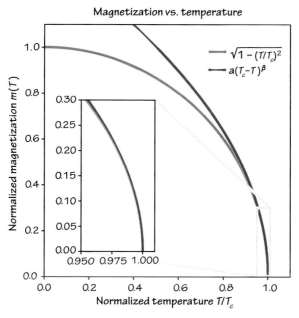

FIGURE W1.1. Full functional dependence vs. power-law description near phase boundary.

With the correct values for a and β, the power-law description very accurately describes the behavior near the phase boundary, but deviates quickly away from it. Moving forward, we must remember that many descriptions of critical phenomena are (approximately) correct only near phase boundaries!

4. Below is a plot of the superconducting gap energy of Tantalum close to the critical temperature as a function of the rescaled temperature $t = T/T_c$.

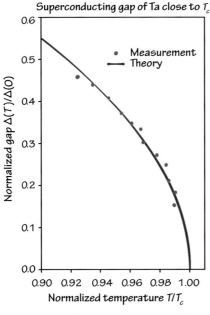

FIGURE W1.2. Superconducting gap of Tantalum close T_c.

TABLE 1.5. Tantalum gap energy temperature dependence.

T/T_c	$\Delta(T)/\Delta(0)$
0.9904	0.1833
0.9895	0.1522
0.9858	0.2112
0.9843	0.2485
0.9781	0.2718
0.9689	0.3013
0.9674	0.3339
0.9612	0.3478
0.9542	0.3711
0.9458	0.4084
0.9349	0.4394
0.9249	0.4580

(a) Judging from the plot, what is the form of the normalized gap $\Delta(T)/\Delta(0)$ close to $t = 1$?

(b) Plotting the data on log-log axes, estimate the critical exponent, as well as the constant of proportionality.

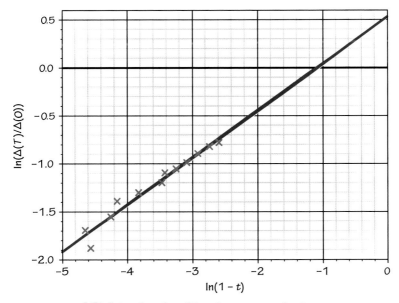

FIGURE W1.3. Log-log plot of Tantalum superconducting energy gap.

Is the theory a good description close to criticality? Close to T_c the theoretical value is $\Delta(T)/\Delta(0) = 1.74\sqrt{1 - T/T_c}$.

2 BACKGROUND: FUNDAMENTALS OF STATISTICAL MECHANICS

According to Douglas Adams, "There is a theory which states that if ever anyone discovers exactly what the Universe is for and why it is here, it will instantly disappear and be replaced by something even more bizarre and inexplicable." One thing is sure, however: the second law of thermodynamics will still apply.

In this chapter, we summarize the formal relations between statistical mechanics and thermodynamics. It is assumed that the student has already encountered these concepts in a previous course, so this section serves primarily as a review, and to establish notational conventions and vocabulary. In particular, we will remind the reader of the relation between the free energy and the partition function, defined as the sum of the Boltzmann weight over all possible microstates of the system. More generally we will give formal definitions of the various quantities we would like to compute in order to "solve" a problem in statistical mechanics. In some sense, the remainder of the book is devoted to developing methods to carry out (or to approximate) the calculations indicated here, and to develop physically meaningful intuitions for the formal expressions.

2.1 Review of Statistical Mechanics

Statistical mechanics is the study of systems in equilibrium. The equilibrium state appears unchanging from a macroscopic perspective, but on a microscopic level there are many possible configurations. The core idea of statistical mechanics is to model the equilibrium state as a suitable average over all microscopic configurations (microstates).

The canonical ensemble is a description of a system which has reached thermal equilibrium with a surrounding heat bath of temperature T. At equilibrium the probability of micostate s is

$$P(s) = e^{-\beta E(s)}/Z, \tag{2.1}$$

where $E(s)$ is the energy of state s, $\beta = 1/k_B T$ is the inverse temperature, and the normalization constant, known as the partition function,

$$Z = \sum_s e^{-\beta E(s)}, \tag{2.2}$$

ensures that the probability sums to 1, $\sum_s P(s) = 1$. Henceforth we will work in units such that the Boltzmann constant $k_B = 1$.

One of the tenets of statistical mechanics is that macroscopic observables are given by thermal averages i.e., an average where each of the microstates is weighted by its Boltzmann probability. If an observable A has value $A(s)$ in state s, then its thermal average is denoted by

$$\langle A \rangle = \sum_s P(s)A(s). \tag{2.3}$$

In particular, the average energy, E, also known as the "internal energy," is

$$E \equiv \langle H \rangle = Z^{-1} \sum_s E(s)e^{-\beta E(s)}, \tag{2.4}$$

where H is the Hamiltonian, or the energy.[1]

2.1.1 The Link between Thermodynamics and Statistical Mechanics

Despite decades of study of free energy, physicists have yet to invent a perpetual motion machine.

According to thermodynamics, the equilibrium state is the one which minimizes a thermodynamic potential, such as the Gibbs free energy or the Helmholtz free energy.[2] But thermodynamics makes no reference to the composition of matter on a microscopic scale. In order to actually calculate potentials, some sort of microscopic description is necessary—and this is where statistical mechanics comes in. It

---

1. In a general quantum mechanical formulation, we need to deal with the fact that some observables A do not commute with the Hamiltonian. In this case, if s labels the eigenstates of H, then A will not have a well-defined value in some or all states s. This issue can be important in the limit $T \to 0$ for understanding the properties of quantum phases of matter and quantum phase transitions. The issue will not arise in any of the problems we consider here, but in the interest of completeness, it is worth reviewing the correct fully quantum mechanical version of Eqs. 2.2 and 2.3:

$$Z = \mathrm{Tr}\left[e^{-\beta H}\right], \tag{2.5}$$

$$\langle A \rangle = \mathrm{Tr}\left[\frac{e^{-\beta H}}{Z} A\right]. \tag{2.6}$$

It is straightforward to show that these expressions are equivalent to Eqs. 2.2 and 2.3 if we interpret $A(s)$ as the expectation value of A in the energy eigenstate s.

2. The specific potential which is minimized at equilibrium is determined by which macroscopic variables are held fixed. For instance, under conditions of constant temperature and pressure, the Gibbs free energy is minimized, whereas under conditions of constant temperature and volume, the Hemlholtz free energy is minimized—see any standard text on thermodynamics.

links the macroscopic potentials to microscopic structure, in the form of a relation between the Helmholtz free energy and the partition function,

$$F = -T \log Z. \tag{2.7}$$

We will henceforth refer to F as simply "the free energy."

From the free energy, all other thermodynamic quantities can be computed. Of note, the free energy is related to the entropy as

$$F = E - TS, \tag{2.8}$$

where E is the internal energy (Eq. 2.4). From this, you can derive further relations (see Worksheet W2.1):

$$E = -\frac{1}{Z}\frac{\partial Z}{\partial \beta} = T^2 \frac{\partial \ln(Z)}{\partial T}, \tag{2.9}$$

$$S = -\frac{\partial F}{\partial T} = \ln\left[Z e^{\beta E}\right]. \tag{2.10}$$

What entropy means in thermodynamics is unclear,[3] but in statistical mechanics entropy has a very powerful interpretation—it is *the logarithm of the number of accessible microstates* (Problem 2.1). With this in mind, considering how $F = E - TS$ is minimized yields an interesting intuition concerning how temperature influences equilibrium phases of matter. At low T, energetic considerations dominate, and the equilibrium state consists predominantly of those relatively special microstates with low energies. Often these involve some sort of ordering. At high T, entropy dominates. Since there are many more states with high energy, thermal averages are dominated by these (largely "disordered") microstates. In Problem 2.2 you will see how this intuition comes into play at a discontinuous phase boundary—the high-temperature phase has greater entropy whereas the low-temperature phase has lower energy.

2.1.2   The Thermodynamic Limit

A phase of matter is a collective effect involving a large number of degrees of freedom. As such, we are generally interested in the limit where the system size, V, is large compared to all microscopic lengthscales. The limiting case $V \to \infty$ is known as the thermodynamic limit.

For observables to be well defined in the thermodynamic limit, we have to distinguish between extensive and intensive properties. An extensive quantity, such as

---

3. The history of entropy is remarkable; it is one of the great triumphs of science that nineteenth century physicists were able to make sense of the notion of entropy and apply it to thermodynamics, even without knowing what it means. It was Boltzmann who correctly identified its microscopic interpretation, even before the atomic theory was widely accepted.

energy or entropy, is one that grows in proportion to the volume, V. In contrast, an intensive quantity, such as temperature or particle density, is independent of V (for large enough V). In the thermodynamic limit, it does not make much sense to think about extensive quantities; instead, we typically talk about the corresponding intensive quantities, defined by dividing by the volume. For instance, the free energy F diverges in the thermodynamic limit, but the free energy density, $f \equiv F/V$, approaches a well-defined value. Of course f will depend on properties such as the type of material, temperature, and pressure—but it is independent of the system size.[4]

2.1.3 Derivatives of the Free Energy

Derivatives of the free energy are generally related to measurable properties. If a variable A appears in the Hamiltonian as

$$H = E(s) + \lambda A(s), \tag{2.11}$$

with a linear coupling to a conjugate field λ, then there is a simple relation for the thermal average of this observable,

$$\langle A \rangle = \frac{\partial F}{\partial \lambda}. \tag{2.12}$$

This relation between conjugate variables A and λ turns out to be very useful, and its derivation is straightforward (Worksheet W2.1). We will refer to Eq. 2.12 as the "derivative theorem" going forward.

As an example, a magnetic dipole moment $\vec{M}$ in a magnetic field $\vec{B}$ has an energy of $-\vec{M} \cdot \vec{B}$. Since the two variables form a conjugate pair, the derivative theorem recapitulates the thermodynamic identity

$$\frac{\partial F}{\partial \vec{B}} = -\langle \vec{M} \rangle. \tag{2.13}$$

Other examples of conjugate pairs include the volume and pressure, or the chemical potential and particle number.

The second derivatives of F describe the susceptibility of a variable to outside forces. For instance, the heat capacity, which describes the change in the internal energy that results from a small change in temperature, is

$$C_V \equiv \left(\frac{\partial E}{\partial T} \right)_V = T \left(\frac{\partial S}{\partial T} \right) = -T \left(\frac{\partial^2 F}{\partial T^2} \right). \tag{2.14}$$

---

4. There are other ways in which quantities can depend on the system size. For instance, the partition function grows exponentially with V since it is the exponential of the free energy. This can make it unwieldy (see Section 3.2.2).

The compressibility, i.e., the change in the density in response to a small change in pressure, is

$$\kappa = -\frac{1}{V}\left(\frac{\partial V}{\partial P}\right) = \frac{1}{V}\left(\frac{\partial^2 F}{\partial P^2}\right). \tag{2.15}$$

As a final example, the magnetic susceptibility tensor is

$$\chi_{a,b} = \frac{1}{V}\left(\frac{\partial m_a}{\partial B_b}\right)_T = -\frac{1}{V}\left(\frac{\partial^2 F}{\partial B_a \partial B_b}\right)_T, \tag{2.16}$$

which describes the magnetization[5] induced by a small external magnetic field. Since in general any component of the magnetization can be induced by any component of the applied field, the susceptibility is a tensor, where $\chi_{a,b}$ relates the ath component of magnetization to the bth component of the applied field; as it is a second derivative of the free energy, it is necessarily a symmetric tensor, $\chi_{a,b} = \chi_{b,a}$.

2.1.4   Variance and Fluctuations

It is useful at this point to review some statistics (see also Appendix A.7). If $x_1, x_2, \ldots, x_N$ are the values of a random variable x taken from a given distribution, then the mean is computed as

$$\bar{x} = \frac{1}{N}\sum_i^N x_i \tag{2.17}$$

and the variance σ_x^2 as

$$\sigma_x^2 = \frac{1}{N}\sum_i^N (x_i - \bar{x})^2 = \overline{(x_i - \bar{x})^2} = \overline{x^2} - \bar{x}^2. \tag{2.18}$$

Arguably, these two quantities give us the most basic statistical properties of x— $\bar{x}$ tells us what we expect x to be, and σ_x^2 tells us the degree to which a single value of x is likely to deviate from $\bar{x}$.

In statistical mechanics, the mean corresponds to the thermal average, Eq. 2.3, with $\bar{x} \to \langle x \rangle$. The variance corresponds to another quantity, Δ_x^2,

$$\sigma_x^2 = \overline{(x_i - \bar{x})^2} \quad \longrightarrow \quad \Delta_x^2 = \left\langle (x - \langle x \rangle)^2 \right\rangle \tag{2.19}$$

$$= \left\langle x^2 \right\rangle - \langle x \rangle^2, \tag{2.20}$$

which describes the *thermal fluctuations* in x.

---

5. As a reminder, the total magnetic moment $\vec{M}$ and the magnetization $\vec{m}$ are related by $\vec{M} = \int dV \vec{m}(\vec{r})$. In other words, the magnetization is the local magnetic moment density.

As an example, consider the ath component of the magnetic moment, M_a. Its mean squared fluctuation about the mean is

$$\Delta^2_{M_a} \equiv \left\langle \left[M_a - \langle M_a \rangle \right]^2 \right\rangle = \left\langle M_a^2 \right\rangle - \langle M_a \rangle^2 . \tag{2.21}$$

The first term in Eq. 2.21 can be calculated (see Worksheet W2.1) as

$$\left\langle M_a^2 \right\rangle = \frac{1}{Z} \sum_s M_a(s)^2 e^{-\beta H(s)} = T^2 \frac{1}{Z} \frac{\partial^2 Z}{\partial B_a^2}, \tag{2.22}$$

and we already know the second term is

$$\langle M_a \rangle = -\frac{\partial F}{\partial B_a} = T \frac{1}{Z} \frac{\partial Z}{\partial B_a}. \tag{2.23}$$

Comparing these results to Eq. 2.16, we see that fluctuations and the susceptibility are related as

$$\left\langle |M_a - \langle M_a \rangle|^2 \right\rangle = TV \chi_{a,a}. \tag{2.24}$$

It may similarly be shown that the heat capacity is related to fluctuations of the energy as

$$\left\langle \left[E - \langle E \rangle \right]^2 \right\rangle = T^2 C_V. \tag{2.25}$$

The relationship between susceptibilities and fluctuations is a profound and ubiquitous feature of statistical mechanics, to which we return in Section 7.6.

2.1.5   Covariance and Correlation Functions

The variance of a random variable measures the degree of deviation from its average. A related concept called covariance can be applied to two random variables x and y,

$$\sigma_{xy} = \frac{1}{N} \sum_i^N (x_i - \bar{x})(y_i - \bar{y}) \tag{2.26}$$

$$= \overline{xy} - \bar{x}\bar{y}. \tag{2.27}$$

The covariance is a measure of the statistical correlation between two variables. If x and y are statistically independent, then $\overline{xy} = \bar{x}\bar{y}$ and $\sigma_{xy} = 0$.[6] If $\sigma_{xy} > 0$, then larger values of x tend to be associated with larger values of y. Conversely, if $\sigma_{xy} < 0$, then larger values of x tend to be associated with smaller values of y.

In the context of statistical mechanics, x and y are two observables. Often x represents a measurable quantity at one location $\vec{r}$, and y is the same quantity measured

---

6. Keep in mind that independence is a stronger condition than $\sigma_{xy} = 0$: it is possible for two variables to be uncorrelated but still not independent. An example is x uniformly distributed betweeen -1 and 1 and $y = x^2$.

at another location $\vec{r}\,'$. In this case, σ_{xy} is known as a correlation function. For example, suppose in a magnetic material we pick two positions, $\vec{r}$ and $\vec{r}\,'$, and calculate the correlation between the magnetization at these two locations, $m(\vec{r})$ and $m(\vec{r}\,')$,

$$G(\vec{r},\vec{r}\,') = \langle m(\vec{r})m(\vec{r}\,') \rangle - \langle m(\vec{r}) \rangle \langle m(\vec{r}\,') \rangle \,. \tag{2.28}$$

$G(\vec{r},\vec{r}\,')$ measures the correlation between magnetization fluctuations about the mean at two different positions in space; if it is positive, it means that a microstate with a larger than average magnetization at position $\vec{r}$ is also likely to have a larger than average magnetization at position $\vec{r}\,'$.

2.1.6   Correlation Functions and Phases of Matter

Correlation functions are a powerful way to describe phases of matter. Many phases can be characterized by the manner in which $G(\vec{r},\vec{r}\,')$ depends on $\vec{r}$ and $\vec{r}\,'$. It is physically clear that if positions $\vec{r}$ and $\vec{r}\,'$ are sufficiently far separated, the behavior of the system at these two points must be essentially uncorrelated, i.e., $G(\vec{r},\vec{r}\,') \to 0$ as $|\vec{r} - \vec{r}\,'| \to \infty$. Conversely, if $\vec{r}$ and $\vec{r}\,'$ are 'close together', then we expect the fluctuations to be noticeably correlated. As a scale of length to measure what is "close" or "far," we will introduce the notion of a correlation length, represented by the Greek letter ξ. In many situations, the correlation function decays exponentially with the distance between $\vec{r}$ and $\vec{r}\,'$, with a characteristic lengthscale ξ:[7]

$$G(\vec{r},\vec{r}\,') \sim \exp\left(-\frac{|\vec{r} - \vec{r}\,'|}{\xi} \right). \tag{2.31}$$

The full correlation function $G(\vec{r},\vec{r}\,')$ is a function of six degrees of freedom: three coordinates of $\vec{r}$ and three of $\vec{r}\,'$. In many phases of matter, however, symmetries constrain the correlation function so that it does not independently depend on all six variables. For instance, consider the correlation function between $\rho(\vec{r})$ and $\rho(\vec{r}\,')$, the density of particles at point $\vec{r}$ and at nearby point $\vec{r}\,'$. If the phase of matter is homogeneous, such as a liquid or a gas, then the average density $\langle \rho(\vec{r}) \rangle$ is independent of $\vec{r}$. This reflects our understanding that a fluid is invariant under a

---

7. The fact that the correlation function decays rapidly with distance has important consequences for the susceptibility. Since the net magnetic moment is an integral of the magnetization,

$$M_a = \int d\vec{r}\, m_a(\vec{r}), \tag{2.29}$$

the susceptibility (Eq. 2.24) can be expressed in terms of the correlation function (Eq. 2.28) as

$$\chi_{a,b} = \frac{1}{TV} \int d\vec{r} d\vec{r}\,' \, G_{a,b}(\vec{r},\vec{r}\,'). \tag{2.30}$$

Naively, we would expect the double integral to grow as V^2, meaning that overall $\chi_{a,b} \sim V$. However, the fact that the correlation is rapidly decaying means that the double integral goes as $V\xi^d$; the factor of V cancels out, giving a χ that is properly intensive.

translation: $\langle \rho(\vec{r}) \rangle = \langle \rho(\vec{r} + \delta \vec{r}) \rangle$ for any $\delta \vec{r}$. The same considerations imply that the density-density correlation function,

$$G(\vec{r}_1, \vec{r}_2) \equiv \Big\langle \big(\rho(\vec{r}_1) - \langle \rho(\vec{r}_1) \rangle \big) \big(\rho(\vec{r}_2) - \langle \rho(\vec{r}_2) \rangle \big) \Big\rangle, \qquad (2.32)$$

must be invariant under $\vec{r}_1 \rightarrow \vec{r}_1 + \delta \vec{r}$ and $\vec{r}_2 \rightarrow \vec{r}_2 + \delta \vec{r}$, i.e., the correlation function cannot depend on the absolute locations $\vec{r}_1$ and $\vec{r}_2$; it can only depend on their relative displacement $\vec{r} \equiv \vec{r}_2 - \vec{r}_1$.

If the system is furthermore isotropic (i.e., invariant under a spatial rotation), the correlation function must depend only on the *distance* between the points $r \equiv |\vec{r}_2 - \vec{r}_1|$, and not on the direction of the displacement between the points.[8] By contrast, in a crystal G depends independently on both $\vec{r}_1$ and $\vec{r}_2$, and is sensitive to the direction of the displacement between the two points.

To simplify the discussion, consider the radial-distribution function, defined by averaging G over space and over the orientation of $\vec{r}_1 - \vec{r}_2$, and normalizing by the average density $\bar{\rho}$:

$$g(r) \equiv \int \frac{d\vec{r}_1}{V} \int d\vec{r}_2 \, \frac{\delta(r - |\vec{r}_1 - \vec{r}_2|)}{4\pi r^2} \left[\frac{\langle \rho(\vec{r}_1) \rho(\vec{r}_2) \rangle}{\bar{\rho}^2} \right]. \qquad (2.33)$$

This quantity measures the probability that if there is a particle at some point in space, there is another particle a distance r away. It is defined such that $g(r) = 1 + \bar{\rho}^{-2} G(\vec{r}_1, \vec{r}_2)$ with $r = |\vec{r}_1 - \vec{r}_2|$ for an isotropic and homogeneous system (a fluid). If there are no correlations anywhere, then $g(r) = 1$ for all r. If $g(r) > 1$, then the probability that any two particles are separated by a distance r is greater than you would expect if they were uncorrelated. If $g(r) < 1$, then the opposite is true.

Shown in Figure 2.1 are cartoons of $g(r)$ for a few states of matter. In an ideal gas, the interactions between particles are negligible, so the locations of the particles are statistically independent. Accordingly, the radial distribution is constant. This corresponds to a situation in which the molecules in the gas are sparse, so it is rare to find pairs of molecules that are close enough to interact significantly. However, with increasing density, the interactions between the molecules become increasingly important, leading to short-range correlations between them. Specifically, if a molecule is located at one location, it is impossible for another molecule to be too close due to the strong short-distance repulsion between molecules. Accordingly, the radial distribution function of an interacting gas must essentially vanish for distances smaller than twice the molecular radius r_0. At slightly larger distances, there is typically a peak in the pair distribution function at $r \approx 2r_0$; this peak accounts

---

8. Gases and liquids are isotropic phases of matter; their properties do not change when they are rotated. However, solids are not isotropic, in that the axes of a crystal point in a particular direction in space. Nematic liquid crystals are anisotropic like a crystal, but as homogeneous as any other liquid (Section 5.2.4).

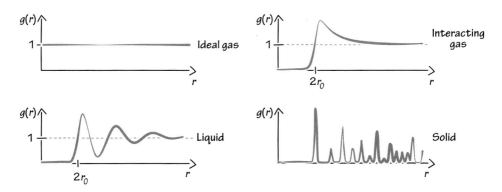

FIGURE 2.1. The radial distribution function sketched out for a few phases of matter as described in the text.

for pairs of particles that would have had smaller separations in the absence of interactions.

In a liquid, the correlations are still more structured. Each molecule has a shell of closest neighbors, then another shell of second neighbors around that, and so forth. The presence of these shells leads to oscillations in the radial distribution. However, with increasing distance, these shell effects become increasingly less well defined; they extend out only to a finite correlation length beyond which the positional correlations between molecules become rapidly small.

In a solid, the molecules arrange into a regular crystalline lattice. Accordingly, the correlations are very regular: there is a series of sharp peaks reflecting the periodic nature of the lattice.[9]

2.2   *Other Fundamental Issues Concerning Statistical Mechanics

This and later starred sections covers "supplemental" material that is not needed to understand subsequent discussions.

It is worth considering what statistical mechanics has to say about the three laws of thermodynamics. The first law of thermodynamics—the conservation of energy—implies that in any process, the change in the internal energy is equal to the heat flowing into the system minus the work done by the system. Applied to processes that are always in instantaneous equilibrium (i.e., for "quasi-static" processes), this statement is equivalent to Eq. 2.8. (Of course, energy conservation holds under much more general circumstances.) However, the second law of thermodynamics—the increase in entropy—intrinsically goes beyond equilibrium considerations. It is a dynamical statement that ensures that, over sufficiently long times, the entropy of the universe always increases to a state of maximum possible entropy—which is the state of thermal equilibrium. (Of course, this statement is not

---

9. The lack of periodicity in the sharp peaks in Figure 2.1 is because g is plotted as a function of radial distance; for a regular crystal the spacing between peaks must decrease in proportion to $1/r^2$.

necessarily true of the subset of the universe that most interests us at any given time.) Because the second law is a dynamical statement, it is beyond the scope of the this book—but it provides the legal framework that justifies our focus on equilibrium phenomena.[10]

It remains to discuss the third law—the vanishing of the entropy density as $T \to 0$—from the perspective of statistical mechanics. Since statistical mechanics provides explicit expressions for the free energy and entropy in terms of the quantum mechanical energy levels of the system, it is straightforward to see why this law holds. In general,[11]

$$Z \to N_{gs} e^{-\beta E_{gs}} \quad \text{as} \quad T \to 0, \tag{2.34}$$

where E_{gs} is the ground-state energy and N_{gs} is the number of degenerate ground states. Under most circumstances, this means that

$$F \sim E_{gs} - T \ln \left[N_{gs} \right] \quad \text{as} \quad T \to 0. \tag{2.35}$$

Hence, $E(T) \to E_{gs}$ (as it must) and

$$S \to \ln \left[N_{gs} \right] \quad \text{as} \quad T \to 0. \tag{2.36}$$

As $T \to 0$, the only accessible states are the ground states, so this is in line with our interpretation of the entropy as the logarithm of the number of accessible microstates. As promised, $S/V \to 0$ in the thermodynamic limit as $V \to \infty$ (assuming that N_{gs} does not diverge exponentially with volume).[12]

2.2.1 Ergodicity

There are deep issues concerning the applicability of statistical mechanics that we have not discussed at all. After all, any actual system at a given point in time is presumably in a specific microstate. The microstate evolves according to some (in principle known) dynamics; it is not a statistical distribution of many

---

10. The existence of a uniquely defined temperature for a system in thermal equilibrium with a heat bath is often referred to as the zeroth law of thermodynamics. The reasoning that leads to this law is already embedded in our formulation of statistical mechanics, where we define the equilibrium state in terms of a Boltzmann weight with a fixed inverse temperature, β.

11. There are a few pathological model systems with an unphysically large density of low-energy states. It turns out that certain "holographic models" motivated by quantum gravity have a density of microstates that diverges so rapidly that the third law is violated, but this is thought to be unlikely to occur in any physically reasonable circumstance.

12. An example of a system that appears to violate the third law of thermodynamics is ordinary ice—crystalline H_2O; here, as was shown by Pauling, there are a large number of nearly degenerate configurations of the H atoms bridging the O atoms, with $N_{gs} \sim e^{\gamma_H N}$ where $\gamma_H \approx \sqrt{3/2}$ and N is the number of water molecules. While these states are not truly exactly degenerate, the energy differences are so small that under reasonable experimental circumstances they act as if they were. Thus, as $T \to 0$, the entropy of ice approaches $S/V \to n \ln(\gamma_H)$ as $T \to 0$ where n is the density of water molecules.

time-independent microstates. The statistical mechanical approach must be justified in some manner. We would be remiss not to at least mention some of the ideas
that have been explored.

The classic rationale rests on the notion of ergodicity. The ergodic hypothesis
posits that for a sufficiently complex system (i.e., a system with chaotic dynamics), and for generic initial conditions, the time-averaged properties of the system
are equal to a thermal average over microstates. Since the timescales characterizing microscopic processes (such as collisions between atoms) are typically fast
compared to the timescales over which macroscopic systems are observed, the
implicit additional assumption here is that macroscopic measurements are well
approximated as averages over all time.

Ergodicity can be proven for classical systems only in very limited circumstances.
It certainly agrees with our experience that most macroscopic material systems, if
left to their own devices, rapidly lose information about the state in which they were
prepared at some initial time—i.e., they relax to their equilibrium state. However,
even where ergodicity can be formally established, there are few theoretical results
that offer an understanding of the *rate* at which measurable quantities approach
their equilibrium values, and how large errors one should expect if the system is
not allowed sufficient time to equilibrate. Indeed, in some systems, such as in very
viscous liquids and glasses, the equilibration process is observed to be arbitrarily
slow. In these situations an equilibrium description is not suitable.

While the ergodic hypothesis sounds intuitively reasonable for classical systems,
it relies on notions from classical mechanics, especially on notions of classical chaos.
Thus, its application to closed quantum systems is even more vexing. Remarkably recently, a fully quantum argument justifying the applicability of statistical
mechanics has been developed, known as the Eigenstate Thermalization Hypothesis
(ETH).[13] The essential insight here—which applies to classical systems as well—is
that if we look at a subregion of a sufficiently large system, then the rest of the system
acts as a heat bath coupled to the subregion of interest. In the context of ETH, the
proposal is that the expectation value of any operator acting in a small subregion
is the same in almost any energy eigenstate with a given energy. Equivalently, this
implies that the expectation value in a given energy eigenstate is equal to its value
in the microcanonical ensemble.

Interest in these fundamental issues is having a renaissance now as experimental techniques allow new ways to measure the properties of macroscopic systems
over a large range of time and lengthscales, and new ways of constructing material systems are permitting studies of closed quantum systems of controllable size.
However, for the purposes of this book, we shall accept without further ado the
ensemble approach as the central tenet of statistical mechanics: measurable properties of macroscopic systems in equilibrium are given by the appropriate averages
over microstates.

---

13. Some recent reviews include Deutsch, *Rep. Prog. Phys.* **81** 082001 (2018) and D'Alessio, Kafri,
Polkovnikov, and Rigol, *Advances in Physics*, **65**(3) 239–362 (2016).

An understanding of the properties of matter in thermal equilibrium allows an understanding of a large range of observable phenomena. Even when systems are slightly perturbed away from equilibrium, statistical mechanics can predict the response properties of the systems. For example, the conductivity of a metal is given by Ohm's law, $\vec{j} = \sigma \vec{E}$, where $\vec{E}$ is an (assumed weak) electric field whose application drives the system slightly out of equilibrium, and $\vec{j}$ is the resulting nonequilibrium current density that arises as a consequence. Importantly, it turns out that the conductivity, σ, can be expressed in terms of an equilibrium correlation function, even though σ itself is a non-equilibrium quantity. The conductivity σ is an example of a linear response function, as it determines the response (in this case, the current) of a system initially in equilibrium to the application of an external perturbation (in this case, the electric field).

In Chapter 7, we will see explicitly how the linear response of a system to an external perturbation can be expressed in terms of equilibrium correlation functions. While the specific example of the conductivity is probably the most familiar example of a linear response function to most students, it is a rather subtle one and computing it is beyond the scope of the present book. This is because it is a dissipative response in the sense that entropy is produced in the nonequilibrium state. Thus identifying the appropriate correlation function, known as the Kubo formula, involves dynamical aspects of the quantum problem that we will not treat explicitly.[14]

Finally, it is worth mentioning that plenty of interesting physics cannot be treated from a thermodynamic perspective. Clearly, this caveat applies to microscopic systems, since a statistical description can only be justified in the limit of a large number of degrees of freedom. But even for macroscopic systems, equilibrium considerations are of limited usefulness for systems which are far from equilibrium.

2.3 Symmetry

In his famous article, "More is different," P. W. Anderson wrote, "it is only slightly overstating the case to say that physics is the study of symmetry." While this may be more than a slight exaggeration, it is certainly not an understatement that symmetry is exceedingly important.

The reader will have previously seen examples in which an analysis of the symmetries of a problem can be used to simplify a calculation. However, in the analysis of phases and phase transitions, the role of symmetry takes on another level of significance. This section is meant to establish some basic understanding of the concepts, so when we re-encounter them later in the text there will be some sense of understanding already. We will first review the meaning of symmetry, then put it in to the context of statistical mechanics. We will conclude the section with a look at the meaning of "broken symmetries."

---

14. If you are interested, see Chapter 7 of the classic graduate-level textbook, Chaikin and Lubensky's *Principles of Condensed Matter Physics* (Cambridge University Press, Cambridge, 2000).

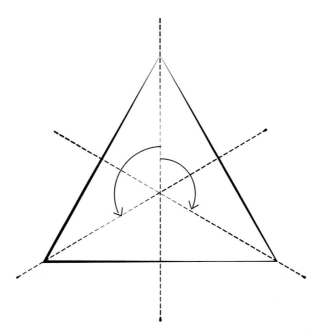

FIGURE 2.2. For an equilateral triangle, the shape is identical to its starting configuration if reflected across any of the dashed lines, or rotated by 120 degrees either clockwise or counterclockwise.

2.3.1   Meaning of "Symmetry"

At a young age, we encounter visual representations of geometric shapes for which the symmetries are summarized by statements such as "it stays the same when you flip it left to right" (e.g., Figure 2.2). A five-blade ceiling fan has a symmetry under "rotation about its axis by 72 degrees." In analysis, we learn to classify even and odd functions, where $f_{even}(x) = f_{even}(-x)$, $f_{odd}(x) = -f_{odd}(-x)$; in other words, even functions are invariant under[15] $x \to -x$ and odd functions are invariant under $x \to -x$ followed by negation. More abstract symmetries such as time-reversal symmetry (e.g., Newton's Laws are the same under $t \to -t$) and Lorentz symmetry (e.g., Maxwell's equations are the same under a Lorentz transform of space-time) also play a key role in elementary physics.

More formally, any "transformation" in which the transformed object is indistinguishable from the initial object is a symmetry transformation of that object. Here, an object means anything that we can characterize precisely, e.g., functions, geometric shapes, graphs, etc.[16] We use the word *invariant* to mean that the object is the same before and after a transformation. An object possesses a symmetry if it is invariant under a transformation.

---

15. "Under" meaning "having gone through the operation."

16. The word "transformation" may evoke a more visual or physical interpretation, whereas the word "operation" may feel more mathematically abstract. However, both mean the same thing in the context of symmetries, and are used interchangeably.

It may seem esoteric at first, but this language allows us to quickly describe symmetries. Using some of the examples from before:

- The equilateral triangle (object) in Fig. 2.2 is invariant under:
  - reflection across any of the three central lines (transformation),
  - rotation by 0, ±120 degrees about its center (transformation).
- An even function $f_{even}(x)$ (object) is invariant under $x \to -x$ (transformation).

The formal mathematical structure for studying symmetries is called group theory; the set of symmetry transformations that leave a given object invariant forms a group. The basics of group theory are sketched in Appendix A.10. However, the discussion in the main text is framed in such a way that familiarity with group theory is not assumed.

2.3.2 Symmetries in Statistical Mechanics

In the context of statistical mechanics, we are interested in answering the following questions: For a given system, what are the symmetries of the underlying physics that govern the behavior of the system? What are the symmetries exhibited by the macroscopic properties of the same system? How do the differences in these symmetries arise?

We are familiar with the idea that the apparent symmetries in the macroscopic world are a subset of the fundamental symmetries of the microscopic physics. For instance, the principle of relativity is a statement concerning the underlying symmetries of the laws of physics—they are translationally invariant, time-translationally invariant, rotationally invariant, and boost invariant. However, in our everyday life, we experience a world that is spatially structured, that changes with time, in which up and down are differentiated, and in which we know perfectly well when we are moving and when we are standing still. Indeed, we age in a way that makes the notion of time-reversal symmetry seem absurd.

At a more concrete level, one of the key issues we will confront here is the nature and origin of spontaneously broken symmetries. The Hamiltonian, H, encodes the set of microscopic physical processes that govern the microstates of the system. The set of symmetry transformations that leave H invariant is the group of microscopic symmetries of the system in question. Moreover, all statistical mechanical averages are defined to have a Boltzmann weight defined by H. Thus, one might reasonably expect all thermodynamic quantities to exhibit the same group of symmetries as H. Surprisingly—or perhaps unsurprisingly?—they do not. The exact manner in which phases of matter can have less symmetry than the underlying physics will be discussed with some care in Chapter 5. More generally, since symmetry is a precise concept—an object is either invariant under a given transformation or it is not—classifying the symmetries of different phases of matter plays a central role in all the ensuing discussion.

2.4 Problems

2.1. *Equivalence of canonical and microcanonical ensembles in the thermodynamic limit.* In the canonical ensemble, the entropy is defined as in Eq. 2.10, but in the microcanonical ensemble, it is defined as $S = \log \Omega(E)$, where $\Omega(E)$ is the number of microstates with energy E. Show that the two definitions are equivalent in the thermodynamic limit. *Hint:* In large systems, microstates that contribute significantly to the free energy have energy very close to $E(T)$. Write down the partition function as an integral over E and use the saddle point approximation.

2.2. *Free energy at a 1st-order transition.* Consider a system exhibiting two phases of matter with free energies $F_1(T)$ and $F_2(T)$. Suppose that the system undergoes a 1st-order transition at a temperature T_c.

   (a) What condition determines T_c?

   (b) In general, without a special symmetry, there is no reason to think that any derivatives of F_1 and F_2 have anything to do with each other. Suppose that phase 1 is the phase of higher entropy at the transition, i.e., $S_1(T_c) > S_2(T_c)$. Which phase is the equilibrium phase at T slightly above T_c? *Hint:* $S = -\partial F/\partial T$.

   (c) The latent heat is defined as the discontinuity in the internal energy at T_c. Write an expression for the latent heat in terms of $S_1(T_c)$ and $S_2(T_c)$. *Hint:* $F = E - TS$.

   (d) Conclude that the phase stable at the lower temperature has the smaller internal energy.

2.3. **Moments, cumulants, and generating functions.* Let $p(x)$ be the probability distribution over some variable x. The mth moment of the distribution is defined $\langle x^m \rangle = \int dx\, p(x) x^m$.

   (a) Consider the moment-generating function, $\phi(t) \equiv \langle e^{tx} \rangle$. Write down a Taylor expansion of $\phi(t)$ about $t = 0$ and find the relation between the moments and the derivatives of $\phi(t)$ at $t = 0$.

   The cumulants $\langle x^m \rangle_c$ are related to the moments. The first cumulant is the mean, $\langle x \rangle_c = \langle x \rangle$, and the second cumulant is the variance, $\langle x^2 \rangle_c = \langle x^2 \rangle - \langle x \rangle^2$. Cumulants are related to derivatives of the cumulant-generating function, $f(t) \equiv \log \langle e^{tx} \rangle = \log \phi(t)$, in the same way that moments are related to derivatives of the moment-generating function.

   (b) Use the Taylor expansion of $f(t)$ to find the third cumulant in terms of the first three moments. *Hint:* $\log(1 + x) = x - x^2/2 + x^3/3 + \dots$.

   Consider a system with Hamiltonian $H = E + \lambda A$. Up to some constants, the partition function is a moment generating function, and the free energy is a cumulant generating function.

(c) Write down an expression for $\langle A^m \rangle$ in terms of derivatives of the partition function.

(d) Write down an expression for the $\langle A^m \rangle_c$ in terms of derivatives of the free energy.

2.5 Worksheets for Chapter 2

W2.1 Calculating Thermal Averages

While we assume some of these exercises are just a review of prerequisites, it is never a bad idea to refresh our memory on these things.

Free energy and the partition function

The definition of the (Helmholtz) free energy is

$$F \equiv E - TS \tag{2.37}$$

and that of the partition function is

$$Z = \sum_s e^{-\beta E_s}, \tag{2.38}$$

where E_s is the energy of state s, and $\beta = 1/T$, such that the probability of state s occurring is:

$$P(s) = \frac{e^{-\beta E_s}}{Z}. \tag{2.39}$$

1. The energy term E in the free energy is the expectation value of E_s. Find an expression for $E \equiv \langle E_s \rangle$ in terms of the partition function Z and inverse temperature β.

2. We assume the energy of each microstate is a function of generalized coordinates q_j, such that $E_s = E_s(q_j)$. Associated with each is a generalized force $f_{sj} = -\frac{\partial E_s}{\partial q_j}$. Find an expression for $f_j \equiv \langle f_{sj} \rangle$ in terms of Z, β, and q_j.

3. Since the partition function should depend on both temperature and the generalized coordinates, we can write it as $Z = Z(\beta, q_j)$. Start with the expression for the total differential of $\ln Z$,

$$d \ln Z = \frac{\partial \ln Z}{\partial \beta} d\beta + \frac{\partial \ln Z}{\partial q_j} dq_j, \tag{2.40}$$

and prove that $F = -T \ln Z$ follows from the definition Eq. 2.37.
    You may find the first law of thermodynamics useful:

$$dE = TdS - f_j dq_j. \tag{2.41}$$

The derivative theorem

We make use of this throughout the book, so understand it well!

4. Given that the Hamiltonian of a system is

$$H(s) = E(s) + \lambda A(s), \qquad (2.42)$$

show that

$$\langle A \rangle = \frac{\partial F}{\partial \lambda}. \qquad (2.43)$$

Hint: Use the definition of the partition function and $F = -T \ln Z$.

Thermal averages: Part two

Using Eq. 2.3, show that

5.

$$\left\langle [M_a - \langle M_a \rangle]^2 \right\rangle = \left\langle M_a^2 \right\rangle - \langle M_a \rangle^2. \qquad (2.44)$$

6.

$$\langle [M_a - \langle M_a \rangle][M_b - \langle M_b \rangle] \rangle \qquad (2.45)$$

$$= \langle M_a M_b \rangle - \langle M_a \rangle \langle M_b \rangle. \qquad (2.46)$$

Fluctuation and susceptibility

7. Show that

$$\langle M_a M_b \rangle = T^2 \frac{1}{Z} \frac{\partial^2 Z}{\partial B_a \partial B_b}. \qquad (2.47)$$

8. Given that

$$\chi_{a,b} = -\frac{1}{V} \left(\frac{\partial^2 F}{\partial B_a \partial B_b} \right), \qquad (2.48)$$

use the product rule to show that:

$$\chi_{a,a} = \frac{1}{TV} \Delta_{M_a}^2, \qquad (2.49)$$

where the fluctuation of M_a is

$$\Delta_{M_a}^2 = \langle M_a M_a \rangle - \langle M_a \rangle \langle M_a \rangle. \qquad (2.50)$$

W2.2 Identifying Symmetries

The Ising ferromagnet is an example of a system with an ordered state that breaks a discrete symmetry. In the fully ordered ($T = 0$) state, there are two symmetry-related ground states, one with all spins up and the other with all spins down. In other systems, the symmetry broken can be a continuous symmetry. Below, we enumerate some common symmetry classes.

Symmetry	Form of the order parameter
Ising	$\sigma = \{\pm 1\}$
XY [O(2)]	$\vec{s} = (\cos\theta, \sin\theta); 0 \le \theta < 2\pi$
Heisenberg [O(3)]	$\vec{s} = (\cos\phi\sin\theta, \sin\phi\sin\theta, \cos\theta); |\vec{s}|^2 = 1$
q-state clock model	$\vec{s} = (\cos\theta, \sin\theta); e^{iq\theta} = 1$

For the following physical situations, indicate the model that fits their symmetry (breaking). We are introducing you to the language of symmetries, so don't worry if not every word makes sense yet:

1. Ferromagnetism of spins subject to an easy-axis magnetic anisotropy (i.e., spins prefer to point along a single axis).

2. Ferromagnetism of spins subject to an easy-plane magnetic anisotropy (i.e., spins prefer to lie in a 2D plane).

3. Ferromagnetism of spins without any magnetic anisotropy.

4. Picking a preferred bond orientation on a honeycomb lattice.

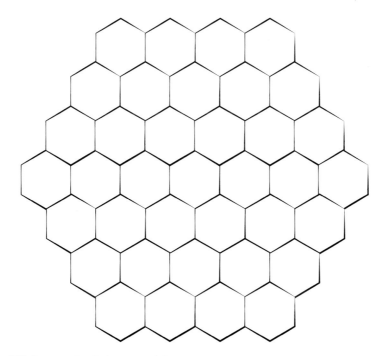

FIGURE W2.1. Example of a honeycomb lattice.

5. Distortion of a tetragonal lattice (which has two equivalent axes).

6. A superfluid transition, where the order parameter is a "condensate wave-function," specified by a complex number. As is true of all wavefunctions in quantum mechanics, the wavefunction can be multiplied by an overall phase without changing the expectation value of any operator.

3 THE ISING MODEL

> *The Ising model is to statistical mechanics as the fruit fly is to genetics, but without the larvae and the gooey stuff they eat.*

Here we introduce a paradigmatic model of statistical physics: the Ising model. It is a simple model of a system with many interacting parts. Despite its simplicity, the Ising model captures the essence of a variety of rich phenomena exhibited by interacting systems, such as broken-symmetry phases and thermodynamic phase transitions.

The Ising model is an abstract mathematical model; it has no pretense of accurately representing any real physical system. What is the motivation for studying it? Alas, for any realistic system, it is impossible to reliably calculate the thermodynamic properties; even the Ising model has no exact solution in three dimensions. Rather than tackling realistic Hamiltonians, our strategy is to begin with simplified models that capture some essential behavior.

Abstraction has an additional advantage of allowing the Ising model to emulate a wide variety of phase transitions. A solution of the Ising model lends insight into many problems at once. In one fell swoop it teaches us about ferromagnetic transitions, liquid-gas transitions, order-disorder transitions, and more. In a deeper sense, by being divorced from a concrete interpretation, the Ising model gives a universal framework for understanding a range of phenomena, beyond the quirks of any particular system. Nevertheless, for intuitive clarity, we will tend to view the Ising model as a depiction of a ferromagnet, as it was originally conceived.

3.1 Definition

The Ising model consists of N variables, referred to as "spins," arranged on a lattice of sites (Figure 3.1). Each spin can take on two possible values, ± 1. The sites of the lattice are labeled with an index $j = 1, 2, \ldots, N$, and the spin of the jth site is denoted by σ_j, with $\sigma_j = +1$ for spin "up" ($\uparrow$) and $\sigma_j = -1$ for spin "down" ($\downarrow$). There are 2^N distinct ways the spins can be configured; each configuration is a microstate of the system. The energy of a state reflects interactions between spins on nearby lattice sites.

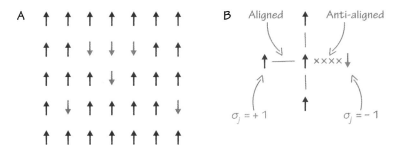

FIGURE 3.1. The Ising model consists of spins on a lattice. Each spin can take on one of two values. There is a bond between each pair of neighboring spins. The model may be defined for arbitrary dimensions; shown above is a depiction in $d = 2$ dimensions on a square lattice.

To draw the correspondence between the Ising model and a ferromagnet, think of each site as an atom arranged in a crystal lattice. The spin on each site represents a magnetic dipole moment associated with each atom. Each spin can point either up or down. If the spins tend to align in the same direction, then the magnetic effect of each spin adds up to yield an overall net magnetic moment, known as ferromagnetism. Conversely, if the directions of the spins are scrambled in a way that their magnetic moments cancel out, then there is no net macroscopic moment, known as paramagnetism.

The simplest version of the Hamiltonian of the Ising model is

$$H = -h \sum_{j=1}^{N} \sigma_j - J \sum_{\langle ij \rangle} \sigma_i \sigma_j. \tag{3.1}$$

The first term represents the tendency of a spin to align with an externally applied magnetic field h; a spin-up has energy $-h$, and a spin-down has energy $+h$. The greater the magnitude of h, the greater the energetic preference for one spin direction over the other. Each individual spin contributes separately to this sum. The second term is the interaction between spins on neighboring sites of the lattice. The notation $\langle ij \rangle$ means a sum over pairs of nearest-neighbor sites. When the spins on two adjacent sites are aligned ($\uparrow\uparrow$ or $\downarrow\downarrow$), the energy is $-J$; when they are anti-aligned ($\uparrow\downarrow$ or $\downarrow\uparrow$), the energy is $+J$. Worksheet W3.1 gives some practice using this Hamiltonian.

The sign of J is physically significant:

- If $J > 0$, then an aligned pair of adjacent spins is lower in energy than an anti-aligned pair. If $h = 0$, there are two degenerate ground states: (1) all the spins are pointing up[1] ($\uparrow\uparrow\uparrow\uparrow$), and (2) all the spins pointing down ($\downarrow\downarrow\downarrow\downarrow$). This is known as a ferromagnet, and interactions with $J > 0$ are known as ferromagnetic interactions.
- In contrast, if $J < 0$, then neighboring spins have an energetic incentive to anti-align. Such a system is known as an antiferromagnet. The nature of its

1. Sometimes we call this fully polarized.

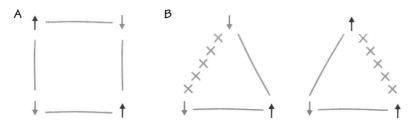

FIGURE 3.2. The ground-state of the Ising antiferromagnet depends on the nature of the lattice. (A) The square lattice is unfrustrated. (B) A triangular lattice is frustrated: it is impossible to simultaneously minimize all the interaction energies.

$h = 0$ ground state depends on how the sites are arranged in the lattice (see Figure 3.2). On a square lattice, it is possible for all adjacent pairs of spins to be simultaneously anti-aligned, and the ground state is a checkerboard of alternating up and down spins. In contrast, on a triangular lattice, this is impossible and there is necessarily at least one (energetically unfavorable) aligned pair of neighboring spins on each triangle. Such systems are "frustrated" (see Worksheet W3.2).

Since frustration introduces a host of additional complexities, we will from here on out restrict our attention to Ising ferromagnets, with $J > 0$.

Most systems of interest are three-dimensional, but it is also valuable to study the Ising model in $d = 2$ (such as a square lattice) or $d = 1$ (such as a chain of spins). As a mathematical abstraction, the Ising model can also be studied in $d > 3$. The type of lattice structure (e.g., square or triangular) can sometimes lead to interesting effects, but for concreteness, we will focus on the d-dimensional hypercubic lattice: a square lattice for $d = 2$, a cubic lattice for $d = 3$, etc.

Did you know? At times, it is useful to think about d as a *continuous* parameter—i.e., to think about systems whose dimensionality is non-integer. Indeed, one of the great breakthrough papers in statistical physics, which we will touch upon in the final chapter, was entitled "Critical phenomena in 3.99 dimensions."

3.1.1   What We Would Like to Know

To "solve the Ising model" means to calculate the free energy and its various derivatives. These include the following thermodynamic quantities:

- The magnetic moment $M \equiv \sum_j \langle \sigma_j \rangle$ is the sum of all the spins.
- The magnetization $m \equiv M/N$ is the average spin direction. It is the intensive quantity associated with the extensive quantity M. If the spins are nearly all aligned up (down), then m is close to $+1$ (-1); if the spins are randomly

oriented with equal numbers up and down, then $m = 0$. In a uniform state, the average spin on every site is the same, i.e., $m = \langle \sigma_j \rangle$ for all j.

- The (zero-field) susceptibility $\chi \equiv \frac{\partial m}{\partial h}\big|_{h=0}$ measures how sensitive the magnetization is to an external magnetic field. If the susceptibility is large, a small field causes a big change in the magnetization.

- The two-site "connected" correlation function[2]

$$G(i,j) \equiv \left\langle \left(\sigma_i - \langle \sigma_i \rangle \right) \left(\sigma_j - \langle \sigma_j \rangle \right) \right\rangle = \langle \sigma_i \sigma_j \rangle - \langle \sigma_i \rangle \langle \sigma_j \rangle \qquad (3.2)$$

measures the extent to which the ith and j^{th} spins are correlated. Because interactions are local, we expect that the correlation is larger between nearby spins and smaller if i and j are further apart. This encodes the answer to the question: if we know that the spin at site i is up, how does that affect the probability (what is the conditional probability) that the spin at site j is also up?

- The correlation length ξ is the characteristic length over which spins are correlated; in many cases, the correlation function falls off as $G(i,j) \sim e^{-|i-j|/\xi}$. Two spins separated by less than ξ are substantially correlated; two spins separated by much more than ξ are essentially uncorrelated. A patch of aligned spins has a typical size of ξ.

We are interested in determining the manner in which these quantities depend on the temperature T, the coupling strength J, and the strength of the external magnetic field h. Typically we consider the thermodynamic limit $N \to \infty$.

One key question of the Ising model is whether local interactions can lead to a collective effect—whether the tendency for nearby spins to align can lead to more global alignment of a macroscopic number of spins. Long-range order of this sort would be reflected in an average spin value, m, which deviates from zero. Of course, m is automatically nonzero if an external magnetic field is applied ($h \neq 0$) to bias one spin direction over the other. The more interesting question is whether it could be possible for $m \neq 0$ even in the *absence* of an external magnetic field. Such a collective ordering of spins is an example of a spontaneously broken symmetry—without any external field, the system has no impetus to prefer $m > 0$ or $m < 0$, but does so anyway. This such an important concept that we will devote more attention to it in Chapter 5.

3.2 Qualitative Properties

Despite the simple appearance of the Hamiltonian, Eq. 3.1, it is challenging to directly calculate quantities such as m as a function of T and h. We will thus begin with qualitative considerations and simple approximations.

---

2. Note that this is the same correlation function from Chapter 2, now using site indices instead of position vectors. "Connected" means that we have subtracted the second term.

If a broken-symmetry phase occurs in an Ising model, it is expected at low temperatures, since thermal excitations tend to scramble the spins. At exactly zero temperature, the system is in the lowest energy state where all spins point in the same direction, aligned with the external magnetic field. Thus, $m = \text{sign}(h)$ at $T = 0$. (That is, $m = 1$ if $h > 0$ and $m = -1$ if $h < 0$.) As the temperature is increased from zero, thermal excitations cause some spins to flip, so the average magnetization is closer to zero. In the limit of infinite temperature, all configurations are equally likely regardless of their energy, and so $m(h) = 0$ for $T = \infty$.

An important feature of the zero-temperature limit is that the magnetization at $h = 0$ depends on how the $h \to 0$ limit is approached. If h approaches 0 from positive values, then the magnetization approaches 1, i.e., $\lim_{h\to 0^+} m(h) = 1$. In contrast, $\lim_{h\to 0^-} m(h) = -1$. In both cases, the magnitude of the magnetization is nonzero, i.e., $|m(h)| \to 1 \neq 0$ as $h \to 0$. This nonzero magnetization is the defining feature of a broken symmetry phase, explored more formally in Chapter 5.

Given that the symmetry is broken at $T = 0$ and not at $T = \infty$, there are three natural possibilities for the thermal phase diagram. It is possible that the broken symmetry exists only at $T = 0$ (which turns out to be the case in $d \leq 1$). It is possible that the symmetry is restored only as $T \to \infty$ (which is never the case for physically reasonable circumstances). There can be a nonzero critical [3] temperature, T_c, separating the two phases such that such that the symmetry is broken below T_c and restored above T_c, i.e.,

$$\begin{cases} \lim_{h\to 0} |m(h)| \neq 0, & T < T_c \\ \lim_{h\to 0} |m(h)| = 0, & T > T_c. \end{cases} \tag{3.3}$$

3.2.1 The Droplet Argument in $d = 2$

In two dimensions, there is[4] a clear line of analysis showing that the broken-symmetry phase survives to nonzero T. In Worksheet W3.3 you have the chance to work through this logic in detail.

Consider the Ising model on a square lattice with a small but positive h so that the ground state is the configuration with all spins up and $m = 1$. At $T = 0$, the thermal ensemble contains only the ground state. At low but nonzero T, the thermal ensemble includes excited states with an energy greater than the ground state but still relatively low energy.

The lowest-energy excitation is the flip of a single spin, as illustrated in Figure 3.3B; flipping a single spin creates a domain wall of length 4 (in dotted lines,

---

3. Note that there must be a precise critical temperature, because the magnetization is either zero or nonzero. There is no in-between!

4. A rigorous version of this argument was originally constructed by Peierls before it was generally accepted that statistical mechanics could really account for phase transitions and for broken-symmetry phases.

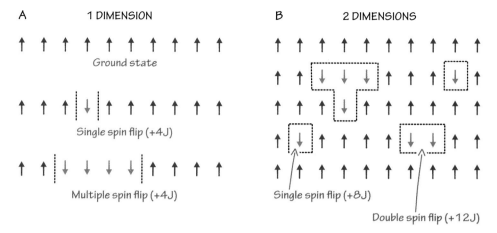

FIGURE 3.3. The excitations of the Ising model are domains of flipped spins. Walls between domains are shown as dotted lines. (A) In one dimension, the energy of a domain is independent of its size. (B) In two (or more) dimensions, the energy of a domain depends on the length (or hyperarea) of its border.

between neighboring spins). A domain wall is a boundary between a region of up spins and a region of down spins. At each segment of the domain wall, the interaction energy is $+J$ instead of $-J$, giving an overall cost of $2J$ units of energy per segment. As there are 4 segments here, the energy of this excitation is $8J$ above the ground-state.

The next highest excitation (also shown in the figure) is the flip of two adjacent spins, which has a domain wall of length 6, and costs $12J$ energy.

In general, the excitations are domains of spin-downs ("droplets") in a "sea" of spin-ups (see Figure 3.4A). To determine whether it is favorable to form these "droplets," consider the change in the free energy[5] as a result of creating a domain with a wall of length L,

$$\Delta F(L) = \Delta E(L) - T\Delta S(L). \tag{3.4}$$

If the overall $\Delta F(L)$ is positive, it is thermodynamically unfavorable to form droplets; if it is negative, it is favorable. The change in energy from such a droplet is $\Delta E(L) = 2LJ$. The change in entropy is $\Delta S(L) = \log \mathcal{N}(L)$, where $\mathcal{N}(L)$ is the number of ways to form a droplet with domain wall length L. There is an inherent trade-off between these two terms; larger droplets (with longer perimeters) are less energetically favorable, but they are more entropically favorable because they have more possible shapes (higher multiplicity).

---

5. This is not in the strict sense a calculation of the free energy, which as a thermodynamic potential requires considering all the possible states and their associated Boltzmann weights. Rather, the quantity ΔF reflects the relative thermodynamic likelihood for one state to occur in reference to some other state, i.e., $\Delta F > 0$ means it is less likely and $\Delta F < 0$ means it is more likely.

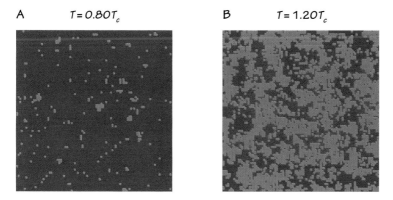

FIGURE 3.4. (A) At low temperatures, a typical configuration of the Ising spins consists of small domains of down spins (represented by blue squares) embedded in a sea of up spins (red squares). (B) At high temperatures, the up and down domains are equivalently distributed. These configurations are taken from a Monte Carlo simulation on a 100×100 square lattice using codes written by the authors.

In order to count $\mathcal{N}(L)$, we first observe that a domain wall must trace a closed path on the lattice with no self-intersections. Since at every step we have between 1 to 3 possible choices of direction, the total number of possibilities after L steps must satisfy

$$1^L < \mathcal{N}(L) < 3^L. \tag{3.5}$$

For large L, $\mathcal{N}(L)$ asymptotically approaches α^L for a particular value of α between 1 and 3 (see Worksheet W3.3).[6] This gives a useful approximation for the entropy, $\Delta S(L) = \log \mathcal{N}(L) \simeq L \log(\alpha)$. For large L, the number of ways to form a droplet grows exponentially, so the entropy grows linearly with L. We thus obtain

$$\Delta F(L) = \Delta E(L) - T \Delta S(L)$$

$$\cong L \left\{ 2J - T \log \alpha \right\}. \tag{3.6}$$

The sign of the quantity in braces depends on T. For $T < 2J/\log \alpha$, it is positive, so forming large droplets is thermodynamically unfavorable. As such, at low T, the excitations are small islands of overturned spins, and these thermal excitations produce only small corrections[7] to the all-spin-up ground state. This is illustrated in Figure 3.4A. However, for $T > 2J/\log \alpha$, it becomes thermodynamically favorable (decreases the free energy) to form arbitrarily large droplets—so the picture of "a few droplets of spin-downs in a sea of spin-ups" is no longer valid, as in Figure 3.4B. This suggests that there is a phase transition at $T_c \approx 2J/\log \alpha$, from a phase where spins are overall aligned but with small pockets of misalignment, to a phase where

---

6. In fact, the value of α has been determined to be ≈ 2.638. See Jensen and Guttman, *J. Phys. A: Math. Gen.* **32**, 4867 (1999), and Clisby and Jensen, *J. Phys. A: Math. Theor.* **45**, 115202 (2012).

7. Self-consistently, so long as ΔF is large compared to T, the concentration of droplets of size L (i.e., their contribution to the partition function) must be small in proportion to $e^{-\beta \Delta F(L)}$.

the excitations are so large and numerous that the distinction between island and sea, up and down, is lost. Snapshots of computer simulations of these two phases are illustrated in Figure 3.4. In the latter case, exactly half of the spins are up and other half are down in the limit $h \to 0^+$.

Droplet argument in arbitrary dimension

The argument for $d = 2$ actually works for all dimensions $d \geq 2$. To see this, consider a domain of down spins with a mean radius R. In general, in d dimensions, the number of sites in the domain scales as a "hypervolume," $V \sim R^d$, and the number of bonds on the boundary of the domain scales as a "hyperarea," $A \sim R^{d-1}$. For large R, the free energy[8] associated with such a domain is

$$F_R \sim A \left\{ 2J - T \log[\alpha] \right\}, \tag{3.7}$$

where the number of distinct domains scales as α^A.

For all $d > 1$, the situation is very similar to what we just discussed in $d = 2$: at temperatures $T \ll 2J / \log[\alpha]$, the free energy cost of a large domain grows strongly with increasing R, and therefore the concentration of large domains of overturned spins (roughly proportional to $\exp[-F_R / T]$) is exponentially small and thus negligible. This corresponds to the intuitive picture of small islands of flipped spins in a large area.

However, for $d = 1$ the situation is completely different. In this case the boundary of a domain is the same regardless of how large it is, $A \sim R^0$. This reflects an important feature of the excited states in $d = 1$, illustrated in the middle and lower rows of Figure 3.3A. The single spin-flip state has an excitation energy of $4J$ relative to the ground state—but any other state with multiple contiguous flipped spins also has the same excitation energy. Thus, at nonzero temperature arbitrarily large domains are no less likely than small domains. We will see below why this precludes the possibility of an ordered phase.

We reach the conclusion that, for all $d > 1$, the Ising ferromagnet exhibits a broken-symmetry phase below a nonzero critical temperature, T_c.[9]

3.2.2   Low-*T* Series Expansion in Arbitrary *d*

To confirm that $d = 2$ is the lowest dimension in which the Ising model exhibits a phase transition, let us turn to a more precise way to formulate the above discussion. Our goal, here, is to obtain an expansion for the magnetization—or any other intensive thermodynamic property—as a Taylor series in powers of $e^{-\beta J}$. (Note that $e^{-\beta J}$ is small if T is close to zero.) If this series is convergent for some value of $e^{-\beta J}$,

---

8. At this stage, we can also justify why the infinitesimal magnetic field may be ignored in the above analysis: the magnetic field adds a term $2Vhm$ to the domain free energy, where $m = -1$ is the magnetization inside the domain, but so long as $h \ll J/R$, this term is negligible.

9. This statement only applies to systems with short-range interactions. If a system has interactions with long enough range, it can indeed manifest a nonzero T phase transition even in $d = 1$.

we can infer from it that the results at this temperature must differ little from those found at $T=0$. Conversely, if the series diverges then the phase is quantitatively different than at $T=0$. More formally, the $T=0$ phase persists at least for the range of temperatures over which the series is convergent. The details are worked out in Worksheet W3.4.

Again, let us consider the Ising model at $T=0$ with $h=0^+$ on a d-dimensional hypercubic lattice with N total sites. As before, the unique ground-state is the configuration with all spins up[10] and $m=1$. We will measure energies relative to the energy of the ground state, $E_G = -2dNJ$[11].

At low temperatures, the only significant microstates are those with an energy close to the ground-state energy. Let us sum up the Boltzmann weights of these microstates to see how powers of $e^{-\beta J}$ naturally arise in the expression for Z. We group the states according to their energy. The first contribution comes from the ground state. The lowest-energy excitations are single spin-flips which break $2d$ bonds and cost $4dJ$ in energy with a Boltzmann factor $e^{-4d\beta J}$. The multiplicity of such states is N, since there are N total sites where the single spin-flip can occur. The next highest energy excitation, which flips two neighboring spins, breaks $4d-2$ bonds, and costs $2(4d-2)J$ in energy, with a Boltzmann factor $e^{-2(4d-2)\beta J}$. The multiplicity is Nd, since there are d orientations for the two adjacent spins, and N sites to place the spins per orientation.

Continuing in this manner, enumerating the higher-energy excited states term by term, we arrive at an expression for the partition function in orders of the excitation energy,

$$Z = e^{2dN\beta J}\left[1 + Ne^{-4d\beta J} + Nde^{-2(4d-2)\beta J} + \mathcal{O}\left(e^{-8d\beta J}\right)\right]. \tag{3.8}$$

The notation $\mathcal{O}(\epsilon)$ means corrections of order ϵ or smaller, where ϵ is presumed to be a small quantity. There is a subtlety here: while we use a system of finite size N here to facilitate calculations, eventually we are interested in the thermodynamic limit $N\to\infty$. The implication here is that the second term (which is $\propto N$) is already infinitely larger than the first, no matter how small T is. To make matters worse, higher order terms are superextensive.[12] For instance, there are $N(N-d-1)/2$ ways to flip two non-neighboring spins, meaning that they make a contribution $N(N-d-1)/2\ e^{-4d\beta J}$ to Z. This grows in proportion to N^2. There is no way that this expansion converges for $N\to\infty$.

---

10. To make the discussion technically precise, we take h to be positive and let $h\to 0$ as $N\to\infty$ in such a way that $Nh\to\infty$. This condition is imposed such that we only need to consider states where $m>0$. This footnote is discussed in greater detail in Chapter 5.

11. Since an overall offset in energy in the Hamiltonian, and therefore the free energy, does not affect the physics of the system, we are justified in doing so. Of course, good bookkeeping is important, so this is accounted for by an overall factor of $e^{-\beta E_G} = e^{2dN\beta J}$ in the partition function.

12. Superextensive meaning $\propto N^k$, where $k>1$, i.e., more than just proportional to the size of the system.

The resolution to this conundrum is to instead calculate a physically meaningful, intensive quantity which is well behaved in the limit $N \to \infty$ (see Worksheet W3.4 for an explicit example). An example of a such a quantity is the average magnetization. This can be calculated as the expectation of any particular spin,

$$m = \langle \sigma_1 \rangle = \frac{1}{Z} \sum_{\{\sigma_i\}} \sigma_1 e^{-\beta E(\{\sigma_i\})}$$

$$= \frac{e^{2dN\beta J}}{Z} \left[1 + (N-2)e^{-4d\beta J} + (N-4)de^{-2(4d-2)\beta J} + \mathcal{O}\left(e^{-8d\beta J}\right) \right]$$

$$= 1 - 2e^{-4d\beta J} - 4de^{-2(4d-2)\beta J} + \mathcal{O}\left(e^{-8d\beta J}\right), \tag{3.9}$$

where we used the expansion $(1+x)^{-1} = 1 - x + \mathcal{O}\left(x^2\right)$ in the last step. Note that terms proportional to N from the numerator and denominator have canceled each other, so all dependence on N has vanished in the final expression.

Equation 3.9 is the low-temperature expansion of the magnetization. Qualitatively, we can understand each term as follows: The first term is the zero-temperature magnetization. At low T, the single spin flip excitations will cause a small fraction $\sim e^{-4d\beta J}$ of the sites to be flipped. This gives a leading-order correction of $-2e^{-4d\beta J}$ to the magnetization (with the prefactor of 2 resulting from going from a $+1$ to a -1 spin). Similarly, the double spin-flip excitations will give a correction on the order of $e^{-2(4d-2)\beta J}$, with the prefactor accounting for the spin-flip as well as the multiplicities possible due to geometry. The low-temperature expansion is discussed in more depth and rigor in Appendix A.2.

In one dimension, the series in Eq. 3.9 will never converge. This is already clear from comparing the T-dependence of the second and third term.[13]

In two dimensions, however, the terms in the expansion become successively smaller if $e^{-\beta J}$ is small enough:

$$m = 1 - 2e^{-8\beta J} - 8e^{-12\beta J} + \mathcal{O}\left(e^{-16\beta J}\right). \tag{3.10}$$

This corroborates our expectation that the broken-symmetry phase persists to T nonzero for T small compared to J. Moreover, the prefactors of successive terms— which reflect the multiplicity of droplets from the droplet analysis—get increasingly larger. Therefore the series does not converge if $e^{-\beta J}$ is too large. This is consistent with our expectation that the phase of matter is qualitatively different for $T > T_c \sim J/\log[\alpha]$; namely, the high-temperature phase is a mixture of relatively small patches of up and down spins, such that $m \to 0$ as $h \to 0$.

---

13. Contrast $d = 1$ with the case of the droplets in $d \geq 2$ in which all low-energy excitations involve a small number of spins, so they do not destroy order.

3.3 Solution in Zero Dimensions

Having completed our qualitative discussion, we now turn to exact solutions, beginning with the simple case of $d = 0$.

The zero-dimensional Ising model is a single spin on a single site, with Hamiltonian

$$H = -h\sigma. \tag{3.11}$$

Since the spin is isolated with no neighbors to interact with, the interaction terms from Eq. 3.1 are missing. Given some finite value for h, the energy of the system depends on the spin state, either $\sigma = +1$ with energy $-h$, or $\sigma = -1$ with energy h. The partition function is

$$Z = \sum_{\sigma = \pm 1} e^{-\beta H(\sigma)} = 2\cosh[\beta h], \tag{3.12}$$

and the free energy is

$$F = -T\log Z = -T\log[2\cosh(\beta h)]. \tag{3.13}$$

The magnetization m can be calculated directly as the thermal average of σ. Since the probability of spin-up ($\sigma = +1$) is $e^{+\beta h}/Z$ and the probability of spin-down ($\sigma = -1$) is $e^{-\beta h}/Z$, the thermal average of σ is

$$m = \langle \sigma \rangle = \frac{1}{Z} \sum_{\sigma = \pm 1} \sigma e^{-\beta H(\sigma)} = \tanh(\beta h). \tag{3.14}$$

The magnetization can also be calculated as a derivative of the free energy with respect to the conjugate field h, yielding the same result,

$$m = -\left(\frac{\partial F}{\partial h}\right)_T = \tanh(\beta h). \tag{3.15}$$

The magnetic susceptibility is found by taking another derivative with respect to h:

$$\chi = \frac{\partial m}{\partial h} = \beta \operatorname{sech}^2(\beta h). \tag{3.16}$$

The magnetization as a function of the external field is shown in Figure 3.5 at various nonzero temperatures. It exhibits the following properties. When $h = 0$, the average spin is zero, as required on the basis of symmetry. As h increases, the energy gap between the two spin states grows, so the lower-energy spin state is more and more favored. For higher temperatures, a greater field strength is needed to achieve the same magnitude of m. If $h \to \pm\infty$, then correspondingly $m \to \pm 1$.

Of particular interest is the $T \to 0$ limit. At exactly zero temperature, $m = \operatorname{sign}(h)$. If the external field is taken to zero, there is ambiguity as to which state the spin will occupy because both states are equal in energy. It will depend on whether $h \to 0^+$ or $h \to 0^-$.

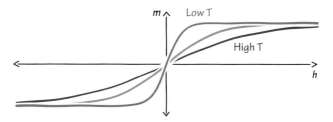

FIGURE 3.5. The magnetization of an isolated Ising spin plotted as a function of external magnetic field, for a few different temperatures.

3.3.1 $J = 0$ limit in d dimensions

The zero-dimensional solution of a single Ising spin may be extended to arbitrary dimensions if the spins are noninteracting ($J = 0$). The only difference is that there are N independent copies rather than one. Thus, the free energy per site and the magnetization are given by the single site expressions, Eqs. 3.13 and 3.15, respectively. Moreover, since there are no interactions, there are necessarily no statistical correlations between spins on different sites; mathematically, this means the spin-spin correlation function factorizes:

$$\langle \sigma_i \sigma_j \rangle = \langle \sigma_i \rangle \langle \sigma_j \rangle, \quad (i \neq j). \tag{3.17}$$

As with the zero-dimensional case, there are no distinct phases in this limit: $m \to 0$ as $h \to 0$ for all $T > 0$.

3.4   Solution in One Dimension

The Ising model in one dimension is also readily solvable. The solution gives us a chance to explicitly see the method involved in summing over all the states (i.e., all possible configurations of spins), so it is instructive to go through the details in the worksheets.

In one dimension, the Hamiltonian can be written as

$$H = \sum_{j=1}^{N} \left[-h\sigma_j - J\sigma_j \sigma_{j+1} \right]. \tag{3.18}$$

This form of the Hamiltonian assumes periodic boundary conditions $\sigma_{N+1} \equiv \sigma_1$, where the spins live on a ring with sites 1 and N bonded to each other. In this case each site has exactly two neighbors, simplifying some of the mathematics.

3.4.1 $h = 0$ Solution–The Combinatorics Approach

In the simpler case with $h = 0$, there is a straightforward method for calculating Z based on combinatorics. (Worksheet W3.5 follows this derivation step-by-step.) The first step is to identify the microstates by specifying the positions of all domain walls between regions of up and down spins. Because of the periodic boundary conditions, the number of domain walls k must be even. Each spin configuration corresponds to a unique configuration of domain walls, but each configuration of domain walls corresponds to exactly two spin configurations—related by flipping all the spins. The partition function can then be expressed as

$$Z = 2e^{N\beta J} \sum_{k \, \text{even}}^{N} \mathcal{N}(k,N)e^{-2\beta Jk} = 2e^{N\beta J} \sum_{k \, \text{even}}^{N} \frac{N!}{(N-k)!k!}e^{-2\beta Jk}, \qquad (3.19)$$

where $\mathcal{N}(k,N)$ is the number of ways there can be k domain walls in a 1D system of size N, $e^{-2\beta Jk}$ is the Boltzmann weight associated with k domain walls, and the ground-state energy is accounted for by the overall factor of $e^{N\beta J}$. The overall factor of 2 is due to the degeneracy in energy from a global spin-flip.[14] With the binomial theorem, the summation can be directly performed to yield

$$Z = e^{N\beta J} \left[(1 + e^{-2\beta J})^N + (1 - e^{-2\beta J})^N \right]. \qquad (3.20)$$

We see from this that the partition function is an analytic function of T for all nonzero T. Therefore, the free energy is also analytic everywhere and there can be no phase transition (see Section 1.1.4).

3.4.2 The General Solution Using the Transfer Matrix Approach

For the general solution with $h \neq 0$, we adopt a different approach using the "transfer matrix." The idea is to consider the energy $K(\sigma_a, \sigma_b)$ between pairs of adjacent spins σ_a and σ_b,

$$K(\sigma_a, \sigma_b) = -\frac{h}{2}(\sigma_a + \sigma_b) - J\sigma_a\sigma_b, \qquad (3.21)$$

such that the Hamiltonian can be re-expressed as

$$H = \sum_{j=1}^{N} K(\sigma_j, \sigma_{j+1}). \qquad (3.22)$$

It is helpful to define a matrix where the entries, $t_{\sigma,\sigma'}$, are the Boltzmann weights of the four possible bond types,

$$t = \begin{bmatrix} e^{-\beta K(\uparrow,\uparrow)} & e^{-\beta K(\uparrow,\downarrow)} \\ e^{-\beta K(\downarrow,\uparrow)} & e^{-\beta K(\downarrow,\downarrow)} \end{bmatrix}. \qquad (3.23)$$

This is known as the transfer matrix.

14. This is the manifestation of the spin-flip symmetry discussed in Section 5.1. It is applicable here due to the absence of h, the "symmetry-breaking field."

Much of what we wish to know can be derived simply from the properties of t. These calculations are worked through in Worksheet W3.6. The first important observation is that the Boltzmann weight associated with a given spin configuration is

$$e^{-\beta H(\{\sigma\})} = t_{\sigma_1,\sigma_2} t_{\sigma_2,\sigma_3} \cdots t_{\sigma_L,\sigma_1} = \prod_j t_{\sigma_j,\sigma_{j+1}}. \tag{3.24}$$

Thus the sum over spin configurations necessary to compute Z becomes simply the sums over repeated indices implied by matrix multiplication. As a result, the partition function can be expressed as

$$Z = \text{Tr}\left[t^N\right]. \tag{3.25}$$

Did you know? It is interesting to note that for finite N, this analysis relates the classical statistical mechanics of the 1D Ising model to the quantum statistical mechanics of a single quantum qubit (see Problem 3.5).

To proceed further, we need to find the eigenvalues $\lambda_\pm$ and eigenvectors $|\psi_\pm\rangle$, with the result (as shown in the worksheet)

$$\lambda_\pm = e^{\beta J}\left[\cosh(\beta h) \pm \sqrt{\sinh^2(\beta h) + e^{-4\beta J}}\,\right]. \tag{3.26}$$

It thus follows that

$$Z = \sum_{a=\pm} \langle\psi_a|\, t^N\, |\psi_a\rangle = [\lambda_+]^N + [\lambda_-]^N. \tag{3.27}$$

In the limit $N \to \infty$, since $\lambda_+ > \lambda_-$, the second term is exponentially smaller than the first, and so can be ignored. Thus $\lim_{N\to\infty} Z = [\lambda_+]^N$, and consequently the free energy is

$$F = -T \log Z = -NT \log \lambda_+, \tag{3.28}$$

the magnetization is

$$m = -\frac{1}{N}\frac{\partial F}{\partial h} = \frac{\sinh(\beta h)}{\sqrt{\sinh^2(\beta h) + e^{-4\beta J}}}, \tag{3.29}$$

which vanishes as $h \to 0$ for all T, and the connected correlation function between two spins on sites i and j is

$$G(i,j) = \langle\sigma_i\sigma_j\rangle - \langle\sigma_i\rangle\langle\sigma_j\rangle = \left(\frac{\lambda_-}{\lambda_+}\right)^{j-i} |\langle\psi_+|\tau_z|\psi_-\rangle|^2, \tag{3.30}$$

where τ_z is the Pauli-z matrix.

Defining

$$g_0(h) \equiv \langle \psi_+ | \tau_z | \psi_- \rangle, \tag{3.31}$$

$$\xi \equiv \left| \log \left(\frac{\lambda_-}{\lambda_+} \right) \right|^{-1}, \tag{3.32}$$

the correlation function takes on the simple form

$$G(i,j) = |g_0(h)|^2\, e^{-|i-j|/\xi}, \tag{3.33}$$

where $|g_0(h)|^2$ characterizes the strength of correlations, and ξ is the characteristic correlation length of the system. In the limit $h \to 0$, these expressions simplify to

$$g_0(0) = 1, \quad \xi = |\log(\tanh \beta J)|^{-1}. \tag{3.34}$$

Eq. 3.33 implies that the correlation between two spins fall off exponentially with distance, becoming negligible for spins separated by distances much greater than the correlation length ξ. As $T \to 0$, the correlation length diverges as $\xi \sim (1/2)e^{2J/T}$, meaning that the domains get larger and larger as $T \to 0$. This agrees with the intuitive conclusion from Section 3.2.1 that there is no broken-symmetry state for $T > 0$.

Did you know? The transfer matrix method is related to the transition matrix of Markov chains in probability theory. In this view, the configurations are obtained by walking down the chain step-by-step and sampling each successive spin. The probability that any particular spin is up or down depends only on the previous spin.

3.5 Solution in Two Dimensions

Exact solutions of interacting models in $d > 1$ are rare. In a feat of mathematical gymnastics, the great statistical physicist, Onsager, first solved the 2D Ising model on the square lattice in 1944. His solution is famously inscrutable, but others after him have come up with simpler ones.[15] To this day, however, the analytic expression for $h \neq 0$ is unknown. Solutions in three or more dimensions are also unknown. It is humbling that the simplest model of statistical mechanics cannot be solved in three dimensions!

The exact solution confirms (as expected) that there exists a ferromagnetic phase with $m \neq 0$. The critical temperature is found to be the solution of the implicit

---

15. See Baxter's book, *Exactly Solved Models in Statistical Mechanics* (Academic Press, London, 1982).

equation[16] $\sinh(2J/T_c) = 1$, or in other words,

$$T_c = 2J/\log(1 + \sqrt{2}) \simeq 2.27J. \tag{3.35}$$

Furthermore, for $T < T_c$, an expression for the mean value of the spin $m = \langle \sigma_j \rangle$ was announced by Onsager, but only later derived (in print) by Yang:

$$m = \left[1 - \text{csch}^4(2\beta J)\right]^{1/8} \sim \left[\frac{T_c - T}{T_c}\right]^{1/8} \left[1 + \cdots\right], \tag{3.36}$$

where ... refers to higher order terms that vanish as $T \to T_c$. For T near T_c, the leading-order behavior is $m \sim (T_c - T)^\beta$ with a critical exponent of $\beta = 1/8$. (Note that this is different than the experimentally observed exponent of $\beta \simeq 1/3$ in three-dimensional magnetic systems!)

Remarkably, the exact transition temperature of the square-lattice Ising model can be derived even without knowledge of the exact solution. The square lattice is actually quite special in this respect because it has a property known as self-duality. This is discussed further in Section 3.7.3.

3.6   Variations on the Ising Model

The essence of the Ising model is a lattice of interacting up/down spins. We have discussed a particular form of the model—one where the lattice is cubic and the interactions occur between the nearest neighbors—but beyond this, one can conceive a wide range of variations which retain the same Ising spirit. We could consider lattices of different geometries in 2D, in 3D, or in even higher dimensions (Figure 3.6A). We could also consider more complicated couplings that allow each spin to interact with nearby spins beyond the nearest neighbor (Figure 3.6B). The interactions might be ferromagnetic (favoring alignment) or antiferromagnetic (favoring anti-alignment), and they might extend over a short range or a long range.

It is important to understand which details matter. We have seen that the dimensionality of the lattice plays an important role, but we have not fully explored the effects of lattice geometry or interaction structure. As we will soon appreciate, many "microscopic details" have no impact on the nature of the phases or phase transitions. (See Problem 4.4.) It is pretty clear that if an Ising system has interactions where nearby spins tend to align, then at low enough temperatures, it will exhibit a ferromagnetic ordered phase in $d > 1$.[17] The transition temperature, however, will

---

16. For an anisotropic version of the same model with interactions J_x and J_y between nearest-neighbor spins in the x and y directions, T_c is found to be the implicit solution of the equation $\sinh(2J_x/T_c)\sinh(2J_y/T_c) = 1$. See Problem 4.1.

17. In $d = 1$, if the interactions range over a long enough distance, then there is still a possibility of an ordered phase—see Appendix A.11.

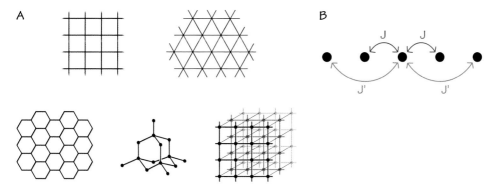

FIGURE 3.6. (A) Various lattice structures. Clockwise from upper left: square, triangular, cubic, diamond, honeycomb. (B) A situation with nearest-neighbor couplings J and next-nearest-neighbor couplings J'.

certainly depend on the range of interactions and the geometry of the lattice. In Chapters 5 and 6 we will further discuss why certain features are universal and others are dependent on microscopic details.

3.7 *Low-T Expansion, High-T Expansion, and Duality

3.7.1 Low-Temperature Expansion to All Orders

It is informative to think about the formal structure of the low-temperature series for the partition function—not just keeping the first few terms in powers of $e^{-\beta J}$ as we did in Section 3.2.1, but for all orders. Formally, we can start from a ground-state configuration in which all the spins are up ($\sigma_j = +1$) and then can sum over all patterns of domains of overturned spins:

$$Z = \sum_{\substack{\text{domain} \\ \text{configurations}}} \exp[-\beta E_0 - \beta 2 J A], \qquad (3.37)$$

where E_0 is the ground-state energy, $2J$ is the energy cost of having a bond connecting two sites with opposite spins, and A is the total number of bonds that cross the boundary between regions of up and down spins.[18] At low T, where βJ is large, the important terms in this sum have a low concentration of overturned spins—or equivalently, a relatively low concentration of generally small domains.

---

18. As we saw above, there will always be troubles with convergence of this sum in the thermo-dynamic limit since the number of terms grows exponentially with the size of the system, but for any finite-size system there is only a finite (even if very large) number of terms in this sum (i.e., all possible ways of making domains of up and down spins), so we need not worry about the convergence of this sum.

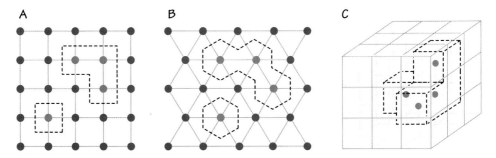

FIGURE 3.7. Structure of domain boundaries for (A) square lattice, (B) triangular lattice, and (C) cubic lattice. Red circles represent up spins and blue circles represent down spins. The boundaries between domains are shown in dotted black lines.

In Eq. 3.37 the spin configurations are enumerated by summing over all possible domain structures. The geometric meaning of this depends on the system at hand. In one dimension, the boundary between a region of up spins and a region of down spins is the "bad bond" between them—as shown in Figure 3.3A. Thus, the sum is over all sets of locations of bad bonds, and A is the total number of bad bonds. In two dimensions, the boundary between domains is a closed loop, hence the sum is over all possible ways of drawing nonoverlapping closed loops, and A is the total length of the loops. This is shown in Figures 3.7A (for the square lattice) and 3.7B (for the triangular lattice). In three dimensions, the boundary between domains is a closed surface, and hence A is the total surface area of all the bounding surfaces enclosing regions of down spins (Figure 3.7C).

In all these cases, the domain boundaries live on the "dual lattice." To be concrete, we focus on two dimensions, where the dual lattice consists of sites that sit at the centers of the plaquettes of the original lattice—i.e., at the centers of the squares of the square lattice and of the triangles of the triangular lattice, as illustrated in Figures 3.7A and 3.7B. Note that the dual lattice of the square lattice is itself a square lattice, while the dual of the triangular lattice is a hexagonal lattice. Conversely, the dual of the hexagonal lattice is the triangular lattice. A configuration of loops corresponds to a set of occupied bonds on the dual lattice which forms closed paths. (This is equivalent to the condition that an even number of bonds emanates from each dual lattice site.)

Crucially, there is a one-to-one correspondence between each configuration of spins on the original lattice and each pattern of closed loops on the dual lattice. Therefore, we can define an equivalent problem with an identical partition function,

$$Z = e^{-\beta E_0} \sum_{\substack{\text{loops on} \\ \text{dual lattice}}} e^{-\beta E_{loop}}. \tag{3.38}$$

Comparing this to Eq. 3.37, the energy associated with a set of loops should be $E_{loop} = 2JA$, with A being the total length of loops—i.e., the number of occupied bonds.

We thus find an equivalence between the Ising ferromagnet on the original lattice, Eq. 3.37, and a model of loops on the dual lattice, Eq. 3.38.

What we have shown is a rather remarkable concept: there exist pairs of statistical mechanics models that are equivalent in the sense that both models give the same result for the partition function—and hence for the free energy. With a little more work, we could have shown corresponding equivalences for various derivatives of the free energy. From a thermodynamic perspective, the problem of an Ising ferromagnet on a square lattice is equivalent to the problem of nonoverlapping loops on the dual square lattice. The low-temperature phase can be described either in terms of the original variables as consisting of a high concentration of spin-up sites, or in terms of the dual variables as having a low concentration of loops.

One might wonder if this rewriting of the original model in terms of new variables is useful, or just a bit of mathematical gymnastics. A remarkable result is that this duality allows us to derive an exact expression for the critical temperature of the Ising model on a square lattice—which we will return to below in the discussion of high-temperature series. But there is a more fundamental point of perspective that is worth noting. We began our discussion in this book by focusing on the Ising ferromagnet as our go-to illustrative model, but we already were sensitive to the idea that the spins could be taken to have various possible physical meanings (see Section 5.2). But suppose we could just as well have taken—as a natural problem in statistical mechanics on which to base our discussion—the problem of closed surfaces on a lattice with an energy proportional to A. Then, somewhere along the way we would have discovered the equivalence between our "physical" model and a mathematical abstraction, the Ising ferromagnet. Who is to say a priori which is the natural physical model and which a mathematical construct?

3.7.2 High-T Series Expansion in Arbitrary d

At high temperatures, the typical configuration of spins looks like a random mixture of up and down spins. This is a very different situation than at low temperatures, where the important configurations involve relatively rare (dilute) spin-flips. But there is a clever way to reformulate the problem at high temperatures so that it is amenable to a systematic high-temperature expansion.

We start with the observation that the piece of the Boltzmann weight that comes from the interaction between two neighboring spins can be expressed as

$$\exp[\beta J \sigma_i \sigma_j] = X[1 + Y \sigma_i \sigma_j] \ \text{ with } \ X = \cosh(\beta J) \ \& \ Y = \tanh(\beta J). \qquad (3.39)$$

(This follows from the fact that the only two allowed values of $\sigma_i \sigma_j$ are ± 1, as you can check for yourself.) This allows us to write an expression for the partition function,

$$Z = \sum_{\{\sigma\}} \exp\left[-\beta J \sum_{<i,j>} \sigma_i \sigma_j\right] = X^{N_b} \sum_{\{\sigma\}} \prod_{<i,j>} \left[1 + Y \sigma_i \sigma_j\right], \qquad (3.40)$$

where N_b is the number of bonds, e.g., on a d-dimensional hypercubic lattice $N_b = dN$. As usual, the sum is over all spin configurations, i.e., $\sum_{\{\sigma\}} \equiv \prod_j \sum_{\sigma_j=\pm 1}$. It is important to note that $Y \to 0$ as $T \to \infty$, i.e., in the high-temperature limit, Y is small.

We now perform a bit of abstract analysis to give a graphical representation of the product over bonds. For every bond, we can choose either the factor 1 or the factor proportional to Y. In expanding out the product, any given term contains a factor of Y for an "occupied" bond, and a factor of 1 for an unoccupied bond. The full product, therefore, corresponds to a sum over all configurations of occupied bonds:

$$Z = X^{N_b} \sum_{\{\sigma\}} \sum_{\substack{\{\text{bonds} \\ i_n,j_n\}}} \left[(Y\sigma_{i_1}\sigma_{j_1})(Y\sigma_{i_2}\sigma_{j_2})\ldots \right]. \tag{3.41}$$

Next, for each configuration of occupied bonds, we can evaluate the sum over spins. Consider the sum over the spin on one site, σ_j. If there are n occupied bonds that touch site j, then the term is proportional to σ_j^n; thus, for n odd, the sum over σ_j vanishes, while for n even, the sum over σ_j gives a factor of 2. After performing the sums over spins on all sites, the only terms which remain are the terms where each site is touched by an even number of bonds. This is a constraint on the configurations of bonds—only configurations which correspond to nonoverlapping loops on the lattice are allowed. Thus,

$$Z = Z_\infty \sum_{\text{loops}} Y^L, \tag{3.42}$$

where $Z_\infty = X^{N_b} 2^N$, the sum runs over all configurations of occupied bonds that form nonoverlapping loops on the lattice, and L is the total length of loop in the given configuration. Clearly, for small Y, the configurations that contribute significantly to the partition function have relatively few loops of typically small extent—the high-temperature phase can be characterized as a dilute system of small loops.

3.7.3 Duality

Does this remind you of something? On the square lattice, we wrote the low-temperature series in terms of a sum over domain walls separating domains of up and down spins. But we also noted that the sum over domains is equivalent to a sum over nonoverlapping loops on the dual lattice. The Boltzmann weight associated with each particular configuration of loops, as expressed in Eq. 3.42, was proportional to $\left[e^{-2\beta J}\right]^A$, where A—the number of bonds crossing the boundary of spin-down domains—is simply the length of the loops. Since the dual lattice of the square lattice is itself a square lattice, this means that the sum over loops has exactly the same form in the low- and high-temperature series, with only the change of variables, $Y \leftrightarrow \left[e^{-2\beta J}\right]$.

This is elegant, but it is also practical. We expect the sum over loops that appears in the low-temperature expansion to be well behaved so long as the weight, $e^{-2\beta J}$, is sufficiently small. Conversely, we expect the loop expansion to diverge as $T \to T_c$, that is to say, as $e^{-2\beta J} \to e^{-c}$ where $e^{-c} \equiv e^{-2J/T_c}$. We do not, a priori, know the value of $c = 2J/T_c$. However, the same loop sum appears in the high-temperature series. It should also be well behaved for Y small, and should diverge as $Y \to e^{-c}$. Thus, by comparing the low- and high-temperature series, we conclude that[19]

$$e^{-2J/T_c} = \tanh \left[J/T_c \right] \quad \to \quad c = \ln \left[1 + \sqrt{2} \right]. \tag{3.43}$$

As alluded to earlier, this agrees with the exact solution of the Ising model on the square lattice!

Alas, the situation is more complicated on other 2D lattices, since in general the dual lattice has a different structure than the original lattice (see Problem 3.8). It is even more complicated in 3D, where the high-temperature series is (again) a series of loops while the low-temperature series involves a sum over closed surfaces. The Ising model on the square lattice is special because it is self-dual. But the insight of considering dual problems is of broader significance. A number of subtle dualities underlie recent theoretical advances[20] in statistical mechanics and quantum field theory.

3.8 Problems

3.1. *Droplets in one dimension.* Try to repeat the droplet argument for the one-dimensional Ising model. Where does it go wrong?

3.2. *Applications.* Explore the internet for applications of the Ising model to different problems.

3.3. *Ising antiferromagnet.* Consider the Ising *anti*ferromagnet on a square lattice in the absence of an external field, with Hamiltonian

$$H = K \sum_{ij} s_i s_j, \tag{3.44}$$

where K is positive so that an anti-aligned bond $\downarrow\uparrow$ is *lower* in energy than an aligned bond $\uparrow\uparrow$.

(a) What are the two ground states of the system? Do they exhibit long-range order?

(b) The magnetization $m = N^{-1} \sum_j \langle s_j \rangle$ is not very useful in measuring the ordering of the antiferromagnet. Why?

---

19. This assumes that there is only one critical temperature at which a transition occurs between the low and high-temperature phases.

20. If you want to be a tourist in the strange land of dualities, take a look at Seiberg, Senthil, Wang, and Witten, *Annals of Physics* **374** 395–433 (2016). You are not expected to understand the article, but if you did, go to graduate school immediately!

(c) Devise a better set of variables, which we will call σ_j, to represent the microstates of an antiferromagnet, such that the *staggered* magnetization $m_s = \langle \sigma_j \rangle$. How is σ_j related to s_j? *Hint:* $m_s = -1$ in one ground state, $m_s = +1$ in the other, and $m_s = 0$ in the disordered state ($T \rightarrow \infty$).

(d) By rewriting the antiferromagnetic Hamiltonian in terms of the σ_j variables, argue why the behavior is equivalent to that in the ferromagnetic case when $h = 0$.

3.4. *Binary alloy.* Consider a solid material composed of two types of elements, A and B, arranged in a regular lattice. This can be modeled as an Ising system, where $\sigma_j = 1$ represents an atom of type A at site j and $\sigma_j = -1$ represents an atom of type B at site j. Suppose that there is an interaction energy between neighboring atoms of ϵ_{aa}, ϵ_{ab}, or ϵ_{bb}, depending on the identities of the atoms.

(a) The Hamiltonian may be rewritten in the form

$$H = E_0 - h \sum_j \sigma_j - J \sum_{\langle ij \rangle} \sigma_i \sigma_j. \qquad (3.45)$$

What are the parameters E_0, h, and J, in terms of ϵ_{aa}, ϵ_{ab}, and ϵ_{bb}?

(b) Comment on the fact that the same Hamiltonian can describe both a ferromagnet and a binary alloy. Interpret the $J > 0$ and $J < 0$ cases.

3.5. *Analogy to the quantum mechanics of spin-half.* The math of the transfer matrices is analogous to the thermodynamics of a quantum mechanical system.

(a) Suppose a quantum mechanical system has Hamiltonian $\hat{H}$ with energy levels E_n. What is its partition function at a temperature T written as a trace?

(b) This expression is reminiscent of the partition function of a 1D Ising chain, Eq. 3.25. What is the quantity corresponding to the transfer matrix t? The length of the chain N?

We have shown that the statistical mechanics of the one-dimensional Ising model has the same mathematics as a zero-dimensional (i.e., a single) quantum mechanical spin. In general, a $(d+1)$-dimensional statistical mechanical model is equivalent to a d-dimensional quantum mechanical one.

3.6. *Classical XY model in one dimension.* Consider a model where each spin, instead of pointing up or down, points at any angle $0 \leq \theta_j < 2\pi$ on the plane. These are called XY spins. Consider a chain of N XY spins with Hamiltonian

$$H = -J \sum_{j=1}^{N} \cos(\theta_{j+1} - \theta_j), \qquad (3.46)$$

where the cosine means that the energy of a bond is minimized when the two neighboring spins are pointing in the same direction. Assume periodic boundary conditions $\theta_1 \equiv \theta_{N+1}$.

(a) Write down the partition function as an integral over each of the θ_j coordinates. Define a transfer matrix t such that $Z = \text{Tr}\left[t^N\right]$. What is $\langle\theta|t|\theta'\rangle$?

(b) Find the eigenvalues of t, $t|m\rangle = \lambda_m|m\rangle$. You may find it helpful that

$$e^{\beta J \cos(\theta-\theta')} = \sum_{n=-\infty}^{+\infty} I_n(\beta J)e^{in(\theta-\theta')},$$

where the I_n are the modified Bessel functions of the first kind.

(c) Evaluate the partition function in terms of λ_m in the thermodynamic limit. Note that $I_n(x) = I_{-n}(x)$ and that $I_0(x) > I_n(x)$ for $n \neq 0$.

(d) Find the two-spin correlator, $\langle e^{\theta_i - \theta_j}\rangle$, in the thermodynamic limit. Show that it is of the form $\exp[-|i-j|/\xi]$. What is the correlation length ξ?

3.7. *Monte Carlo simulation.* Using your favorite programming language, write a simulation of the 2D Ising model using the Monte Carlo metropolis method. Measure various correlation functions as a function of the temperature, especially near the critical temperature $T_c \approx 2.27J$? How does the pattern of spins visually appear near criticality? A nice JavaScript implementation which can run in your web browser can be found at:

http://mattbierbaum.github.io/ising.js/

3.8. *Duality.* Consider the four following problems:

1. Ising spins on a triangular lattice.

2. Ising spins on a hexagonal lattice.

3. Nonoverlapping loops on a triangular lattice.

4. Nonoverlapping loops on a hexagonal lattice.

(a) Up to a constant, write down the form of the low-temperature and high-temperature series expansions of Z for models (1) and (2). Be sure to specify the expansion parameter as well as what lattice the sum of loops runs over.

(b) Among models (1), (2), (3), and (4), which pairs of problems are equivalent?

(c) Let T and T' be a pair of temperatures which are related to each other as $e^{-2J/T} = \tanh(J/T')$, or equivalently, $\sinh[2J/T]\sinh[2J/T'] = 1$. Suppose that we know the solution to model (1) at temperature T. Based on this information, of which other models, and at what temperature(s), can we also determine the solution?

(d) Repeat part (c), but instead suppose we have solved model (2) at temperature T.

(e) Can you figure out how the transition temperatures of models (1) and (2) are related? (In both cases rest assured that there is a unique T_c.)

3.9 Worksheets for Chapter 3

W3.1   The Ising Hamiltonian–A Primer

In Section 3.1, we learned that in the absence of a magnetic field, the Hamiltonian of the Ising model is

$$H = -J_1 \sum_{n.n.} \sigma_i \sigma_j. \tag{3.47}$$

Before moving on to more complicated lattices and an infinite number of sites (i.e., thermodynamic limit $N \to \infty$), let us practice with a 4-site model. We will start simple, with just the nearest-neighbor (n.n.) interactions, J_1. With the above definition, $J_1 > 0$ is a ferromagnetic interaction, whereas $J_1 < 0$ is antiferromagnetic.

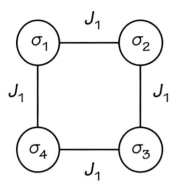

FIGURE W3.1. Ising model with 4 sites coupled by nearest-neighbor interaction, J_1.

1. In a table, enumerate all of the possible spin configurations, and write down the corresponding energy of each spin configuration. You can reduce the size of the table by treating pairs of "flipped" states as equivalent, i.e., $\uparrow\uparrow \atop \uparrow\uparrow$ is equivalent to $\downarrow\downarrow \atop \downarrow\downarrow$.

2. What are the corresponding Boltzmann factors of each configuration? Assuming that $J_1 > 0$, what is the ground-state ($T \to 0$) configuration? What about for $J_1 < 0$?

W3.2   More on the 4-Site Problem

Let us extend the previous question and examine its behavior with next-nearest-neighbor (n.n.n.) (see Figure W3.2) interactions, J_2:

$$H = -J_1 \sum_{n.n.} \sigma_i \sigma_j - J_2 \sum_{n.n.n.} \sigma_i \sigma_j. \tag{3.48}$$

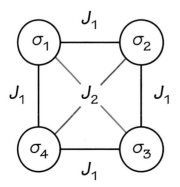

FIGURE W3.2. Ising model with 4 sites coupled by nearest-neighbor interaction J_1, as well as next-nearest-neighbor interaction, J_2.

1. In a table, again enumerate the possible spin configurations, as well as the total energy of each configuration, treating pairs of flipped states as equivalent.

2. Since we now have two parameters to tune, our phase diagram no longer lives on just a line, as before. Describe the nature of the ground state(s) in each region of the phase diagram shown in Figure W3.3, as well as on the phase boundaries.

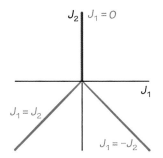

FIGURE W3.3. Semi-completed phase diagram of the next-nearest-neighbor 4-site Ising model.

3. If the four sites are arranged as a tetrahedron such that the interactions for any given pair of spins are identical, where does this situation arise in the phase diagram in Figure W3.3?

W3.3   The Droplet Argument in 2D

We start the 2D Ising system in its ground state (all aligned) on an infinite square lattice. Without loss of generality, let us assume all the spins are pointing up. The Hamiltonian is

$$H = -J \sum_{\langle ij \rangle} \sigma_i \sigma_j. \qquad (3.49)$$

1. What is the change in energy from a single pair of spins that is switched from ↑↑ to ↑↓?

2. Sketch out all the droplet configurations that have domain walls of the following lengths. A droplet is a group of connected spins, and the domain wall length L is the total length of the boundary enclosing the droplet.

   (a) $L = 4$:

   (b) $L = 6$:

   (c) $L = 8$:

3. Assume that inside each droplet the spins are all pointing down, and outside the spins are all pointing up. What is the overall energy change ΔE from the ground state for each of the above cases? What is the general expression for $\Delta E(L)$?

4. For each of the prescribed wall lengths, how many different configurations $\mathcal{N}(L)$ were possible? If we assume there are on average α choices when forming each segment of the wall such that $\mathcal{N}(L) = \alpha(L)^L$, what is $\alpha(L)$ for each L?

L	$\mathcal{N}(L)$	$\alpha(L)$
4		
6		
8		
10	28	1.395

We get the above results by requiring the walls to form a closed, non-self-intersecting polygon. Without these constraints, a contiguous wall of length L will have 3^L ways of forming. From the above exercise, we should gain some intuition for the argument presented in the main text: for each closed wall of length L, we have at least one way and at most 3^L ways to construct it, i.e.,

$$1 \le \mathcal{N}(L) = \alpha(L)^L < 3^L. \tag{3.50}$$

By performing this algorithmically on a computer[21], the current best estimate of the asymptotic value is $\lim_{L \to \infty} \alpha(L) \approx 2.638$. You can get a sense of this from the plot in Figure W3.4.

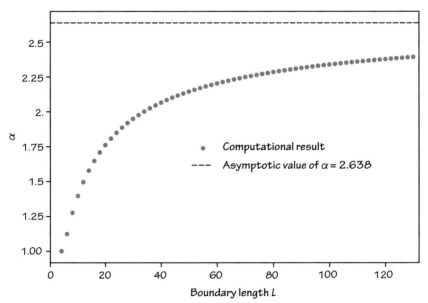

FIGURE W3.4. Value of α evaluated computationally up to $L = 130$, and the asymptotic value.

21. Jensen and Guttman, *J. Phys. A: Math. Gen.* **32**, 4867 (1999), and Clisby and Jensen, *J. Phys. A: Math. Theor.* **45**, 115202 (2012).

5. We can assign a quantity

$$\Delta F(L) = \Delta E(L) - T\Delta S(L) \tag{3.51}$$

for the droplet states of length L, where $\Delta S(L) = \log \mathcal{N}(L)$ represents the change in entropy associated with these states. What is the physical meaning $\Delta F(L)$? *Hint:* The "partition function of droplets" can be written as:

$$Z = \sum_L \mathcal{N}(L) e^{-\beta \Delta E(L)}. \tag{3.52}$$

6. The partition function stops being analytic at a certain temperature T_c, indicating a phase transition. You can make the approximation that $\alpha(L) = \alpha$, i.e., a constant. Find T_c. *Hint:* The geometric series converges iff $|r| < 1$:

$$\sum_{k=0}^{\infty} ar^k = \frac{a}{1-r}. \tag{3.53}$$

W3.4   Ising Model in 0D

The zero-dimensional Ising model is just an isolated spin. Since there are no interactions, there is only one term in the Hamiltonian:

$$H = -h\sigma. \tag{3.54}$$

Although this model won't necessarily teach us much about collective behavior, it is a nice example to see the calculations of statistical mechanics in action!

1. Given that the two possible states are $\sigma = \pm 1$, write down the partition function $Z(\beta, h) = \sum_\sigma e^{-\beta H(\sigma)}$.

2. Calculate the magnetization $m = \langle \sigma \rangle$:

   (a) Directly from the definition:

$$\langle \sigma \rangle = \frac{1}{Z} \sum_\sigma \sigma e^{-\beta H(\sigma)}. \tag{3.55}$$

   (b) From the free energy calculated via the partition function. Recall the derivative theorem from Section 2.1.3.

3. Calculate the susceptibility $\chi = \frac{\partial m}{\partial h}$.

W3.5   Ising Model in 1D–The Combinatorics Approach

For solving the Ising model in 1D, if the external field is $h = 0$, then we can employ a combinatorics method to calculate various quantities.[22] In this case, we have, again, the Hamiltonian:

$$H = -J \sum_j \sigma_j \sigma_{j+1}. \tag{3.56}$$

For our purposes, the spins occupy a 1D chain with N sites with periodic boundary conditions (i.e., $\sigma_{N+1} = \sigma_1$). Since the model has only nearest neighbor interactions, it is convenient to picture the ring equivalently as an ordered set of N bonds, on each of which the spins are either parallel ($E = -J$) or anti-parallel ($E = J$).

1. How many possible ways can k anti-parallel bonds be arranged on N total bonds? Try sketching out a few cases for small N.

2. Can k be even/odd with periodic boundary conditions? Does this depend on the size N of the chain?

3. Knowing that the ground-state (all bonds parallel) energy is $E_G = -NJ$, we can enumerate states based on their excitation energy E_k. How does E_k depend on the number (k) of anti-parallel bonds?

4. Combining the results from the previous parts, express the partition function as a sum over the number of anti-parallel bonds.

5. Evaluate the sum to get an expression for the partition function, and then the free energy $F = -T \log Z$ in the thermodynamic limit $N \to \infty$. Is the expression analytic for all T? What does this say about the existence of a phase transition? *Hint:* The binomial theorem might come in handy here:

$$(a \pm b)^N = \sum_k^N \binom{N}{k} (\pm 1)^k a^{N-k} b^k, \tag{3.57}$$

where $\binom{N}{k} = \frac{N!}{k!(N-k)!}$ is the binomial coefficient.

W3.6   Ising Model in 1D–The Transfer Matrix Approach

Let us now solve the 1D Ising model in full generality, external field h included. The Hamiltonian and some quantities of interest are:

$$H = -J \sum_j \sigma_j \sigma_{j+1} - h \sum_j \sigma_j, \tag{3.58}$$

$$Z = \sum_{\{\sigma\}} \exp\left[-\beta H(\{\sigma\})\right], \tag{3.59}$$

---

22. This is mechanically easier than the general purpose solution via transfer matrices, which is the subject of the next exercise.

$$\langle \sigma_i \rangle = Z^{-1} \sum_{\{\sigma\}} \sigma_i \exp\left[-\beta H(\{\sigma\})\right], \tag{3.60}$$

$$\langle \sigma_i \sigma_j \rangle = Z^{-1} \sum_{\{\sigma\}} \sigma_i \sigma_j \exp\left[-\beta H(\{\sigma\})\right]. \tag{3.61}$$

We will once again assume a 1D chain with N sites and periodic boundary conditions $\sigma_{N+1} = \sigma_1$.

Reformulation of the problem

1. Taking inspiration from the previous approach where we considered the bonds between spins, find a function $K(\sigma_a, \sigma_b)$ such that we can recast the Hamiltonian in the form

$$H = \sum_{j}^{N} K(\sigma_j, \sigma_{j+1}). \tag{3.62}$$

It should satisfy $K(\sigma_a, \sigma_b) = K(\sigma_b, \sigma_a)$ to be interpreted as the energy of a bond between two spins.[23]

2. The partition function, written out explicitly, is

$$Z = \sum_{\sigma_1 = \pm 1} \sum_{\sigma_2 = \pm 1} \cdots \sum_{\sigma_N = \pm 1} e^{-\beta K(\sigma_1, \sigma_2)} e^{-\beta K(\sigma_2, \sigma_3)} \cdots e^{-\beta K(\sigma_N, \sigma_1)}. \tag{3.63}$$

Recall that matrix operations can be expressed as sums:

$$(AB)_{ik} = \sum_{j} A_{ij} B_{jk}, \tag{3.64}$$

$$\text{Tr}[A] = \sum_{j} A_{jj}. \tag{3.65}$$

Reinterpret the partition function as matrix multiplications such that

$$Z = \text{Tr}\left[t^N\right]. \tag{3.66}$$

What is the matrix t? *Hint:* Start by interpreting the terms as matrix elements of t:

$$e^{-\beta K(\sigma_a, \sigma_b)} = \langle \sigma_a | t | \sigma_b \rangle = t_{\sigma_a \sigma_b}. \tag{3.67}$$

3. Using the cyclic property of the trace,[24] show that, for any values of i and j:

---

23. Mathematically, the choice of $K(\sigma_a, \sigma_b)$ is not unique, but the criterion we require allows for a physical interpretation, and more convenient algebra.

24. Which is $\text{Tr}[AB] = \text{Tr}[BA]$. Combining this with the associative property of matrices $(AB)C = A(BC)$, the cyclic property can be extended to arbitrary number of matrices multiplied together, e.g. $\text{Tr}[ABCD] = \text{Tr}[BCDA] = \text{Tr}[CDAB] = \text{Tr}[DABC]$.

(a) The expectation value of any spin, which we will define as the magnetization m, is

$$\langle \sigma_i \rangle = \langle \sigma_j \rangle = Z^{-1} \mathrm{Tr} \left[\tau_z t^N \right] \equiv m, \qquad (3.68)$$

where $\tau_z = \begin{bmatrix} 1 & 0 \\ 0 & -1 \end{bmatrix}$ is the standard Pauli-z matrix.

(b) the expected two-site correlation, is

$$\langle \sigma_i \sigma_j \rangle = Z^{-1} \mathrm{Tr} \left[\tau_z t^{j-i} \tau_z t^{N-j+i} \right]. \qquad (3.69)$$

Calculations

Having reformulated the problem as matrix operations, we will perform the calculations. Let us use the notation:

$$t | \psi_\pm \rangle = \lambda_\pm | \psi_\pm \rangle. \qquad (3.70)$$

So $\lambda_\pm$ are the two eigenvalues of t, and $|\psi_\pm\rangle$ are the two corresponding eigenvectors.

1. In its eigenbasis, $t' = U t U^{-1}$ is diagonal (by construction) with the form

$$t' = \begin{bmatrix} \lambda_+ & 0 \\ 0 & \lambda_- \end{bmatrix}. \qquad (3.71)$$

   Express the partition function in terms of $\lambda_\pm$.

2. The trace can be computed in any basis, so we could write

$$\mathrm{Tr}\,[A] = \langle \psi_+ | A | \psi_+ \rangle + \langle \psi_- | A | \psi_- \rangle. \qquad (3.72)$$

   Express the following quantities in terms of $\lambda_\pm$, $|\psi_\pm\rangle$, and τ_z:

   (a) $\langle \sigma_i \rangle$.

   (b) $\langle \sigma_i \sigma_j \rangle$.

   Hint: The identity matrix can be expressed as:

$$I = | \psi_+ \rangle \langle \psi_+ | + | \psi_- \rangle \langle \psi_- |. \qquad (3.73)$$

3. Now, to do some heavy lifting, find the eigenvalues $\lambda_\pm$ and the corresponding normalized eigenvectors $|\psi_\pm\rangle$. *Hint:* If you need a refresher on how to handle 2×2 matrices, Appendix A.1 is a good place to be.

4. Let us put it all together. Calculate explicitly, in the thermodynamic limit[25] $N \to \infty$:

---

25. Eliminate terms that are infinitely smaller than other terms, but retain the N dependence in the terms you keep.

(a) The partition function Z. Does this agree with the answer derived using combinatorics for $h = 0$?

(b) The free energy $F = -T \log Z$.

(c) The magnetization from the free energy:

$$m = -\frac{1}{N} \frac{\partial F}{\partial h}. \tag{3.74}$$

(d) The magnetization as the expectation value:

$$m = \langle \sigma_i \rangle. \tag{3.75}$$

Does this agree with the previous part?

(e) The expected two-site correlation:

$$\langle \sigma_i \sigma_j \rangle. \tag{3.76}$$

(f) The connected correlation function, which is defined as

$$G(i,j) = \langle \sigma_i \sigma_j \rangle - \langle \sigma_i \rangle \langle \sigma_j \rangle, \tag{3.77}$$

and recast the expression into the form:

$$G(i,j) \propto e^{-\frac{|i-j|}{\xi}}. \tag{3.78}$$

What is the expression for the correlation length ξ?

(g) In the case that $h = 0$, how does the correlation length behave as a function of temperature?

W3.7 Low-Temperature Expansion in 2D[26]

We can further understand the effect of dimensionality on the critical behavior of the Ising model by performing a quantitative expansion at low temperatures. While the text discusses this in general for any dimensionality d, here we will work out the details for $d = 2$ on a square lattice with N total sites. The Hamiltonian is

---

26. This is a tricky exercise; feel free to try your best on each step, then check your answer for correctness and understanding before moving on to the next step!

$$H = -J \sum_{\langle ij \rangle} \sigma_i \sigma_j, \tag{3.79}$$

and we will assume ferromagnetic interactions ($J > 0$).

The partition function

As a reminder, the partition function is

$$Z = \sum_{\{\sigma\}} e^{-\beta H(\{\sigma\})} = e^{4N\beta J} \sum_{E_i} g(E_i) e^{-\beta E_i}, \tag{3.80}$$

where the first form is expressed as a sum over microstates, but we can equivalently express the partition function as an infinite series in which the states with excitation energy E_i are weighted by their multiplicities $g(E_i)$. As before, our starting point is the ground state, with all spins pointing up. The overall factor of $e^{4N\beta J}$ comes from accounting for the energy of the ground state, $E_G = -4NJ$, so we can use the notation E_i to indicate the excitation energy relative to the ground state.

1. The lowest energy state is the ground state, with $E_0 = 0$ and therefore Boltzmann factor of $e^0 = 1$. The first excited state is a single spin-flip, with $E_1 = 8J$. What are the next **two** possible excitation energies E_2 and E_3, and what are the types of spin-flips associated with them? *Hint: Excited spins are not necessarily adjacent to each other.*

2. Now we will need to count the multiplicity of states associated with each excitation energy. The ground state is unique, so $g(0) = 1$. The first excited state is a single spin, with N possible sites for this excitation, so $g(8J) = N$. Find $g(E_2)$ and $g(E_3)$. *Hint: You may find it helpful to look at how we enumerated the droplet shapes in the previous exercise.*

3. Now that we have all the ingredients, write down the next two orders in the expansion for Z:

$$Z e^{-4N\beta J} = 1 + N e^{-8\beta J} + \qquad + \qquad\qquad + \mathcal{O}\left(e^{-\beta E_4}\right). \tag{3.81}$$

The magnetization

Since the lattice is translationally invariant, calculating the magnetization is the same as calculating the expectation value of a spin on any specific site, which we will call σ_1. So:

$$m = \langle \sigma_1 \rangle = \frac{1}{Z} \sum_{\{\sigma\}} \sigma_1 e^{-\beta H(\{\sigma\})} \tag{3.82}$$

$$= \frac{e^{4N\beta J}}{Z} \sum_{E_i} \left[g_+(E_i) - g_-(E_i) \right] e^{-\beta E_i}, \tag{3.83}$$

where the first line is the definition of the expectation value, and the second line is the form it takes as an infinite series. For each possible excitation energy E_i, there are $g_+(E_i)$ configurations in which $\sigma_1 = +1$, and $g_-(E_i)$ configurations in which $\sigma_1 = -1$, with $g_+(E_i) + g_-(E_i) = g(E_i)$.

1. The unique ground-state has all spins pointing up, so $g_+(0) = 1$ and $g_-(0) = 0$. The first excited state has exactly one configuration where σ_1 is flipped, so $g_-(E_1) = 1$ and $g_+(E_1) = N - 1$. Find $g_-(E_2)$ and $g_-(E_3)$. *Hint:* Again, you might want to revisit the droplet enumeration we did before. Think about how many unique ways in which you can place the droplets on the lattice such that they contain a specific site.

2. From the definitions above, we see $g_+(E_i) - g_-(E_i) = g(E_i) - 2g_-(E_i)$. Use your results from previous parts to find the next two orders:

$$m = \frac{e^{4N\beta J}}{Z}\left[1 + (N-2)e^{-8\beta J} + \right.$$

$$\left. + \mathcal{O}\left(e^{-\beta E_4}\right)\right]. \tag{3.84}$$

3. To find the final answer for the magnetization, we will now incorporate the $1/Z$ normalization. Use the expression for Z found previously, and fully expand the expression for m up to order $\mathcal{O}\left(e^{-\beta E_3}\right)$. *Hint:* Multiply out the series completely, keeping terms up to the required order. You may find the following useful:

$$\frac{1}{1+x} = 1 - x + x^2 + \mathcal{O}\left(x^3\right). \tag{3.85}$$

When the dust settles, you will find that we are left with an expansion with no terms proportional to any power of N. Since m is an intensive quantity, the expansion (rightfully) reflects this property. Remarkably, as the expansion is carried out to higher orders, no further powers of N will appear!

4. *The exact solution by Onsager and Yang[27] for the 2D Ising model gives the order parameter to be:

$$m = \left[1 - \frac{1}{\sinh^4(2\beta J)} \right]^{1/8}. \tag{3.86}$$

Does it agree with the expansion we just found?

The free energy, entropy, and heat capacity

Having found the expansion for the partition function, we can also find the free energy, entropy, and heat capacity. For all calculations below, keep terms up to order $\mathcal{O}\left(e^{-\beta E_3}\right)$.

---

27. L. Onsager, *Phys. Rev.* **65**, 117 (1944), and C. N. Yang, *Phys. Rev.*, **85**, 808 (1952)

1. Starting with the expansion for Z, find the expansion for the free energy $F = -T \log Z$. *Hint:* You may find the following useful:

$$\log(1 + x) = x - \frac{x^2}{2} + \mathcal{O}\left(x^3\right). \tag{3.87}$$

2. Calculate the entropy $S = -\partial F / \partial T$.
3. Calculate the heat capacity per site:

$$c = \frac{C}{N} = \frac{T}{N} \frac{\partial S}{\partial T}. \tag{3.88}$$

Since this is also an intensive quantity, the correct answer should not have any terms proportional to any power of N.

4 MEAN-FIELD THEORY

Enough of generalities. It is time to roll up our sleeves and calculate!

In the last chapter, we solved the Ising ferromagnet in one dimension. This is a rare exception. Most problems—even extremely simplified paradigmatic models—cannot be solved exactly. It is thus necessary to develop approximate methods. Ideally, these methods should give qualitatively correct accounts of phases and phase transitions, even if they are only semi-quantitatively reliable.

A powerful and useful method of approximation is known as "mean-field theory." This approximation is broadly applicable across disciplines; the Hartree-Fock approximation from atomic physics is a prominent example.

4.1 A Headfirst Introduction

In a mean-field treatment, the interactions between microscopic variables (e.g. spins) are approximated by a suitable average potential, or "mean field." To get a sense of what this means, consider the Ising ferromagnet of Eq. 3.1. In any given microstate, the spin on a particular site, j, experiences a local magnetic field that has two contributions: one from the applied magnetic field, and one from each of its nearest neighbors (n.n.):

$$b_j^{\text{eff}} = h + J \sum_{k=\text{n.n.}j} \sigma_k. \qquad (4.1)$$

Crucially, the local magnetic field, b_j^{eff}, depends on the configuration of spins on neighboring sites, σ_k. The intuitive essence of the mean-field method is to ignore this dependence, and to replace Eq. 4.1 by its average,

$$b \equiv \left\langle b_j^{\text{eff}} \right\rangle = h + zJm, \qquad (4.2)$$

when performing the sum over the values of σ_j. Here $m \equiv \langle \sigma_k \rangle$ is the average spin of a neighbor and z is the number of neighbors.

Under this approximation, the spin at site j can be thought of as an isolated spin in a magnetic field of strength b. Assuming that all sites are equivalent, this viewpoint reduces the problem from N interacting spins to N identical, noninteracting spins.

We have solved this problem already (Eq. 3.15), so we know that m is a function of b ($m = \tanh(b/T)$). The only new ingredient here is that b, in turn, must be simultaneously determined "self-consistently" from Eq. 4.2 as a function of m.

Below, once we have derived mean-field theory in a more systematic manner, we will see that this simple analysis gives a remarkably useful way of addressing key questions of interest. In particular, the self-consistent solution of Eq. 4.2 has a critical temperature $T_c = zJ$, such that for $T > T_c$, the only self-consistent solution has $m \to 0$ as $h \to 0$, while for $T < T_c$, of m approaches a nonzero value, indicating a broken-symmetry phase.

4.2   The Variational Principle

It is typically not possible to exactly compute the thermodynamic properties of a system described by any physically reasonable Hamiltonian H. Instead, in carrying out a mean-field theory of the given problem, we introduce a "trial Hamiltonan" H_{tr} which is simple enough to be solvable, but which still captures relevant features of the original problem. Typically, we choose a trial Hamiltonian with one or more adjustable parameters, and then determine the parameter values that best approximate the original problem. We focus on the free energy F to serve as the point of comparison—after all, F is the root of all thermodynamic properties.

There is a simple inequality which underlies all such approaches. Let us define the variational free energy, F_{var}, as

$$F_{var} = F_{tr} + \langle H - H_{tr} \rangle_{tr}, \tag{4.3}$$

where the subscript tr means to calculate the corresponding value using the trial Hamiltonian, i.e.,

$$F_{tr} = -T \log Z_{tr}; \quad Z_{tr} = \mathrm{Tr}\left[e^{-\beta H_{tr}}\right]; \quad \langle O \rangle_{tr} = \mathrm{Tr}\left[\frac{e^{-\beta H_{tr}}}{Z_{tr}} O\right]. \tag{4.4}$$

The variational free energy is useful since it is an upper bound on the actual free energy:

$$F_{var} \geq F. \tag{4.5}$$

It stands to reason that the smaller the value of F_{var}, the closer it approximates F, and the more closely the trial ensemble represents the original problem (see Problem 4.3).

Equation 4.5 follows from a very general result in probability theory, Jensen's inequality. Let ϕ be a discrete or continuous random variable, and let $P(\phi)$ be the associated probability distribution. The expectation value for a function $f(\phi)$ under this distribution is

$$\langle f(\phi) \rangle = \mathrm{Tr}\left[P(\phi)f(\phi)\right], \tag{4.6}$$

where the trace indicates a sum or integral over all possible values of ϕ. Jensen's inequality states that for any convex[1] function f, $\langle f(\phi) \rangle \geq f(\langle \phi \rangle)$. We prove the relevant case of this inequality in Worksheet W4.2.

From here, the full proof of Eq. 4.5 is straightforward.[2] We express the partition function as

$$Z = \mathrm{Tr}\left[e^{-\beta H} e^{\beta H_{tr}} e^{-\beta H_{tr}} \right] = Z_{tr} \left\langle e^{-\beta(H - H_{tr})} \right\rangle_{tr}. \qquad (4.7)$$

Since $e^{-\beta(H - H_{tr})}$ in Eq. 4.7 is convex in $H - H_{tr}$, it follows from Jensen's inequality that Z is bounded below as

$$Z \geq Z_{tr} e^{-\beta \langle H - H_{tr} \rangle_{tr}}. \qquad (4.8)$$

To get the free energy, take the logarithm of both sides of this inequality and multiply by $-T$; this leads to the stated inequality, Eq. 4.5. F is hence bounded above by F_{var}.

4.2.1   Usefulness of the Variational Approach

The variational principle gives a criterion for constructing a simplified (solvable) problem that "best" approximates the true problem of interest. The strategy is to define an H_{tr} that contains a set of "variational" parameters, $\{b_1, b_2, \ldots, b_n\}$, and then to choose the parameter values which minimize F_{var}. This criterion picks the optimal H_{tr} out of the class of trial Hamiltonians parameterized by $\{b\}$.

Needless to say, it is a separate question whether "best" is good enough. In general, making H_{tr} simple enough to be solvable also means that its thermal ensemble lacks many aspects of the original problem, H. Thus, each time we apply a mean-field approach, we must assess whether our variational ansatz (i.e., the form of H_{tr}) is sufficient for the problem at hand. The good news is that even simple mean-field theories can yield surprisingly rich results.

4.2.2   An Illustration of the Principle

Our old friend the Ising model,

$$H = -J \sum_{\langle ij \rangle} \sigma_i \sigma_j - h \sum_j \sigma_j, \qquad (4.9)$$

serves as a fine example of how this all works. To begin, let us consider a case in which the variational approach would (formally) produce a result that is exact.

---

1. A function is convex if any line segment between two points on its graph lies above the graph. A sufficient (but not necessary) condition for convexity if the second derivative is well defined is that it be nonnegative everywhere.

2. The proof given here is valid for the class of models—including the Ising model—in which H and H_{tr} commute. For the case where they do not commute (i.e., where $H H_{tr} \neq H_{tr} H$), see Appendix A.8.

Suppose we took as our trial Hamiltonian a version of the Ising model itself:

$$H_{tr} = -\sum_{\langle ij \rangle} K_{ij}\sigma_i\sigma_j - \sum_j b_j\sigma_j, \tag{4.10}$$

where the coupling strengths K_{ij} and field strengths b_j are allowed to be different for different sites i and j. Notice that H_{tr} becomes the model we wish to solve, H, if we choose the variational parameters to be $K_{ij} = J$ and $b_j = h$ for all i and j. From a practical perspective, this is not a terribly useful exercise, since H_{tr} itself is not solvable—in fact, it is harder to solve than the original problem, since K_{ij} and b_j now vary from site to site. However, let us proceed formally and see where it gets us.

The variational approach states that the optimal values of the parameters K_{ij} and b_j are the ones which minimize the variational free energy (Eq. 4.3). To find these values, we set the derivatives with respect to these parameters equal to zero,

$$\frac{dF_{var}}{dK_{ij}} = \frac{dF_{var}}{db_j} = 0. \tag{4.11}$$

Note that this condition does not guarantee that we find the global minimum of the variational free energy (or a minimum at all); it just solves for stationary points where small changes to the parameters do not change F_{var}. In evaluating derivatives of this sort, we are greatly helped by the derivative theorem discussed in Section 2.1.3. In the present context this implies that

$$\frac{\partial F_{tr}}{\partial K_{ij}} = -\langle\sigma_i\sigma_j\rangle_{tr} \quad \& \quad \frac{\partial F_{tr}}{\partial b_j} = -\langle\sigma_j\rangle_{tr}. \tag{4.12}$$

With a little straightforward algebra it follows that

$$\frac{\partial F_{var}}{\partial K_{ij}} = -\sum_{\langle kl \rangle}[J - K_{kl}]\frac{\partial\langle\sigma_k\sigma_l\rangle_{tr}}{\partial K_{ij}} - \sum_k[h - b_k]\frac{\partial\langle\sigma_k\rangle_{tr}}{\partial K_{ij}}, \tag{4.13}$$

$$\frac{\partial F_{var}}{\partial h_j} = -\sum_{\langle kl \rangle}[J - K_{kl}]\frac{\partial\langle\sigma_k\sigma_l\rangle_{tr}}{\partial h_j} - \sum_k[h - b_k]\frac{\partial\langle\sigma_k\rangle_{tr}}{dh_j}. \tag{4.14}$$

This might appear to be of limited usefulness since we have already noted that we do not know how to compute $\langle\sigma_k\sigma_l\rangle_{tr}$ and $\langle\sigma_k\rangle_{tr}$; fortunately, we do not need to, since we can see that the derivatives are zero when

$$K_{ij} = J \quad \text{and} \quad b_j = h \quad \text{for all } i, j. \tag{4.15}$$

Of course, since $H_{tr} = H$ for this choice of parameters, the variational principle has given us exactly the right answer. But this example was rather contrived because we chose a class of trial Hamiltonians which was general enough to include the actual

Hamiltonian of the problem. Still, what we have seen is that the strategy of minimizing F_{var} is a promising criterion for picking out the best solution from the set of all trial Hamiltonians.

4.3 Mean-Field Theory of the Ising Model

Now we make practical use of the variational method to find an approximate solution to our good pal, the Ising model. We will pick an H_{tr} which is simple enough to be soluble, yet captures a good amount of the physics:

$$H_{tr} = -\sum_j b_j \sigma_j. \tag{4.16}$$

We pretend that instead of an interaction term each spin j experiences an "effective magnetic field" of b_j. Our task is to adjust the variational parameters b_j so as to minimize F_{var}.

The trial Hamiltonian is a problem of N noninteracting spins, which we have solved already in Section 3.3.1, with the following results:

$$F_{tr} = -T \sum_j \log[2\cosh(\beta b_j)] = \frac{T}{2} \sum_j \log\left[\frac{1-m_j^2}{4}\right], \tag{4.17}$$

$$m_j \equiv \langle \sigma_j \rangle_{tr} = -\frac{\partial F_{tr}}{\partial b_j} = \tanh(\beta b_j), \tag{4.18}$$

$$\langle \sigma_i \sigma_j \rangle_{tr} = \begin{cases} 1 & \text{if } i=j \\ m_i m_j & \text{if } i \neq j. \end{cases} \tag{4.19}$$

Notice that, since our H_{tr} has no interactions between different spins, the trial ensemble has no correlations between different spins; $\langle \sigma_i \sigma_j \rangle_{tr}$ factorizes into $m_i m_j$. This aspect of the variational approach is generically wrong, but we will see that it still manages to capture much of the relevant physics.

Substituting these expressions into the variational free energy (Eq. 4.3) gives F_{var} as a function of the parameters $\{b_j\}$,

$$F_{var} = F_{tr} - J \sum_{<ij>} m_i m_j + \sum_j (b_j - h) m_j. \tag{4.20}$$

The details of this calculations are worked out in Worksheet W4.4. To minimize F_{var}, we set its derivatives to zero,

$$\frac{\partial F_{var}}{\partial b_j} = 0 = \sum_i \frac{\partial m_i}{\partial b_j}\left[b_i - h - J \sum_{k=\text{n.n.}i} m_k\right] \quad \text{for all } j, \tag{4.21}$$

or equivalently,

$$b_i = h + J \sum_{k=\text{n.n.}i} \tanh(\beta b_k) \quad \text{for all } i, \tag{4.22}$$

where the sum runs over the nearest neighbors (n.n.) of i. The expression for b_i represents the average local field experienced by spin i—the same expression we encountered in Eq. 4.1 by heuristic reasoning.

Eq. 4.22 forms a coupled set of equations for b_j. Solutions correspond to stationary points of F_{var} (which might be local minima, local maxima, or saddle points). Our goal is to find the global minimum—the best variational approximation to the Ising model.

For the reminder of this chapter, we will study the simplest case, where b_j takes on the same value, b, on all N sites. Such a solution is spatially uniform, or translationally invariant. It turns out that for the ferromagnetic case ($J > 0$), the global minimum of F_{var} is spatially uniform—we will see why shortly.[3] Taking $b_j = b$ on all sites j leads to a single equation,

$$b = h + zJ \tanh(\beta b), \tag{4.23}$$

where z is the number of nearest-neighbor sites.

4.3.1   The Mean-Field Phase Diagram

The mean-field equation above can be analyzed to obtain an approximate description of the phase diagram of the Ising ferromagnet in the $h - T$ plane. To do this, for each value of h and T, we first determine the value of the variational parameter $b = b_{min}$ which minimizes the variational free energy $F_{var}(b, h, T)$. From here, we compute the mean-field approximation to the free energy,

$$F_{mf}(h, T) \equiv F_{var}(b, h, T)\big|_{b=b_{min}}, \tag{4.24}$$

and the magnetization,

$$m_{mf}(h, T) = -\frac{1}{N}\frac{dF_{mf}}{dh} = \tanh(\beta b_{min}). \tag{4.25}$$

Phase boundaries are then indicated by interesting features such as discontinuities or singularities as functions of h and T. Though this process is conceptually simple, the difficulty is that the self-consistency relation (Eq. 4.23) is a transcendental equation with no closed-form solution. Nevertheless, even if we cannot directly solve this equation, we can still figure out overall features of the $h - T$ phase diagram.

---

3. We address situations with nonuniform states (i.e., states where b_j varies from site to site) in Chapter 7.

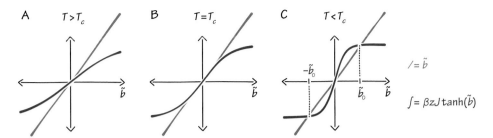

FIGURE 4.1. The self-consistency relation may be solved graphically by plotting both sides of the equation and looking for intersections. Nonzero solutions exist only when the blue line has a shallower slope than the red curve at the origin.

Determining the critical temperature T_c

We first consider the case with $h = 0$. In this case, the self-consistency relation can be rewritten in terms of a dimensionless quantity, $\tilde{b} \equiv \beta b = b/T$:

$$\tilde{b} = \beta z J \tanh \tilde{b}. \tag{4.26}$$

Only one (dimensionless) parameter remains, $\beta z J$. The nature of the solutions depends on whether $\beta z J$ is greater or less than 1. This is easy to see graphically: in Figure 4.1, we plot the left and right hand expressions of Eq. 4.26 for various values of $\beta z J$. Solutions correspond to points at which the two curves intersect. There is always a solution at $\tilde{b} = 0$. For $\beta z J < 1$ (left panel), this is the only solution; for $\beta z J > 1$ (right panel), there is an *additional* pair of solutions at $\tilde{b} = \pm \tilde{b}_0$. These nonzero solutions turn out to be thermodynamically favorable (i.e., have lower variational free energy). The middle panel shows the critical case $\beta z J = 1$, where the nonzero solutions first appear.

We have found two qualitatively distinct regimes, separated by a precise value of $\beta z J = z J / T = 1$. This defines a critical temperature, $T_c \equiv z J$, which separates two phases! For $T > T_c$, the symmetry is preserved ($m = 0$) and any given spin is equally likely to be up or down. In contrast, for $T < T_c$ the symmetry is broken ($m = \pm \tanh \tilde{b}_0 \neq 0$), as one spin direction is more likely than the other.

The way in which m depends on T is shown in the green curve of Figure 4.2A. As T is lowered past T_c, the magnetization m rises smoothly from zero. Thus, there is a continuous phase transition at $T = T_c$, $h = 0$.

The discontinuous phase boundary

Below the critical temperature for $h = 0$, there are *two* equivalent solutions, $m = \pm m_0$, with the same variational free energy (green curve in Figure 4.2A). This twofold ambiguity can be resolved with an externally applied magnetic field h that lowers the (free) energy of one spin orientation compared to that of the other. For $h > 0$, the

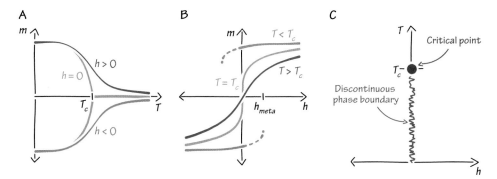

FIGURE 4.2. (A) The temperature dependence of the magnetization $m(h, T)$ in the absence of a field $h = 0$ (green) and in the presence of positive/negative fields (blue and red). (B) The field dependence of $m(h, T)$ for fixed $T > T_c$ (red), $T = T_c$ (green), and $T < T_c$ (blue). The dashed portions of the blue curves correspond to metastable solutions of the mean-field equations (local but not global minima of F_{var}; see Section 4.4.3). (C) The mean field phase diagram of the Ising model.

solution with $m > 0$ is preferable (red curve), and conversely for $h < 0$ (blue curve). This implies that when h changes sign below T_c, the magnetization *jumps discontinuously*. Thus, there is a discontinuous phase boundary lying along $h = 0$ for $T < T_c$, as shown in the phase diagram in Figure 4.2C.

This discontinuity can also be seen by plotting m as a function of h at fixed T, as in Figure 4.2B. The solid blue curve illustrates that for $T < T_c$, the function $m(h)$ has a discontinuity at $h = 0$, i.e., it approaches a different limiting value depending on whether $h \to 0$ from the left or the right. This is the essence of a broken-symmetry phase.

Susceptibility to magnetic fields near T_c

The discontinuity in $m(h)$ below T_c affects the behavior of the susceptibility, χ. Under normal circumstances, if an external magnetic field h is applied to a magnet, a small field induces a proportionally small change to the magnetization, $m \simeq \chi h$. Here χ is the slope of $m(h)$ at $h = 0$. The red curve in Figure 4.2B illustrates this linear behavior for $T > T_c$. However, just below T_c, we have seen that even an infinitesmal field produces a nonzero m. Correspondingly, it stands to reason that the susceptibility is very large for T just above T_c. In fact, χ diverges as T approaches T_c from above. This can be seen in the green curve in Figure 4.2B, which has a vertical slope at $h = 0$.

What this means is that an Ising model near its critical point is extremely susceptible to magnetic fields. As it is cooled from above to below T_c, the symmetric state becomes increasingly sensitive to the slightest magnetic fields—even for T above T_c. One can "smell" that a transition is approaching even before encountering it!

Beyond qualitative results

With barely any calculation, we have arrived at some rather remarkable results, summarized in Figure 4.2. Notably, the solutions of the mean-field equations give rise to a sharp transition temperature, T_c. To obtain a clearer understanding, we would like to supplement this general picture with more quantitative predictions. Since an analytic solution of Eq. 4.23 is impossible, our next best hope is an asymptotic expression—a series expansion which is accurate in certain limiting regimes. In the low-temperature limit, $T \ll T_c$, we can derive an asymptotic expression for m (see Worksheet W4.5)—an expression that agrees with the first few terms in the low-T series from the previous chapter. Another interesting region of the phase diagram is the region near the critical point, where T is close to T_c and h is close to 0. We now focus on this parameter regime.

4.4 Variational Free Energy as Function of m_j

We will first shift perspective to make calculations more intuitive. Namely, we restate the problem in terms of a free energy that is a function of magnetization at each site, $F_{var}(m_j)$. Later we will see that this forms the basis of the Ginzburg-Landau theory of phase transitions.

Consider the invertible relationship between b_j (the mean field at site j) and m_j (the average spin at site j):

$$b_j(m_j) = T \operatorname{arctanh}\left[m_j\right] = \frac{T}{2} \ln\left[\frac{(1+m_j)}{(1-m_j)}\right]. \tag{4.27}$$

To each value of b corresponds exactly one value of m, and vice versa. It is thus equivalent to view the variational problem as minimizing F_{var} over the variables $\{m\} = \{m_j \text{ for all } j \text{ in lattice}\}$ and to view it as minimizing over the $\{b\}$, in the sense that setting

$$\frac{\partial F_{var}(\{m\})}{\partial m_j} = 0 \tag{4.28}$$

leads to the same self-consistency equation as Eq. 4.21. This means that we can treat $\{m\}$ instead of $\{b\}$ as the variational parameters for F_{var}. In Worksheet W4.3, you will prove this to be a general feature of mean-field theories. Specifically for the Ising model, you can work out the expression for the variational free energy in terms of the $\{m\}$ by using Eq. 4.27 to substitute $b(m_j)$ for b_j in Worksheet W4.6. You will find that

$$F_{var}(\{m\}) = \sum_j f_{var}(m_j) + \frac{J}{2} \sum_{<ij>} \left[m_i - m_j\right]^2 - h \sum_j m_j, \tag{4.29}$$

where

$$f_{var}(m_j) = -T \log 2 + \frac{T}{2} \log(1 - m_j^2) - \frac{zJ}{2} m_j^2 + T m_j \operatorname{arctanh} m_j \tag{4.30}$$

is the free energy per site, i.e., the free energy density.

The first term in Eq. 4.29 is the local contribution to the free energy due to the spin at each site j, summed over all sites independently. The second term represents the tendency for neighboring spins to align; it is the energy penalty if the average magnetizations of two neighboring sites i and j are unequal. The third term represents the tendency for spins to align with the external magnetic field; it is the average energy of spin j in a field h.

We can now see why the global minimum of the free energy is necessarily spatially uniform, with m_j the same on all sites.[4] The only term which is sensitive to spatial variation (i.e., situations where neighboring sites have unequal values of m_j) is the second term. It is minimized when $m_i = m_j$ for all neighboring pairs $\langle ij \rangle$ i.e., when $m_j = m$ is the same for all j. Thus, the global minimum is necessarily spatially uniform. By restricting our attention to uniform states ($m_j = m$), Eq. 4.29 simplifies considerably to

$$\frac{1}{N}F_{var}(m, h, T) = f_{var}(m) - hm. \tag{4.31}$$

Eq. 4.29 expresses the (variational) free energy as a functional of the magnetization. This idea forms the essence of an insightful approach to phases and phase transitions known as Landau theory (or, more generally, Ginzburg-Landau theory). Here, in our first encounter with the approach, we have computed $F_{var}(\{m\})$ via a laborious calculation from a Hamiltonian. In subsequent chapters, we will arrive at much the same result more simply, from symmetry arguments.

4.4.1 Expansion for Small m

The minima of F_{var} still cannot be found in closed form. However, we can obtain a much simpler expression by taking a Taylor expansion in powers of m (Worksheet W4.6):

$$\frac{1}{N}F_{var} = f_0 - hm + \frac{\alpha}{2}m^2 + \frac{u}{4}m^4 + \mathcal{O}(m^6), \tag{4.32}$$

with coefficients

$$f_0 = -T \log 2, \tag{4.33}$$

$$\alpha = T - zJ = T - T_c, \tag{4.34}$$

$$u = T/3. \tag{4.35}$$

The crucial parameter is α, which measures the "distance" from the critical temperature—it is positive for $T > T_c$ but negative for $T < T_c$.[5] (We have chosen these symbols and funny-looking parameters for future convenience.)

---

4. Remember that we are restricting our attention to the ferromagnetic case $J > 0$, where adjacent spins prefer alignment over anti-alignment.

5. Note that this is *not* the same as the critical exponent, α—there are unfortunately only so many letters in the English and Greek alphabets. We hope it is clear from context which α is being referred to.

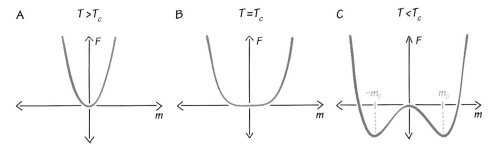

FIGURE 4.3. The shape of the function $f_{var}(m) \approx \alpha m^2/2 + um^4/4$ for α positive, zero, and negative.

This approximation is valid so long as $|m|$ is small—so it is perfect for studying what happens close to a continuous phase transition. (For $T > T_c$, the condition is satisfied so long as h is small, since $m \propto h$. For $T < T_c$, this is valid for small h and T close enough to T_c, since m rises continuously from 0 as T decreases.)

The simple form of Eq. 4.32 gives a rather nice way to see how the symmetry is broken below T_c. In Figure 4.3 we have plotted $f_{var}(m)$ as a function of m at various temperatures. The overall shape of the curve depends on α: for $\alpha > 0$, there is one global minimum at $m = 0$, whereas for $\alpha < 0$, there are two global minima at $m = \pm m_0$. In the broken-symmetry phase, there is nothing which favors one or the other of the minima—hence the "spontaneous" nature to the symmetry breaking.

4.4.2 Critical Exponents

Our approximate free energy can yield predictions for how various thermodynamic quantities depend on h and T. We expect these expressions to be asymptotically correct in the regime where $|m|$ is small, such as near the critical point.

For example, the equilibrium values of m can be determined by solving for the minima,

$$\frac{\partial f_{var}}{\partial m} = \alpha m + um^3 = 0 \implies m = \pm\sqrt{|\alpha|/u} \text{ when } \alpha < 0. \tag{4.36}$$

This is the mean-field prediction for the temperature dependence of m: as T is cooled past T_c, the magnetization grows according to $m \sim |T - T_c|^\beta$, with $\beta = 1/2$.

We have encountered the exponent β before. Experiments on 3D systems (Section 1.3) give a value of $\beta \approx 1/3$, and the Onsager solution of the 2D Ising model (Section 3.5) gives $\beta = 1/8$. With the mean-field approximation we find $\beta = 1/2$, independent of spatial dimension d. In later chapters, we will address the issue of why mean-field theory often produces wrong values of critical exponents.

Susceptibility

As discussed earlier, we also expect interesting critical behavior in the (zero-field) susceptibility, $\chi \equiv \lim_{h \to 0} \frac{dm}{dh}$, because near the critical point, a small change in the external field can induce a disproportionately large change in the magnetization. In

Worksheet W4.7 we derive the expression

$$\chi = \frac{\partial m}{\partial h}\bigg|_{h=0} = \frac{1}{\alpha + 3um^2} = \begin{cases} 1/\alpha, & T > T_c, \\ 2/|\alpha|, & T < T_c. \end{cases} \tag{4.37}$$

The susceptibility approaches infinity upon approach to the critical point as $\chi \sim |\alpha|^{-1} \sim |T - T_c|^{-\gamma}$. This divergence agrees with the qualitative discussion from earlier.

This analysis predicts another exponent, $\gamma = 1$. It is worth noting that mean-field theory predicts the same exponent regardless of whether one approaches T_c from above or below. Furthermore, there is a "universal amplitude ratio" of the susceptibility just above and just below the critical point:

$$\lim_{\delta T \to 0^+} \frac{\chi(T_c + \delta T)}{\chi(T_c - \delta T)} = 2. \tag{4.38}$$

Nonlinear response to h

There is one more critical exponent that governs how m depends on h. At temperatures above T_c there is a linear response; i.e., a slight magnetic field causes a proportionate change in magnetization.[6] On the other hand, below T_c, the magnetization changes discontinuously as a function of h. It is interesting to see how these two behaviors interpolate: at exactly $T = T_c$, the small-h response is not linear in h, but follows yet another power-law dependence,

$$0 = -h + um^3 \implies m = (h/u)^{1/3} \quad \text{at } T = T_c. \tag{4.39}$$

This striking nonlinearity is illustrated in the green curve in Figure 4.2B. It is characterized by another critical exponent, $m \sim h^{1/\delta}$, whose mean-field value is $\delta = 3$.

Singularities and universality

Our first approximate glimpse into critical phenomena has revealed some remarkable facts. We have encountered a thermodynamic quantity (the susceptibility) which *diverges* at the critical point. This is surprising, since the functions encountered in physics are generally finite and, indeed, analytic. Typically, if we vary control parameters (such as the temperature) by a small amount, we expect the thermodynamic variables (such as susceptibilities) to also change smoothly by a small amount in a way amenable to Taylor expansion. Critical points are an important exception, where functions have singularities: divergences, cusps, discontinuities, etc.

The singularities predicted by mean-field theory are power laws with specific exponents. These "critical exponents" are given Greek letters by (accidents of) convention. We have seen three of these so far:

---

6. That is, $m = \chi h + \ldots$, where the $\ldots$ represents higher order terms in h.

- as the temperature is lowered toward T_c, the zero-field susceptibiliity goes as $\chi \sim (T - T_c)^{-\gamma}$, with $\gamma = 1$;
- at exactly $T = T_c$, as the external magnetic field is varied, the magnetization goes as $m \sim h^{1/\delta}$, with $\delta = 3$;
- as T is lowered past T_c, the magnetization grows as $m \sim |T - T_c|^{\beta}$, with $\beta = 1/2$.

Although the mean-field values of the exponents are not correct, the theory correctly predicts that, near a critical point, there is singular behavior characterized by a set of fractional exponents. The most striking feature of these exponents is that they are universal—they are pure numbers independent of microscopic details of the model. If we were to repeat the same analysis for a different version of the Ising model—perhaps a different sort of lattice—we would still find exactly the same exponents! (See Problem 4.4.) Mean-field theory demonstrates that certain aspects of critical behavior are robust and completely independent of microscopic considerations—a profound notion which underlies much of our understanding of emergent phenomena.

4.4.3 *Metastability and Two-Phase Coexistence

For a certain range of parameters, there is an additional "metastable" solution to the mean-field equations, as illustrated in Figure 4.4A. Specifically, for $T < T_c$ and h small, there is a secondary minimum of the variational free energy at $m = -m_0(T) + \chi(T)h + \ldots$ in addition to the global minimum at $m = m_0(T) + \chi(T)h + \ldots$. The secondary local minimum is known as the metastable solution. The metastable solution is plotted in dotted lines in Figure 4.2B; it only exists if h is smaller than a critical value $h_{meta}(T)$ (see Problem 4.7). The region of the phase diagram which admits a metastable solution is bordered by the "spinodal" lines, indicated by the dashed curves in Figure 4.4B.

The meaning of metastability can be thought of in analogy with the problem of a ball in a potential with two unequal minima, as in Figure 4.4A. No matter where we start the ball, in a short time—due to friction and wind and other random forces—the ball is likely to settle into one of the two minima. If it finds the global minimum, it is likely to remain there indefinitely, but if it initially settles into the local minimum, it will likely stay there for a long time, until a big gust of wind (or possibly a child with a passion for kicking balls) comes along and displaces it. Similarly, if we prepare a physical system in a local minimum of the free energy, we expect it to remain in this metastable state long enough that it can be treated almost as if it were in equilibrium (see Section 5.1.2 and Problem 4.7).

Metastability turns out to be a rather general property of discontinuous phase transitions. In many cases it is possible to "superheat" or "supercool" a material past a discontinuous phase boundary without causing it to change phase. For example, if you pour very pure water into a smooth container and carefully chill it below its

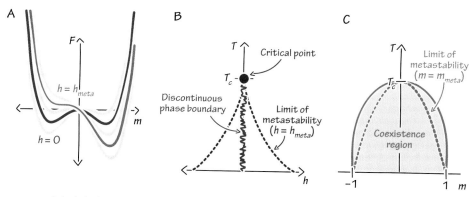

FIGURE 4.4. (A) The variational free energy density as a function of the magnetization, m, for fixed h at $T < T_c$. There is a metastable minimum of the free energy for $0 < h < h_{meta}$. (B) The corresponding h–T phase diagram with region of metastability in between the dotted and solid line. (C) The same phase diagram in the m–T plane.

freezing point, you will find that it remains in a liquid state even below $0°C$. This is because ice crystals do not usually form out of nowhere—they only grow after being seeded, or nucleated, from a small speck of dust or an imperfection in the walls of the container. (Sometimes a sharp disturbance can nucleate crystal growth, as you can find by googling "supercooled water video"!) If the water is very pure and there are no nucleation sites, it can remain in a metastable liquid state indefinitely—despite the fact that crystal growth is spontaneous (lowers the free energy).

Another intriguing property of the Ising system below T_c is a "forbidden region" of the phase diagram in the $m - T$ plane. (Until now we have only considered the $h - T$ plane). For $T > T_c$, any value of m between -1 and 1 is possible, but for $T < T_c$, there is a forbidden range $-m_0(T) < m < m_0(T)$. This is illustrated in Figure 4.4C. In the forbidden region there is no value of h corresponding to an equilibrium magnetization m. This is reminiscent of what happens at a 1st-order transition between a liquid and a gas: at any given $T < T_c$, there is no homogeneous phase with a density intermediate between that of the liquid and the gas. Rather, if a fluid of this density is placed in a sealed vessel of fixed volume at $T < T_c$, it spontaneously separates into a liquid portion and a vapor portion; the thermodynamic state is a state of macroscopic two-phase coexistence. Analogously, if an Ising system were held at a fixed total magnetization $M = Nm$ in the forbidden region, it would separate into two portions. A fraction g of the system would have $m = m_0$ and a fraction $1 - g$ would have $m = -m_0$, where $g = \frac{1}{2}(m + m_0)/m_0$. This is the meaning of the label "coexistence region."

4.5 Validity of Mean-Field Theory

There is a simple metric to assess the validity of the mean-field method. The approximation amounts to replacing the local effective field b_j^{eff} by its mean value b. Therefore, if b_j^{eff} does not tend to stray far from b, then the approximation is good.

Any inaccuracies incurred must be due to the fluctuations of the local field away from the mean field. If the root mean square magnitude of a typical fluctuation,

$$|\Delta b| \equiv \left\langle \left(b_j^{\text{eff}} - b\right)^2 \right\rangle^{1/2}, \tag{4.40}$$

is small compared to the temperature, $|\Delta b| < T$, then we expect mean-field theory to be reasonably accurate.

To estimate the magnitude of the fluctuations, observe that the local field as defined in Eq. 4.2 involves the sum of z random variables. If there are enough neighbors and the neighbors are sufficiently independent of one another, then the central limit theorem (Problem 4.2) implies that

$$|\Delta b| \sim \sqrt{z}J, \tag{4.41}$$

where z is the number of neighbors. For T near $T_c \simeq zJ$, the criterion $|\Delta b|/T \ll 1$ is satisfied if $z^{-1/2} \ll 1$. This result suggests that mean-field theory is exact in the limit $z \to \infty$: if the number of neighbors is large, the relative magnitude of fluctuations in the local field is small and the approximation is good.

This line of reasoning turns out to be only partially true. In deriving Eq. 4.41, we assumed that neighboring spins are uncorrelated with one another, but as we will see in Chapter 7, the correlations between sites become stronger and longer ranging near the critical point. Thus, the validity of mean-field theory upon close approach to T_c is a more subtle issue.[7]

Dependence on dimensionality

Another shortcoming of our current mean-field treatment is that it predicts a phase transition at $T_c = zJ$ irrespective of the number of spatial dimensions. This is incorrect. The exact solution from Chapter 3 showed that no transition exists in $d = 1$, and the droplet argument establishes that $d \geq 2$ is necessary.

Nevertheless, the mean-field method gives the correct general picture in $d \geq 2$. There is a high-temperature phase with $m = 0$, a low-temperature phase with $m \neq 0$, and a continuous phase transition between the two phases. However, the predicted temperature of the transition $T_c^{mf} = zJ$ is not quite right. It is too high. This is a general feature of a mean-field method. It arises because fluctuations are ignored. Thermal fluctuations have the effect of scrambling spins, making it harder for them to align. By neglecting fluctuations, the mean-field approximation overestimates the tendency for alignment and thereby predicts too high a transition temperature.

The extent to which T_c is overestimated is consistent with the expected errors of order $\sqrt{z}J$. To illustrate this, in Table 4.1 we compare the exact value of T_c (typically

---

7. The discussion here can be generalized and formalized to models with a longer range of interaction—see Appendix A.11.

Table 4.1. Comparison of the exact T_c and the mean-field $T_c^{mf} = zJ$ of the Ising ferromagnet on various regular lattices, given in units of J. The square lattice result is from the Onsager solution; the remaining numbers are from numerical studies of these models.

Lattice	square	triangular	honeycomb	cubic	diamond	4D hypercubic
T_c/J	2.27	3.64	1.52	4.51	2.7	6.68
T_c^{mf}/J	4	6	3	6	4	8

obtained from numerical simulations) with the mean-field values for Ising models on a variety of different lattices. As expected, the mean-field value is more accurate the larger z is.

> **Did you know?** The mean-field method is widely used in many-body physics: the self-consistent field of electronic structure calculations, the Debye-Huckel theory of electrolytes, the van der Waals theory of interacting gases, etc. It is employed in other fields, such as cellular automata, machine learning, game theory, and more.

Near the phase transition, mean-field theory predicts that thermodynamic observables follow power laws with universal exponents. This prediction is only partially right: critical exponents do not depend on lattice geometry, but they *do* depend on dimension. In Table 1.4, we listed the values of critical exponents extracted from exact solutions (where they are available) or from large-scale numerical simulations. There is one set of exponents for Ising models in $d = 2$, another set in $d = 3$, and yet another set in $d \geq 4$. One remarkable fact (not shown in the table) is that in *all* $d \geq 4$, the critical exponents have exactly the same values as predicted by mean-field theory!

Clearly, there is important physics to explore beyond mean-field theory. For instance, in future chapters, we will certainly explore the origin of this peculiar dependence on d. Still, we should be suitably impressed by how much this simple treatment captures of the qualitative physics of phases and phase transitions.

4.6   Problems

4.1. *Variational treatment of the anisotropic 2D Ising model.* A simple generalization of the Ising model allows for anisotropy, where the coupling is stronger along some axes than along others. In this problem, you will explore the anisotropic 2D Ising model to varying degrees of variational approximation by

- treating both axes approximately (Weiss mean field),
- treating one axis exactly and the other approximately (1D ansatz), and
- treating both axes exactly (Onsager's exact solution).

The Hamiltonian may be written as

$$H = -\sum_{i,j} \left\{ J_X \sigma_{i+1,j} \sigma_{i,j} + J_Y \sigma_{i,j+1} \sigma_{i,j} + h\sigma_{i,j} \right\}, \tag{4.42}$$

where $\sigma_{i,j}$ is the spin in the ith column and j^{th} row, and J_X and J_Y are the couplings along the x and y directions, respectively. (Note that the notation is slightly different from that in the text.)

(a) First, consider a trial Hamiltonian of the familiar form:

$$H_{tr}^{(1)} = -b\sum_{i,j} \sigma_{i,j}, \tag{4.43}$$

where the parameter b is the mean field. Find the value of b which minimizes the variational free energy

$$F_{var} = F_{tr} + \left\langle H - H_{tr} \right\rangle_{tr}, \tag{4.44}$$

and compare the resulting self-consistent relation to the one obtained in Eq. 4.23. Interpret the results. *Hint:* Much of the calculation is analogous to that in Section 4.3. Remember that there are $\sum_{i,j} 1 = N$ sites total, and that $\langle \sigma_{i,j} \rangle$ is the same on all sites.

(b) With a graphical argument, find the transition temperature $T_c^{(1)}$ predicted by the trial Hamiltonian $H_{tr}^{(1)}$ when $h = 0$. What happens in the limit $J_X = J_Y$? How about $J_Y = 0$?

(c) Now consider a variational ansatz which accounts for the interactions in the x-direction,

$$H_{tr}^{(2)} = -K\sum_{i,j} \sigma_{i+1,j} \sigma_{i,j} - b\sum_{i,j} \sigma_{i,j}. \tag{4.45}$$

For a variational ansatz to be useful, we need to be able to exactly compute thermodynamic averages in the corresponding trial ensemble. Fortunately, this problem is equivalent to one we have already solved. To what model is this trial Hamiltonian equivalent? Why? *Hint:* There is no algebra required here.

(d) From the explicit form of $F_{tr} = -T\log Z_{tr}$, show that

$$\frac{\partial F_{tr}}{\partial b} = -Nm \quad \text{and} \quad \frac{\partial F_{tr}}{\partial K} = -NC_x, \tag{4.46}$$

where $m = \left\langle \sigma_{i,j} \right\rangle_{tr}$ is the average spin and $C_x = \left\langle \sigma_{i,j} \sigma_{i+1,j} \right\rangle_{tr}$ is the correlation between a spin and its neighbor in the x direction, all in the trial ensemble. What is $m = \langle \sigma_{i,j} \rangle_{tr}$ in terms of the parameters K, b, and β? What about $C_y = \langle \sigma_{i,j} \sigma_{i,j+1} \rangle_{tr}$ in terms of m? You do not need to give an explicit expression for C_x. *Hint:* Which sites are coupled to which sites, and which sites are independent?

(e) With Eq. 4.45 as a trial Hamiltonian, minimize the variational free energy with respect to the variational parameters b and K. Show that this leads to the mean-field equations

$$K = J_X \quad \text{and} \quad b = h + 2J_Y m(K, b), \quad (4.47)$$

where $m(K, b)$ is your expression from part (d). Interpret your results. *Hint:* Use the results from earlier parts, and leave the algebra in terms of quantities such as $\frac{\partial m}{\partial K}$ and $\frac{\partial C_x}{\partial K}$. You will find that two quantities need to vanish when you demand that $\frac{\partial F_{var}}{\partial b} = \frac{\partial F_{var}}{\partial K} = 0$.

(f) Once again, use a graphical argument to find an (implicit!) expression for the predicted transition temperature $T_c^{(2)}$ predicted by the trial Hamiltonian $H_{tr}^{(2)}$ when $h = 0$.

In Onsager's exact 2D Ising model solution, he found that the exact(!) transition temperature T_c satisfies

$$\sinh\left(\frac{2J_X}{T_c}\right) \sinh\left(\frac{2J_Y}{T_c}\right) = 1. \quad (4.48)$$

(g) In the highly anisotropic limit where the y-couplings are very weak ($J_Y \ll J_X$), how does the exact T_c compare to the results obtained from the two trial Hamiltonians you have just analyzed? Interpret your findings. *Hint:* In this limit, you will find that $J_X \gg T_c \gg J_Y$, which means that you can use a small argument approximation for $\sinh(2J_Y/T_c)$ and a large argument approximation for $\sinh(2J_X/T_c)$.

(h) In the isotropic limit where both x and y directions are equivalent ($J_X = J_Y$), use a computer to give numeric estimates of $T_c^{(1)}$, $T_c^{(2)}$, and T_c in units of J. How do the three respective transition temperatures compare? Reason about the differences.

In general, the usefulness of a variational method depends on the *suitability* of the ansatz for that particular problem. The ansatz of part (c) is more effective when the underlying problem is highly anisotropic.

4.2. *More neighbors, more mean.* This problem justifies why mean-field theory is more accurate when the number of neighbors z is large. Consider the molecular field around an Ising spin with z neighbors, $-Jzs - h$, where

$$s = \frac{1}{z} \sum_{j=1}^{z} \sigma_j \qquad (4.49)$$

is the average spin of the neighbors. (The sum runs over the neighbors). The central spin will preferentially adopt the average spin of its neighbors. In the noninteracting trial ensemble, the neighbors are statistically independent.

(a) In the trial ensemble, what is the probability that a neighbor σ_j is up, as a function of the external field? Let $P(\uparrow) = p$ and $P(\downarrow) = 1 - p$ for the rest of this problem.

(b) When $z = 4$ and $h = 0$, what are the possibilities for the neighbors' average spin, and what are their probabilities? Sketch out this probability distribution as a bar graph. *Hint:* There are $2^4 = 16$ configurations.

(c) Mean-field theory presumes that s always takes on its average value $\langle s \rangle$, that is, it pretends that the probability distribution over s is a single peak at $\langle s \rangle$! How accurate is this assumption for part (b)?

(d) The size of deviations from the mean can be captured by the variance, $\langle (s - \langle s \rangle)^2 \rangle = \langle s^2 \rangle - \langle s \rangle^2$. What is the variance of a single spin σ_j?

(e) What is the variance of the *average spin of neighbors, $s = \frac{1}{z} \sum_j \sigma_j$?* *Hint:* Remember that the neighbors are not correlated with one another, so $\langle \sigma_i \sigma_j \rangle = \langle \sigma_i \rangle \langle \sigma_j \rangle$ for $i \neq j$.

(f) The square root of the variance, the standard deviation, is the typical measure for the spread of a probability distribution. How does the standard deviation of s depend on the number of neighbors z?

(g) What happens to the shape of the probability distribution from part (b) as the number of neighbors increases? What does this say about the validity of mean-field theory?

4.3. *Variational principle and K-L divergence.* The variational inequality is useful not only in statistical mechanics, but also in many branches of statistics and machine learning. In these contexts, we often compare different probability distributions. For instance, if we observe a set of data points with distribution P and fit a model with distribution Q, we would like to have a measure of how different P and Q are. A standard measure of "distance" between two probability distributions is the K-L divergence,

$$D_{KL}(P\|Q) \equiv \sum_x P(x) \log \frac{P(x)}{Q(x)}. \qquad (4.50)$$

The smaller the K-L divergence, the more similar the probability distributions $P(x)$ and $Q(x)$. In fact, the K-L divergence is always nonnegative, and it is zero only when $P(x)$ and $Q(x)$ are identical.

The K-L divergence is a nice way to understand what the variational principle is doing. Let P and Q represent the trial and actual thermal ensembles, respectively:

$$P(x) = \frac{e^{-H_{tr}(x)/T}}{Z_{tr}}, \quad Q(x) = \frac{e^{-H(x)/T}}{Z}. \tag{4.51}$$

Using the facts stated above, prove the inequality

$$F_{tr} + \langle H - H_{tr}\rangle_{tr} \geq F \tag{4.52}$$

and justify the variational approach. *Hint:* Start with $D_{KL}(P\|Q) \geq 0$.

4.4. *Universality of critical exponents.* Apply the mean-field approach to the nearest-neighbor Ising model on a 2D *triangular* lattice, and predict a phase diagram, the transition temperature T_c, and the leading-order behavior of m and χ near the critical point. What is different and what is same compared to the square lattice?

4.5. **Mean-field XY.* Apply the mean-field approach to the classical XY model on a d-dimensional hypercubic lattice with $z = 2d$ neighbors per site. Use the Hamiltonian

$$H = -J \sum_{\langle ij\rangle} \cos(\theta_i - \theta_j) - h \sum_j (\theta_j - \phi_0), \tag{4.53}$$

where $0 < \theta_j < 2\pi$ is the angle of the j'th spin and the external magnetic field has magnitude h and direction ϕ_0. Predict a phase diagram, a transition temperature, and the leading-order behavior of $\mathbf{m} = \langle\cos\theta\rangle\hat{\mathbf{e}}_1 + \langle\sin\theta\rangle\hat{\mathbf{e}}_2$ near the critical point.

4.6. *3-state clock model.* Consider a model where each spin takes on one of three possible values. We can represent this with a two-dimensional spin vector, $\mathbf{S_j} = S_j^1\hat{\mathbf{e}}_1 + S_j^2\hat{\mathbf{e}}_2$, where the three orientations are unit vectors spaced apart by $120°$:

$$\mathbf{S_j} = \hat{\mathbf{e}}_1, \quad \mathbf{S_j} = -\frac{1}{2}\hat{\mathbf{e}}_1 + \frac{\sqrt{3}}{2}\hat{\mathbf{e}}_2, \quad \mathbf{S_j} = -\frac{1}{2}\hat{\mathbf{e}}_1 - \frac{\sqrt{3}}{2}\hat{\mathbf{e}}_2. \tag{4.54}$$

We would like to solve the Hamiltonian

$$H = -J \sum_{\langle ij\rangle} \mathbf{S_i} \cdot \mathbf{S_j} = -J \sum_{\langle ij\rangle} S_i^1 S_j^1 + S_i^2 S_j^2 \tag{4.55}$$

with the help of the trial Hamiltonian $H_{tr} = -b\sum_j S_j^1$.

(a) Give a qualitative description of the disordered phase and of the broken-symmetry phase.

(b) Sketch out the three possible spin states on a 2D plane. Which of the three broken-symmetry states is captured by our trial Hamiltonian?

(c) Let's first solve the trial Hamiltonian. Show that

$$F_{tr} = -T \log \left(e^{b/T} + 2e^{-b/2T} \right); \quad \langle S_j^1 \rangle_{tr} = \frac{e^{b/T} - e^{-b/2T}}{e^{b/T} + 2e^{-b/2T}}.$$

Can you argue why $\langle S_j^2 \rangle_{tr} = 0$?

(d) Now let's use our solution to tackle the original, interacting problem. Calculate the variational free energy, and set its derivative w.r.t b to zero to derive the mean-field equation $b = zJm$, where $m = \langle S_j^1 \rangle_{tr}$.

(e) Based on an argument of slopes, what is the temperature T^* at which a nonzero solution to the mean-field equation first appears? (In contrast to the Ising case, it turns out that the nonzero solutions are not necessarily lower in free energy than the zero solution!)

(f) Derive the following expression for the variational free energy solely in terms of m:

$$f_{var} = -T \log 3 + \frac{2}{3} T (1 - m) \log(1 - m)$$

$$+ \frac{1}{3} T (1 + 2m) \log(1 + 2m) - \frac{Jz}{2} m^2. \tag{4.56}$$

You will find it helpful to first show that

$$e^{-3b/2T} = \frac{1 - m}{1 + 2m} \implies b = -\frac{2}{3} T \log \left(\frac{1 - m}{1 + 2m} \right). \tag{4.57}$$

(g) If m is small, we can take a Taylor expansion of f_{var} to figure out where the minima are. Using a computer, find the first few terms of the expansion. Do you notice anything different in this expansion compared to the Ising model? (The presence of this extra term causes the transition to be discontinuous, as we will explore further in Chapter 6.)

4.7. *Metastability and hysteresis.* Consider an Ising-like system.

(a) Pick a temperature T_0 below T_c, and sketch the graph of $f_{var}(m)$ for two values of h: one slightly positive, and one slightly negative.

(b) Suppose the system were initially prepared with $h = 0^-$. What is the equilibrium state?

(c) After the system has equilibrated, Jack decides to change the external magnetic field to be slightly positive, $h = 0^+$. What is the new equilibrium state?

(d) What do you expect the state of the magnet to be
   i. immediately after Jack changes the magnetic field?
   ii. an infinitely long time after Jack changes the magnetic field?

(e) Jack is impatient and decides to turn the field stronger. What happens to the metastable solution once the field is sufficiently strong?

(f) Determine how strong the magnetic field needs to be, $h > h_{meta}$, for what you said in part (e) to happen. *Hint:* At exactly $h = h_{meta}$, there is one global minimum and one inflection point in the free energy.

4.7 Worksheets for Chapter 4

W4.1   Weiss Mean-Field Theory

Let us start with the now (hopefully) familiar Hamiltonian of the Ising model:

$$H = -J \sum_{\text{n.n.}} \sigma_i \sigma_j - h \sum_j \sigma_j. \tag{4.58}$$

We assume the spins exist on an N-site lattice with translational symmetry, and the spins interact only with their z nearest neighbors. A uniform external field h is also allowed. In this exercise we derive one of the simplest mean-field models.

1. Begin by considering the value of each spin in relation to the expectation value of the spins $\langle \sigma_i \rangle = m$, such that each individual spin can be written as:

$$\sigma_i = \langle \sigma_i \rangle + \delta \sigma_i,$$

where $\delta \sigma_i$ is the fluctuation of the spin relative to the average. Use this change of variables to rewrite the interaction term in the Hamiltonian such that fluctuations exist only as a 2nd-order term.

2. Let us also make the nearest-neighbor sum more tractable by showing that:

$$\sum_{\text{n.n.}} = \frac{1}{2} \sum_i \sum_{j \text{ n.n. of } i}.$$

3. So far everything is exact, only with the interaction re-expressed, including a 2nd-order fluctuation term $\delta \sigma_i \delta \sigma_j$. The mean-field approximation entails ignoring second (and higher) order fluctuations,[8] such that $\delta \sigma_i \delta \sigma_j = 0$. Using this approximation and the rewritten form for nearest-neighbor summation, express the Hamiltonian as a single sum over all sites. *Hint:* Keep in mind that each site has z nearest neighbors.

4. By now we have approximated the original interacting problem with N identical noninteracting effective problems. Solving the problem now amounts to solving any single copy. Note that we did not completely throw away the interaction—the thermodynamics of the interacting system is carried implicitly by the m and m^2 terms.

---

8. The reason that this is justified (to an approximate extent) is that while the term $\delta \sigma_i \delta \sigma_j$ itself may be large compared to m, the average of this term over neighboring pairs is much smaller.

Using the new Hamiltonian, calculate the partition function and the free energy, and obtain an expression for the average magnetization $m = \langle \sigma_i \rangle$. *Hint:* Note that the conjugate variable of σ is $-h$ in the Hamiltonian.

5. *Often we may write the quantity $zJ = J_{\text{eff}}$ to indicate that this is the effective interaction strength after the mean-field procedure. What would happen to the mean-field Hamiltonian if we considered interactions beyond nearest neighbor, i.e., z_1 nearest-neighbor spins with interaction J_1, z_2 next-nearest-neighbor spins with interaction J_2, and so on? Can we still express the problem as N noninteracting spins?

The expression for the magnetization is a transcendental self-consistent equation. Even though we cannot solve for the magnetization explicitly, this does give a tractable description of its behavior, which we will analyze in greater detail in following exercises.

W4.2   Variational Principle

Here we prove a number of the assertions involved in our discussion of the variational principle in statistical mechanics.

1. Prove that $\langle e^{-\lambda\phi} \rangle \geq e^{-\lambda\langle\phi\rangle}$ for a random variable ϕ. This is a special case of Jensen's inequality relevant for statistical mechanics. *Hint:* Note that $1 = e^{\lambda\langle\phi\rangle - \lambda\langle\phi\rangle}$, and $e^x \geq 1 + x$ for all real x.

2. Now consider a Hamiltonian H and its partition function $Z = \text{Tr}\left[e^{-\beta H}\right]$, as well as a trial Hamiltonian H_{tr} and its partition function $Z_{tr} = \text{Tr}\left[e^{-\beta H_{tr}}\right]$, assuming that $[H, H_{tr}] = 0$. Use a similar approach to show that its Helmholtz free energy satisfies

$$-T \log Z = F \leq F_{tr} + \langle H - H_{tr} \rangle_{tr},$$

Where $\langle\rangle_{tr}$ denotes an expectation value calculated using Boltzmann weights determined by the trial Hamiltonian. *Hint:* Start with the expression of the partition function Z, and note that the identity $\mathbb{1} = e^{\beta H_{tr}} e^{-\beta H_{tr}}$.

3. The usefulness of the variational principle lies with the parameters $\{b_i\}$ in the trial Hamiltonian $H_{tr}(\{b_i\})$. Describe a general approach to optimize the parameters so that F_{var} provides the best possible approximation to F.

W4.3   *Variational Mean-Field Theory–General Features

You can do this exercise before or after tackling the mean-field theory solution of the Ising model. In either case, this somewhat abstract exercise is meant to help you better understand the variational mean-field reasoning, without being tied down to a specific physical model.

Consider a Hamiltonian of the form:

$$H = H_0 - \sum_j h_j \sigma_j. \tag{4.59}$$

This is a very general form. The second term is a bilinear coupling of the system, here denoted by a spin σ_j, to some external field h_j applied at site j. Everything else, including interaction terms, are relegated to the first term H_0, with the only requirement that it has no dependence on $\{h\} = \{h_j \text{ for all } j\}$ (this notation refers to the entire set of h_j). The free energy due to this Hamiltonian will be a function of the set of fields $\{h\}$ and the temperature T. From the derivative theorem, we know

$$\frac{\partial F(\{h\}, T)}{\partial h_k} = -\langle \sigma_k \rangle (\{h\}, T), \tag{4.60}$$

and higher order correlation terms can be calculated by taking successively higher order derivatives, which we will look at in greater detail in Chapter 7.

The quantity $\langle \sigma_k \rangle$ is the order parameter, which reflects symmetry-broken states of the system, and is therefore usually the first quantity we are interested in. However, the functional dependence of $F(\{h\}, T)$ is often difficult to calculate analytically or even numerically. This is where the variational method comes in.

The mean-field trial Hamiltonian

Let us consider a trial Hamiltonian that we can solve of the form:

$$H_{tr} = H_{tr}^0 - \sum_j b_j \sigma_j, \tag{4.61}$$

where $\{b\}$ are "variational parameters" to be determined. While H_{tr} looks very similar to the original Hamiltonian at first glance, we emphasize two things: first, the term $-\sum_j b_j \sigma_j$ is what leads to the commonly named "mean field," since b_j takes on the role of "field" conjugate to σ_j (parallel to b_j in H); second, we should *be able to solve* the thermodynamics (free energy, expectation values, . . .) of H_{tr} in its entirety, so H_{tr}^0 is shorthand for terms in H_{tr} that are not directly responsible for the "mean fieldness," but are within our ability to calculate nonetheless. H_{tr}^0 does not contain any $\{b\}$ dependence.[9]

We can use H_{tr} to calculate F_{tr}, which also follows the derivative theorem and gives us:

$$\frac{\partial F_{tr}}{\partial b_k} = -\langle \sigma_k \rangle_{tr}. \tag{4.62}$$

---

9. Though it may contain other variational parameters, for the sake of brevity, here we assume it does not, but the result here generalizes to where additional parameters exist.

Here we use $\langle \, \rangle_{tr}$ to mean taking the expectation value in the trial ensemble. Note that $\langle \sigma_k \rangle_{tr}$ is a function of $\{b\}$ and T. Let us give it a new name (and just to make the $\{b\}$ dependence clear):

$$\langle \sigma_k \rangle_{tr} \equiv m_k(\{b\}, T) = m_k(b_j, T). \tag{4.63}$$

Here and onward, whenever variables with subscripts appear as arguments to functions, the set symbol $\{\}$ is implied.

1. Show that, given the form of Eqs. 4.59 and 4.61, the variational free energy can always be written in the form:

$$F_{var} = F_{tr} + F_{var}^0 - \sum_j (h_j - b_j) m_j, \tag{4.64}$$

where

$$F_{tr} = F_{tr}(b_j, T) \tag{4.65}$$

has no explicit dependence on $\{h_j\}$ and

$$F_{var}^0 = F_{var}^0(m_j, T) = \left\langle H_0 - H_{tr}^0 \right\rangle_{tr} \tag{4.66}$$

has no explicit dependence on $\{h\}$, and depends only on $\{b\}$ implicitly through $\{m\}$.

2. We have argued previously that the best approximation to the original free energy F is obtained by solving for

$$\frac{\partial F_{var}}{\partial b_k} = 0. \tag{4.67}$$

Show that, given Eq. 4.64, this leads to the self-consistency condition:

$$\sum_j \left(\frac{\partial F_{var}^0}{\partial m_j} - h_j + b_j \right) \frac{\partial m_j}{\partial b_k} = 0. \tag{4.68}$$

Hint 1: If a function depends on a variable implicitly, partial differentiation requires the chain rule:

$$\frac{\partial f[g_\alpha(s_\beta)]}{\partial s_k} = \sum_\gamma \frac{\partial f}{\partial g_\gamma} \frac{\partial g_\gamma}{\partial s_k}. \tag{4.69}$$

Hint 2: $\{h\}$ are independent variables with $\partial h_j / \partial h_k = \delta_{jk}$. Derivatives of other variables with respect to h_j are 0.

Since $\partial m_j / \partial b_k \neq 0$ in general, the condition becomes, for each j:

$$\frac{\partial F_{var}^0}{\partial m_j} - h_j + b_j = 0. \tag{4.70}$$

3. The condition Eq. 4.70 means that formally there is a set of solutions $\{b_j = b_j(h_j, T)\}$ that minimizes F_{var}, so we can define an optimized variational free energy F_{mf} as:

$$F_{mf}\left(h_j, T\right) \equiv F_{var}\left(b_j, h_j, T\right)\Big|_{\frac{\partial F_{var}}{\partial b_j}=0}, \qquad (4.71)$$

where F_{mf} is a function of only $\{h\}$ and T. Show that, subject to Eq. 4.70:

$$\frac{\partial F_{mf}}{\partial h_k} = -m_k(h_j, T), \qquad (4.72)$$

where

$$m_k(h_j, T) = m_k(b_j, T)\Big|_{\frac{\partial F_{var}}{\partial b_j}=0}.$$

Approximating the order parameter

For Hamiltonians of the form of Eq. 4.59, the order parameter is defined as

$$\frac{\partial F(h_j, T)}{\partial h_k} = -\langle \sigma_k \rangle (h_j, T). \qquad (4.73)$$

We can therefore obtain an approximate order parameter $\langle \tilde{\sigma}_k \rangle$ from any approximation to the free energy $\tilde{F}$ by:

$$\frac{\partial \tilde{F}}{\partial h_k} = -\langle \tilde{\sigma}_k \rangle. \qquad (4.74)$$

Note that this procedure is general and does not depend on how we get the approximate free energy $\tilde{F}$.

4. Using the variational procedure with a trial Hamiltonian of the form of Eq. 4.61, we find the variational approximation $\tilde{F} = F_{mf}$. Show that the approximate order parameter is:

$$\langle \tilde{\sigma}_k \rangle = \langle \sigma_k \rangle_{tr}\Big|_{\frac{\partial F_{var}}{\partial b_j}=0} = m_k(h_j, T). \qquad (4.75)$$

That is, the approximate order parameter *is* the expectation value in the trial ensemble $\langle \sigma_k \rangle_{tr} = m_k(h_j, T)$, evaluated at the self-consistence conditions!

This result is not apparent a priori, and is a feature of the form of Eqs. 4.59 and 4.61. Fortunately, physical systems frequently have Hamiltonians of the form of Eq. 4.59, which makes picking Eq. 4.61 as the trial Hamiltonian a useful general method of approximation.

This particular variational approximation is called the "mean-field theory." We have previously mentioned that the parameter b_j takes on the role of field, and is what we solve for in the self-consistency equation Eq. 4.70. Its conjugate variable (in the thermodynamic sense) $\langle \sigma_j \rangle_{tr} = m_j$ turns out to be the order parameter in this approximation. Therefore, the physical interpretation of the parameter b_j is indeed a thermodynamically averaged field, or "mean field," experienced by σ_j.

Equivalence of b_j and m_j as variational parameters

Eqs. 4.62 and 4.63 define a new set of functions $\{m\}$ by

$$m_k(b_j, T) \equiv -\frac{\partial F_{tr}}{\partial b_k}, \qquad (4.76)$$

meaning that we have the explicit functional dependence of $\{m\}$ on $\{b\}$. Formally, this implies that we can invert the relationship and write:

$$b_k = b_k(m_j, T). \qquad (4.77)$$

We can then substitute this into the expression for the variational free energy,

$$F_{var}\left[b_k(m_j, T), h_k, T\right] = F_{var}(m_k, h_k, T), \qquad (4.78)$$

meaning that we replace all dependence on $\{b\}$ explicitly with $\{m\}$.

While we can define an arbitrary set of functions that gives a correspondence:

$$g_k(b_j, T) \iff b_k(g_j, T),$$

only the set of functions $\{m\}$ defined by Eq. 4.76 satisfies the following specific condition that we will now prove.

5. Show that

$$\frac{\partial F_{var}}{\partial m_k} = 0 \qquad (4.79)$$

   leads to the same self-consistency equations as Eq. 4.70.

Physical interpretation of $F_{var}(m_k, h_k, T)$

Proving that Eq. 4.79 leads to the same self-consistency equations tells us that $\{m\}$, the variational order parameters, can be used as the variational parameters in F_{var} instead of $\{b\}$.

While mathematically equivalent to Eq. 4.67, minimizing F_{var} with respect to $\{m\}$ gives a more transparent physical interpretation: the variational free energy is minimized with respect to the *order parameter* of the system. This also gives us an intuitive phenomenological understanding[10] of phase transitions, which we will

---

10. This means that it is based on observed phenomena pertaining to the system, such as the order parameter and other macroscopic observables.

explore in Chapters 6 and 7. While historically the name "mean-field theory" stems from the $\{b\}$ formulation of the variational theory, it is applicable also to this phenomenological variant, owing to the fact that formulating F_{var} either in $\{b\}$ or in $\{m\}$ are mathematically equivalent.

W4.4   Mean-Field Solution of the Ising Model

You can do this exercise before or after tackling the general features of mean-field theory. In either case, this concrete exercise gives you a physically meaningful example of variational mean-field theory in practice.

Consider the Hamiltonian of the Ising ferromagnet with field h_j applied at each site σ_j:

$$H = -J \sum_{\langle i,j \rangle} \sigma_i \sigma_j - \sum_j h_j \sigma_j. \tag{4.80}$$

The derivative theorem tells us that the (sitewise) magnetization is:

$$\langle \sigma_j \rangle = -\frac{\partial F}{\partial h_j}. \tag{4.81}$$

Let us employ the variational method to approximate the behavior of the system, using the trial Hamiltonian:

$$H_{tr} = -\sum_j b_j \sigma_j, \tag{4.82}$$

where $\{b_j\}$ are the variational parameters we will use to minimize F_{var}. We know from the derivative theorem that:

$$-\frac{\partial F_{tr}}{\partial b_j} = \langle \sigma_j \rangle_{tr} \equiv m_j, \tag{4.83}$$

where the last step is a definition for m_j.

Calculating averages with respect to the trial ensemble

1. As a small first step, calculate Z_{tr} from H_{tr} for only two spins. Simplify your answer to a single product.

2. Now consider an arbitrary number of spins. From H_{tr}, calculate Z_{tr}, then F_{tr}, $\langle \sigma_i \rangle_{tr}$, and $\langle \sigma_i \sigma_j \rangle_{tr}$. Show that $\langle \sigma_i \rangle_{tr} = \tanh(\beta b_i)$ and $\langle \sigma_i \sigma_j \rangle_{tr} = \langle \sigma_i \rangle_{tr} \langle \sigma_j \rangle_{tr}$ for $i \neq j$. *Hint:* The σs are independent in H_{tr}, so you can manipulate the sums and products easily. The previous part should come in handy.

3. Using these results, calculate $F_{var} = F_{tr} + \langle H - H_{tr} \rangle_{tr}$. It is not exactly an elegant expression, but will be useful soon. *Hint:* The notation $\langle\ \rangle_{tr}$ means to calculate expectation values with the trial ensemble, which means we can apply results of expectations we have already calculated above.

Derive the mean-field equations

To minimize the variational free energy F_{var}, we compute the derivative of F_{var} with respect to each b_k (we will use a different dummy index to not interfere with the existing dummy indices), and then set this derivative equal to zero:

$$\frac{\partial F_{var}}{\partial b_k} = 0. \tag{4.84}$$

1. Show that this gives the result:

$$b_j = h_j + J \sum_{i\,\text{n.n.\,of}\,j} \tanh(\beta b_i). \tag{4.85}$$

2. Now, show that:

$$\frac{\partial F_{var}}{\partial h_k} = -m_k. \tag{4.86}$$

   Hint: We know $\partial F_{tr}/\partial b_k = -\langle \sigma_k \rangle_{tr}$; what's $\partial F_{tr}/\partial h_k$? Use the chain rule for partial differentiation.

Consequences of the mean-field solution with uniform h

Let us briefly examine the behavior of the system for a uniform applied field, i.e., $h_j = h$ for all j. The self-consistency equation, Eq. 4.85, then becomes:

$$b_j = h + J \sum_{i\,\text{n.n.\,of}\,j} \tanh(\beta b_i). \tag{4.87}$$

3. Explain how we could motivate the solution of Eq. 4.87 to be translationally invariant, i.e., $b_j = b$ for all j.
   Hint: What can you say about minimizing the ground-state ($T = 0$) variational free energy?

   Once we have assumed the solution is translationally invariant, we will see that there are two phases: a disordered phase at $T > T_c$ in which $\langle \sigma_j \rangle \to 0$ as $h \to 0$ and an ordered phase at $T < T_c$ in which $\langle \sigma_j \rangle \to \text{sign}(h)m(T)$ with $m(T) > 0$ as $h \to 0$.

4. From Eq. 4.87, find T_c. *Hint:* Make a sketch of the graphs of b and $\tanh(\beta b)$; what happens when you increase β? Set $h = 0$.

5. From Eq. 4.87 show that $m(T) = \tanh[\beta(zJm(T) + h)]$. Remember that $m = \langle \sigma \rangle_{tr}$ when the variational parameters minimize F_{var}.

6. We have argued above that for $T > T_c$, $m = 0$. For T below that near T_c (i.e., $T_c \gg (T_c - T) > 0$), m is very small. Find an expression for $m(T)$ near T_c. Again set $h = 0$.

W4.5   Low-Temperature Analysis of Mean-Field Equations

Even though we can't solve the mean-field equations analytically, we can still get approximate expressions in certain limiting cases. This is known as asymptotic analysis, since the results will not be exact but "asymptotic." You'll get a chance to think carefully about what that means. For now, let's take a look at the low-temperature limit of the equation

$$\tilde{b} = \beta z J \tanh \tilde{b} \qquad (4.88)$$

where $\tilde{b} \equiv \beta b$.

1. How does $\tilde{b}$ behave as $T \to 0$? *Hint:* Remember that $-1 \le \tanh x \le 1$ for all x.

2. With this in mind, write down the first few terms in a series approximation of Eq. 4.88. Remember that $\tanh(x) = (1 - e^{-2x})/(1 + e^{-2x})$. *Hint:* What is the appropriate small parameter to expand in?

3. Write down an expression for $\tilde{b}$ in terms of T and $T_c = zJ$. You only need the leading-order term of $\tilde{b}$ for this and the rest of this worksheet.

4. Using your answer from (3), come up with a series expression for $m = \pm \tanh(\tilde{b})$. (The $\pm$ reflects the two possible states of broken symmetry.) Write down the dominant term and the leading-order correction. What is the size of the next-order correction?

5. Which term(s) are important as $T \to 0$?

6. What happens to the relative importance of the higher order corrections in your expression as $T \to 0$?

It is intriguing to relate your result with what we got in the low T expansions for m from the previous chapter:

7. At precisely $T = 0$, what is the ground state? Compare this to the dominant term in m.

8. At low nonzero T, what are the lowest energy excitations? How much of an energy cost are they? What is the corresponding Boltzmann factor? Compare this to the 1st-order term in your expression for m.

It is a triumph for mean-field theory that it agrees with the m computed in a low-T expansion.

W4.6   The Variational Free Energy F_{var} as a Function of the Order Parameter

We are now going to express the variational free energy (F_{var}) as a function of $\{m\}$ instead of $\{b\}$ by using the definition:

$$-\frac{\partial F_{tr}}{\partial b_j} = m_j = \tanh(\beta b_j). \tag{4.89}$$

This allows us to use the order parameters $\{m\}$ as the variational parameters. Now is a good time to do W4.3 if you have not yet done so.

1. Derive the following expression:

$$F = \sum_j V(m_j) + \frac{J}{2}\sum_{<i,j>}(m_i - m_j)^2, \tag{4.90}$$

where

$$V(m_j) = \frac{T}{2}\ln\left[\frac{1-m_j^2}{4}\right] + m_j T\tanh^{-1}(m_j) - \frac{zJ}{2}m_j^2 - h_j m_j. \tag{4.91}$$

Hint: The following identity may come in handy:

$$\cosh^2 x - \sinh^2 x = 1 \implies 1 - \tanh^2 x = 1/\cosh^2 x. \tag{4.92}$$

If you can't get the answer going forward from F_{var}, try also working backward.

2. As long as $|m_j|$ are all small, we can expand this in powers of m_j. Using the identity

$$\tanh^{-1}(x) = \frac{1}{2}\ln\left[\frac{1+x}{1-x}\right], \tag{4.93}$$

show that

$$V(m_j) = V_0 - h_j m_j + \frac{\alpha}{2}m_j^2 + \frac{u}{4}m_j^4 + \ldots, \tag{4.94}$$

where

$$V_0 = -T\ln(2), \tag{4.95}$$

$$\alpha = T - zJ = T - T_c, \tag{4.96}$$

$$u = \frac{T}{3}. \tag{4.97}$$

We can see a lot from Eqs. 4.90 and 4.91. Firstly, any spatially non-uniform solution of m_j will always have higher free energy than a uniform solution so long as $J > 0$. Secondly, for $T > T_c$ there is at least a local minimum of the free energy with all $m_j = 0$ (which is in fact a global minimum). For $T < T_c$, we can see that $m_j = 0$ is a local maximum of the free energy. If we use the expansion Eq. 4.94, we can determine that there are two minimum free energy solutions by solving $\partial\tilde{V}/\partial m_j = 0$:

$$m_j = m = \pm\sqrt{|\alpha|/u} \sim (T_c - T)^{1/2}. \tag{4.98}$$

This is approximate, but it is valid as long as $|m|$ is small, which is to say $|\alpha|$ is small, which in turn is to say $(T_c - T) \ll u$. However, the symmetry of the solution is exact; since it contains only even powers of m, V is an even function of m, so if there is a minimum of the free energy with $m_j = m$, there is an equally good minimum of the free energy for $m_j = -m$.

W4.7   The Mean-Field Critical Exponents β, γ, δ

From the previous part, we have:

$$F_{var} = V_0 - hm + \frac{\alpha}{2}m^2 + \frac{u}{4}m^4 + \cdots, \tag{4.99}$$

where

$$V_0 = -T \ln 2,$$

$$\alpha = T - zJ$$

$$= T - T_c,$$

$$u = \frac{T}{3}.$$

Since we have now proven that the value of the order parameter m minimizes the free energy, the phase of the system is determined by the equation:

$$\frac{\partial F_{var}}{\partial m} = 0$$

$$\implies -h + \alpha m + um^3 = 0. \tag{4.100}$$

We ignore higher order terms in m since the first terms necessarily dominate the behavior of the system.

Critical exponent β

The first critical exponent concerns the power-law behavior of the order parameter m near the critical temperature. We are looking for the form $m \propto (T_c - T)^\beta$. This case requires that $h = 0$, since any nonzero h leads to a symmetry-broken state at any given temperature. Solve Eq. 4.100 and show that $\beta = 1/2$.

Critical exponent γ

The next critical exponent concerns the susceptibility of the order parameter, evaluated at zero applied field:

$$\chi = \left.\frac{\partial m}{\partial h}\right|_{h=0} \propto |T_c - T|^{-\gamma}. \tag{4.101}$$

Solve for χ from Eq. 4.100 and show that $\gamma = 1$. *Hint:* Remember implicit differentiation?

Critical exponent δ

The third critical exponent we will study for now is the field dependence of the order parameter on the phase boundary, i.e., $T = T_c$. We are looking for the form $m \propto h^{1/\delta}$. Again, from Eq. 4.100, show that $\delta = 3$.

BROKEN SYMMETRIES

Symmetries of nature, like promises of undying love, oft are broken.

This chapter builds on what we learned from our study of the Ising model to develop a more general perspective on the theory of phase transitions. Specifically, we will delve more deeply and precisely into the concepts of "spontaneously broken symmetry" and "order parameters," and into the "universal" properties of systems in the vicinity of certain critical points.

We have seen that the Ising ferromagnet at $T < T_c$ can exist in two distinct thermodynamic states. The two states are distinguished by the sign their magnetization m. More generally, in any system with a spontaneously broken symmetry, there is an analogously defined observable, known as the order parameter, which encodes the information needed to specify the thermodynamic state. The perspective of order parameters and broken symmetries is powerful because it allows us to develop a description of phases and phase transitions which generalize across different systems.

5.1 Spontaneously Broken Symmetry and Order Parameters

In Section 2.3, we alluded to the central role of symmetry in the study of phases and phase transitions: the Hamiltonians of physical systems often possess a set of symmetries, but their equilibrium phases may not respect all of those symmetries. For instance, in the absence of a symmetry-breaking field ($h = 0$), the Ising ferromagnet has the up-down symmetry of $\sigma_j \to -\sigma_j \; \forall \; j$; the energy is invariant under a global spin-flip.

At high temperatures, the equilibrium phase of matter always exhibits all the microscopic symmetries, meaning that the thermodynamic state is invariant under any symmetry transformation. However, it often is the case that below a critical temperature, T_c, the thermodynamic state no longer longer manifests all these symmetries. In the case of the Ising model, the paramagnetic phase is invariant under the spin-flip operation, but the ferromagnetic phase (with nonzero m) is not. Below T_c, the $\sigma_j \to -\sigma_j \; \forall \; j$ symmetry is broken.

The concept of broken symmetry is a powerful way to characterize equilibrium states of matter. Different ordered phases break different symmetries—and the broken symmetries determine much of the interesting physics of the phases and their transitions. Typically, but not universally, low-temperature phases are more ordered than higher temperature phases; correspondingly, low-temperature phases typically have a smaller set of symmetry transformations (see Section 2.3 and Appendix A.10 for more detailed discussion on symmetry groups).

The order parameter ϕ generalizes the notion of m from the Ising ferromagnet. It is zero in the fully symmetric phase, and it is nonzero in the ordered phase in a manner which captures the nature of the broken symmetry. For different types of order, the order parameter could take on three or four (as opposed to two) discrete values, or could even be a multicomponent vector. Furthermore, even though we have only seen examples where the order parameter is uniform throughout a material, it is in general a local quantity which can vary in space. We will see some explicit examples of this soon.

> **Did you know?** It can happen in some cases that a low-temperature phase is *more* symmetric than a higher temperature one. For example, helium alone among the elements remains a (quantum) liquid down to the lowest temperatures, $T \to 0$, at ambient pressure, although it does crystallize at sufficiently high pressure. For a range of pressures, however, the helium isotope ^{3}He is a liquid at absolute zero, crystallizes upon heating above a nonzero temperature, T_{c1}, only to melt again above a still higher temperature, $T_{c2} > T_{c1} > 0$. If you find this interesting and mysterious, you can look up the Pomarenchuk effect.

5.1.1   Why the Thermodynamic Limit Is Necessary

The possibility of a spontaneously broken symmetry is not immediately apparent in the formal expressions we summarized in the Chapter 2. When we compute thermal averages by summing over microstates, the states which are related by symmetry all have the same probability weight (since they have the same energy). From this it is easy to prove that any thermodynamic observables must be symmetric under all symmetry transformations—broken symmetries should not be possible!

Let us look for the loophole in this argument, using as example the now familiar case of the Ising ferromagnet. We begin by applying this symmetry argument to construct a "proof" that $m = 0$ at all T. The microstates of the system are specified by the configuration of all the spins, $\{\sigma\}$. In terms of this basis, the average magnetization is

$$\langle \sigma_j \rangle = \sum_{\{\sigma\}} \frac{e^{-\beta H(\{\sigma\})}}{Z} \, \sigma_j. \tag{5.1}$$

Alternatively, if we perform the sum in a "flipped" basis $\{\tau\}$ where $\tau_j = -\sigma_j$, we see that

$$\langle \sigma_j \rangle = \sum_{\{\tau\}} \frac{e^{-\beta H(\{\tau\})}}{Z} (-\tau_j). \tag{5.2}$$

But since $H(\{\sigma\}) = H(\{\tau\})$, it follows that

$$\langle \sigma_j \rangle = -\langle \sigma_j \rangle. \tag{5.3}$$

Since the only solution to $x = -x$ is $x = 0$, we are forced to conclude that $\langle \sigma_j \rangle = 0$. (This argument can be readily generalized to complicated problems with richer sets of symmetries.) The only flaw in this proof is that it applies only to finite systems, where we can truly enumerate all of the states. Finite-size systems cannot exhibit spontaneous symmetry breaking—only infinite-size systems can.

Another way to see this fact is by considering the analyticity of the free energy. In any finite-size system, F is necessarily an analytic function of T. This is because F is composed of a (finite number of) analytic functions: the partition function is a sum of exponentials, and since $Z > 0$ for all $T > 0$, it follows that $F = -T \log Z$ is also analytic for all nonzero temperatures. Since phase transitions correspond to non-analyticities of F, we conclude that no finite system can undergo a phase transition (at nonzero T). However, since an *infinite* sum of analytic functions is not necessarily analytic everywhere,[1] an infinite system may undergo a phase transition.

Formally, spontaneously broken symmetries can only arise in the "thermodynamic limit," i.e., when the number of sites $N \to \infty$. To see how this occurs, one has to take the limit in a particular manner. Consider the example of the Ising ferromagnet again, but now in the presence of a symmetry-breaking field, h. Here the magnetization, $m(h; N) = \langle \sigma_j \rangle$, is a function of both h and N, and there is no symmetry-related reason for it to vanish for $h \neq 0$. To define the order parameter, we first take the thermodynamic limit, $N \to \infty$, and only then take the limit as the symmetry-breaking field vanishes, $h \to 0$:

$$m \equiv \lim_{h \to 0} \left[\lim_{N \to \infty} m(h; N) \right]. \tag{5.4}$$

(We have implicitly been working in this limit in previous chapters, but without being quite so explicit.)

As an illustration of this limiting procedure, let us think about the spin configurations which are significant at low T. The configurations can be divided into two groups, one in which spins are mostly up, and another in which spins are mostly down. The thermal average $\langle \sigma_j \rangle$ includes contributions from both groups. If $h = 0$, the two groups contribute equally, but if $h > 0$, then the "mostly up" group dominates

---

1. Consider, for instance, the Fourier series of a square wave. The square wave is non-analytic since it is discontinuous, and yet it can be expressed as an infinite sum of (analytic) sinusoids. The partial sums are analytic, but they converge to a nonanalytic function.

because it is lower in energy. Since the difference in energy between a mostly up and mostly down configuration is roughly $2hN$, the Boltzmann weight of the "mostly down" group is smaller by a factor of $e^{-2hN/T}$. Consequently, if $h \gg T/N$, the contribution of the mostly down configuration is exponentially small. If we take the limits $N \to \infty$ and $h \to 0$ in such a way that $hN/T \to \infty$, we will approach a unique and well-defined value for the order parameter $m(T)$ which is nonzero for all $T < T_c$.

There is another way to look at the same problem that avoids the need to introduce a symmetry-breaking field. It involves looking at the correlations between the value of the order parameter at two points, $\vec{R}_i$ and $\vec{R}_j$, far apart in space. So long as all interactions are local in space, we would expect that points far apart are uncorrelated, so that

$$\lim_{|\vec{R}_i - \vec{R}_j| \to \infty} \langle \sigma_i \sigma_j \rangle = \langle \sigma_i \rangle \langle \sigma_j \rangle. \tag{5.5}$$

In a disordered phase, this implies that the correlation function vanishes at long distances; in any phase in which it approaches a nonzero constant, i.e., if there is "long-range order," we can identify the long-distance behavior with the value of the order parameter. Notice that the thermodynamic limit is still required for this definition to make sense, because in a finite-size system, the separation between any two points, $|\vec{R}_i - \vec{R}_j|$, is necessarily finite. Formally, this second definition of long-range order involves first taking the limit $N \to \infty$, and then the limit $|\vec{R}_i - \vec{R}_j| \to \infty$. In general, the two methods for identifying states with broken symmetries are mutually compatible.

5.1.2   Phase Transitions in the Real World

Infinity is an abstraction.

In the real world, all systems are finite. Nevertheless, the number of atoms in a typical macroscopic piece of matter is large, on the order of Avogadro's number $N \sim 10^{26}$. Intuitively, this large number might seem close to infinity, but conceptually it is still important to think about what "close to infinity" means, and whether any corrections arise from the finiteness of N. Indeed, in an age of microfabrication, where the physics of increasingly small material systems are of interest, it is of practical interest to understand the properties of "mesoscopic" systems, in which $N \gg 1$ but the fact that N is finite still plays a role in the physics.

There are two ways to understand why phase transitions are observed in the real world where N is finite:

Small but nonzero *h*

The conceptually simpler approach is to consider the problem in the presence of a very small symmetry-breaking field—after all, in the case of a ferromagnet, unless we go to heroic efforts to screen it out, there is always the presence of earth's magnetic field. In the majority of situations, this small but nonzero h biases the energies enough to choose a preferred direction of the magnetization below T_c, but not enough to have any other significant effect on thermodynamic averages. As we have seen, for $T < T_c$, so long as $hN m(T) \gg T$, only states with one orientation of the net magnetization contribute to the sum over states, while so long as $T \gg h$, the symmetry-breaking field has little other effect on thermodynamic averages. Hence, so long as $T \gg h \gg T/[N m(T)]$, the magnetization of a finite sample in a weak symmetry-breaking field will behave similarly to the $m(T)$ of an infinite sample in the absence of a field (Eq. 5.4). For N sufficiently large, almost any small h will satisfy this condition.[2]

Timescales

Another way to view the problem is in terms of a hierarchy of timescales. Recall that our use of statistical ensembles is ultimately justified by the ergodic hypothesis. A key assumption of this is that the system explores all the relevant configurations during the observation period. In situations where symmetry is broken, however, this assumption is not valid; only a portion of configurations are accessible over typical laboratory timescales.

Consider the time evolution of a finite-size Ising ferromagnet cooled below T_c. The important microstates can be divided into two groups, ones with $M \simeq Nm(T) > 0$ and ones with $M \simeq -Nm(T) < 0$. Statistical mechanics assumes that the system has enough time to sample all relevant configurations—both the $M > 0$ group and the $M < 0$ group. In a large system, however, the amount of time it takes to get from one set of configurations to the other is extremely long—maybe longer than the lifetime of the universe, and certainly longer than any scientist measuring the system is willing to wait! Effectively, then, the system is stuck in one of the two sets of states.

To be more precise, suppose that the finite-size Ising ferromagnet is initially prepared in a small field $h > 0$. Once it settles into an equilibrium state with $M \simeq Nm(T) \approx N$, we set $h \to 0$. For a while, the system will (with high probability) only access microstates in the $M > 0$ group. However, if we wait for a very long time $t \gg \mathcal{T}$, the system will eventually explore configurations in the other $M < 0$ group. We can estimate the timescale $\mathcal{T}$ as follows: in going from net positive to net negative magnetization, the system must pass through an intermediate state with $M = 0$. Any configuration with $M = 0$ necessarily has an energy that is larger by an amount

---

2. Note that no matter how large N is, this condition will not be satisfied near enough to T_c, since $m(T) \to 0$ as $T \to T_c$. Hence, close to T_c, we will always encounter a regime in which finite-size corrections lead to a "rounding" of the critical phenomena—another manifestation of the fact that phase transitions are ultimately precisely defined only in the thermodynamic limit.

ΔE than the ground-state energy, and hence the probability of finding the system in such a "transition state" is small in proportion to $\exp[-\Delta E/T]$. Correspondingly, the length of time one expects to wait before the system evolves to such a state is $\mathcal{T} \sim \tau_0 \exp[\Delta E/T]$, where τ_0 is a microscopic time which characterizes the transition from one microstate to a nearby state.

Naturally, ΔE depends on the nature of the transition state. The lowest energy state with $M = 0$ (and hence the one that is exponentially more likely to arise than any other) is one in which one half the system has $m \approx 1$, and the other has $m \approx -1$. There is a "domain wall" between the two ordered regions, something we will study in detail in Chapter 7. Since the system is in local equilibrium everywhere away from the domain wall, the energy cost comes entirely from the domain wall itself. In a three-dimensional system, the energy of such a domain wall is proportional to the cross-sectional area, $\Delta E \propto A$, of the finite system—consequently, $\mathcal{T}$ is grows exponentially with the size of the system and is typically extremely long.[3] For times small compared to $\mathcal{T}$ but large compared to τ_0, the system behaves as if in the thermodynamic limit.

5.2   Examples of Broken Symmetries

To make these formal considerations more tangible, we take a survey of some phase transitions. Some involve simple and familiar ideas such as rearranging atoms. Others involve quantum mechanical properties such as ferromagnetism and superconductivity.

5.2.1   Order-Disorder Transitions

In solid alloys (materials with multiple elements), a type of solid-to-solid phase transition known as an order-disorder transition may occur. The classic example is β-brass, a 50-50 binary alloy of copper and zinc in a body-centered cubic (BCC) lattice. The BCC lattice is bipartite—it can be decomposed into two interpenetrating but otherwise identical cubic lattices (which we will call A and B) such that all the nearest neighbors of any site on the A sublattice are on the B sublattice, and vice versa. This is depicted in Figure 5.1. At relatively high temperatures (but still lower than the melting temperature), the two types of atom are randomly located. The probability of finding a Cu or a Zn atom on any given lattice site is the same for any lattice site; there is a "sublattice symmetry" in that the A and B sublattices are indistinguishable. By contrast, at $T = 0$, the atoms form an ordered pattern with Cu atoms on one sublattice—which we will identify as the A sublattice—and Zn on the other—the B sublattice—as shown schematically in Figure 5.1A. Importantly,

---

3. In d dimensions, the domain wall has a "hyperarea" L^{d-1}, where L is the spatial extent of the system.

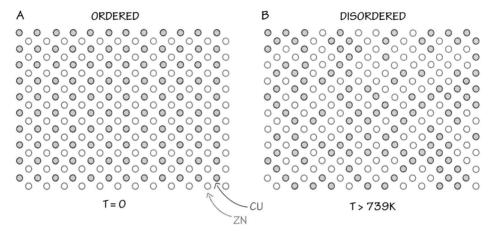

FIGURE 5.1. Beta-brass undergoes an order-disorder transition at $T_c = 739\ K$. The Zn and Cu atoms tend to interleave alternately in the ordered phase, but arrange randomly in the disordered phase. Note the BCC lattice remains intact; only the positions of the atoms change.

at $T = 0$ the two sublattices are not equivalent; the symmetry between the sublattices is spontaneously broken. At low but nonzero temperature, the perfect order that characterizes the ground state is lost in that there is a chance of finding a Zn atom on the predominantly Cu A sublattice, and vice versa. However, so long as T is smaller than a critical temperature T_c, the two sublattices remain distinguishable in the sense that one sublattice has more than half the Zn atoms while the other has less than half. The sublattice symmetry is still spontaneously broken for all $T < T_c$.

The fact that there are two possible broken-symmetry ground states (one with the Zn on the A sublattice and one with the Zn on the B sublattice) is reminiscent of the Ising model. Taking each possible arrangement of atoms to be a microstate, we represent the states of this "lattice gas" by defining Ising variables, σ_j, such that $\sigma_j = 1$ if there is a Zn atom on site j and $\sigma_j = -1$ if there is a Cu atom. The only difference here is that the problem in question is an Ising *anti*-ferromagnet: from the fact that the ground state has alternating spin orientations on the two sublattices, we can infer that the effective interaction favors opposite sign spins on neighboring sites; i.e., J in Eq. 3.1 is negative.

Because the lattice is bipartite, even this difference is of no great importance. We can define a new set of Ising variables τ_j such that $\tau_j = \sigma_j$ for j on the A sublattice, and $\tau_j = -\sigma_j$ for j on the B sublattice. The sublattice symmetry that interchanges the A and B sublattices thus transforms $\tau_j \rightarrow -\tau_j$. The state in which all the Zn atoms are on the A sublattice corresponds to $\tau_j = 1\ \forall\ j$, while the state in which they are all on the B sublattice has $\tau_j = -1\ \forall\ j$. Thus, at the formal level, this problem is equivalent to an Ising ferromagnet in which $m = N^{-1}\sum_j \langle \tau_j \rangle$ represents the excess concentration of Zn on the A sublattice. Accordingly, it is possible to define an order parameter ϕ that behaves in exactly the same manner as the magnetization of a ferromagnet.

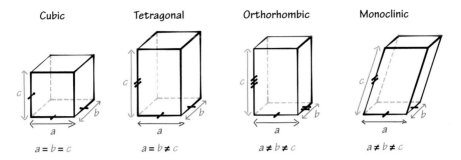

FIGURE 5.2. Nature exhibits crystals whose lattices have varying degrees of symmetry. Shown above are unit cells of four possible Bravais lattices, with the most symmetric one on the left, and successively less symmetric ones on the right.

5.2.2 Structural Transitions

The cubic-to-tetragonal transition we introduced in Section 1.2.1 is another example of how symmetry is broken at a phase transition. In the high-temperature cubic phase, all three axes are equivalent, whereas in the lower-temperature tetragonal phase, one of the three axes is unique. In this case there are three distinct choices for the broken-symmetry state—corresponding to the three crystal axes—which are all equivalent in the high-temperature cubic phase. Worksheet W5.3 lets you think about how to describe this type of broken symmetry.

Some tetragonal crystals can undergo further symmetry-breaking structural transitions as they are cooled. If the two equivalent axes were to become distinguishable, then the crystal would become orthorhombic (Figure 5.2): when viewed along the c axis, the shape in the $a - b$ plane distorts from a square to a rectangle. The tetragonal-to-orthorhombic transition must occur at a precise critical temperature T_c such that

$$a = b, \quad T > T_c,$$
$$a \neq b, \quad T < T_c.$$

Either the a or the b axis becomes the longer one below T_c. Since there are two equivalent choices for the broken-symmetry phase, this is an Ising-like symmetry. Accordingly, we can define an order parameter in terms of the lattice parameters, $\phi \equiv (a - b)/(a + b)$, such that ϕ is zero in the entire tetragonal phase and nonzero in the orthorhombic phase. In analogy with m in the Ising ferromagnet, ϕ can be either positive or negative; the sign indicates which of the two equivalent directions in the tetragonal lattice spontaneously becomes the longer one.[4]

There are even more structural transitions, as you could imagine. For example, in an orthorhombic lattice the c axis is perpendicular to the $a - b$ plane, so there is

---

4. By convention, the axes of an orthorhombic crystal are defined such that $a < b < c$.

a symmetry of reflection across the $a - b$ plane. If the c axis were to tilt away from the perpendicular, this mirror symmetry would be spontaneously broken, leading to a monoclinic crystal. In general, crystals can exhibit a wide variety of structural symmetries and undergo structural transitions at points where the symmetry changes.

Such transitions allow us to see why symmetry is a precise, clear-cut way of classifying the phases of matter. Most properties of matter change continuously as a function of temperature. For instance, most crystals expand under heating; their lattice constants a, b, and c vary with T in a continuous manner. However, the symmetry of the crystal cannot change continuously. A crystal either has a symmetry or not; two lattice constants are either equal or unequal. There is no in-between. Thus, if there is a symmetry-breaking structural transition, we can identify a *precise* critical temperature at which the symmetry changes sharply.

5.2.3   Crystalline Solids

We have already alluded to the fact that the atoms in a crystal are arranged in a regular, repeating manner (see Section 1.2.2). One subtlety we have glossed over is that the positions of the atoms do not perfectly coincide with the abstract sites of the lattice. At nonzero temperatures, the atoms experience random, thermal motions, and even at $T = 0$, there is quantum-mechanical uncertainty due to the zero-point motion. Nevertheless, in a crystal the *average* atomic positions have a regular, periodic pattern.

In a crystal, the translational symmetry is broken since an atom is more likely to be found at some points in space than at others. The manner in which the symmetry is broken is defined by the periodicity of the thermally averaged density, $\langle \rho(\vec{r}) \rangle$. Because $\langle \rho(\vec{r}) \rangle$ has a set of regular peaks centered around the lattice sites, it is only invariant under a special subset of translations $\{\vec{R}\}$, called the Bravais lattice:

$$\langle \hat{\rho}(\vec{r} + \vec{R}) \rangle = \langle \hat{\rho}(\vec{r}) \rangle \ \ \forall \, \vec{r} \ \& \ \forall \, \vec{R} \in \{\vec{R}\}. \tag{5.6}$$

As discussed in textbooks on solid state physics,[5] any Bravais lattice vector (in 3D) can be written as a linear combination $\vec{R} = n_a \vec{r}_a + n_b \vec{r}_b + n_c \vec{r}_c$, where $\vec{r}_\alpha$ are the basis vectors of the unit cell, and n_α are integer coefficients. In addition to translational symmetry, the crystalline state also breaks rotational symmetry, since the Bravais lattice is only invariant under a smaller subset of all possible rotations. Some crystals break inversion symmetry as well.[6]

5.2.4   Liquid Crystals

The power of the broken-symmetry perspective is, perhaps, nowhere clearer than when considering liquid crystals. From a broken-symmetry perspective, a liquid

---

5. The standard, definitive book is Ashcroft and Mermin's *Solid State Physics* (Saunders College Publishing, Philadelphia, 1976).

6. The inversion transformation is the operation $(x, y, z) \rightarrow (-x, -y, -z)$.

A   B

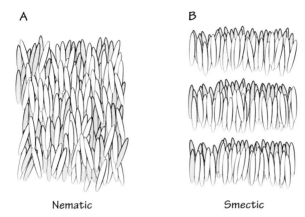

Nematic   Smectic

FIGURE 5.3. The symmetries of liquid crystals are intermediate between those of liquids and solids. (A) In a nematic liquid crystal, the positions of the molecules are homogeneously distributed, but the long axis of the molecules is oriented preferentially in one direction. (B) In a smectic liquid crystal, the molecules stack into orderly layers, but remain liquid-like within the layers.

crystal is a phase with a pattern of broken symmetry somewhere between a liquid and a crystal. Accordingly, liquid crystals tend to arise in a range of temperatures above the melting temperature of a crystalline solid phase but below the critical temperature of an isotropic fluid phase. Their defining characteristic is that only a *subset* of translational and rotational symmetries are broken. (Worksheet W5.1 gives you a chance to explore and understand what this means.)

For example, a nematic liquid crystal retains the full translational symmetry of a liquid, but spontaneously breaks its rotational symmetry. This phase of matter can be thought of as a liquid of long rod-like molecules where the positions of the molecules are randomly distributed, but the orientations generally lie along a common axis in space, as illustrated in Figure 5.3A. Because the molecules tend to be oriented in a specific (arbitrary) direction—say, the x direction—the rotational symmetry is spontaneously broken. This directional asymmetry can be observed, for instance, by measuring how the material responds to an electric field applied in different directions.[7] In an isotropic fluid, the response is independent of the direction of the field, but in a nematic liquid crystal, the response is different when the field is applied along the x direction than when a field of equal strength is applied along the y or z direction.[8]

---

7. The degree to which polarized light passes through a nematic liquid crystal depends on the direction of its polarization; this property is exploited in liquid crystal display technologies.

8. To be precise, the dielectric tensor, ϵ_{ab}, of a liquid crystal is anisotropic. To linear order, the relation between an applied electric field and the total displacement field is given by the dielectric tensor

$$D_a = \epsilon_{ab} E_b$$

(where a, $b = x$, y, z, and summation convention is assumed). In an isotropic system such as a liquid, this relation reduces to the familiar relation $\vec{D} = \epsilon \vec{E}$, which is to say that $\epsilon_{ab} = \epsilon\, \delta_{a,b}$. However in a nematic

In a smectic liquid crystal, the translational symmetry is broken—but only in one direction. Roughly speaking, a smectic looks like a crystal along one direction, but looks like a liquid along the others. More precisely, the density in a smectic is periodic in the x direction and but uniform in the y or z direction,

$$\langle \hat{\rho}(\vec{r}) \rangle = \langle \hat{\rho}(x) \rangle = \langle \hat{\rho}(x + n\lambda) \rangle, \tag{5.7}$$

where n is any integer and λ is the period of the smectic state. (Of course, such a state also breaks rotational symmetry: the choice of x in the above is arbitrary.) At an intuitive level, one can picture a smectic liquid crystal as consisting of stacked layers, as in Figure 5.3B. The layers are stacked in a regular, periodic fashion, but molecules within a layer have no positional order and can flow, as in a two-dimensional fluid.[9] Such a situation might arise, for instance, if a crystal of stacked planes undergoes a partial melting process in which the order within each plane is lost but the stacking of the planes remains intact.

Beyond nematic and smectic phases, there are a host of other possible liquid crystalline phases with different patterns of spatial symmetry breaking. In electronic systems where quantum effects are important, there are a still richer variety of possible "electronic liquid crystalline" phases, which combine the spatial symmetry-breaking patterns of classical liquid crystals with various sorts of quantum behaviors, such as metallicity or superconductivity.

5.2.5   Ferromagnetism

Even though we have formally studied the Ising ferromagnet as an abstract problem in statistical mechanics, we still have not discussed its relation to *real* ferromagnets. Compared to the other examples we have seen, the ferromagnet is trickier to understand because its broken symmetry is not immediately apparent and because its existence is deeply quantum mechanical.

To begin with, consider the thermodynamic changes that arise as a crystalline solid is cooled through its Curie temperature T_c. Above this temperature, the crystal is nonmagnetic. When the crystal is cooled below this temperature, it becomes a ferromagnet with nonzero magnetic moment $\vec{M}$. Assuming that the material is a

---

phase, one direction—which we will take to be x—is different from the others, which is to say that

$$\epsilon_{ab} = [\epsilon - \mathcal{N}]\delta_{a,b} + 3\mathcal{N}\delta_{a,x}.$$

Here $\mathcal{N}$ (which is the traceless part of ϵ and is zero by symmetry in an isotropic phase) can be viewed as the nematic order parameter.

9. There is a technical point that is not quite right in this discussion of the smectic phase—in fact, true smectic long-range order is possible only in spatial dimension $d > 3$. In $d = 3$ the smectic phase is, instead, characterized by quasi-long-range order, akin to the order we will find for the XY model in $d = 2$.

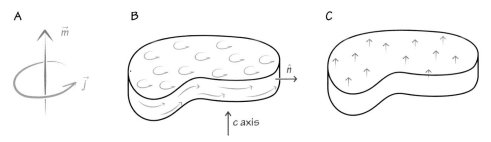

FIGURE 5.4. (A) In classical electromagnetism, magnetic dipoles result from current loops. (B) The magnetization of a ferromagnet implies a persistent bound current flowing along the edge of the material. This bound current can be thought of as the sum of many small current loops in the bulk of the ferromagnet. (C) The microscopic origin of ferromagnetism is the quantum mechanical spins of electrons in the material, which play the role of the current loops shown in panel (B).

crystal for a range of T that extends above T_c, the "high-temperature" phase already exhibits preferred directions in space (namely, the axes of the crystal). Thus, when the "low temperature" phase develops a magnetic moment, it will lie along one of the crystal's axes. Furthermore, if the three axes of the crystal are all nonequivalent (say it is an orthorhombic crystal with $a \neq b \neq c$), $\vec{M}$ will lie along one of these axes—say, the c-axis, to be concrete. In other words, the orientation of the magnetic moment is already largely determined by the symmetries of the crystal.

In the orthorhombic situation, the only remaining ambiguity about $\vec{M}$ in the ferromagnetic phase is whether it points "up" or "down" along the c-axis. Now the connection to the Ising model is apparent: the m of the Ising model corresponds to the component of the magnetization, $\vec{m} = \vec{M}/V$, along the c-axis, and the symmetry of the Ising Hamiltonian $\sigma_j \rightarrow -\sigma_j$ corresponds to an operation which takes $\vec{m}$ to $-\vec{m}$.[10] But what exactly is the symmetry which reverses the direction of magnetization? It is the rather abstract symmetry under reversal of time, $t \rightarrow -t$, as if a video is viewed in reverse.

To see why a time-reversal transformation also reverses the direction of magnetization, recall that, in classical electromagnetism, a magnetic moment $\vec{m}$ is the result of current flowing around a loop (Figure 5.4A). The direction of $\vec{m}$ is determined from the direction of the current via the right-hand rule. When the arrow of time is reversed, the direction of the current is reversed, and the direction of the magnetic moment is reversed.[11]

---

10. To be more precise, the average spin is unitless whereas the magnetization has units of magnetic dipole moment per volume—so the full correspondence requires multiplying by the magnetic dipole moment of a spin and dividing by the unit cell volume.

11. In quantum mechanics, it is still true that the magnetic dipole moment picks up a minus sign under time-reversal.

Thus, time-reversal symmetry is spontaneously broken when a crystal is cooled through T_c and develops a magnetization. The microscopic equations of motion (whether classical or quantum mechanical) are time-reversal invariant, and the state of the crystal above T_c respects this symmetry. However, as the time-reversal operation transforms the state with $m > 0$ into a macroscopically distinct state with $m < 0$, the ferromagnetic phase is not time-reversal invariant.

The origin of ferromagnetism

From the viewpoint of classical electromagnetism, a block of ferromagnet possesses a persistent electrical current that flows around its edge. Recall that a static magnetization $\vec{m}(\vec{r})$ can be related to a "bound" current

$$\vec{j}(\vec{r}) = \vec{\nabla} \times \vec{m}(\vec{r}). \qquad (5.8)$$

In a block of ferromagnet where $\vec{m}(\vec{r})$ is a constant everywhere in the bulk of the material, and zero everywhere outside the material, Eq. 5.8 implies that there is zero current in the bulk, but there is a net *equilibrium* current $\vec{I} = \hat{n} \times \vec{m}$ along the edge of ferromagnet, where $\hat{n}$ is a unit normal vector pointing outward from the edge.

To visualize the meaning of this "bound current," imagine that at each site of the crystal there is a small loop of current rotating counterclockwise in a plane perpendicular to the *c*-axis—as shown in Figure 5.4B. Each current loop produces a magnetic field that points up within the loop, leading to a net magnetization. In the bulk of the sample, there is no net macroscopic current, because the currents from any one site are cancelled out by the opposite currents at a neighboring site. However, at the edge of the sample, the currents are uncompensated, meaning that a net current flows counterclockwise around the edge of the sample.

We thus conclude that a ferromagnet contains equilibrium persistent currents.[12] In a classical world, this would be impossible! Charges travelling in loops are accelerating, which classically implies they should radiate energy, and the current should thus decay.[13] This tells us that the origin of ferromagnetism is profoundly quantum mechanical. In general, the thermodynamics of a classical system is always independent of the magnetic field (Problem 5.1)—so without quantum effects, it is impossible for phases of matter to exhibit magnetic properties.

In reality, ferromagnetism arises from the spins of electrons, as schematically shown in Figure 5.4C. Spin is the quantum mechanical version of a magnetic dipole moment and behaves unlike any classical analogue. The orientation of an electron's spin can only be fully specified along one axis at a time and is quantized to be either

---

12. Remember that the ferromagnet is an *equilibrium* phase of matter, which means that its currents (or any other macroscopic observable) do not change in time.

13. This is reminiscent of the issues with a classical picture of an atom: when a classical electron orbits a nucleus, it will lose energy due to radiation and eventually spiral into the nucleus.

"up" or "down." The quantized nature of spin means that its magnetic moment does not decay.

Beyond Ising ferromagnets

So far we have discussed the situation where a ferromagnetic transition occurs in an orthorhombic crystal: the ordered moment must lie either "up" or "down" along one axis of the crystal. However, if ferromagnetic order develops in a liquid phase—as happens in a ferrofluid—the ordered moment can point in any direction in space. In this situation, the ferromagnetic transition breaks the rotational symmetry of space along with time-reversal symmetry, and the order parameter is unavoidably the three-component pseudo-vector[14] quantity, $\vec{m}$. This sort of ferromagnet is often referred to as a Heisenberg ferromagnet.[15]

Between the Ising case (where $\vec{m}$ is restricted to lie on a line) and the Heisenberg case (where $\vec{m}$ can point in any direction on a sphere), there can be a host of other symmetries possible. If the crystal in its nonmagnetic state is tetragonal, then depending on the nature of the microscopic interactions, the ordered moment could preferentially lie either along the tetragonal c-axis or along the equivalent a- and b-axes. In the former case, there are two possible patterns of broken symmetry ($-c$ or $+c$) and the system is an Ising ferromagnet. However, in the latter case, there are four possible broken-symmetry states: $+a$, $-a$, $+b$, and $-b$. This situation is referred to as a 4-state clock ferromagnet. The ordered state now not only breaks time-reversal symmetry, but it also breaks the equivalence between the a- and b-axes because the magnetization must point along one of the two axes. In fact, since the a- and b-axes are no longer equivalent in the ordered phase, the crystal must become orthorhombic below T_c. Symmetry is precisely defined, and if it is broken in any way, then it is not a symmetry.

5.2.6 Superfluids

When Helium-4 is cooled to 2.17 degrees above absolute zero, it undergoes a transition to a superfluid phase, which is an exotic quantum phase. A superfluid can be described as a mixture of two components, an ordinary fluid component and a superfluid component which has zero viscosity. The superfluid component can flow indefinitely without dissipating any energy. Another strange property of superfluid helium is that it creeps up the sides of its vessel, forming a film which is 30

---

14. The prefix "pseudo" means that, unlike an ordinary vector, the direction of a psuedo-vector does not change under inversion $(x, y, z) \rightarrow (-x, -y, -z)$.

15. Even in the solid state, it is often true that the spin-orbit coupling is weak (since it is a relativistic effect), in which case the underlying microscopic physics has an approximate spin-rotational symmetry, suggesting that the system might be modeled as nearly a Heisenberg ferromagnet. We will consider more carefully the role of approximate symmetries in Section 7.5.2.

nanometers thick. The same lack of resistance to flow also makes a superfluid prone to leaking out of holes.

The superfluid transition can be identified by a singularity in the specific heat, with a small negative critical exponent of $\alpha \approx -0.15$.[16] For Helium-4, T_c is 2.17 K, but it is much lower for Helium-3, $T_c \approx 0.001$ K at atmospheric pressure.

The difference in T_c between the two isotopes gives a clue about the origin of superfluidity. Helium-4 is a boson, but Helium-3 is a fermion.[17] From earlier courses in statistical mechanics, you may remember that non-interacting bosons can condense into their quantum mechanical ground state (known as Bose condensation). Roughly, this describes the superfluid transition of Helium-4. The ordering in Helium-3 is more complicated. In a crude approximate sense, Helium-3 atoms become weakly bound into pairs, which are effectively bosons; these pairs can then condense. It is because the pairing energy is so small that the T_c of Helium-3 is so much smaller than that of Helium-4.

The pertinent order parameter of a superfluid is a complex number, Φ, referred to as the "macroscopic wavefunction." It can be thought of as the wavefunction of all the bosons in the ground state, but the exact definition is quite involved (see Appendix A.12). The magnitude of Φ is related to the density of the superfluid component, whereas the complex phase of Φ is not directly observable. Recall from elementary quantum mechanics that observable properties are left invariant if Φ is shifted by an arbitrary complex phase, $\Phi \rightarrow e^{i\theta}\Phi$, where $e^{i\theta}$ is a complex number of unit magnitude but arbitrary angle between 0 and 2π. As it turns out, this is the broken symmetry which distinguishes the normal fluid phase from the superfluid phase. We can represent $\Phi = |\Phi|e^{i\theta}$ as a two-component real vector with components $\Phi_1 = \mathrm{Re}[\Phi]$ and $\Phi_2 = \mathrm{Im}[\Phi]$. Since θ can take any value between 0 and 2π, (Φ_1, Φ_2) can point in any direction in an abstract 2D plane (one which has no relation to real space). The symmetry with respect to multiplying Φ by a complex phase is an "XY-symmetry," since it is equivalent to a global rotational symmetry in 2D. We refer to it as an "internal symmetry" because it is unrelated to any symmetry of space. Still, in the ordered state, the order parameter "points" in the same direction throughout the system—so we can describe it as a form of XY ferromagnet, where the "spin" on each site represents some complex quantum mechanical quantity. Strikingly, we can come to understand various properties of this most quantum liquid solely by representing it as a classical XY ferromagnet!

5.2.7 Superconductors

The superconducting phase is another remarkable quantum phase of matter. A superconductor has zero electrical resistance. Therefore, a superconducting ring

---

16. The specific heat itself does not diverge, but has a divergent first derivative at T_c, $dC/dT \sim |T - T_c|^{-(1-|\alpha|)}$.

17. Helium-4 contains an even number of subatomic particles whereas Helium-3 contains an odd number.

can host an electrical current indefinitely, without needing any battery or external drive. Another defining property is the ability to expel magnetic fields from the interior of a superconductor, known as the Meissner effect. More practically, superconductors can support the large electrical currents required to produce powerful magnetic fields in MRI machines. You may have also seen demonstrations that a superconductor can be stably levitated above a permanent magnet.

Most metals become superconductors when cooled down to within a few degrees of absolute zero; mercury, for instance, has a T_c of 4.2K at atmospheric pressure. The superconductor used in most MRI machines is an alloy of niobium and titanium with a transition temperature of $T_c = 9.4$K; a bath of liquid helium is required to cool the alloy below this temperature. Other materials can become superconductors, too. Notably, certain brittle ceramic materials have particularly high transition temperatures, such as YBCO, with $T_c = 92$K. This is certainly not high-temperature in the everyday sense (it is still $-181°$C!) but at least it is above the temperature of liquid nitrogen, 77K. These materials are referred to as "high-temperature superconductors."

There are similarities between a superfluid and a superconductor. You can think of a supercurrent as the flow of a charged superfluid of electrons. Many properties about superconductors can be described with an order parameter field, $\Phi(\vec{r})$, which can be thought of as the wavefunction of the condensate. One subtlety is that electrons are fermions, so they cannot directly condense into their ground state. However, as is the case with Helium-3, the electrons can pair up, into pairs known as "Cooper pairs," which in turn condense to form the superconducting state.[18]

5.3 Order Parameters and Symmetry

There is a general mathematical structure that underlies all the examples of broken symmetries we have discussed. Here we briefly sketch this more general structure.

We consider the set of all symmetry operations, $g \in G$, that leave the system invariant. In the mathematical sense, G is a group. In the absence of any symmetry-breaking fields, the Boltzmann weight governing the thermal averages is invariant under any symmetry transformation of the Hamiltonian. This implies that the thermal average of any quantity, $\hat{O}$, that is not a "scalar" (i.e., invariant) under the operations of the symmetry group must vanish. An example of this was given in Section 5.1.1, where we observed that $\langle \sigma_j \rangle$ would be expected to vanish in the symmetric phase. More generally, this statement can be proven as follows: Let $U(g)$

---

18. One key difference between superfluids and superconductors is that Cooper pairs are charged. Gauge invariance implies that all physical quantites must be invariant under the gauge transformation, $\vec{A} \to \vec{A} + \vec{\nabla}\chi$, $A_0 \to A_0 - \partial_t \chi$, and $\Phi(\vec{r}) \to \Phi(\vec{r})e^{i(2e/\hbar)\chi}$. Thus, the phase of the superconducting order parameter is related in an intimate fashion to gauge invariance. This makes the discussion of superconductivity on the basis of notions of broken symmetry somewhat less direct. We will return to this problem in Section 10.5.

be the unitary operator that implements the symmetry operation g. For instance, if the symmetry in question is rotational symmetry, then $U(g)$ is the operator that rotates the coordinates of all the particles.[19] As this is a symmetry of the Hamiltonian, the Boltzmann weight must be invariant, $U^\dagger(g)e^{-\beta H}U(g) = e^{-\beta H}$. It is then straightforward to see that

$$\langle \hat{O} \rangle = Z^{-1}\mathrm{Tr}\left[e^{-\beta H}\hat{O}\right] = Z^{-1}\mathrm{Tr}\left[U^\dagger(g)U(g)e^{-\beta H}\hat{O}\right] \tag{5.9}$$

$$= Z^{-1}\mathrm{Tr}\left[e^{-\beta H}U(g)\hat{O}U^\dagger(g)\right] = \langle g[\hat{O}]\rangle,$$

where $g[\hat{O}]$ is the symmetry transformed quantity. If we wish, we can sum this relation over all g, meaning that $\langle \hat{O} \rangle$ is equal to its average over all symmetry transformations. For a scalar, this average is trivial; for any quantity that is not a scalar under the symmetries, (by definition) its average over all symmetry transformations is 0.

With this in mind, we define an order parameter, ϕ, as the expectation value of any quantity that has no scalar component. This way it would be expected to vanish identically in the symmetric phase, but develop a non-zero expectation value in the broken-symmetry phase. Typically we further choose the order parameter so that its transformation properties under the symmetry group are simple—technically, we would like it to transform according to an "irreducible representation" of the symmetry group. The order parameter must be defined either by applying a symmetry-breaking field and taking the limit as it vanishes, as in Eq. 5.4, or from looking at the long-range behavior of the two-point correlation of the same quantity, as in Eq. 5.5. Beyond these symmetry considerations, there is considerable flexibility in the precise definition of the order parameter.

One benefit of this level of abstraction is that it allows us to appreciate the similarities in the behavior of otherwise dissimilar phenomena. In any phase transition which breaks an Ising symmetry—the order-disorder transition in β-brass, various ferromagnetic transitions, and more—the broken-symmetry phase is characterized by a single Ising-like order parameter, ϕ. In the case of β-brass, ϕ can be construed to be the difference between the concentration of Zn atoms on the A and B sublattices in the neighborhood of point $\vec{r}$, while in the ferromagnet, ϕ is the local magnetization, i.e., the difference in the number of up and down spins. All such systems share similar properties; for example, at a given temperature in the broken-symmetry phase, the order parameter can take on two values of opposite sign and equal magnitude.

5.3.1   Thermodynamic Role of the Order Parameter

In a broken-symmetry phase, the order parameter identifies which of the possible symmetry-related ordered states is realized. For an Ising ferromagnet, the two

---

19. There is a subtlety, which we will gloss over here, that not all symmetries can be represented by a unitary operator—for example, in quantum mechanics, time-reversal symmetry is represented by an anti-unitary operator.

possible states are distinguished by the sign of ϕ. For a Heisenberg ferromagnet, at any given pressure and temperature below T_c, there is a whole family of possible equilibrium states with the same magnitude of magnetization, but which are distinguished by the orientation of the magnetization. To fully describe the thermodynamic state, the three-component vector ϕ must be specified.

In the symmetric phase which exhibits the same symmetries as the underlying Hamiltonian, ϕ vanishes identically. The phase boundary between the symmetric and broken-symmetry phases is then precisely defined as the boundary between regions of a phase diagram in which $\phi = 0$ and $\phi \neq 0$. As shown in Figure 5.5, the transition can be either continuous or discontinuous; in the former, ϕ rises continuously from zero upon entering into the broken symmetry phase, while in the latter it jumps discontinuously to a nonzero value. Just as the magnetization is (minus) the derivative of the free energy with respect to magnetic field, the order parameter is generically the derivative with respect to an appropriate symmetry-breaking field. Thus, this definition of a "discontinuous" transition agrees with the one in Chapter 1—it is a transition at which a first derivative of the free energy changes discontinuously.

The order parameters and symmetry groups of various phase transitions is listed in Table 5.1.

Did you know? It seems plausible that any global symmetry of nature can be spontaneously broken in some suitably ordered equilibrium state of matter. A topic of current interest[a] is whether this includes time-translation symmetry. In other words, is it possible for a system in equilibrium (or quasi-equilibrium) to exhibit properties that oscillate as a function of time?

———

a. The current state of knowledge concerning this notion—referred to colloquially as a "time crystal"—has recently been reviewed in Khemani, Moessner, and Sondhi, "A brief history of time crystals," arXiv:1910.10745 (2019).

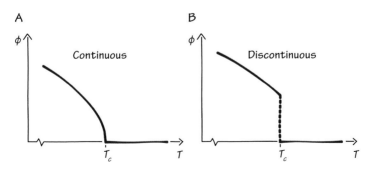

FIGURE 5.5. The order parameter, ϕ, is a generalization of the magnetization in the Ising model. Near a continuous transition, there is a regime in which the order parameter is small, and hence it is reasonable to expand the free energy in powers of ϕ—the Landau expansion. On the other hand, near a discontinuous transition, under conditions in which there is a large jump in ϕ, such an expansion is clearly uncontrolled.

Table 5.1. A list of various physical systems exhibiting phase transitions and their corresponding order parameters and symmetry groups, inspired by Ken Wilson's article in *Scientific American* titled "Problems in Physics with Many Scales of Length," **214.2**, 158–179 (1979).

Symmetry	Dimension	System	Order parameter
Ising	$d = 3$	Ferromagnet in an orthorhombic crystal	magnetization
Ising	$d = 3$	Fluid near a liquid-gas transition[a]	density difference between phases
Ising	$d = 3$	Binary alloy near order-disorder transition	concentration difference between sublattices
3-state clock	$d = 2$	Xenon adsorbed on graphite	occupation of superlattice
4-state clock	$d = 4$	Ferromagnet in a tetragonal crystal	magnetization
XY	$d = 3$	Helium-4 near superfluid transition	amplitude of superfluid condensate
XY	$d = 2$	Helium-4 films on Mylar	amplitude of superfluid condensate
Heisenberg	$d = 3$	Ferrofluid transition	magnetization

[a]In the liquid-gas transition, there is no broken symmetry, so in that sense it is even more amazing that it behaves like other Ising systems.

5.4 Some Paradigmatic Models

Since models are abstractions, their relation to reality is often subtle. Conversely, one abstract model can capture the essential physics of a variety of physical systems. As we have seen, the nature of the Ising model is the same regardless of the physical meaning of the variable σ_j so long as there are exactly two symmetry related ground-states. In this section we discuss some simple paradigmatic models which represent systems with other symmetries.

5.4.1 Models with Discrete Symmetries

An Ising spin has two possible values. A simple generalization is a spin which has more than two possible values. One such model is known as the q-state clock model. In this model, each spin s_j is a two-dimensional vector which can point in one of q discrete, evenly spaced directions on a unit circle (think of it as the minute hand on a clock). For instance, in the 4-preferred-state clock model, the spin s_j can point to 3:00, 6:00, 9:00, or 12:00. More generally, a q-state clock spin can point to one of q angles, $n2\pi/q$, where $n = 1, 2, \ldots, q$. This is illustrated in the right panels of Figure 5.6.

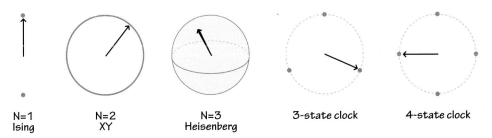

| N=1
Ising | N=2
XY | N=3
Heisenberg | 3-state clock | 4-state clock |

FIGURE 5.6. Schematics of a few different types of symmetries.

In the absence of a symmetry-breaking field, the q spin states are all equivalent and the energy depends only on the relative orientation of spins, not on the absolute orientation. More precisely, the Hamiltonian is invariant under a transformation where all the spins are rotated through any multiple of $2\pi/q$ radians. Clearly, the Ising model can be thought of as a 2-state clock model.

Much like the Ising model, the clock model has multiple ground states. If the interactions are ferromagnetic, there are q ground states in which all the spins point in the same direction. In the low-temperature phase, there is a preferred direction, and it is most likely to find a spin in a particular one of the q spin states. In the high-temperature phase, there is no preferred direction, and any spin state is as likely as any other. Because the symmetry of the clock model is discrete, it shares many properties with the Ising model; for instance, there is a nonzero T_c in $d \geq 2$, but not in $d = 1$. In other respects, however, the clock model behaves quite differently; for instance, the 3-state clock model exhibits a *discontinuous* transition into the broken-symmetry phase for $d \geq 3$.

In the limit of $q \to \infty$, the spin states become more closely spaced together and approach a continuum along the rim of the clock face, approximating what is known as the XY model.

5.4.2 Models with Continuous Symmetries

In an XY model, the spins lie at any angle $0 \leq \theta_j < 2\pi$ on a unit circle. We can represent each spin as a two-dimensional unit vector, $\mathbf{S}_j = \cos(\theta_j)\hat{\mathbf{e}}_1 + \sin(\theta_j)\hat{\mathbf{e}}_2$ (Figure 5.6). Note importantly that $\mathbf{S}_j$ is a vector in an abstract "spin space" which may have nothing to do with real space—for instance, we could very well define the model on a lattice of $d = 3$ spatial dimensions, where on each site on the lattice there is an XY spin, $\mathbf{S}_j$.

The symmetry of the XY model is that all directions in spin space are equivalent. There is a global rotational symmetry in spin space; i.e., the energy of a configuration does not change if the spins on all the lattice sites are uniformly rotated by an angle $\Delta\theta$:

$$\mathbf{S}_j \to R[\mathbf{S}_j] = \cos(\theta_j + \Delta\theta)\hat{\mathbf{e}}_1 + \sin(\theta_j + \Delta\theta)\hat{\mathbf{e}}_2, \tag{5.10}$$

where $R[\mathbf{v}]$ signifies rotation by an angle $\Delta\theta$ in spin space, with $0 \leq \Delta\theta < 2\pi$.

A standard model of an XY ferromagnet is

$$H = -J \sum_{\langle ij \rangle} \mathbf{S}_i \cdot \mathbf{S}_j \tag{5.11}$$

$$= -J \sum_{\langle ij \rangle} \cos(\theta_i - \theta_j). \tag{5.12}$$

As with the other models we have seen, the Hamiltonian has a local preference for the relative direction of neighbors but no global preference for any particular spin direction. For $J > 0$, there is an energetic tendency for neighboring spins to point in similar directions. The ground states are manifestly the configurations in which the spins are all pointing parallel to one another, $\phi_j = \phi \ \forall \ j$.

Unlike the Ising model or the clock model, the XY model has a continuous symmetry, which implies that there is a *continuum* of possible broken-symmetry states, one for every $0 < \theta < 2\pi$. This leads to some interesting new properties. For instance, any two such states can be continuously deformed into each other by smoothly rotating around the circle—something which we will delve into further in Chapter 7.

It is important to keep in mind that the essential feature of this problem is its symmetry, not the exact form of its Hamiltonian (see Worksheet W5.2). If we included interactions between further neighbors or with higher powers of $\vec{s}$, H would still be considered an XY magnet so long as the global rotational symmetry was preserved. For example, even if the interaction between neighboring spins were generalized such that

$$H = -\sum_{\langle ij \rangle} \left[J \mathbf{S}_i \cdot \mathbf{S}_j + J'(\mathbf{S}_i \cdot \mathbf{S}_j)^2 \right], \tag{5.13}$$

the neighbors would still tend to align and the nature of the ground-states would be unaffected so long as $J > 0$ and $J' > -J$.

The most important example of a system with XY symmetry is the superfluid, discussed earlier in this chapter. In this case, there are no "spins," and in fact there is no lattice either—there is just a complex order parameter, $\Phi(\vec{r})$, i.e., the macroscopic wavefunction. The only resemblance between the XY ferromagnet and a superfluid is in symmetry; the free energy of the superfluid is invariant under the transformation $\Phi(\vec{r}) \to e^{i\theta}\Phi(\vec{r})$ for any $0 \leq \theta \leq 2\pi$. However, this connection is sufficient such that many properties of superfluids—including all universal critical properties of the superfluid transition—can be understood by studying the XY ferromagnet.

Beyond the XY model, we can also imagine a model where the spins lie on the unit sphere, i.e., $\mathbf{S}_j$ is a 3D unit vector (Figure 5.6). This is called the Heisenberg model. Again, the interactions involve only the relative angle between nearby spins and the Hamiltonian is invariant under any global rotation in spin space (i.e., transformations where the spins on all the lattice sites are rotated through the same angle). As mentioned before, the Heisenberg model has the appropriate symmetries to describe the magnetic order in a ferrofluid or in a crystal in which spin-orbit coupling is negligible.

It is not hard to generalize this further to situations where the spins S_j are N-dimensional unit vectors and the symmetry group is the set of all rotations in this N-dimensional spin space. This is called the $O(N)$ model.[20] A Hamiltonian which is invariant under this set of rotations (such as Eq. 5.11 or 5.13) is said to have $O(N)$ symmetry. For $N = 1$, this is the familiar Ising model, for $N = 2$, it is the XY model, and for $N = 3$, it is the Heisenberg model. For $N > 3$, it is unimaginatvely referred to as the $O(N)$ model.

In the realms of condensed matter physics, examples of $O(N)$ symmetries with $N > 3$ are rare and generally require a degree of fine tuning. In high-energy physics, large symmetry groups are thought to play a significant role in the interactions among elementary particles. On the theoretical front, it is useful to study models of arbitrary N, similarly to how we have considered lattices of arbitrary d. In Chapter 8, we will see circumstances in which it is possible to solve such models in the limit $N \to \infty$, where corrections are small in proportion to $1/N$. This may not sound all that interesting. Nonetheless, $1/3$ is pretty small—so perhaps it is not surprising that this approach can teach us quite a bit about the Heisenberg ($N = 3$) problem. However, even for the Ising ($N = 1$) problem, the large N approximation yields some valid qualitative understanding.

5.5 Problems

5.1. *Magnetism in classical thermodynamics.* Consider a classical system of N charged particles in the presence of electric and magnetic fields, with Hamiltonian

$$H = \sum_i (p_i + e_i A(r_i))^2 / 2m_i + V(\{r_i\}),$$

where $A(r)$ is the vector potential and $B = \nabla \times A$. Show that the partition function, $Z = \int dr_1 \ldots dr_N dp_1 \ldots dp_N \, e^{-\beta H}$, is independent of A. Conclude that the thermodynamics of a classical system is independent of magnetic field. *(Hint: Make a coordinate transformation.)*

5.6 Worksheets for Chapter 5

W5.1 Liquid Crystals and Broken Symmetries

Liquid crystals are fascinating phases with properties (and symmetries!) between those of a liquid and a solid. Many liquid crystals are made of molecules with a rigid,

---

20. Rotations in N dimensions are represented by $N \times N$ orthogonal matrices, hence the name of the symmetry group $O(N)$.

rod-like structure. Because of the long shape, these molecules can have a tendency to point along the same axis in space. Here is an example. It is called N-(heptoxy-benzylidene)-p-n-pentylaniline.

All liquid crystal phases have orientational order. Some liquid crystal phases also exhibit positional order (regularities in the positions of molecules). This worksheet illustrates how orientational order and positional order are related to rotational and translational symmetries.

phase	symmetry			
	arbitrary rotations around the x-axis	arbitrary rotations around y- or z-axis	arbitrary translations along the x-axis	arbitrary translations along y- or z-axis
liquid				
solid				
nematic				
smectic A				
smectic C				

1. In the liquid phase, the position and orientation of each molecule is *random*. Fill out the first row of the table with a ✓ if the symmetry is preserved, otherwise put a × if the symmetry is broken in any way.
2. In the (crystalline) solid phase, molecules are arranged with a defined orientation in a regular 3D lattice. Fill out the second row of the table.
3. In the nematic phase, the molecules are oriented so that the long axis tends to point in a shared direction (known as the director), but they are otherwise unordered. Their motion is unhindered, as in a liquid. Fill out the third row. (Define $\hat{x}$ as the orientation of the director.)

A smectic liquid crystal phase maintains the orientational order of a nematic phase, but also exhibits positional ordering. Specifically, the molecules in a smectic phase tend to arrange themselves into regularly spaced layers of planes. It is like a liquid because separate smectic planes can freely slide over one another.

4. In the smectic A phase, the molecules within a smectic plane form a two-dimensional fluid, with random positions. The director is perpendicular to the smectic plane. Fill out the fourth row.

5. The smectic C phase is identical to the smectic A phase, except that the director is tilted away from the smectic plane normal. Fill out the fifth row. (Define the smectic plane as the $y - z$ plane here; the director is *not* in the $\hat{x}$ direction.)

6. Depending on the temperature, N-(heptoxy-benzylidene)-p-n-pentylaniline can exhibit a whole bunch of different phases, including two additional phases called smectic H and smectic B!

$$\boxed{?} \xleftrightarrow{23°C} \boxed{H} \xleftrightarrow{58°C} \boxed{B} \xleftrightarrow{64°C} \boxed{?} \xleftrightarrow{68°C} \boxed{?} \xleftrightarrow{80°C} \boxed{?} \xleftrightarrow{83°C} \boxed{?}$$

Five of the boxes are marked "?"—these correspond to the five phases in the table you just filled out. Based on the symmetries, can you figure out which phase belongs in which box?

W5.2   Interpolating between the XY and Clock Models

As we have stressed in the text, the symmetry of a model is of supreme importance, often more so than the exact form of the Hamiltonian. In this worksheet we will think about the symmetries of the Hamiltonian

$$H = -J \sum_{\langle ij \rangle} \cos(\theta_i - \theta_j) - h \sum_j \cos(q\theta_j), \tag{5.14}$$

where q is an integer, and the spins on each site are an angle $0 \le \theta_j < 2\pi$.

1. For $h = 0$, what is this model equivalent to?

2. Consider a symmetry transformation of the form $\theta_j \to \theta_j + \Delta\theta \ \forall j$. If $h = 0$, what is the set of $\Delta\theta$ which leave the Hamiltonian invariant?

Now let's turn on the "clock anisotropy" term with $h > 0$. (Isotropy means all spin directions are equivalent; anisotropy means they are not.)

3. In the case $q = 1$, what is the ground state of the model? What are the remaining symmetries (if any) of the Hamiltonian?

4. What about $q = 2$? Argue that this case is similar in spirit to an Ising model.

Note that even though the spins are not Ising spins, the $q = 2$ case is still fundamentally an Ising model because of its symmetry.

5. More generally, for $q > 2$, how many ground-states are there, and what is the symmetry of the model?

6. Explain the following phrase to someone at your table: "the clock anisotropy term reduces the continuous rotational symmetry to a q-fold discrete rotational symmetry."

7. For each of the following situations, give a qualitative description of the phase of matter and sketch out the distribution over the angle θ_j of all spins on the lattice. Use a mean-field approximation.

   (a) Very high temperatures, $T \gg h$ and $T \gg J$
   (b) $h \gg T$ and $J = 0$
   (c) $h \gg T$ and $0 < J \ll T$
   (d) $h \gg T$ and $J \gg T$
   (e) $J \gg T$ and $h = 0$
   (f) $J \gg T$ and $0 < h \ll T$

It is possible to extrapolate between the $h \gg J$ and the $h \ll J$ cases. We will revisit this idea of approximate symmetry in more detail in Section 7.5.2.

W5.3   Determining the Symmetry and Order Parameter

Each of the scenarios below describes a phase transition where a discrete symmetry is broken. For each one, please (a) state the number of distinct broken-symmetry phases, (b) give an appropriate paradigmatic model which could be used to describe the system, and (c) describe a suitable order parameter, ϕ.

> *Hint 1:* Remember that your order parameter should be zero in the symmetric state, but nonzero in the broken-symmetry state—and should be different for each of the possible states of broken symmetry!
>
> *Hint 2:* For the 3-state clock model, the three states of broken symmetry have unit vectors of $\mathbf{S}_1 = (1, 0)$, $\mathbf{S}_2 = (-1/2, \sqrt{3}/2)$, and $\mathbf{S}_3 = (-1/2, -\sqrt{3}/2)$. You can leave your expression for the order parameter in terms of $\mathbf{S}_1, \mathbf{S}_2$, and $\mathbf{S}_3$.

1. A ferromagnetic transition in a tetragonal crystal, where the spontaneous magnetization arises in one of the $-a$, $+a$, $-b$, and $+b$ directions.

2. A structural transition from a cubic to a tetragonal crystal. (Remember that in a tetragonal unit cell, all three axes are perpendicular and two have the same length.)

When a gas is placed in contact with a surface, the gas molecules can adsorb to the surface if the interactions are favorable. This is how many important industrial chemical reactions are catalyzed. (Is adsorption energetically favorable or unfavorable? How about entropically?)

3. Adsorption of atoms onto the sites of a 2D square lattice. The atoms are large enough that each adsorbed atom excludes all others from the adjacent sites, but not the diagonals (see figure). *Hint:* How many ground states are there if the lattice is full?

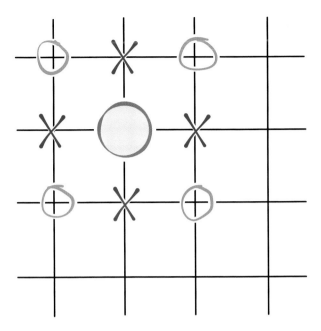

4. Same as part 3 except on a triangular lattice. Each adsorbed atom excludes exactly six directly adjacent sites.

In the following chapter we will see why (at mean-field level) the $q = 3$ clock model has a discontinuous transition while the $q = 4$ clock model has a continuous transition. Knowing this, which of these transitions do you expect to be continuous and which discontinuous?

6 THE LANDAU FREE ENERGY

Landau taught us that whoever understands spontaneously broken symmetries can make multiple wise pronouncements about many complex systems without the need for any further knowledge.

Until now, we have taken a "bottom-up" approach to understanding the emergent properties of complex systems: we started from a detailed knowledge of the interactions between constituent parts (i.e., a microscopic Hamiltonian), and worked up from there to calculate or approximate various thermodynamic quantities. We have begun to appreciate, however, that the resulting phases and phase transitions exhibit similar properties which are rather independent of microscopic details. This suggests that a "top-down" approach may be equally profitable.

In this chapter we adopt this alternative strategy, the Ginzburg-Landau (G-L) approach. It is a theory which starts with a few phenomenological pieces of information, such as the symmetry of the order parameter. From there, the G-L theory uses general arguments to determine as much as possible about thermodynamic properties—all without having to know anything about microscopics! For now we present a simplified version of the G-L theory (usually referred to simply as Landau theory) where the order parameter is spatially uniform. In the following chapter we will further develop G-L theory to account for spatially varying order parameter "textures."

6.1 Once More, the Ising Model

To start with a problem that is by now highly familiar, let us approach the Ising ferromagnet again—but now from the top down. We will not assume any knowledge about the microscopic interactions; we start from general facts of thermodynamics. We will work at an appropriate mean-field level of approximation where the equilibrium magnetization, ϕ, takes on the value which minimizes a variational "Landau" free energy, F_L. Since we are restricting our attention to spatially uniform states (i.e., states where ϕ is constant in all parts of a material), the equilibrium state is specified by a single value of ϕ. At this level of approximation, the Landau free energy is a function of the order parameter ϕ and the control parameters T and h (temperature and external magnetic field, respectively):

$$F_L(T, h; \phi) = \int d\vec{r} \left[\mathcal{F}_L(T; \phi) - h\phi \right]. \tag{6.1}$$

Here the quantity in brackets is the free energy per unit volume, or the free energy density. It is a local property of the material. We focus on this quantity because it is intensive, and we have separated it into an intrinsic part in the absence of an externally applied magnetic field, $\mathcal{F}_L$, and a "magnetic energy," $-h\phi$.

The total free energy is given by integrating the density over the volume of the material. Notice that we are no longer talking about a lattice—we are using a continuum description that is valid regardless of the details of structure on the atomic scale. This is in the spirit of developing a more general theory. We will discuss the continuum limit further in Chapter 7.

We wish to write down an expression for $\mathcal{F}_L$. Generally, this is impossible, but near critical points, there is hope. We have already seen that for small $|h|$ and temperatures above or not too much below T_c, the magnetization is "small."[1] Thus, under many circumstances the full ϕ-dependence of $\mathcal{F}_L$ is irrelevant and we only need to know its values for small ϕ. Since generically the free energy is analytic,[2] we are invited to express it as a Taylor series in powers of ϕ as

$$\mathcal{F}_L(T; \phi) = \mathcal{F}_0 + b\phi + \frac{\alpha}{2}\phi^2 + \frac{s}{3}\phi^3 + \frac{u}{4}\phi^4 + \cdots, \tag{6.2}$$

where we have chosen a funny set of symbols to conform with typical conventions (and for future convenience).

Each of the coefficients ($\mathcal{F}_0, \alpha, u$, etc.) in this expression is a function of T. In general, if there are control parameters other than the temperature (such as pressure or chemical composition), the Landau coefficients will be (analytic) functions of *all* the control parameters. The bottom-up approach aims to calculate (or approximate) these functions directly; the top-down approach employed here will use general features of the coefficients to deduce properties of the phase transition, as we will see shortly.

To proceed further, we appeal to the $\phi \to -\phi$ symmetry of the Ising ferromagnet. In the absence of a magnetic field, the free energy must be invariant under this symmetry operation, i.e.,

$$F_L(T, h = 0; \phi) = F_L(T, h = 0; -\phi). \tag{6.3}$$

---

1. Specifically, it is small compared to its maximum value, ϕ_0, in the broken-symmetry phase at $T = 0$, when all the spins are aligned.

2. Except at critical points.

From this it follows that $\mathcal{F}_L$ can contain only even powered terms, i.e., $b = s = 0$, and hence that

$$\mathcal{F}_L(T; \phi) = \mathcal{F}_0(T) + \frac{\alpha(T)}{2}\phi^2 + \frac{u(T)}{4}\phi^4 + \cdots. \qquad (6.4)$$

This is a powerful step. By demanding that the free energy respect the microscopic symmetries of the Ising ferromagnet, we constrained the form of the free energy considerably. In general, the symmetry of the order parameter dictates which terms are allowed and which are forbidden in the Landau free energy.

Importantly, our expression for the free energy agrees with the one we derived from a microscopic mean-field theory in Eq. 4.32.[3] However, now we have arrived at this expression in a rather different manner, without needing to know anything much about the underlying problem other than the fact that it has an order parameter ϕ with a $\phi \to -\phi$ symmetry. This is remarkable in and of itself. Furthermore, we now recognize that the lack of odd powers of m in Eq. 4.32 was not an accident; from the present perspective, it is an obligatory consequence of symmetry.

Deriving critical phenomena

Although the Landau free energy, Eq. 6.4, recapitulates the same results we derived in Chapter 4, we should approach this problem without preconceived notions of the sign or magnitude of the various coefficients. That is, we will study what happens for general $\alpha(T)$ and $u(T)$ and impose necessary restrictions along the way.

If $u(T) < 0$, then the expression in Eq. 6.4 would not make much sense, because it would become arbitrarily large and negative for large values of $|\phi|$. In this situation higher order terms need to be included; the relevant values of ϕ that minimize F_L are large enough that terms of order ϕ^6 necessarily come into play. We will return to this point when we discuss tricritical points in Section 6.3.1.

For now, let us consider a situation in which $u(T) > 0$. As before, we wish to determine how the equilibrium value of ϕ depends on the control parameter T. Now we recognize that there are two distinct situations. If $\alpha(T) > 0$, then in the absence of a magnetic field, the free energy is minimized by $\phi = 0$; this corresponds to the symmetry-preserving ("disordered") phase. However, if $\alpha(T) < 0$, the free energy is minimized by a nonzero value of ϕ:

$$\phi = \pm\sqrt{|\alpha|/u} \quad \text{for} \quad \alpha(T) < 0; \qquad (6.5)$$

this corresponds to the broken-symmetry phase. We thus infer that the critical temperature (if it exists) is the temperature T_c at which $\alpha(T_c) = 0$.

---

3. In making this correspondence, we identify ϕ with m/ν, where m is the magnetization as defined in Chapter 4 and ν is the volume of a unit cell (which comes into play when going from a lattice to a continuum description).

To explore what happens near the critical point, we expand the Landau coefficients in powers of the dimensionless "distance" from the critical point, $t = (T - T_c)/T_c$:

$$\mathcal{F}_0(T) = \mathcal{F}(T_c) + a_0 t - \frac{c_0}{2} t^2 + \cdots,$$

$$\alpha(T) = \alpha_1 t + \frac{\alpha_2}{2} t^2 + \cdots, \tag{6.6}$$

$$u(t) = u_0 + u_1 t + \cdots.$$

By expanding in this manner, we are keeping with our assumption that physical quantities are analytic—a central assumption which underpins all the analysis in this chapter. We also assume that none of the coefficients is zero;[4] indeed, we know that $\alpha_1 > 0$ since α must be positive for $T > T_c$ and negative for $T < T_c$.

From here, we can analyze the critical properties of the system just as we did in Section 4.4.2. In fact, since the free energy is identical in form to Eq. 4.32, the results are identical to what we already found for the Ising model in a mean-field treatment (see end of Worksheet W6.1). Specifically, for $T > T_c$, the susceptibility $\chi = 1/\alpha$ diverges as $t^{-\gamma}$ to leading order in $|t|$, with an exponent of $\gamma = 1$. For $T < T_c$, we get the same critical exponent $\beta = 1/2$ that we derived before, i.e., $|\phi| \sim |t|^{1/2}$.

An interesting quantity that we have not yet considered is the specific heat per unit volume,

$$c = \frac{T}{V} \frac{dS}{dT} = -T \frac{d^2 \mathcal{F}}{d^2 T}. \tag{6.7}$$

For $T > T_c$, where $\phi = 0$, we find that $c = c_0 + \mathcal{O}(t)$. However, since the equilibrium free energy for $T < T_c$ is

$$\mathcal{F}_L\left(T; \phi = \pm\sqrt{|\alpha|/u}\right) = \mathcal{F}_0 - \frac{\alpha^2}{4u}, \tag{6.8}$$

it follows that

$$c = c_0 + \frac{\alpha_1^2}{2u_0} + \mathcal{O}(t) \text{ for } T < T_c. \tag{6.9}$$

The Landau theory predicts that the specific heat jumps discontinuously as the temperature goes from above to below T_c—a new sort of critical singularity. (Because c is a second derivative of $\mathcal{F}_L$, the discontinuity of c at T_c is consistent with a description of a "continuous" phase transition.)

Implications

The above analysis is rather general: any system (with the requisite symmetry) in which u is positive will exhibit these Ising critical behaviors at a point at which α

---

4. Generically, all the coefficients $a_0, c_0, \alpha_1, \ldots$ in Eq. 6.6 will be nonzero, unless we are "doubly" fine-tuned. See discussion of multicritical points at the end of the chapter.

passes through 0, independent of microscopic details. At no point did we have to make any big assumptions, other than those based on symmetry.

What the Landau theory approach makes transparent is that phase transitions can be understood in a unified manner for many different physical systems. Certain aspects are universal, whereas others are sensitive to microscopic details. From general arguments about the form of the free energy, one infers that there is a critical temperature below which a broken-symmetry phase appears. The theory also predicts the values of critical exponents. Other quantities, such as the value of T_c or the prefactor α_1, can only be understood based on microscopic considerations.

6.2 Other Symmetries

The top-down approach allows us to readily analyze all sorts of different problems without much more work. Different forms of broken-symmetry states are describable in terms of different order parameters and subject to different symmetry-imposed constraints. These constraints alone are often sufficient to pin down information about phases and phase transitions.

As another example, instead of the one-component order parameter, ϕ, we might have an N-component vector, $\boldsymbol{\phi} = (\phi_1, \phi_2, \ldots, \phi_N)$. As mentioned previously, $N = 2$ describes a superfluid and $N = 3$ describes a ferrofluid. We also need to specify the microscopic symmetries of the problem. By requiring that F_L respect these symmetries, we will control which terms in the Taylor series expansion of $\mathcal{F}_L$ are allowed and which terms vanish by symmetry. Our task is to identify the combinations of ϕ which are invariant under all microscopic symmetry transformations (known as "scalars"). We then expand F_L in powers of these scalars.

To illustrate this approach, we will consider systems with a two-dimensional order parameter,

$$\boldsymbol{\phi} = \phi_1 \hat{\mathbf{e}}_1 + \phi_2 \hat{\mathbf{e}}_2 = |\phi| \left[\hat{\mathbf{e}}_1 \cos(\theta) + \hat{\mathbf{e}}_2 \sin(\theta) \right]. \tag{6.10}$$

Depending on what is most convenient, we can specify ϕ in Cartesian coordinates (in terms of ϕ_1 and ϕ_2) or in polar coordinates (in terms of $|\phi|$ and θ). We can also take ϕ to be a complex number $e^{i\theta}|\phi|$, such that the real part of ϕ is $\phi_1 = |\phi| \cos(\theta)$ and the imaginary part is $\phi_2 = |\phi| \sin(\theta)$.

6.2.1 The XY Model: O(2) Symmetry

If there is a full rotational symmetry, i.e., if the free energy is symmetric under $\theta \to \theta + \Delta\theta$ for any $0 \le \Delta\theta < 2\pi$, then the system is said to have O(2) symmetry. In this case, the only scalar quantities we can make are powers of $\boldsymbol{\phi} \cdot \boldsymbol{\phi} = \phi_1^2 + \phi_2^2 \equiv |\phi|^2$. The free energy can thus only depend on the magnitude of the order parameter:

$$\mathcal{F}_L(T; \boldsymbol{\phi}) = \mathcal{F}_0 + \frac{\alpha}{2}|\phi|^2 + \frac{u}{4}|\phi|^4 + \cdots. \tag{6.11}$$

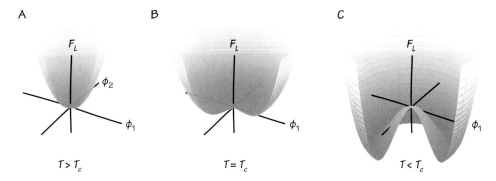

FIGURE 6.1. Illustration of the Landau free energy of an XY model, at temperature above, at, and below T_c.

The analysis of this free energy is almost the same as that in the Ising case. The key difference is that ϕ is now a two-dimensional vector. Below T_c, there is a continuum of degenerate minima lying on a circle; the free energy fixes the magnitude of ϕ but not its angle. This is illustrated in Figure 6.1.

In the presence of an applied field $\mathbf{h} \cdot \phi$, there is an energetic preference for ϕ to lie parallel to $\mathbf{h}$; a unique point on the circle becomes the global minimum. Other than this, the remaining analysis is the same as that in the Ising case, with $h = |\mathbf{h}|$. For $T < T_c$ (where $\alpha < 0$), the Landau free energy is minimized by

$$\phi \to \sqrt{|\alpha|/u}\, \mathbf{h} \ \ \text{as} \ \ |\mathbf{h}| \to \mathbf{0}. \tag{6.12}$$

Since the functional form of the free energy is the same as in the Ising case, all critical exponents will necessarily be the same in this mean-field treatment, despite the differences in the nature of the broken symmetry involved.

6.2.2 q-State Clock Models

In the case of a q-state clock model, the set of symmetries is smaller. As described in Section 5.4.1, this model is invariant only under rotations of the order parameter by multiples of $\Delta\theta = 2\pi/q$. To simplify the discussion, we will also assume that the model has a reflection symmetry—i.e., it is invariant under $\phi_2 \to -\phi_2$ or equivalently under $\phi \to \phi^*$ (where ϕ^* is the complex conjugate of ϕ).

As before, we start with the most general form of the Landau free energy, and then we retain only the allowed terms, i.e., the ones that are invariant under all symmetry operations. Since q-fold symmetry is a smaller set of symmetries than $O(2)$, we will find additional terms in the Taylor series of $\mathcal{F}_L$ that are no longer forbidden.

Without imposing symmetry restraints, the most general expansion of the Landau free energy is

$$\mathcal{F}_L(T; \phi) = \mathcal{F}_0 + b_i\phi_i + \frac{\alpha_{ij}}{2}\phi_i\phi_j + \frac{s_{ijk}}{3}\phi_i\phi_j\phi_k + \frac{u_{ijkl}}{4}\phi_i\phi_j\phi_k\phi_l + \cdots, \tag{6.13}$$

where summation over repeated indices, $i = 1, 2$, etc., is implicit. The coefficients can be interpreted as partial derivatives, and so the indices must be symmetric, e.g. $\alpha_{ij} = \alpha_{ji}$.

$q = 2$ state clock model

The case $q = 2$ is just an awkward way of representing the Ising model. The discrete rotational symmetry forbids all odd powers of ϕ_j, leaving a Landau free energy of the form

$$\mathcal{F}_L(T; \phi) = \mathcal{F}_0 + \frac{\alpha}{2}|\phi|^2 + \frac{u}{4}|\phi|^4 + \frac{\alpha'}{2}\left[\phi_2^2 - \phi_1^2\right] + \frac{v'}{2}\phi_1^2\phi_2^2 + \frac{u'}{4}\left[\phi_1^4 - \phi_2^4\right] + \cdots, \quad (6.14)$$

where ... represents higher order terms in powers of ϕ_j. We expect to find results similar to those from the Ising model. If $\alpha' > 0$, ordering with non-zero ϕ_1 is clearly favored over nonzero ϕ_2; the role of ϕ_1 and ϕ_2 is interchanged if $\alpha' < 0$. Assuming the quartic terms are such that the free energy is bounded below, it is easy to see that the free energy is minimized by $\phi = 0$ only so long as $\alpha \pm \alpha' > 0$. Thus, T_c is determined by the implicit equation $\alpha(T_c) - |\alpha'(T_c)| = 0$. If $\alpha'(T_c) > 0$, then just below T_c the free energy is minimized by $\phi_2 = 0$, and hence the Landau free energy reduces precisely to that for the Ising model, where ϕ_1 is the Ising order parameter, and the effective parameters that control its behavior can be identified as $\alpha - \alpha' \leftrightarrow \alpha$ and $u + u' \leftrightarrow u$.

However, if we consider the case in which $|\alpha'|$ is small, then the regime of temperatures near T_c in which the second component of the order parameter can be neglected is likely to be correspondingly small. Indeed, in the limit in which all the primed coefficients vanish, this problem approaches the O(2) problem discussed above. We will further explore what the LG theory tells us to expect in cases with such "near symmetries" in Section 7.5.2.

$q = 3$ state clock model

For $q = 3$, an entirely new feature of the Landau theory appears. Now there is a cubic term, $\phi^3 + (\phi^*)^3 = 2|\phi|^3 \cos(3\theta)$, that is invariant under the symmetries of the problem (namely, rotations by $120°$). Thus,

$$\mathcal{F}_L(T; \phi) = \mathcal{F}_0 + \frac{\alpha}{2}|\phi|^2 + \frac{s}{3}|\phi|^3 \cos(3\theta) + \frac{u}{4}|\phi|^4 + \cdots. \quad (6.15)$$

Since it is only the cubic term that depends on θ, we can minimize first with respect to θ and then separately with respect to $|\phi|$. For $s < 0$, the free energy is minimized by $\theta = 0°, 120°$, or $240°$, while for $s > 0$ it is minimized by $\theta = 60°, 180°$, or $300°$. In

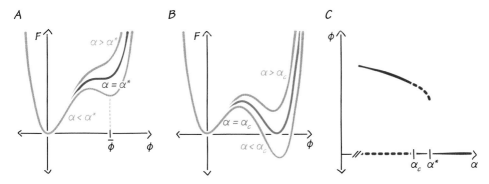

FIGURE 6.2. Landau theory with a ϕ^3 term describes a 1st-order (discontinuous) phase transition. (A) When α crosses from above to below α^*, a new minimum in $F(\phi)$ is born at $\phi = \bar{\phi}$. (B) When α crosses from above to below α_c, the $\phi = \bar{\phi}$ minimum becomes thermodynamically favored. (C) There is a discontinuous transition to a broken-symmetry state (with $\phi = \bar{\phi} \neq 0$) at $\alpha = \alpha_c$. The minima of $F(\phi)$ are plotted as a function of α, with solid lines representing global minima (equilibrium values of ϕ) and dotted lines representing local minima (metastable values of ϕ).

both cases, after minimizing with respect to θ, the Landau free energy has the form

$$\mathcal{F}_L(T; \phi) = \mathcal{F}_0 + \frac{\alpha}{2}|\phi|^2 - \frac{|s|}{3}|\phi|^3 + \frac{u}{4}|\phi|^4 + \cdots. \qquad (6.16)$$

This is shown for various values of α in Figure 6.2.

From here, Worksheet W6.2 works out the phase diagram by considering the shape of the curve for varying values of ϕ. For $\alpha > \alpha^* \equiv s^2/4u$, there is a single minimum of $\mathcal{F}_L$ with $\phi = 0$. However, for $\alpha < \alpha^*$, there are two local minima (Figure 6.2A)—one at $\phi = 0$ and one at $\phi = \bar{\phi}$ with

$$\bar{\phi} = \frac{|s|}{2u} + \sqrt{\frac{s^2 - 4\alpha u}{4u^2}}. \qquad (6.17)$$

The one with the lower free energy describes the equilibrium phase of matter, while the one with higher free energy is a metastable phase. Specifically, there is a critical value of α,

$$\alpha_c \equiv \frac{|s|}{3}\bar{\phi} = \frac{2}{9}\frac{s^2}{u}, \qquad (6.18)$$

at which the two minima have equal free energy. Above this critical value, $\phi = 0$ is lower in free energy; below, the $\phi = \bar{\phi}$ minimum is lower (Figure 6.2B). Consequently, for $\alpha(T) > \alpha_c$, the equilibrium state is symmetric, while for $\alpha(T) < \alpha_c$, it is not. This is illustrated in the solid lines in Figure 6.2C. In the state with spontaneously broken symmetry, there are three possible values of θ which minimize the free energy, each spaced 120° apart. When $\alpha(T)$ crosses α_c, there is a discontinuous (1st-order) transition to one of these three equivalent broken-symmetry states;

the magnitude of the order parameter jumps discontinuously from 0 for $T > T_c$ to $\bar{\phi}$ for T infinitesimally smaller than T_c. Correspondingly, there is a latent heat (i.e., T times the discontinuity in the entropy)

$$L = \frac{1}{2}\frac{d\alpha}{dT}\bar{\phi}^2 - \frac{1}{3}\frac{d|s|}{dT}\bar{\phi}^3 + \frac{1}{4}\frac{du}{dT}\bar{\phi}^4. \tag{6.19}$$

We see that the Landau theory has recapitulated some general properties of 1st-order transitions, including metastability and a latent heat.

As this is a new observation, it is worth analyzing it a bit more carefully. Firstly, the essential feature of this argument follows from the existence of a cubic term in the Landau free energy. From our present top-down perspective, if there is no reason forbidding the existence of a cubic term (e.g., if there is no symmetry with respect to $\phi \to -\phi$), then we expect such a term to exist. It thus follows from Landau theory that if a transition involves a pattern of broken symmetry that allows a cubic invariant, then the transition must be discontinuous. Only if cubic invariants are forbidden by symmetry is a continuous transition possible.[5]

There is also an inherent issue with the line of analysis we have gone through here. We have justified our approach, in part, under the assumption that the magnitude of the order parameter is small. If the transition is continuous, then close enough to the transition the order parameter takes on values that are arbitrarily small. However, for a discontinuous transition, even arbitrarily close to T_c in the ordered phase, the order parameter takes on a nonzero value, $\bar{\phi}$. If $\bar{\phi}$ is small (that is, if the discontinuities at the transition are small), then the transition is "almost continuous," and this is not a problem. But nothing guarantees this. Thus, while our conclusion that the transition cannot be continuous in the presence of a cubic term is unambiguous, the description of the transition is approximate at best.

$q = 4$ state clock model

Up through quartic order, there is only one new term that is allowed here that is excluded in the presence of O(2) symmetry, $\left[\phi^4 + (\phi^*)^4\right] = 2|\phi|^4 \cos(4\theta)$. Thus,

$$\mathcal{F}_L(T; \phi) = \mathcal{F}_0 + \frac{r}{2}|\phi|^2 + \frac{u}{4}|\phi|^4 + \frac{v}{2}|\phi|^4 \cos(4\theta) + \cdots. \tag{6.20}$$

There are two possible patterns of broken symmetry here. If $-u < v < 0$, then in the ordered state the order parameter vector is aligned with either $\hat{\mathbf{e}}_1$ or $\hat{\mathbf{e}}_2$ (i.e., $\theta = 0°, 90°, 180°,$ or $270°$). Conversely, if $v > 0$, the order parameter in the ordered

---

5. We do need to remember that this is still a result of mean-field theory. There are very few counterexamples to this result in the form of transitions that are continuous where Landau theory would predict them to be discontinuous—but they do exist, especially in low dimensions. A famous case, in fact, is the 3-state Potts model (which is equivalent to the 3-state clock model) in two dimensions.

Table 6.1. Summary of q-state clock model for various values of q.

	Invariants	Transition type	Equivalent models
$q = 2$	ϕ_1^2, ϕ_2^2	continuous	Ising
$q = 3$	$|\phi|^2, |\phi|^3 \cos(3\theta)$	discontinuous	N/A
$q = 4$	$|\phi|^2, |\phi|^4 \cos(4\theta)$	continuous	N/A
$q \to \infty$	$|\phi|^2$	continuous	XY

phase lies $45°$ from the axes. In both cases the magnitude of the order parameter behaves identically to that of the Ising model.

$q > 4$ state clock model

Now the only difference between the clock model and the O(2) model will be encoded in higher than quartic terms in the Landau free energy, proportional to $[\phi^q + (\phi^*)^q]$. The effects of these terms are subtle so long as the magnitude of the order parameter is small, i.e., the behavior may be expected to be similar to that of the XY model despite the fact that the symmetry involved is discrete.

A summary of the cases for different values of q is listed in Table 6.1.

6.3 Multicritical Points

Critical points occur at points on the phase diagram when something "special" happens to the coefficients of the Landau expansion. The examples we have studied so far are all "singly fine-tuned," wherein something special happen to exactly *one* of the coefficients (e.g., where α vanishes or, in the case of a 1st-order transition, where it reaches a special value). In these situations, only one control parameter needs to be fine-tuned to reach a critical point. The set of all such points on the phase diagram forms a phase boundary, or "critical manifold," whose dimension is one less than the dimension of the phase diagram—it is said to have co-dimension 1. For example, in a 1D phase diagram (e.g., where the only parameter varied is T), the system is critical at a 0D point (i.e., at $T = T_c$); in a 2D phase diagram (say, in the $P - T$ plane), the phase boundary between two phases (say, a liquid and a solid) is a 1D curve.

A multicritical point is a circumstance where multiple "special things" happen simultaneously—for instance, two Landau coefficients vanish simultaneously. Correspondingly, *multiple* control parameters need to be adjusted in order to reach a multicritical point. The number of parameters which need to be tuned is the co-dimension of the critical manifold. The examples we give below are multicritical points of co-dimension 2, that is, involving two special parameters, occurring at a single point on a 2D phase diagram or on a line on a 3D phase diagram, and so forth.[6]

6.3.1 Tricritical Points

In some circumstances, the phase boundary separating two phases can be continuous at some places and discontinuous at others. The point on the boundary at which a continuous transition turns discontinuous is known as a tricritical point.[7] This situation is doubly fine-tuned because one control parameter needs to be tuned to reach the phase boundary, and another needs to be tuned to reach the point on the boundary where the transition goes from continuous to discontinuous. Specifically, consider an Ising-like system which depends on two control parameters T and x. For example, x might represent the pressure or the concentration of some atomic species. Suppose the system undergoes a symmetry-breaking transition upon cooling T past T_c. The transition temperature $T_c(x)$ will, in general, depend on the value of x. In the situation we are describing, there is a special value of x, x_{tri}, such that the transition at $T_c(x)$ is continuous for $x < x_{tri}$ and discontinuous for $x > x_{tri}$. The tricritical point occurs at $T = T_{tri} \equiv T_c(x_{tri})$ and $x = x_{tri}$.

In the vicinity of the tricritical point where the order parameter is small, it is permissible to perform a Landau expansion.[8] As we will see below, the tricritical point occurs where the Landau coefficients α and u vanish simultaneously. The sign of α roughly (but not exactly—see below) determines whether T is above or below T_c, and the sign of u corresponds to whether x is below or above x_{tri}. The transition is continuous for $u > 0$ (as before) but discontinuous for $u < 0$ (as we will now show).

---

6. There is a subtlety in counting parameters. For instance, if we consider only parameters that respect time-reversal symmetry (i.e., $\sigma_j \to -\sigma_j$), then the critical point of the Ising model is reached by tuning a single parameter, e.g., the temperature. This remains true even if we consider a much more complex version of the Ising model with second and further neighbor interactions—just so long as we stay in the range of parameters in which there is a direct transition from a disordered high-T phase to a ferromagnetic low-T phase. However, if we consider the possibility of including terms that break this symmetry, the number of parameters that need to be tuned increases. In terms of parameters in the Landau expansion, the critical point of interest arises when the coefficients of the linear (ϕ), quadratic (ϕ^2), and cubic (ϕ^3) terms all vanish simultaneously (and assuming $u > 0$), i.e., it involves three parameters to be fine-tuned. In the absence of a magnetic field, the two odd-powered terms vanish by symmetry in the case of the Ising ferromagnet, so they do not enter the parameter count. For the model in the presence of a magnetic field, the symmetry is broken but in a special way, such that only a linear term is generated—so that in the $h - T$ plane the critical point has co-dimension 2, in analogy with the liquid-gas critical point we discussed in Chapter 1.

7. The origin of the nomenclature governing multicritical points is arcane—we will not attempt to motivate it.

8. Even though this situation involves a discontinuous transition, the magnitude of the discontinuity (that is, the amount by which ϕ jumps as T crosses T_c for $x > x_{tri}$) becomes increasingly small upon approach to the tricritical point.

Rather than directly analyzing the phase diagram in the $T - x$ plane, it is simpler to use Landau theory to think about the $\alpha - u$ plane. In the neighborhood of the tricritical point, the two descriptions are generically[9] related by the linear, invertible mapping

$$\alpha(T, x) = \alpha_t \ (T - T_{tri}) + \alpha_x \ (x - x_{tri}) + \cdots, \tag{6.21}$$

$$u(T, x) = u_t \ (T - T_{tri}) - u_x \ (x - x_{tri}) + \cdots, \tag{6.22}$$

where ... represents higher order terms in powers of $(x - x_{tri})$ and $(T - T_{tri})$.[10]

For $u > 0$, we can repeat the analysis of ϕ^4 theory to find a continuous transition as α crosses 0. For $u < 0$, we can no longer truncate the Landau expansion at quartic order. In order to deal with a free energy that is bounded below, we are forced to keep terms to the next order allowed by symmetry, i.e., the sixth order:

$$\mathcal{F}_L(T; \phi) = \mathcal{F}_0 + \frac{\alpha}{2}\phi^2 + \frac{u}{4}\phi^4 + \frac{v}{6}\phi^6 + \cdots, \tag{6.23}$$

where ... signifies still higher order terms in powers of ϕ, and for stability we necessarily assume that $v > 0$.[11] To find the discontinuous transition in the $u < 0$ regime, we follow the same line of reasoning as we employed in studying the 3-state clock model. We look for stable or metastable solutions by solving for values of ϕ that correspond to local minima of $\mathcal{F}_L$. For $\alpha > \alpha^* \equiv u^2/4v$, there is only a single such solution—the symmetric solution with $\phi = 0$. For $\alpha^* > \alpha > 0$, there are two metastable solutions, one with $\phi = 0$ and the other with $\phi = \pm\bar{\phi}$, where

$$\bar{\phi}^2 = \frac{|u|}{2v} + \sqrt{\left(\frac{u}{2v}\right)^2 - \left(\frac{\alpha}{v}\right)}. \tag{6.24}$$

To determine the point at which the 1st-order transition occurs, we compare the free energy for $\phi = 0$ with that at $\phi = \bar{\phi}$. As shown in the blue curve of Figure 6.3A, the latter has lower free energy for

$$\alpha < \alpha_c \equiv \frac{3}{16}\frac{u^2}{v}. \tag{6.25}$$

---

9. Eqs. 6.21 and 6.22 contain the linear-order terms in a Taylor expansion about T_{tri} and x_{tri}, where $\alpha_t = \frac{\partial \alpha}{\partial T}$, etc. Generically, these partial derivatives will not be zero unless we are extremely (un)lucky (i.e., we would need an even more fine-tuned situation of one higher co-dimension).

10. In keeping with our expectation that the ordered phase is favored at lower T, we will assume that $\alpha_t > 0$. The fact that we have defined the sense of x such that $u(T_0, x) > 0$ for $x < x_{tri}$ further implies the inequality $(u_t\alpha_x + u_x\alpha_t) > 0$.

11. In the neighborhood of the tricritical point, any small variation of v with T or x is relatively unimportant.

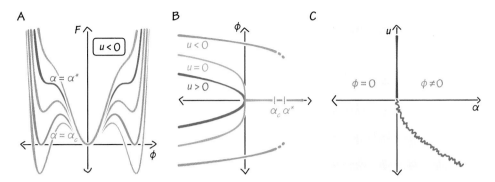

FIGURE 6.3. ϕ^6 theory of the tricritical point. (A) The Landau free energy (Eq. 6.23) for $u < 0$. A pair of broken-symmetry minima is born at $\alpha = \alpha^*$ and becomes thermodynamically favored at $\alpha = \alpha_c$. (B) For $u \geq 0$, there is a continuous transition at $\alpha = 0$; for $u < 0$, the transition is discontinuous and happens at a positive value of α. (C) The phase diagram in the $\alpha - u$ plane, with the continuous phase boundary in magenta and the discontinuous boundary in red.

Consequently, there is a 1st-order transition from the ordered to the disordered phase that occurs when α is still positive, as illustrated by the blue curve in Figure 6.3B.

Figure 6.3C summarizes the phase diagram in the $\alpha - u$ plane: for $u > 0$, there is a continuous transition as α crosses 0; for $u < 0$, there is a discontinuous transition as α crosses α_c. The corresponding phase boundary in the $T - x$ plane lies at

$$T_c(x) = \begin{cases} T_0(x) & \text{for} \quad x < x_{tri} \\ T_0(x) + 3u^2/16v\alpha_t & \text{for} \quad x > x_{tri}, \end{cases} \tag{6.26}$$

where $T_0(x)$ is the temperature at which α vanishes, $\alpha(T_0(x), x) = 0$. Note that there is a new sort of nonanalyticity here—one that shows up in the shape of the phase boundary $T_c(x)$ itself. In this case it is a rather subtle non-analyticity, a discontinuity in $d^2 T_c/dx^2$ at the tricritical point. For $x > x_{tri}$, the order parameter jumps discontinuously from 0 to

$$\bar{\phi}_c = \sqrt{3|u|/4v} \sim (x - x_{tri})^{1/2}, \tag{6.27}$$

which, since it gets smaller the closer we are to the tricritical point, is a measure of how weakly 1st-order the transition is. The manner in which the latent heat vanishes as $x \to x_{tri}$ is calculated in Problem 6.3.

6.3.2 Bicritical and Tetracritical Points

Again, we think of a two-dimensional phase diagram, now of the sort shown in Figure 6.4. We now allow for the possibility of two different types of ordering, described with order parameters ϕ_1 and ϕ_2. An interesting question to ask here is whether the two types of orderings can coexist, i.e., whether there is a region of

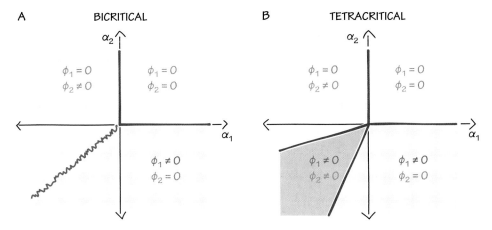

FIGURE 6.4. When two different types of ordering, with order parameters ϕ_1 and ϕ_2, are possible, they can either compete (left, bicritical) or coexist (right, tetracritical). The red region is where ϕ_1 is nonzero; the blue region is where ϕ_2 is nonzero. Magenta solid lines represent a continuous phase boundary; red squiggly lines represent a discontinuous boundary.

the phase diagram where both ϕ_1 and ϕ_2 are nonzero. Consider a scenario where the high-temperature phase is symmetric. For $x < x_c$, upon lowering the temperature, we encounter a transition (assumed to be continuous) at a critical temperature, $T_1(x)$, to a spontaneously broken-symmetry phase in which one order parameter, ϕ_1, develops a nonzero expectation value. For $x > x_c$ the transition we encounter at $T_2(x)$ is to a phase in which a different order paramter, ϕ_2, is nonzero. By construction, these two phase boundaries meet at a multicritical point at x_c, such that $T_1(x_c) = T_2(x_c)$. There are two possibilities for the region of the phase diagram below the multicritical point. If there is a discontinuous phase boundary between the two ordered phases, as shown in Figure 6.4A, then the multicritical point is known as a bicritical point. If, instead, there is an intermediate coexistence phase, as shown in Figure 6.4B, then it is a tetracritical point.

For simplicity, let us treat explicitly the simplest example, in which ϕ_1 and ϕ_2 are each Ising order parameters. Then, we can treat the problem in terms of a two-component vector order parameter, ϕ, with components ϕ_1 and ϕ_2. The Landau free energy to quartic order is then of the same form as what we already introduced in Eq. 6.14. By appropriately scaling ϕ_1 and ϕ_2, we can arrange for the magnitude of the terms proportional to ϕ_1^4 and ϕ_2^4 to be equal, so that

$$\mathcal{F}_L(T; \phi) = \mathcal{F}_0 + \frac{\alpha_1}{2}\phi_1^2 + \frac{\alpha_2}{2}\phi_2^2 + \frac{u}{4}|\phi|^4 + \frac{v}{2}\phi_1^2\phi_2^2 + \cdots. \tag{6.28}$$

The above-mentioned critical temperatures are obtained as the implicit solutions of $\alpha_j(T_j(x), x) = 0$, and the multicritical point occurs at $x = x_c$ where $T_1(x_c) = T_2(x_c)$. The condition of thermodynamic stability requires $u > 0$ and $v > -u$.

For $v > 0$, we are dealing with a bicritical point. As shown in Figure 6.4A, there is a disordered phase everywhere that both $\alpha_1 > 0$ and $\alpha_2 > 0$, and a broken-symmetry phase with $\phi_1 = \pm\sqrt{|\alpha_1|/u}$ and $\phi_2 = 0$ wherever $0 > \alpha_1$ and $\alpha_2 > \alpha_1$ and with $\phi_2 = \pm\sqrt{|\alpha_2|/u}$ and $\phi_1 = 0$ wherever $0 > \alpha_2$ and $\alpha_1 > \alpha_2$. There is a 1st-order phase boundary on the line defined implicitly by the equation $\alpha_1 = \alpha_2 < 0$ along which the order parameter jumps from one type to the other. This transition becomes increasingly weakly first order upon approach to the bicritical point with $\Delta\phi_1 = -\Delta\phi_2 \sim (T_{bi} - T)^{1/2}$. The latent heat at the transition likewise vanishes upon approach to the bicritical point, as shown in Problem 6.4.

For $0 > v > -u$, we are dealing with a tetracritical point, as shown in Figure 6.4B. It is relatively simple to see that $\phi_1 = \pm\sqrt{|\alpha_1|/u}$ and $\phi_2 = 0$ for $0 > \alpha_1$ and $\alpha_2 > -|\alpha_1|[1 - |v|/u]$ and that $\phi_2 = \pm\sqrt{|\alpha_2|/u}$ and $\phi_1 = 0$ for $0 > \alpha_2$ and $\alpha_1 > -|\alpha_2|[1 - |v|/u]$. However, now there is a coexistence regime in which both ϕ_1 and ϕ_2 are nonzero which occurs where both $\alpha_1 < 0$ and $\alpha_2 < 0$ and where $[1 - |v|/u]^{-1} > |\alpha_1|/|\alpha_2| > [1 - |v|/u]$. Moreover, all the transitions are continuous. The details of how various quantities depend on the distance to the critical points is the subject of Problem 6.5.

6.4 Problems

6.1. *Cubic invariant with a symmetry-breaking field.* In Section 6.2.2 we saw how a ϕ^3 term generically causes a transition to be discontinuous. Now consider the additional effect of a symmetry-breaking field h,

$$\mathcal{F}_L = \mathcal{F}_0 - h\phi + \frac{\alpha}{2}\phi^2 - \frac{s}{3}\phi^3 + \frac{u}{4}\phi^4 + \cdots.$$

Assume that $s > 0$.

(a) Does $h > 0$ increase or decrease the transition temperature? How about $h < 0$?

(b) *Determine the phase diagram in the $h - \alpha$ plane. Draw in solid lines the discontinuous phase boundary and in dotted lines the limit of metastability.

6.2. *Vanishing ϕ^4 term.* Suppose we have a system with a broken Ising symmetry, but fine-tuned such that the ϕ^4 term vanishes:

$$\mathcal{F}_L = \mathcal{F}_0 + \frac{\alpha}{2}\phi^2 + \frac{v}{6}\phi^6 + \cdots,$$

where $v > 0$ and $\alpha \propto T - T_c$. Determine the mean-field critical exponents β, γ, and δ, and the critical amplitude ratio $\chi(T + dT)/\chi(T - dT)$. What is the same and what is different compared to the more generic Ising case?

6.3. *Tricritical point.* Calculate the leading-order dependence of the latent heat on $x - x_{tri}$ near a tricritical point. *Hint:* The discontinuous phase boundary involves a jump in ϕ from 0 to $\bar{\phi}_c$ (Eq. 6.27), so the latent heat is

$$L = T\Delta S = T\frac{\partial \mathcal{F}_L}{\partial T}\bigg|_{\phi=\bar{\phi}_c} - T\frac{\partial \mathcal{F}_L}{\partial T}\bigg|_{\phi=0}. \tag{6.29}$$

6.4. *Bicritical point.* Calculate the leading-order dependence of the latent heat on $T - T_c$ near a bicritical point.

6.5. *Tetracritical point.* Calculate how the magnitude of ϕ_1 varies as you enter the coexistence phase from the $\phi_1 \neq 0$, $\phi_2 = 0$ phase.

6.5 Worksheets for Chapter 6

W6.1   Deducing the Landau Free Energy from Symmetry Considerations

Whereas we built up our intuition for the Ising model by considering microscopic interactions, universality implies a powerful general approach to describe the behavior near a critical point in terms of a free energy F expressed as an expansion in orders of ϕ, where the allowed powers of ϕ are determined by the symmetry (breaking) associated with the phase transition. This is known as Landau theory.

1. Let $F_L(T, h; \phi)$ be the free energy of an arbitrary(!) Ising system with volume V, which is a function of the temperature T and symmetry-breaking field h, as well as an order parameter ϕ (which we assume is spatially uniform for now). Why are we justified to claim

$$F_L(T, h; \phi) = \int_V d\vec{r}\ [\mathcal{F}_L(T; \phi) - h\phi]?$$

   What is the interpretation of the two terms inside the brackets?

2. The most general possible expansion in ϕ up to fourth order is

$$\mathcal{F}(T; \phi) = \mathcal{F}_0(T) + b(T)\phi + \frac{\alpha(T)}{2}\phi^2 + \frac{s(T)}{3}\phi^3 + \frac{u(T)}{4}\phi^4.$$

   In what regime of the phase diagram would you expect this expansion to be accurate, and why?

3. Because this is an Ising system, the free energy must be invariant under the symmetry $\phi \to -\phi$ if there is no symmetry-breaking field. Based on this, which of the terms in $\mathcal{F}$ are allowed? Which must be zero?

4. What is the restriction on u to obtain a physically realistic free energy? (Remember that the equilibrium values of ϕ are the minimums of the free energy.)

5. At what value of α will the phase transition occur?

6. Remember that $\mathcal{F}_0$, α, and u each depend on T. Since we are near the critical point, let's expand in powers of a dimensionless "distance" from the

critical point,

$$t \equiv \frac{T - T_c}{T_c},$$

which is sometimes known as a "reduced temperature." If a function $g(T)$ is analytic and amenable to Taylor expansion around $T = T_c$, we can say

$$g(T) = g_0 + g_1 t + \frac{g_2}{2} t^2 + O(t^3). \tag{6.30}$$

How are the coefficients g_0, g_1, and g_2 related to (derivatives of) $g(T)$?

7. Near the critical point, we only need to keep the lead-order term of the expansion in t. What is the leading-order term for $\mathcal{F}_0$, α, and u?

8. Compare your generic expression for $\mathcal{F}(T, \phi)$ to the variational free energy density you got from the mean-field Ising model,

$$\frac{F_{var}}{N} = f_{var} = -T \log 2 + \frac{T - zJ}{2} m^2 + \frac{T}{12} m^4 + O(m^6).$$

What about the critical exponents? Remember that β, γ, and δ are defined by:

$$m \propto (T_c - T)^\beta \qquad \frac{\partial m}{\partial h}\bigg|_{h=0} \propto |T_c - T|^{-\gamma} \qquad m|_{T=T_c} \propto h^{1/\delta}. \tag{6.31}$$

Turns out, since the Landau free energy has the exact same form as the mean-field Ising model, all of the critical exponents are the same! So $\beta = 1/2$, $\gamma = 1$, and $\delta = 3$. This is not a coincidence, as we will explore much later in the book.

W6.2   Understanding the Discontinuous Transition with a ϕ^3 Term

Depending on the symmetry of the system, the Landau free energy can have various permitted and forbidden terms which can affect the properties of the phase transition!

1. As a warm-up, suppose that we were studying a system where the graph of the free energy looks something like Figure W6.1. What is the equilibrium value of ϕ? Are there any metastable solutions (local minima)?

For a 3-clock model, the Landau free energy permits a ϕ^3 term:[12]

$$\mathcal{F}_L = \frac{\alpha}{2} \phi^2 - \frac{s}{3} \phi^3 + \frac{u}{4} \phi^4 + \cdots. \tag{6.32}$$

---

12. For the purposes of this worksheet, we'll take $s > 0$. The symbol ϕ represents the magnitude of the order parameter—which, since we're talking about a 3-clock model, can point in the $0°$, $120°$, or $240°$ direction on a clock face.

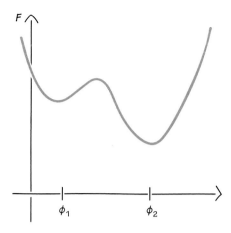

FIGURE W6.1. A potential with two unequal minima.

Before we do any calculations, let's take a look at the graph of $\mathcal{F}_L(\phi)$ for various values of the parameter α. Each curve is a different value of α.

2. There is one value of ϕ which is a (local) minimum of $\mathcal{F}_L$, regardless of α here. What is it?

3. For some values of α this is the only minimum. For some values of α there is an *another* minimum at a different value of ϕ—let's call it $\bar{\phi}$. Separating these cases is a special value of $\alpha = \alpha^*$. Trace over this special curve with your pencil and label it α^*.

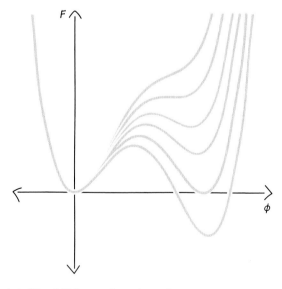

FIGURE W6.2. A plot of Eq. 6.32 for varying values of α.

4. For $\alpha < \alpha^*$, is $\bar{\phi}$ necessarily the equilibrium state? Why or why not?

5. There is another special value of $\alpha = \alpha_c$ which separates the case where $\bar{\phi}$ is the equilibrium state and where it isn't. Trace over this curve and label it α_c (with a different color if you have colored markers!)

6. Sketch a graph of how the equilibrium value of ϕ depends on α, making sure to label α^* and α_c on your x-axis. Is this transition continuous or discontinuous?

Now that we understand the overall behavior of this model, let's calculate the values of α^* and α_c.

7. By setting $d\mathcal{F}_L/d\phi = 0$, solve for the values of ϕ which extremize $\mathcal{F}_L$. Should you take the + or the − sign for $\bar{\phi}$? What does the other sign correspond to?

8. For what values of α is your expression for $\bar{\phi}$ valid? Use this fact to determine α^*.

9. Give an equation in terms of $\bar{\phi}$ which you can use to solve for α_c.

10. It's possible but a little tedious to solve for α_c directly with this, so instead let's use another way. If you look back to the curve you labeled α_c, you may notice an interesting feature of quartic equations: the local maximum is exactly halfway between $\phi = 0$ and $\phi = \bar{\phi}$. This is the only value of α where this is the case! Use this fact to determine α_c.

GINZBURG-LANDAU THEORY

One scale to rule them all!

Up until now, we have explored (at mean-field level) the physics of spatially uniform states. However, important additional insights into the nature of equilibrium phases can be obtained by considering nonuniform states. Such states arise in various ways. For instance, if a system is perturbed in a spatially localized fashion, the equilibrium state becomes altered in the nearby region; the extent of this region is given by the correlation length.

In Chapters 4 and 6 we examined the same set of problems from two different perspectives: a bottom-up approach, in which we began with a microscopic model and then constructed a mean-field approximation; and a top-down approach, in which we began with an assumption about a broken symmetry and approached the problem phenomenologically. This chapter employs both approaches. Specifically, we will carry through the mean-field analysis allowing for a spatially varying order parameter, and we will derive a continuum expression for the free energy—now not only in powers of the order parameter but also in powers of spatial derivatives of the order parameter.

7.1 The Thermodynamic Correlation Length

The notion of a thermodynamic correlation length, ξ, is a central concept in characterizing spatial variation in an equilibrium phase of matter. We have already seen ξ appear in the context of equilibrium fluctuations in correlation functions: correlations are only substantial between points that are separated by distances shorter than the correlation length. In this chapter, the correlation length will appear again in other situations with spatial variation, such as the response of a system to a localized perturbation, or the size of a domain wall. All these properties are characterized by a single scale.[1]

---

1. There are a number of special circumstances in which exceptions to this general rule arise. One of the most interesting has to do with the physics of quantum phases of matter in the limit $T \to 0$; here, far from any quantum critical points, the correlation length as usually defined from the properties of thermodynamic correlation functions will typically approach a finite limit as $T \to 0$, but there

As we have already mentioned in Chapter 4, the correlation length diverges upon approaching the critical point. This fact is of profound significance—it is the ultimate reason why critical behavior is universal. Features shorter than the correlation length get averaged out. When the correlation length is long compared to the scale of the lattice, the microscopic details of the lattice are irrelevant to the thermodynamic description. Therefore, near a critical point, the behavior at long distances can be well approximated by a continuum description and the structure of the lattice may be disregarded.

Furthermore, the correlation length plays a key role in justifying the thermodynamic limit. Any discrepancies between a finite-size and an infinite-size system are known as "finite-size effects." The importance of the correlation length is that any two points separated by distances longer than ξ can be treated as approximately independent of one another. Hence, when $L \gg \xi$, finite-size effects are expected to be quite small.[2] You have the chance to familiarize yourself with the correlation length in Worksheet W7.1.

7.2   Spatial Correlations in the Ising Model

As always, we begin by considering our friendly neighborhood Ising ferromagnet. Although the trial Hamiltonian we introduced in Eq. 4.16 allowed for spatially varying fields, b_j, until now we restricted our attention to uniform Hamiltonians, $h_j = h$, and uniform solutions, $b_j = b$. This was sufficient for studying the spatially uniform states that arise in the absence of any applied fields, or when the applied field is uniform. Now we are going to consider inhomogeneous solutions—those that arise from an underlying inhomogeneous problem (in this case, in the presence of a spatially varying field h_j) or those that spontaneously break translational symmetry (i.e., where h_j is spatially uniform but the equilibrium state is not).

Unsurprisingly, the steps we take are parallel to those in Chapter 4 (see Worksheet W7.2 for details). We introduce the trial Hamiltonian,

$$H_{tr} = -\sum_j b_j \sigma_j \,, \tag{7.1}$$

---

can be a distinct scale, $\xi_{coh}(T)$, over which quantum coherence is maintained, such that $\xi_{coh} \to \infty$ as $T \to 0$. Quantum mechanics has a nonlocal character; for instance, the spectrum of a given Hamiltonian depends on the boundary conditions. One consequence of this is that "mesoscopic systems," with $\xi_{coh} > L \gg \xi$, can exhibit novel properties that are intermediate between those of microscopic and macroscopic systems. Another example is disordered systems, where under some circumstances there can be a wide range of lengthscales such that the suitable averages that enter measurements of different physical properties lead to wildly different correlation lengths.

2. Herein also lies the reason why sufficiently close to a critical point, finite-size effects will necessarily play a role in the behavior: for T close enough to T_c, the correlation length grows so large that it becomes comparable to the system size, $\xi \sim L$.

and we use this to compute the value of the variational free energy, $F_{var}(\{h\}; \{b\})$, as in Eq. 4.20. Finally, to obtain the mean-field description, we minimize F_{var} with respect to all the b_j, leading to the mean-field equations

$$b_j = h_j + J \sum_{i=\text{n.n. } j} m_j, \qquad (7.2)$$

with $m_j = \langle \sigma_i \rangle_{tr} = \tanh(\beta b_j)$. Again, however, it will be more convenient to consider $\{m\}$ as our variational parameters as in Eq. 4.29. In terms of these variables, the variational free energy is given by

$$\tilde{F}_{var}(\{h\}; \{m\}) = \sum_j \tilde{f}_{var}(m_j) + \frac{J}{2} \sum_{\langle ij \rangle} \left[m_i - m_j \right]^2 - \sum_j h_j m_j, \qquad (7.3)$$

where $\tilde{f}_{var}(m)$ is given in Eq. 4.30. Minimizing this with respect to m_j yields the same mean-field equation as Eq. 7.2, although now interpreted as an implicit equation for m_j. The parameters $\{b\}$ (or, equivalently, $\{m\}$) are variational parameters that we have introduced. For a given set of fields $\{h\}$ there is a set $m_j(\{h\})$ that minimizes $\tilde{F}_{var}$. Substituting this set of parameters into the $\tilde{F}_{var}$ is what ultimately gives the (physical) mean-field free energy, as solely a function of $\{h\}$:

$$F_{mf}(\{h\}) = F_{var}(\{h\}; \{b\})\big|_{b_j=b_j(\{h\})} = \tilde{F}_{var}(\{h\}; \{m\})\big|_{m_j=m_j(\{h\})}. \qquad (7.4)$$

7.2.1 Position-Dependent Correlation Functions

Many of the quantities we are interested in are familiar correlation and response functions, but now with an extra dependence on position:

- The magnetization is the first derivative of the free energy,

$$\langle \sigma_j \rangle = -\frac{\partial F(\{h\})}{\partial h_j} = m_j(\{h\}). \qquad (7.5)$$

  In the mean-field approximation, where F is replaced with F_{mf}, this leads to the intuitively reasonable result that $\langle \sigma_j \rangle = \langle \sigma_j \rangle_{tr}$ (see Worksheet W4.3).
- The "connected" correlation between spins at sites i and j is given by a second derivative of the free energy,

$$G_{ij}(\{h\}) \equiv \langle \sigma_i \sigma_j \rangle - \langle \sigma_i \rangle \langle \sigma_j \rangle = -T \frac{\partial^2 F}{\partial h_i \partial h_j}. \qquad (7.6)$$

  We will be analyzing the mean-field result obtained by substituting $F \to F_{mf}$ in this expression. Importantly, the result is *not* the same as it would be if we were to simply compute the corresponding correlation functions in the trial ensemble: $\langle\sigma_i\sigma_j\rangle_{tr} - \langle\sigma_i\rangle_{tr}\langle\sigma_j\rangle_{tr} = \delta_{ij}[1 - m_j^2]$. Even though the trial ensemble has no correlations between spins on different sites, we will find that the mean-field G_{ij} exhibits a rich spatial structure of correlations.

- The nonlocal susceptibility χ_{ij} describes how a small, localized field at site j perturbs the average spin at site i. Intuitively, we expect χ_{ij} to be larger if i and j are closer together; in fact, we will see that "close" or "far" are defined in comparison to the correlation length. χ appears when expanding the magnetization to linear order in $\{h\}$,

$$\langle\sigma_i\rangle = \sum_j \chi_{ij}h_j + \cdots, \tag{7.7}$$

where ... represents higher order terms in powers of h_k. Equivalently, it is the second derivative of the free energy,

$$\chi_{ij} = \frac{\partial m_i}{\partial h_j}\bigg|_{\{h\}=\{0\}} = -\frac{\partial}{\partial h_j}\frac{\partial F}{\partial h_i}\bigg|_{\{h\}=\{0\}}, \tag{7.8}$$

where we have used Eq. 7.5. Comparing this expression to Eq. 7.6 tells us that the equilibrium fluctuations (G_{ij}) are directly related to the susceptibility (χ_{ij}),

$$\chi_{ij} = T^{-1} G_{ij}(\{h\})\big|_{\{h\}=\{0\}}. \tag{7.9}$$

We will return to this point at the end of the the chapter.

7.2.2 Calculating the Nonlocal Susceptibility χ_{ij}

As a first worked example, let us compute χ_{ij} at a temperature $T > T_c = zJ$. Since the techniques required are a bit involved, we will leave out most of the steps. Worksheet W7.2 walks through this calculation in detail.

In this range of temperatures, m_j is small, so we can expand $\tilde{F}_{var}$ in powers of m_j as

$$\tilde{F}_{var} = \sum_j \left[-T\log 2 + \left(\frac{T - T_c}{2}\right)m_j^2 - h_j m_j \right] + \frac{J}{2}\sum_{\langle ij\rangle}[m_i - m_j]^2 + \ldots, \tag{7.10}$$

where ... indicates higher order terms. Minimizing this with respect to m_j (or equivalently, expanding Eq. 7.2 to linear order) yields a set of coupled equations for χ_{ij},

$$\delta_{ik} = \sum_j \left[(T - T_c)\delta_{ij} + J\Delta_{ij}\right]\chi_{jk}$$

where

$$\Delta_{ij} = \begin{cases} z & \text{if} & i = j, \\ -1 & \text{if} & i \text{ and } j \text{ are nearest neighbors}, \\ 0 & & \text{otherwise}. \end{cases} \tag{7.11}$$

This 2nd-order difference equation can be solved via Fourier transform,

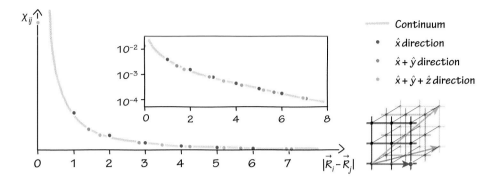

FIGURE 7.1. Correlation function. The gray line shows the continuum approximation (Eq. 7.26) while the dots show the correlation function of a cubic lattice (Eq. 7.12), computed via numerical integration. The colors indicate different directions along the lattice, illustrated in the lower right. The inset shows the same plot with the y-axis on a log scale. Parameters are $\xi = 4a$.

$$\chi_{ij} = \int \frac{d\vec{k}}{(2\pi)^d} \left\{ \frac{\exp[i\vec{k} \cdot (\vec{R}_i - \vec{R}_j)]}{(T - T_c) + Jg(\vec{k})} \right\}, \tag{7.12}$$

where $\vec{R}_i$ is the position of lattice site i, d is the number of dimensions, $g(\vec{k})$ is the Fourier transform of $\Delta(\vec{R}_i - \vec{R}_j) = \Delta_{ij}$, and the range of the integral is the first Brillouin zone (see Worksheet W7.2 and Appendix A.3 for a detailed explanation).

This expression is hard to compute exactly, but it can readily be evaluated numerically. (We will discuss its behavior at more length in Section 8.1.) Figure 7.1 plots χ_{ij} for the case of a cubic lattice as a function of the distance between sites i and j. The key result is that χ_{ij} falls off rapidly with distance; if we perturb a system at a point in space, we alter the equilibrium state in the surrounding region, but far from the perturbation, the effect becomes immeasurably small. The radius of the affected region defines a correlation length, ξ, which you can estimate from Eq. 7.12 in Worksheet W7.3.

7.3 The Effective Field Theory

We have been thinking of the order parameter as a local property, m_j, on each site of the lattice, but if we wish to study more general or universal features, it is helpful to generalize so that we are not beholden to any particular lattice structure. Rather than an order parameter on the sites of a lattice, a natural step is to define ϕ as a function of a continuous position coordinate, $\vec{r}$. We can think $\phi(\vec{r})$ as a physical quantity with a value at every point in space, known as a field.

Different situations will call for different kinds of fields. In the case of the magnetization in an Ising ferromagnet, $\phi(\vec{r})$ is a real scalar field, designating the magnitude of the local magnetization. It is analogous to the electric potential $V(\vec{r})$ and the charge density $\rho(\vec{r})$ of electrodynamics. To describe more complex forms of symmetry breaking, we can introduce fields with multiple components, analogous to the vector and pseudo-vector electric and magnetic fields, $\vec{E}(\vec{r})$ and $\vec{B}(\vec{r})$, of electrodynamics.

One advantage of a field treatment is that it is often easier to carry out calculations with functions defined over a continuous variable. Moreover, the continuum limit naturally arises if we are concerned with what happens on the longest lengthscales. If we zoom out sufficiently, we expect that details on the scale of the lattice do not matter.

The free energy generalizes as

$$F_{var}(\{m\}) \longrightarrow F_{var}[\phi], \tag{7.13}$$

from a function of many variables to a function of a function—a functional. In classical mechanics, you may have encountered the action functional, the time integral of the Lagrangian. From the principle of least action, it is possible to derive a set of differential equations, Newton's laws. In a similar manner, by minimizing the free energy functional, it is possible to derive a set of differential equations for the equilibrium configuration of the fields.

7.3.1 Taking the Continuum Limit

From a bottom-up approach, a continuum description of the variational free energy can be readily derived. We assume that we are interested in field configurations that are smoothly varying on the scale of the lattice spacing. Corresponding to each configuration of fields m_j, we define a continuous function $\phi(\vec{r})$ such that $\phi(\vec{R}_j) = m_j$ on the lattice points and $\phi(\vec{r})$ interpolates as smoothly as possible between the lattice points. A continuum version of the symmetry-breaking field, $h_j \to h(\vec{r})$, may be similarly defined. In this case, we can easily take the continuum limit with

$$\sum_j f(\vec{R}_j) \to \int \frac{d\vec{r}}{\nu} f(\vec{r}), \tag{7.14}$$

$$m_i = \phi(\vec{R}_j) + (\vec{R}_i - \vec{R}_j) \cdot \vec{\nabla}\phi(\vec{R}_j) + \cdots, \tag{7.15}$$

where ν is the volume of a unit cell, and ... signifies higher order terms in derivatives of ϕ (see Appendix A.5). In this way, the continuum version of the variational free energy in Eq. 7.3 becomes

$$\tilde{F}_{var}[h;\phi] = \int d\vec{r} \left\{ V(\phi) + \sum_{\mu,\mu'} \frac{\kappa_{\mu,\mu'}}{2} (\partial_\mu \phi)(\partial_{\mu'}\phi) - h\phi + \cdots \right\}, \tag{7.16}$$

where $\phi(\vec{r})$ and $h(\vec{r})$ inside the integral are evaluated at the position $\vec{r}$, and ... signifies higher order terms in spatial gradients. The first term,

$$V(\phi) = \nu^{-1} \tilde{f}_{var}(\phi), \tag{7.17}$$

is an effective potential which depends only on the value of the order parameter at that point. The second term is the leading-order penalty for local variations of $\phi(\vec{r})$. The coefficient of this term,

$$\kappa_{\mu,\mu'} = \nu^{-1} J \sum_{\vec{R}} \Delta(\vec{R}) R_\mu R_{\mu'} = \nu^{-1} J \left. \frac{\partial^2 g(\vec{k})}{\partial k_\mu \partial k_{\mu'}} \right|_{\vec{k}=\vec{0}}, \tag{7.18}$$

is a symmetric tensor $\kappa_{\mu,\mu'} = \kappa_{\mu',\mu}$ which represents the "stiffness" of the order parameter field measured along various directions. For the case of a cubic lattice, the x, y, and z directions are equivalent, meaning that $\kappa_{\mu,\mu'} = \kappa\delta_{\mu,\mu'}$, with $\kappa = \nu^{-1} J a^2 = J/a$ where a is the lattice spacing. In this case the quadratic term simplifies to $\frac{\kappa}{2}|\nabla\phi|^2$.

Arriving at the same expression with a top-down approach

Before analyzing this continuum theory, let us pause and consider it from a top-down perspective. Suppose we were asked to guess the form of the variational free energy for an Ising ferromagnet in the absence of an external field, $\tilde{F}_{var}[0; \phi]$. We would be able to go a fair distance toward obtaining Eq. 7.16 on the basis of rather general considerations, as we have done in Chapter 6.

To begin, we would expect to be able to express $\tilde{F}_{var}[0; \phi]$ as an integral over space of a free energy density, $\tilde{\mathcal{F}}_{var}$. Since the material is homogenous, $\tilde{\mathcal{F}}_{var}$ cannot have any explicit dependence on position. Furthermore, on the basis of the locality of physics, the free energy density at a point in space must depend only on the value of the order parameter field and its spatial derivatives at that point. When $\tilde{\mathcal{F}}_{var}$ is expanded in spatial gradients of ϕ, we expect the effective potential V to be the leading term, followed by terms which are first order in spatial derivatives, second order, etc. However, most lattices have an inversion symmetry, $\vec{r} \to -\vec{r}$, which forbids any terms that are first order in spatial derivatives. The next-order term, which is quadratic in spatial derivatives, is certainly allowed. In general, its coefficient $\kappa_{\mu,\mu'}$ could be a function of $\phi(\vec{r})$. Worksheet W7.4 provides a chance to work through this reasoning.

In the most general case, the components of κ are unrelated, but in cases where the system has additional symmetries, we could further simplify the problem. If the system we are interested in were rotationally invariant (e.g., a liquid of some sort), then the stiffness tensor must be isotropic and hence can only have one independent component, $\kappa_{\mu,\mu'} = \kappa\,\delta_{\mu,\mu'}$. Interestingly, we have seen that this is also the case for a system with an underlying cubic lattice: to leading order in the gradient expansion, a cubic system has the same rotational symmetry as free space. In crystals with lower degrees of symmetry, there can be more independent components of $\kappa_{\mu,\mu'}$; for example, in a tetragonal system in which the z direction differs from the x and y directions, there are two independent components, parallel and perpendicular to the z-axis: $\kappa_{\mu,\mu'} = \delta_{\mu,\mu'} \left[\kappa_\parallel + \delta_{\mu,z}(\kappa_\perp - \kappa_\parallel) \right]$.

Finally, if we focus on the case in which the order parameter is small, we can expand all the terms in a Taylor series expansion in powers of ϕ. The fact that (for $h = 0$) the free energy must be symmetric under $\phi(\vec{r}) \to -\phi(\vec{r})$ implies that no odd terms in this series arise. Thus, on the basis of locality, symmetry, and analyticity, we can infer the form of the variational free energy functional to be the same as Eq. 7.16, with

$$V(\phi) = V_0 + \frac{\alpha}{2}\phi^2 + \frac{u}{4}\phi^4 + \cdots, \qquad (7.19)$$

$$\kappa_{\mu,\mu'}(\phi) = \kappa_{\mu,\mu'} + \cdots, \qquad (7.20)$$

where ... represent higher order terms in the Taylor expansion. This is known as the Ginzberg-Landau formulation of the theory of phases and phase transition. It is a generalization of the Landau theory in Chapter 6. Admittedly, this approach does not allow us to compute α, u, and other parameters, but it is reasonable to assume that they are all continuous and analytic functions of T and P and any other physical parameters we wish to vary.

7.3.2   The Equilibrium Order Parameter Field

To determine the equilibrium field configuration $\bar{\phi}(\vec{r})$, we find the function which minimizes $\tilde{F}_{var}[h; \phi]$. As explained in Worksheet W7.5 and Appendix A.4, this is done by setting the functional derivative to zero:

$$\left.\frac{\delta F_{var}[h; \phi]}{\delta \phi(\vec{r})}\right|_{\phi = \bar{\phi}} = \frac{\partial V(\bar{\phi})}{\partial \bar{\phi}} - \kappa_{\mu,\mu'}\partial_\mu\partial_{\mu'}\bar{\phi} - h = 0. \qquad (7.21)$$

This is the continuum version of the self-consistency relation in Eq. 7.2.

For concreteness, let us consider a system with cubic symmetry, so $\kappa_{\mu,\mu'} = \kappa\,\delta_{\mu,\mu'}$, and let us restrict ourselves to $T > T_c$, where we can neglect terms in V that are higher than quadratic order. In this case the variational free energy takes on a particularly simple form,

$$F_{var}[h; \phi] = \int d\vec{r} \left\{ V_0 + \frac{\alpha}{2}\phi^2 + \frac{\kappa}{2}|\nabla\phi|^2 - h\phi \right\}, \qquad (7.22)$$

leading to the mean-field equation

$$\left[\alpha - \kappa\nabla^2\right]\bar{\phi}(\vec{r}) = h(\vec{r}). \qquad (7.23)$$

Suppose for simplicity we look for solutions satisfying the boundary conditions $\phi(\vec{r}) \to 0$ as $|\vec{r}| \to \infty$—we will consider other cases later in this chapter. It is clear that if there is no symmetry-breaking field, then $\bar{\phi}(\vec{r})$ is zero everywhere. However, if $h(\vec{r}) \neq 0$ in some region of space, then the equilibrium order parameter nearby would be nonzero.

To understand how the order parameter field responds to an applied symmetry-breaking field, first consider the extreme example of a local perturbation,

$h(\vec{r}) = h_0 \delta(\vec{r} - \vec{r}_0)$: a delta function of magnitude h_0 localized at position $\vec{r}_0$. By definition, the response at position $\vec{r}'$ is given by $\bar{\phi}(\vec{r}) = h_0 \chi(\vec{r}' - \vec{r}_0)$: the nonlocal susceptibility times the size of the perturbation. Since Eq. 7.23 is linear, if two perturbations were simultaneously applied, the response would be the sum of the responses to the individual perturbations. Indeed, any arbitrary $h(\vec{r})$ can be decomposed into the appropriate sum (or integral) of delta-function sources. Therefore, the response of the order parameter field to a general source can be expressed as a convolution integral,

$$\bar{\phi}(\vec{r}) = \int d\vec{r}_0 \chi(\vec{r} - \vec{r}_0) h(\vec{r}_0). \tag{7.24}$$

The nonlocal susceptibility $\chi(\vec{r})$ is the elementary response to an impulse; it is obtained by solving the differential equation Eq. 7.23 for a delta-function source. As before, this can be done via Fourier transform; in Worksheet W7.5 you show that it has the "Ornstein-Zernicke" form

$$\chi(\vec{r}) = \int \frac{d\vec{k}}{(2\pi)^d} \frac{e^{i\vec{k}\cdot\vec{r}}}{\alpha + \kappa k^2}. \tag{7.25}$$

This expression is analogous to the one we obtained on the lattice, Eq. 7.12, with the relation $\alpha \leftrightarrow (T - T_c)$, $Jg(\vec{k}) \leftrightarrow \kappa k^2$, and the integral over the first Brillouin zone replaced by an integral over all $\vec{k}$. Indeed, the large-$|\vec{r}|$ behavior of the two expressions are the same, since at large $|\vec{r}|$ the integral is dominated by small $|\vec{k}|$, where $g(\vec{k}) = k^2 + \mathcal{O}(k^4)$. The advantage of the continuum expression is that the integral can be evaluated analytically. In $d = 3$, the result is

$$\chi(\vec{r}) = \left(\frac{1}{4\pi\kappa}\right) \frac{e^{-|\vec{r}|/\xi}}{|\vec{r}|} \quad \text{with } \xi = \sqrt{\kappa/\alpha}. \tag{7.26}$$

In Problem 7.3 you will derive an exact expression in $d = 1$, and in Appendix A.6, the asymptotic expression in arbitrary dimension d:

$$\chi(\vec{r}) \sim \begin{cases} r^{-(d-1)/2} e^{-|\vec{r}|/\xi} & \text{for} \quad |\vec{r}| \gg \xi, \\ r^{-(d-2)} & \text{for} \quad |\vec{r}| \ll \xi. \end{cases} \tag{7.27}$$

The correlation length

We have obtained an explicit expression for the correlation length, ξ. Notably, it diverges near a critical point as $\alpha \to 0$. From Eq. 7.26 and the usual T dependence of α it follows that this divergence is governed by another critical exponent, ν, where, as $T \to T_c^+$:

$$\xi \sim \xi_0 [(T - T_c)/T_c]^{-\nu} \quad \text{with } \xi_0 \sim a \text{ and } \nu = 1/2. \tag{7.28}$$

Furthermore, recall that ξ not only determines the spatial extent over which variations in the equilibrium state are induced by a local perturbation, it also tells us about the lengthscale over which equilibrium fluctuations are correlated. Specifically, as in Eq. 7.9, the zero-field correlation function is related to the susceptibility as

$$G(\vec{r} - \vec{r}') \equiv \langle \phi(\vec{r})\phi(\vec{r}') \rangle - \langle \phi(\vec{r}) \rangle \langle \phi(\vec{r}') \rangle = T\chi(\vec{r} - \vec{r}'). \tag{7.29}$$

For $T \neq T_c$, the correlations fall off exponentially with distance, such that beyond a lengthscale ξ the correlations are negligible. However, at $T = T_c$, the situation is quite different because ξ is divergent! Instead, the integral in Eq. 7.25 yields

$$\chi(\vec{r}) \sim |\vec{r}|^{-(d-2)}, \quad \text{for } T = T_c, \tag{7.30}$$

which falls (in $d > 2$) with a *power* of distance. This is a striking result: it implies that at criticality, there is no lengthscale that characterizes the return to equilibrium. A small perturbation at the origin produces deviations from equilibrium behavior that decays with distance from the origin, but not in a way that is negligible beyond any particular distance. The same applies to the equilibrium fluctuations at criticality, $G(\vec{r})$: two arbitrarily faraway points are correlated in a nonnegligible way. This sort of power-law decay of correlations—and more generally the lack of any specific longest lengthscale that characterizes the decay of correlations—is another characteristic feature of critical phenomena.

To put this result in context, consider the behavior of $\chi(\vec{r})$ at temperatures above, but not much above, T_c. Here the correlation length is quite large but not infinite. For distances short compared to the correlation length ξ (but still larger than the lattice spacing a), we can ignore α in performing the integral in Eq. 7.25—consequently, for $a \ll |\vec{r}| \ll \xi$, $\chi(\vec{r})$ falls off in a power-law manner as in Eq. 7.30. It is only at distance scales $|\vec{r}| \gg \xi$ that $\chi(\vec{r})$ falls exponentially as in Eq. 7.26. Thus, any experiment which probed the system only at lengthscales short compared to ξ could not determine whether the system is at its critical temperature or only very close to it.[3]

More concerning the continuum limit

The continuum limit is meant to be an approximation of long-distance physics, i.e., for distances much longer than the lattice constant, $|\vec{R}_i - \vec{R}_j| \gg a$. For short distances, because of the discrete nature of the lattice, we have no reason to expect correlations calculated with the continuum theory to agree with those calculated with the lattice theory. For example, Eqs. 7.26 and 7.29 imply that $G(\vec{r})$ diverges as $|\vec{r}| \to 0$; however, on the lattice, it is manifest that $|\langle \sigma_i \sigma_j \rangle| \leq 1$ (since $|\sigma_j| = 1$), which implies that $|G_{ij}| \leq 1$. These two results clearly disagree as $r \to 0$. Thus there is a very tangible sense in which the field theory result is unphysical—i.e., *wrong*.

---

3. It is a peculiarity of $d = 3$ that the power-law prefactor governing the long-distance behavior of χ is the same as that governing its "shorter"-distance behavior near criticality, since $(d - 2) = (d - 1)/2 = 1$ for $d = 3$. This peculiarity does not survive when critical correlations are treated more accurately, by methods that go beyond mean-field theory, as we shall see in the final chapters.

Nonetheless, at long distances compared to the lattice constant, the continuum theory is a reasonable approximation to the lattice theory. In Figure 7.1 we plot χ computed from the continuum theory and the lattice theory, Eqs. 7.26 and 7.12, respectively. The two agree well for $|\vec{R}_i - \vec{R}_j| \gg a$. Interestingly, if we consider the correlation function G_{ij} measured along different directions of the lattice, we find that at long distances the direction does not matter (see the different colored dots in Figure 7.1). The fact that correlations are approximately rotationally invariant is another nontrivial success of the continuum description.

For the continuum approximation to be reasonable, the correlation length must also be long compared to the lattice constant, $\xi \gg a$. This is precisely the situation near a critical point—in fact, the field-theoretic description of long-distance behavior becomes asymptotically more accurate the closer a system is to criticality. The fact that the continuum description does not explicitly refer to a lattice is another basis for understanding how the microscopic (in this case, lattice scale) details of a problem get averaged out near a critical point.

7.3.3   Correlations in the Ordered Phase

So far, we have restricted our attention to $T \geq T_c$. However, the extension to behavior in the ordered phase at $T < T_c$ is straightforward. We can still use Eq. 7.6 to compute the connected correlation function. Now, however, because we are investigating the ordered state, we cannot drop the higher order terms in the variational free energy as $\phi(\vec{r})$ is not small. Since we will be taking the limit as $h \to 0$ at the end of the calculation, we can look for $\phi(\vec{r})$ of the form

$$\phi(\vec{r}) = \bar{\phi} + \delta\phi(\vec{r}), \tag{7.31}$$

where $\bar{\phi}$ is a value of the magnetization that minimizes the variational free energy in the absence of any applied field and $\delta\phi(\vec{r})$ is assumed to be small.

In the case where T is not much below T_c, F_{var} can be approximated by the quartic form in Eq. 7.19, with $\alpha = -(T_c - T)$, $u = T/3$, and $\bar{\phi} = \sqrt{|\alpha|/u}$. Substituting in our ansatz for $\phi(\vec{r})$ gives

$$\tilde{F}_{var}[\phi, h] = F_{var}[\bar{\phi}, 0] + \int d\vec{r} \left\{ |\alpha|(\delta\phi)^2 + \frac{\kappa}{2}|\nabla\delta\phi|^2 - h\delta\phi + \cdots \right\}, \tag{7.32}$$

where ... represents higher order terms in powers of $\delta\phi$. (The absence of any terms linear in $\delta\phi$ other than the term linear in h reflects the fact that $\phi(\vec{r}) = \bar{\phi}$ is a minimum of the free energy for $h = 0$.) You will derive a similar expression in Worksheet W7.6.

Notice that the quantity in braces is identical to the expression for $T > T_c$ in Eq. 7.22, with the substitution $\alpha \to 2|\alpha|$. Thus, for small $\delta T \ll T_c$, the correlation function $G(\vec{r})$ for $T = T_c + \delta T$ has exactly the same form as for $T = T_c - 2\delta T$. Consequently, the correlation length diverges with the same critical exponent $\nu = 1/2$

regardless of whether we approach the critical temperature from above or below. However, there is an asymmetry in the way the correlation length diverges from above and below T_c,

$$\frac{\xi(T_c + \delta T)}{\xi(T_c - \delta T)} \to \sqrt{2} \ \text{ as } \ \delta T \to 0^+ \ . \tag{7.33}$$

This is another example of a universal critical amplitude ratio.

Finally, again paralleling the analysis we went through for $T > T_c$, we can obtain a similar relation between the susceptibility and the equilibrium correlation functions for $T < T_c$:

$$\langle \sigma_j \rangle = m_j + \sum_i \chi_{ji} h_i + \cdots, \tag{7.34}$$

where ... signifies higher order terms in h and $\chi_{ij} = -T^{-1} G_{ij}$.

7.4   Beyond the Ising Model

The same set of considerations we have applied for the Ising ferromagnet can be applied to more general forms of order and more general symmetries. For a change of perspective, we will illustrate this starting with the Landau-Ginzburg formulation. To have an explicit symmetry in mind we will treat the XY ferromagnet—expressed in terms of the complex field $\phi(\vec{r})$ as in Eq. 6.10. The real and imaginary parts of ϕ represent the two components of the order parameter. As we have discussed in Section 6.2.1, in the ordered phase the magnetization $\phi(\vec{r}) = e^{i\theta} \bar{\phi}$ has a fixed magnitude, $\bar{\phi} = \sqrt{|\alpha|/u}$, but can point at any arbitrary angle in the XY plane, $0 \le \theta < 2\pi$.

For $T > T_c$, the situation is not much different than that of the Ising case. For small h we can ignore any terms in the free energy of higher than quadratic order. Minimizing this with respect to $\phi(\vec{r})$ yields exactly the same equation as for the Ising case, Eq. 7.23, but now ϕ has two components. Since each component can be considered separately, we arrive at the conclusion that for $T > T_c$, both $G(\vec{r})$ and $\chi(\vec{r})$ have the same form as for the Ising case. Thus, the correlation length diverges as $T \to T_c$ from above in precisely the same manner. The same results also apply regardless of the symmetry of the order parameter—so long as there is sufficient symmetry to preclude a cubic term in the LG free energy (which would cause the transition to be discontinuous).

For $T < T_c$ $(\alpha < 0)$, however, the situation is more intricate. You have the chance to explore this more fully in Worksheet W7.6. As we did in the Ising case, for the case of small h we will expand the free energy in powers of the deviation of ϕ about its $h = 0$ value:

$$\phi(\vec{r}) = e^{i\theta} \bar{\phi} + \delta\phi(\vec{r}) \ \text{ where } \ \bar{\phi} = \sqrt{|\alpha|/u}. \tag{7.35}$$

From here, using the identity

$$|\phi|^4 = \bar{\phi}^4 + \bar{\phi}^2 \left[4|\delta\phi|^2 + \left(e^{-i\theta}\delta\phi\right)^2 + \left(e^{i\theta}\delta\phi^\star\right)^2 \right] + |\delta\phi|^4 \tag{7.36}$$

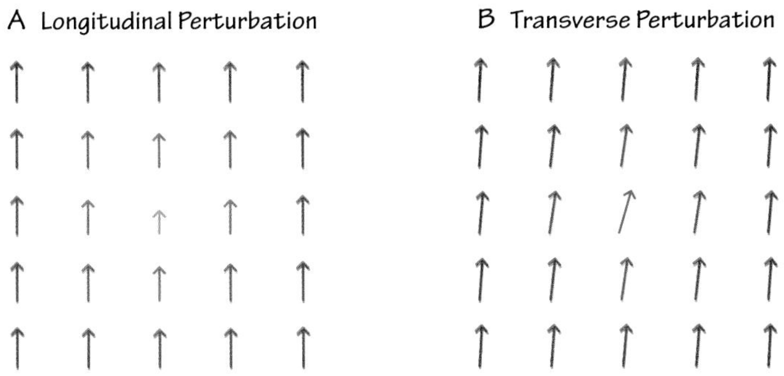

FIGURE 7.2. In the ordered phase of the XY model, the susceptibility of the system is different in the directions parallel (A) and perpendicular (B) to the direction of ordering.

we derive an expression

$$\mathcal{F}_{var}[\phi] = \mathcal{F}_{var}[\bar{\phi}] + \frac{\kappa}{2}|\vec{\nabla}\delta\phi|^2 - \frac{1}{2}\left[h^\star\delta\phi - h\delta\phi^\star\right] \tag{7.37}$$

$$+ \frac{|\alpha|}{4}\left[2|\delta\phi|^2 + (e^{-i\theta}\delta\phi)^2 + (e^{i\theta}\delta\phi^\star)^2\right] + \cdots, \tag{7.38}$$

where ... is higher order terms in $\delta\phi$. Without loss of generality, let us take the direction of magnetization to lie along the real axis, $\theta = 0$. Then, (neglecting the higher order terms) F_{var} is minimized by field configurations that satisfy the equations

$$\left[-\kappa\nabla^2 + 2|\alpha|\right]\delta\phi_R(\vec{r}) = h_R(\vec{r}), \tag{7.39}$$

$$\left[-\kappa\nabla^2\right]\delta\phi_I(\vec{r}) = h_I(\vec{r}), \tag{7.40}$$

where the subscripts R and I signify the real and imaginary parts. These equations determine the components of the order parameter parallel and perpendicular to the direction of ordering, respectively. The first equation is the same as in the Ising case, but the second equation is not. Instead, it looks like the Ising case precisely at $T = T_c$ (where $\alpha = 0$). Thus,

$$\chi_\parallel(\vec{r}) \sim |\vec{r}|^{-(d-1)/2}\exp{-|\vec{r}|/\xi},$$

$$\chi_\perp(\vec{r}) \sim |\vec{r}|^{-(d-2)}, \tag{7.41}$$

where the susceptibilities are defined such that the lowest (linear) order response of the magnetization to an applied field at $T < T_c$ is

$$\langle\phi(\vec{r})\rangle_{\{h\}} = \bar{\phi} + \int d\vec{r}'\left[\chi_\parallel(\vec{r}-\vec{r}')h_R(\vec{r}') + i\chi_\perp(\vec{r}-\vec{r}')h_I(\vec{r}')\right] + \cdots. \tag{7.42}$$

Here, $\xi = \sqrt{\kappa/2|\alpha|}$ is the correlation length familiar from earlier discussions, and the subscripts $\parallel$ and $\perp$ mean "longitudinal" and "transverse" to the direction of ordering, respectively.

This is a new piece of physics. For $T > T_c$, the linear response of the system to an applied field is the same regardless of the direction of the applied field, $\vec{h}$. This is required by symmetry. But below T_c, this symmetry is spontaneously broken—the magnetization points in a particular direction, which we have arbitrarily taken to be $\theta = 0$. Now, if we apply an external field also along the $\theta = 0$ direction—i.e., a longitudinal field—it behaves in much the same way we have seen: it causes a small local change in the magnitude of the order parameter. The region which is changed by the longitudinal perturbation has a finite lengthscale, ξ, which diverges on approach to T_c from below in the same way as in the Ising case. By contrast, if we apply an external field in the *transverse* direction, $\theta = \pi/2$, the response falls slowly, with only a power-law decay with distance. To first order in $h_\perp$, what is going on is that the magnetization rotates away from the $\theta = 0$ direction toward $\theta = \pi/2$ without changing its magnitude. The effects of a longitudinal and a transverse perturbation in the ordered phase are illustrated in the left and right panels of Figure 7.2, respectively.

To see how dramatic this difference is, consider the response of the system to a uniform externally applied field, so that the relevant susceptibility is the spatial integral of $\chi(\vec{r})$ over all space (or, equivalently, the Fourier transform of the susceptibility, Eq. 7.25, evaluated at $\vec{k} = \vec{0}$). The longitudinal susceptibility, $\tilde{\chi}_\parallel(\vec{k} = \vec{0}) = 1/2|\alpha|$, is finite, but the transverse susceptibility diverges, $\tilde{\chi}_\perp(\vec{k}) \to \infty$, as $|\vec{k}| \to 0$. This is physically reasonable: the transverse field causes the direction of magnetization to rotate, but since the initial direction of the magnetization was the result of spontaneous symmetry breaking, there is no restoring force to the transverse field. The fact that the transverse susceptibility falls off as a power law is less obvious—it is codified as "Goldstone's theorem," which we will discuss in Section 8.3.1.

7.5 Order Parameter Textures

There is another new piece of physics that arises in the broken-symmetry state below T_c. It is an ineluctable feature of a state of spontaneously broken symmetry that there is more than one state with the same free energy. Consequently, even a small perturbation in a macroscopic portion of space can have a large effect—forcing the system to chose a unique one of the possible broken-symmetry states. We focused on this notion as one of the defining features of a broken symmetry, where we applied a uniform symmetry-breaking field to the entire system, and then considered the limit as the strength of this field tends to zero. But it is important to also consider what happens if we apply one sort of symmetry-breaking field in one part of a system and another elsewhere. Even if these fields are weak, if they act on macroscopic portions of the system they must locally force the system into the corresponding preferred ordered state. If the preferred order is different in different parts of the

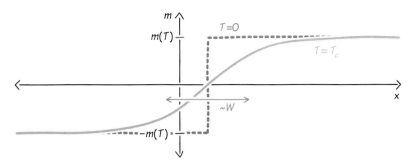

FIGURE 7.3. A schematic depiction of the magnetization as a function of position in conditions in which $m_j \to m(T)$ as $x_j \to -\infty$ and $m_j \to -m(T)$ as $x_j \to +\infty$. The characteristic width of the domain wall is W. The solid blue line is for T close to T_c, where the continuum Landau-Ginzburg theory applies, while the dashed red line is for $T \to 0$, where the domain wall is one lattice constant wide.

system, then there must be some characteristic manner in which the system interpolates between these two states in the intervening regions of space. This can produce various interesting order parameter "textures." Even if we take the limit in which the symmetry-breaking fields tend to zero, we will find that certain such textures remain as metastable states, and can also be important for the physics of phase transitions.

7.5.1   Domain Walls in the Ising Ferromagnet

We start with the case of an Ising ferromagnet, where there are two possible broken-symmetry states, one with $m_j = m(T)$ and the other with $m_j = -m(T)$. Consider a large system in which far to the left (x_j large and negative) we apply a weak positive field $h_j = h > 0$, far to the right (x_j large and positive) we apply a weak negative field $h_j = -h$, and in a large region around the origin ($x_j = 0$) we apply no symmetry-breaking fields. Such a set of fields would favor $m_j = -m(T)$ in the left and $m_j = m(T)$ on the right. This situation is illustrated in Figure 7.3.

   Somewhere in the region of interest there must be a domain wall or kink—a finite region of width W over which the magnetization changes from $-m(T)$ to $m(T)$. To find the exact form of the domain wall in the context of mean-field theory we would need to find the pattern of m_j that minimizes $\mathcal{F}_{var}$ in Eq. 7.21 subject to the given boundary conditions. This is readily done numerically, and even analytically in certain limits (discussed below). However, it is useful to first discuss the qualitative considerations at play.

   Given that the value of m_j that minimizes $\mathcal{F}_{var}$ in a uniform state is $m_j = \pm m(T)$, there must clearly be a free energy cost whenever m_j takes on any other value. This tendency favors textures with W as small as possible. However, the free energy also has a term of order J that disfavors large changes of m_j between neighboring sites. In the Landau-Ginzberg version of the theory, this tendency is represented by the stiffness, κ. This term favors textures that vary slowly in space, and hence tends to favor larger W. The balance between these two terms determines the optimal value of W. Perhaps unsurprisingly, we will find that this leads to $W \sim \xi$, where ξ is the already discussed correlation length.

Near criticality

Near the critical point, the free energy is well approximated by the Landau-Ginzberg free energy, Eq. 7.16, with $V(\phi)$ approximated by Eq. 7.19, including the quartic term because we are considering $T < T_c$. Setting the functional derivative to zero leads to the nonlinear differential equation

$$\left[-\kappa \nabla^2 - |\alpha| + u\phi^2 \right] \phi = 0. \tag{7.43}$$

While systematic methods exist for solving such equations, for now we will simply guess a form, and then prove that it is a solution. Our guess is a simple function that satisfies the requisite boundary conditions:

$$\phi(\vec{r}) = -\bar{\phi} \tanh(x/W) \text{ with } \bar{\phi} = \sqrt{|\alpha|/u}. \tag{7.44}$$

This describes a domain wall of width W which interpolates between $\phi(\vec{r}) \rightarrow \mp \bar{\phi}$ as $x \rightarrow \pm \infty$. Substituting this into Eq. 7.43 yields the condition:

$$W = 2\xi \text{ with } \xi = \sqrt{\kappa/2|\alpha|}. \tag{7.45}$$

In Worksheet W7.7 you will see how this result can be obtained from dimensional analysis, up to a factor of order 1.

We can also evaluate the free energy cost (relative to the uniform solution) of the domain wall by evaluating the integral over all space of $\tilde{F}_{LG}$ with the result

$$\Delta F = \sigma A \text{ with } \sigma = \frac{\alpha^2 \xi}{2u} C, \tag{7.46}$$

where σ, which is referred to as the surface tension, is the energy per unit area of domain wall, and

$$C = \int_{-\infty}^{\infty} dx \left[1 - \tanh^4(x) \right] = 10/3. \tag{7.47}$$

The domain wall thus becomes broader in proportion to ξ, and the surface tension, $\sigma \sim (T_c - T)^{3/2}$, vanishes as $T \rightarrow T_c^-$.

Note that this solution does not depend on the magnitude of the fields we applied far from the domain wall, $|x|/W \gg 1$, to impose the requisite boundary conditions. Indeed, if we imagine turning off these fields once we have used them to induce a domain wall, the domain wall will remain unchanged: it is a local minimum of the variational free energy. Any distortion in which the domain wall in some region moves a bit to the right or to the left will increase the amount of domain wall area

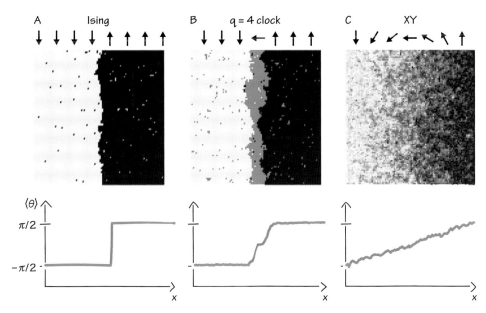

FIGURE 7.4. The effect of boundary conditions $\theta(0) = -\pi/2$ and $\theta(L) = \pi/2$ for a few models. (A) Ising model. (B) 4-state clock model; there is an intermediate region of $\theta = 0$. (C) XY model; there is no domain wall and θ interpolates linearly from $x = 0$ to $x = L$. The temperature is $T = 0.7 \, T_c$, where $T_c \approx 2.27, 1.13,$ and 0.89 for the Ising, 4-state clock, and XY models on a square lattice, respectively.

and hence increase the free energy cost by roughly σ times the change in area. Thus, domain walls are metastable.

Finally, we remark that if we were dealing with problems with richer patterns of discrete symmetry breaking, the character of the domain walls will be similar to those of the Ising case, with the one difference that there may be more than one type of domain wall. For instance, in the q-state clock model, there is one sort of domain wall between regions in which θ differs by an amount $\pm 2\pi/q$. However, the domain wall between regions that differ by $\pm 4\pi/q$ might have a different structure—in fact it will often break up into two distinct domain walls with an intermediate region with a phase shifted by $2\pi/q$. This is illustrated in Figure 7.4B.

Zero temperature

The construction of the domain wall is still easier as $T \to 0$ (red curve in Figure 7.3). Here, there are no thermal fluctuations, so each m_j is equal either to -1 or 1. The optimal domain wall solution in this limit is $m_j = -\text{sgn}(x_j)$ and the corresponding surface tension is easily seen to be $\sigma = 2Ja^{-2}$, where a is a lattice constant. This looks qualitatively similar to the solution of the Landau-Ginzburg equations we have just discussed in the limit that $W \sim a$. Manifestly, as the temperature increases from $T = 0$ to $T \to T_c^-$, the width of the domain wall increases and the surface tension decreases.

7.5.2   Textures in the XY Model

Domain walls are the natural textures that arise when a discrete symmetry is broken. A domain wall separates two regions with distinct orientations of the order parameter. In situations where a continuous symmetry is spontaneously broken, however, it is possible for nearby regions to differ in orientation by arbitrarily small amounts—leading to a totally different kind of "domain wall."

The absence of domain walls in the XY model

We now consider the case in which a continuous symmetry is spontaneously broken. To be concrete, we start with the Landau-Ginzburg field theory description of the XY model. Now we apply external fields so that

$$\phi(\vec{r}) \to |\bar{\phi}| e^{\pm i\theta_0} \text{ as } x \to \pm L/2, \tag{7.48}$$

where $\bar{\phi} = \sqrt{u/|\alpha|}$ is the equilibrium magnitude of the order parameter, θ_0 is arbitrary, and L is the (assumed macroscopic) width of our system. In other words, we take boundary conditions of $\vec{\phi} = \bar{\phi}\left[\cos(\theta_0)\hat{e}_1 \pm \sin(\theta_0)\hat{e}_2\right]$ far to the left or right of the region of space we are studying, respectively, and we want to see how the order interpolates between these two limits. Minimizing the free energy with respect to $\phi(\vec{r})$ is achieved by a solution to the nonlinear differential equation,

$$\left[-\kappa\nabla^2 - |\alpha| + u|\phi|^2\right]\phi = 0. \tag{7.49}$$

Again, let us solve the problem by guessing the answer and showing that our guess solves this equation. Consider the ansatz $\phi(\vec{r}) = |\bar{\phi}| e^{i\theta(\vec{r})}$, i.e, take the order parameter to have the preferred magnitude $|\bar{\phi}|$ at each point in space, but an unknown and possibly spatially varying angle $\theta(\vec{r})$. For $\theta(\vec{r}) = $ a constant, this expression represents an absolute minimum of the free energy. In our case, we would like to find a solution subject to the boundary conditions, $\theta(\vec{r}) \to \pm\theta_0$ as $x \to \pm L/2$. In Problem 7.1, you find the solution to this equation subject to the requisite boundary conditions:

$$\theta(\vec{r}) = \theta_0\left[\frac{2x}{L}\right]. \tag{7.50}$$

The angle of the order parameter linearly interpolates between the left and right halves of the sample, as illustrated in Figure 7.4C. Importantly, the character of this solution is entirely different from the domain wall we found for the Ising ferromagnet. There, we found a domain wall of width ξ, which arose through balancing the gradient energy, which favors a broad domain wall, and the potential energy, which favors a particular magnitude of the magnetization and so is minimized when the wall is as narrow as possible. In the XY case, we found a solution in which the magnitude of the order parameter is everywhere equal to its equilibrium value, so there is only a gradient energy. Consequently, to minimize the gradient energy, the domain wall expands to be as broad as it can be—the order parameter interpolates smoothly between its asymptotic values over the entire width of the system, no matter how wide it may be. Domain walls are not expected in states with spontaneously broken continuous symmetries.

*Systems with an "almost" continuous broken symmetry

One can get a fuller understanding of the difference between discrete and continuous symmetries by considering a problem with an "almost" continuous symmetry. This is a more complicated problem than any we have yet encountered. It is more properly associated with the study of the behavior of systems near a multicritical point.

   To have a concrete example, consider the case of the 2-state clock model in Eq. 6.14, and suppose that the only term that breaks the XY symmetry is the leading-order (quadratic) term α', i.e., we set $v' = u' = 0$. Thus, the Landau-Ginzburg free energy for $\phi(\vec{r}) = |\phi(\vec{r})|e^{i\theta(\vec{r})}$ is

$$\mathcal{F}_{LG} = \frac{\kappa}{2}\left[\left\|\vec{\nabla}|\phi|\right\|^2 + |\phi|^2\left|\vec{\nabla}\theta\right|^2\right] + \frac{\alpha}{2}|\phi|^2 - \frac{\alpha'}{2}|\phi|^2\cos(2\theta) + \frac{u}{4}|\phi|^4. \quad (7.51)$$

Since we wish to study order parameter textures in the broken-symmetry phase, we will consider the case that $\alpha < 0$ (so we are in an ordered phase) and $|\alpha| \gg \alpha' > 0$, corresponding to a weak preference for the order to point in the $\hat{e}_1$ orientation, i.e., for $\theta = 0$ or π. Clearly, the free energy is minimized by $\phi(\vec{r}) = \pm\bar{\phi}$, with $\bar{\phi} = \sqrt{u/(|\alpha| + |\alpha'|)}$.

   Now we look for the form of $\phi(\vec{r})$ that mimimizes $\mathcal{F}$ while satisfying the boundary conditions $\phi(\vec{r}) \to \pm\bar{\phi}$ as $x \to \pm\infty$. This problem is somewhat more complicated than the ones we have treated so far, but it is sufficient for our purposes to solve it approximately. Let us look for the best possible solution of the form $\phi(\vec{r}) = \bar{\phi}e^{i\theta(\vec{r})}$ subject to the boundary condition, $\theta(\vec{r}) \to 0$ as $x \to +\infty$ and $\theta \to \pi$ as $x \to -\infty$. This form has the advantage that in the limit $\alpha' \to 0$ it corresponds to what we already have found for the pure XY case. Minimizing $\mathcal{F}_{LG}$ with respect to $\theta(\vec{r})$ leads to the equation

$$-\nabla^2\theta + m^2\sin(2\theta) = 0 \quad \text{with} \quad m^2 = \frac{\alpha'}{\kappa}. \quad (7.52)$$

This is a famous equation—called the sine-Gordon equation—and the domain wall solution is known. Again, rather than deriving it, we can look up the answer and then verify for ourselves (Problem 7.2) that this indeed is a solution of Eq. 7.52:

$$\theta(\vec{r}) = 2\arctan\left(e^{-x/\ell}\right) \quad \text{with} \quad \ell^{-1} = \sqrt{2}m. \quad (7.53)$$

We see that $\theta \approx \pi$ for $-x/\ell \gg 1$, $\theta \approx 0$ for $x/\ell \gg 1$, and θ interpolates smoothly between those values for $|x| \sim \ell$. As the approximate XY symmetry becomes increasingly close, the domain wall width diverges as $\ell \sim |\alpha'|^{-1/2}$.

We can understand this result heuristically: over the width of the domain wall, the potential term in the free energy density is larger than in the uniform state by an amount of order $\alpha' \bar{\phi}^2$, while the cost in gradient energy is of order $\kappa \bar{\phi}^2 (\pi/\ell)^2$. Integrating this over the full width ℓ of the domain wall gives a total free energy cost (per unit cross-sectional area of the domain wall) of $\sim \ell \left[\kappa \bar{\phi}^2 (\pi/\ell)^2 + \alpha' \bar{\phi}^2 \right]$, which is minimized when $\ell \sim \sqrt{\kappa/\alpha'}$, giving rise to a surface tension $\sigma \sim \bar{\phi}^2 \sqrt{\kappa \alpha'}$. Indeed, if we use the approximate solution from Eq. 7.53 to evaluate the surface tension obtained by integrating the free energy cost from Eq. 7.51 over all space, we obtain

$$\sigma = 4|\alpha'|\bar{\phi}^2 \ell = 2\sqrt{2} \frac{u\sqrt{|\alpha'|\kappa}}{(|\alpha| + |\alpha'|)}, \tag{7.54}$$

indicating that the surface tension vanishes as $\alpha' \to 0$.

7.5.3   Vortices in the XY Model

The domain walls we have been studying are a form of topological order parameter texture. What this means is the following: in the ordered phase of an Ising ferromagnet, if we are told that the magnetization in one region of space is up and that it is down in another, there must be (at least) one domain wall between the two regions. The precise location of the domain walls depends on the sample history, and the domain wall width and surface tension depend on all sorts of details of the system in question. However, it is an unavoidable fact about Ising order that one must cross an odd number of domain walls along any path from the first region to the second.

However, while domain walls are natural topological textures for a system with a discrete broken symmetry, we have seen that they are unstable in the case of a continuous symmetry such as the XY model. As group theory is the mathematical structure that provides the best characterization of the symmetries of a system, "homotopy" provides the general way to understand topological textures.[4] This is beyond the scope of the present treatment. Rather than develop this general theory, we will consider a particularly important second example.

A system with XY symmetry cannot have a domain wall, but can still have another form of topological texture known as a vortex. The name is evocative of a whirlpool or a tornado—it is a line (1D) object in 3D space with something swirling about it. The structure of the vortex can be found by solving for the order parameter texture $\phi(\vec{R}) = |\phi(\vec{R})|e^{i\theta(\vec{R})}$ that minimizes the free energy (Eq. 7.22) subject to a certain set of boundary conditions. To get a feeling for the correct boundary conditions, look, for example, at the vortex in the upper portion of the right-hand panel of Figure 7.5.

---

4. A review of homotopy as applied to topological textures can be found in *Mermin, Rev. Mod. Phys.* 51, 591 (1979).

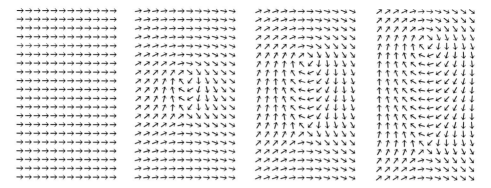

FIGURE 7.5. A pair of vortices. Along a clockwise path around the upper vortex, the change in phase is $+2\pi$, whereas along a clockwise path around the lower vortex, the change in phase is -2π. Along a path which encloses both vortices, the change in phase is zero.

Here, looking down on the system along the z-axis, we see that the XY spins encircle the vortex core in a clockwise sense.

To stabilize such a texture, consider boundary conditions corresponding to the existence of a vortex line along the z-axis. The nature of the vortex is most readily specified in cylindrical coordinates: $\vec{R} = \rho\left[\cos(\gamma)\hat{x} + \sin(\gamma)\hat{y}\right] + z\hat{z}$, where $\rho = \sqrt{x^2 + y^2}$ is the displacement from the z-axis and $\gamma(\vec{\rho}) \equiv \arctan[y/x]$ is the azimuthal angle. Far from the z-axis—i.e., for $\rho \gg \xi$, where $\xi = \sqrt{\kappa/2|\alpha|}$ is correlation length—we will impose the condition $\phi(\vec{R}) = i\bar{\phi}e^{-i\gamma}$. (In vector notation, this corresponds to $\vec{\phi}(\vec{R}) = (\bar{\phi}/\rho)\left[y\hat{x} - x\hat{y}\right]$, which, as in the figure, corresponds to the spins encircling the vortex core in a clockwise manner. We will not use this notation further since it is confusing in the sense that $\vec{R}$ is a vector in 3D real space and $\vec{\phi}$ is a vector in a 2D order parameter space.) Thus, away from the vortex core the system is locally in one of its ground states. Note that although $\gamma(\vec{R})$ is a multi-valued function (mod 2π), this presents no problem since the physical quantity, $\phi(\vec{R})$, is a single-valued function of $\vec{R}$. It is relatively easy to convince oneself that the order parameter texture $\phi(\vec{R})$ that minimizes the LG free energy is translationally invariant in the z direction, has the same angular dependence independent of ρ, and is thus of the form

$$\phi(\vec{R}) = i\bar{\phi}f(\rho/\xi)e^{-i\gamma}, \qquad (7.55)$$

where f is the real-valued solution of a nonlinear differential equation (known as the "nonlinear Schrodinger equation") with the properties

$$f(x) \to 1 \text{ as } x \to \infty, \qquad (7.56)$$

$$f(x) \to 0 \text{ as } x \to 0 .$$

The detailed form of f is of no great qualitative importance, so we will not derive it here. What is significant is that, except in a region of width $\sim \xi$ about the vortex core in which $|\phi|$ is suppressed, the amplitude of the order parameter is approximately constant, $|\phi(\vec{R})| \approx \bar{\phi}$.

This observation permits an approximate approach to the problem that will allow us to treat more general vortex-containing textures. Specifically, if we exclude regions of space associated with vortex cores, we can assume that $|\phi|$ is approximately constant, i.e., $\phi(\vec{R}) \approx \bar{\phi} e^{i\theta(\vec{R})}$. In the context of this approximation,

$$\mathcal{F}_{LG} \approx \mathcal{F}_0 + \frac{\kappa \bar{\phi}^2}{2} |\vec{\nabla}\theta|^2. \tag{7.57}$$

Minimizing this with respect to $\theta(\vec{R})$ leads to the Laplace equation, $\nabla^2\theta = 0$. In a simply connected region of space, there is no single-valued solution to this equation. However, since we have explicitly excluded the regions of vortex cores from consideration, we should imagine we are studying this problem in the (in general, multiply connected) space that excludes regions of size $\sim \xi$ about a set of lines that define the positions of vortices.

The above discussion reminds us of the familiar problem of the magnetic field produced by current carrying wires given by Ampere's law. Firstly, notice that while the phase of the order parameter, $\theta(\vec{R})$, is multivalued,

$$\vec{b}(\vec{R}) \equiv \vec{\nabla}\theta(\vec{R}) \tag{7.58}$$

is a single-valued vector field except at singular points at which $|\phi(\vec{R})| = 0$ (i.e., vortex cores), where θ itself is ill defined. Then, Laplace's equation for $\theta(\vec{R})$ is equivalent to the familiar relation from Maxwell's equations,

$$\vec{\nabla} \cdot \vec{b} = 0. \tag{7.59}$$

However, we also want to impose the condition that the phase changes by 2π upon encircling a vortex. This implies that we also need to satisfy Ampere's law, or in differential form (again in analogy with Maxwell's equations),

$$\vec{\nabla} \times \vec{b} = \vec{j}_v. \tag{7.60}$$

Here $\vec{j}_v(\vec{R})$ is a fictitious "vortex current" that we associate with the vortex line; in the example we have already treated, $\vec{j}_v(\vec{\rho}) = \hat{z}\, \delta^{(2)}(\vec{\rho})$, but this can be generalized to more general vortex configurations consisting of any array of vortex lines.

By taking the divergence of both sides of Eq. 7.60, it is clear that for this equation to be consistent, it is necessary that $\vec{\nabla} \cdot \vec{j}_v = 0$, i.e., that vortex current is conserved. Fortunately, the topological character of the vortex—the fact that the phase winds by 2π along any path that encloses it—implies that this must be the case. Moreover, we see that there is a directionality to the vortex line—in the example we considered, the phase change on any clockwise path about the z-axis is -2π, but we could have looked at solutions with a phase that changed by $+2\pi$—corresponding to $\vec{j}_v(\vec{\rho}) = -\hat{z}\, \delta^{(2)}(\vec{\rho})$—simply by replacing $\gamma(\vec{\rho}) \to -\gamma(\vec{\rho})$ in the above analysis.

More generally, vortices can be characterized by tracing a closed path around a region and tracking the change in phase around the path. Because such a path returns to the same point, $\phi(\vec{R})$ must return to its original value; however, in doing so, the phase can accumulate an overall change $2\pi n$, where n is an integer. If there is a single vortex in the region enclosed by the path, then n is known as the winding number of the vortex. If there are multiple vortices enclosed, n will be the sum of the winding numbers of each vortex. Notably, if there are two vortices with opposite winding numbers, then the phase change around a path enclosing both vortices is zero. The rightmost panel of Figure 7.5 illustrates a pair of vortices, one with $n = 1$ and another with $n = -1$. Viewed from left to right, the panels show how a pair of such vortices can spontaneously emerge from a texture without vortices. Conversely, viewed from right to left, they show how two vortices of opposite winding number can annihilate one another.

The analogy between the XY model and magnetostatics is profound and has many consequences that we will not have time to explore. While there are important differences between the two problems, related to the character of gauge symmetries, dimensionality, etc., the fact that the same mathematics shows up in the two contexts is not accidental. It is a remarkable example of emergence that a system with a spontaneously broken XY symmetry is amenable to a long-distance description in terms of effective magnetic fields and vortex currents.

7.6 Equilibrium Fluctuations

Equation 7.29 is a relation between the connected correlation function, G, and the susceptibility, χ. Here this result has been derived in the context of a mean-field approximation, but the relation between G and χ is an exact identity of the sort already mentioned in Eq. 2.24. This is a remarkable result, given that these are physically distinct quantities: G is a measure of the equilibrium fluctuations of local quantities about their mean, whereas χ tells us about the response of the system to an externally applied perturbation. This relation is a specific instance of a general relation between equilibrium fluctuations (including dynamical fluctuations) and the linear response of a system to an external perturbation (including a time-varying perturbation) known as the "fluctuation-dissipation theorem."

Since we have not treated dynamics, we will only treat the static case, but since this relation is so general it is worth looking a bit more closely into what it means. Consider our good buddy the Ising ferromagnet in the presence of a possibly spatially varying field, h_j. It follows as an algebraic identity starting from the expression for $\langle \sigma_j \rangle$ in terms of the average over microstates, that

$$\frac{\partial \langle \sigma_j \rangle}{\partial h_i} = \beta G_{ij} . \tag{7.61}$$

At the same time, in the limit that $\{h\} \to 0$, the left-hand side of this equation is χ_{ij}. Thus, we obtain Eq. 7.29 not as a mean-field result, but as a fundamental relation between G and $T\chi$. In the continuum limit, the same relation between the equilibrium fluctuations of the order parameter field, $G(\vec{r}) = \langle \phi(\vec{r})\phi(\vec{0})\rangle - [\langle \phi \rangle]^2$, and the corresponding susceptibility can be derived,

$$G(\vec{r}) = T\chi(\vec{r}). \tag{7.62}$$

7.7   Problems

7.1. *No domain wall in the XY model.* Verify that Eq. 7.50 is the solution to Eq. 7.49 with the boundary conditions described in the text.

7.2. *Domain wall with an "almost" continuous symmetry.* Verify that Eq. 7.53 is the solution to the sine-Gordon equation, Eq. 7.52.

7.3. *Correlation function in one dimension.* Show that for $d = 1$, the integral in Eq. 7.25 evaluates to

$$\frac{1}{2\sqrt{\alpha\kappa}}e^{-|r|/\xi} \text{ with } \xi \equiv \sqrt{\kappa/\alpha}. \tag{7.63}$$

7.8   Worksheets for Chapter 7

W7.1   The Meaning of the Correlation Length

As we begin to study how order parameters can vary in space, the concept of the correlation length will repeatedly appear. Let's spend some time to think about this.

1. As a warm-up, let's think back to the exact solution of the 1D Ising model. We had found an expression for the correlation function:

$$\langle \sigma_i \sigma_j \rangle = e^{-|i-j|/\xi}, \quad \xi = |\log(\tanh \beta J)|^{-1}. \tag{3.34}$$

   (a) What is the physical interpretation of $\langle \sigma_i \sigma_j \rangle$?

   (b) Sketch a graph of $\langle \sigma_i \sigma_j \rangle$ as a function of $|i - j|$, and give the intuition behind the shape of the graph.

   (c) Interpret ξ.

2. We will soon be studying a few different situations involving spatial variation in the magnetization, m_j. As a starting point we will use the Ising Hamiltonian

$$H = -J \sum_{\langle ij \rangle} \sigma_i \sigma_j - \sum_j h_j \sigma_j,$$

   and for concreteness, you can have in your mind the picture of a 2D square lattice.

   (a) In this Hamiltonian, each site has its own h_j. What might this represent physically?

(b) Suppose that we are at $T > T_c$, and all the h_j are equal to zero except for $h_i > 0$ on a specific site i. On which sites j would the magnetization m_j be most affected?

(c) Give an interpretation of what is described by the nonlocal susceptibility,

$$\chi_{ji} \equiv \frac{dm_j}{dh_i}.$$

(d) Under what circumstances might χ_{ji} be large? Small? Can you guess how χ_{ji} could depend on i and j? How might the correlation length come into play?

3. The correlation length also plays an important role in justifying the thermodynamic limit.

(a) Let's say we wanted to study a system of size L. Instead of treating it as a finite-size system, let us approximate it as an infinite-size system. Do you expect our approximation to be better if $\xi \ll L$ or if $\xi \gg L$? Why? (There are a couple of reasons here!)

(b) Let's reconsider our expression from part (1). Show that ξ diverges in the limit $T \to 0$ for the 1D Ising system.

(c) What does this mean if we wanted to study our finite-size system at very low T?

W7.2   Correlation Function of the Ising Model with Spatial Variation

To capture situations where the magnetization is not spatially uniform, let us consider the Ising Hamiltonian where there is a separate, local magnetic field for each site j more carefully:

$$H = -J \sum_{\langle ij \rangle} \sigma_i \sigma_j - \sum_j h_j \sigma_j.$$

As before, we'll use a trial Hamiltonian $H_{tr} = - \sum_j b_j \sigma_j$.

1. Let's begin with a short warm-up. In previous chapters, we had considered a spatially uniform field ($h_j = h$ for all j) and we derived the following mean-field equations:

$$b_j = J \sum_{i=\text{n.n. } j} m_i + h \quad \text{for each site } j,$$

where $m_i = \langle \sigma_i \rangle_{tr}$. What is the interpretation of each term in this equation? If we had a spatially varying h_j, what would the equation be?

2. If $T > T_c$ and the h_j are zero on all the sites, then what is m_i?

Let's find out how m_i is affected if h_j is not zero everywhere. To make our lives easier, let's say that h_j is small and that the resulting m_i are also small, so that we can use a linear approximation.

3. Use $m_i = \tanh(b_i/T)$ to eliminate $\{b\}$ from the mean-field equations, and use the expansion $\text{arctanh}(x) = x + x^3/3 + \cdots$ to linearize the equations.

4. Take a derivative with respect to h_k to find a set of linear equations for the nonlocal susceptibility, $\chi_{jk} = dm_j/dh_k$.

5. For later convenience, we'll want to rewrite these equations as

$$\delta_{jk} = (T - T_c)\chi_{jk} + J \sum_i \Delta_{ji}\chi_{ik}$$

so that the sum runs over *all* the lattice sites, not just the nearest neighbors.[5] What should be the definition of Δ_{ji}? (Remember $T_c = zJ$.)

6. Let $\vec{R}_a$ be the position coordinate of lattice site a. Argue why χ_{ij} depends only on $\vec{R}_i - \vec{R}_j$. Show that the same is true of Δ_{ji}.

7. Let $\chi(\vec{r})$ be a function of position where $\chi(\vec{R}_i - \vec{R}_j) = \chi_{ij}$. Define $\Delta(\vec{r})$ similarly. What is the interpretation of $\chi(\vec{r})$? For what values of $\vec{r}$ is $\chi(\vec{r})$ defined?

8. Letting $\vec{R}_k = \vec{0}$, $\vec{R}_j = \vec{r}$, and $\vec{R}_i = \vec{r}\,'$, rewrite your mean-field equations in terms of $\chi(\vec{r})$ and $\Delta(\vec{r} - \vec{r}\,')$.

To proceed further, we'll need some more machinery, the Fourier Transform. Suppose that $\chi(\vec{r})$ can be expressed in the form

$$\chi(\vec{r}) = \int_{\vec{k}} e^{i\vec{k}\cdot\vec{r}} \tilde{\chi}(\vec{k}),$$

where $\int_{\vec{k}}$ means $\frac{1}{(2\pi)^d} \int d^d\vec{k}$. For now, don't worry about the range of $\vec{k}$ we are integrating, or about the normalization of the integral—we will treat this as an ansatz to help us solve for $\chi(\vec{r})$ in the mean-field equations.

9. Substitute this ansatz for $\chi(\vec{r})$, and massage the equation until you get something which looks like $\int_{\vec{k}} e^{i\vec{k}\cdot\vec{r}}[\ldots\ldots] = 0$. Write the resulting expression in terms of $g(\vec{k}) = \sum_{\vec{r}} e^{-i\vec{k}\cdot\vec{r}} \Delta(\vec{r})$.

   Hint 1: $\delta_{\vec{r},0} = \int_{\vec{k}} e^{i\vec{k}\cdot\vec{r}} \cdot 1$.

   Hint 2: You may have to insert $e^{i\vec{k}\cdot\vec{r}} e^{-i\vec{k}\cdot\vec{r}} = 1$ into one of your terms.

10. Since your expression is zero for any $\vec{r}$, it means that the integrand $[\ldots\ldots]$ vanishes. Use this to solve for $\tilde{\chi}(\vec{k})$.

11. Finally, substitute this in your ansatz to arrive at an integral expression for χ_{ij}. Success!

---

5. We also pulled out a $-T_c$ so that the coefficient of the other term vanishes as $T \to T_c$.

We have derived the following expression for the nonlocal susceptibility of an Ising model on a lattice:

$$\chi(\vec{R}_i - \vec{R}_j) = \int_{\vec{k}} \frac{e^{i\vec{k}\cdot(\vec{R}_i - \vec{R}_j)}}{(T - T_c) + Jg(\vec{k})}, \tag{7.64}$$

where $\vec{R}_i$ is the position of lattice site i, $g(\vec{k}) = \sum_{\vec{r}} e^{-i\vec{k}\cdot\vec{r}} \Delta(\vec{r})$, and

$$\Delta(\vec{R}_i - \vec{R}_j) = \begin{cases} z & \text{if} & i = j, \\ -1 & \text{if} & i \text{ and } j \text{ are nearest neighbors}, \\ 0 & & \text{otherwise}. \end{cases} \tag{7.65}$$

The integral cannot be evaluated numerically, but you can argue how it depends on the distance between the lattice sites i and j. For concreteness, let's take a 2D square lattice of spacing a, where

$$\int_{\vec{k}} = \frac{1}{4\pi^2} \int_{-\pi/a}^{\pi/a} \int_{-\pi/a}^{\pi/a} dk_x \, dk_y.$$

1. What is $g(\vec{k})$ for small $|\vec{k}|$?
2. Sketch the integrand without $e^{i\vec{k}\cdot\vec{r}}$ as a function of $|\vec{k}| = k$. (You can use the small $|\vec{k}|$ approximation.) What is the approximate width of the function?
3. Sketch the real part of the numerator. How does this graph depend on $|\vec{r}| = r$?
4. Multiply the two of these together. Sketch a graph of the real part of the resulting function vs. k for the cases where r is large and where r is small. In which case do you expect the integral to be large and in which case small, and why? What does this result imply about how a system responds to local perturbations?
5. How big does r have to be for the integral to be small?

As discussed in the chapter, Ginzburg-Landau theory is an extension of Landau theory where spatial variations are allowed. Instead of considering just the symmetry of the order parameter, we now also have to take into account the spatial symmetry of the system. In this exercise we will demonstrate how to construct the free energy from symmetry arguments alone, using the Ising model as our example.

As before, the free energy of the system is

$$F = \int d\vec{r} \, \mathcal{F}, \tag{7.66}$$

such that the free energy of the system is the volume integral of the free energy density $\mathcal{F}$, which is a local property. When we constructed the free energy density in the previous chapter, we considered only the uniform state. Here, to make the argument as general as possible, we write the free energy density as:

$$\mathcal{F} = \mathcal{F}\left(\vec{r}, \phi(\vec{r}), \partial_i \phi, \partial_{ij}\phi, \dots\right). \tag{7.67}$$

The arguments of the free energy density now reflect the fact that spatial variations are allowed, with explicit dependence on the position $\vec{r}$, and spatially varying order parameter $\phi(\vec{r})$. More importantly (and interestingly), the free energy density can now depend on *spatial derivatives* of the order parameter, denoted by the subscripted partial derivatives.

As before, we assume we are in the vicinity of a phase transition and therefore ϕ is a small quantity. Furthermore, we assume that the spatial variations are slow, such that the spatial derivatives are good quantities to expand on.

1. Given the form of Eq. 7.67, show that the Taylor expansion in terms of the spatial derivatives to second order can be written as

$$\begin{aligned}
\mathcal{F}\left(\vec{r}, \phi(\vec{r}), \partial_i \phi, \partial_{ij}\phi, \dots\right) &= V(\vec{r}, \phi(\vec{r})) \\
&\quad + \sum_i (\partial_i \phi) \nu_i(\vec{r}, \phi(\vec{r})) \\
&\quad + \sum_{ij} (\partial_i \partial_j \phi) \frac{\eta_{ij}(\vec{r}, \phi(\vec{r}))}{2} \\
&\quad + \sum_{ij} (\partial_i \phi)(\partial_j \phi) \frac{\kappa_{ij}(\vec{r}, \phi(\vec{r}))}{2} + \dots,
\end{aligned} \tag{7.68}$$

where

$$\mathcal{F}(\vec{r}, \phi(\vec{r}), 0, 0, \dots) = V(\vec{r}, \phi(\vec{r}))$$

and $\nu_i, \eta_{ij}, \kappa_{ij} \dots$ are general coefficients of expansion (the factors of $1/2$ are included by convention). This is sometimes called a gradient expansion.

2. Given that the system has translational symmetry,

$$\phi(\vec{r}) = \phi(\vec{r} + \vec{R}) \tag{7.69}$$

for any $\vec{R}$, prove that there is no explicit spatial dependence on $\vec{r}$ in any of the terms, i.e.,

$$\mathcal{F}\left(\vec{r}, \phi(\vec{r}), \partial_i \phi, \partial_{ij}\phi, \dots\right) = \mathcal{F}\left(\phi(\vec{r}), \partial_i \phi, \partial_{ij}\phi, \dots\right) \tag{7.70}$$

or, more explicitly,

$$V(\vec{r}, \phi(\vec{r})) \equiv V(\phi(\vec{r})), \qquad \nu_i(\vec{r}, \phi(\vec{r})) \equiv \nu_i(\phi(\vec{r})),$$

$$\eta_{ij}(\vec{r}, \phi(\vec{r})) \equiv \eta_{ij}(\phi(\vec{r})), \qquad \kappa_{ij}(\vec{r}, \phi(\vec{r})) \equiv \kappa_{ij}(\phi(\vec{r})). \qquad (7.71)$$

3. Assume that the system has hypercubic spatial symmetry (e.g., a square lattice in 2D or a cubic lattice in 3D), meaning that rotation by $\pi/2$ along any primary axis does not change the free energy of the system. Eliminate all unnecessary terms of spatial derivative in your previous expression to show that:

$$F = \int d^d\vec{r} \left\{ V(\phi(\vec{r})) + |\nabla\phi|^2 \frac{\kappa(\phi(\vec{r}))}{2} + \cdots \right\}. \qquad (7.72)$$

Hint: This part can be intuitively quick but formally tedious. Don't be afraid to peek at the solutions for the formal justification!

Hint: You might need the vector identity $\nabla \cdot (f\nabla g) = f\nabla^2 g + \nabla f \cdot \nabla g$.

4. Assume that ϕ is small, and that our system obeys the symmetry

$$F[\phi] = F[-\phi]. \qquad (7.73)$$

Expand the first term in the expression B.7.72 to obtain the final Ginzburg-Landau free energy:

$$F[\phi] = \int d^d\vec{r} \left\{ V_0 + \frac{\alpha}{2}\phi(\vec{r})^2 + \frac{u}{4}\phi(\vec{r})^4 + \frac{\kappa}{2}\left[\nabla\phi(\vec{r})\right]^2 + \cdots \right\}. \qquad (7.74)$$

W7.5   *Solving for the Continuum Correlation Length

Let us add to the Landau-Ginzburg free energy we have just derived a term that represents an external perturbation $-h(\vec{r})\phi(r)$. For example, this can be an externally applied magnetic field h coupled to the magnetization so that the free energy respects a global symmetry $F[\phi; h] = F[-\phi; -h]$. The total free energy is now:

$$F[\phi; h] = \int d^d\vec{r} \left\{ V_0 - h(\vec{r})\phi(\vec{r}) + \frac{\alpha}{2}\phi(\vec{r})^2 + \frac{u}{4}\phi(\vec{r})^4 + \frac{\kappa}{2}\left[\nabla\phi(\vec{r})\right]^2 + \cdots \right\}$$

$$= \int d^d\vec{r}\,\mathcal{F}\left[\phi(\vec{r}), \nabla\phi(\vec{r}); h(\vec{r})\right]. \qquad (7.75)$$

1. We are trying to minimize the free energy F with respect to the order parameter ϕ. Since $F[\phi; h]$ is a functional of $\phi(\vec{r})$, the minimization entails finding the functional form of ϕ, which calls for the use of Euler-Lagrange equation

$$\frac{\partial\mathcal{F}}{\partial\phi} - \nabla \cdot \frac{\partial\mathcal{F}}{\partial(\nabla\phi)} = 0, \qquad (7.76)$$

where $\partial/\partial\nabla\phi$ means the partial derivative w.r.t. each component of the spatial derivative $\nabla\phi = (\partial_x\phi, \partial_y\phi, \ldots)$, and is therefore a vector. If you need a recap on the calculus of variations, see Appendix A.4.

Show that the differential equation for ϕ that minimizes F is:

$$\alpha\phi + u\phi^3 - \kappa\nabla^2\phi = h(\vec{r}). \tag{7.77}$$

Correlation length measures the spatial extent of the response to a local perturbation, so we cannot assume a spatially uniform ϕ anymore. We can model a small local perturbation at $\vec{r}_0$ of strength h_0 with a field $h = h_0\delta(\vec{r} - \vec{r}_0)$ where h_0 is small.

2. With out loss of generality, assume that $\vec{r}_0 = 0$ such that the perturbation is at the origin. Use Eq. 7.77 to derive a form of ϕ for both $T > T_c$ and $T < T_c$.

   Hint 1: Assume that the order parameter takes the form

$$\phi(\vec{r}) = \phi_0 + \tilde{\phi}(\vec{r}), \tag{7.78}$$

   where the spatial variation is a small perturbation from the uniform solution ϕ_0 (i.e., when $h = 0$).

   Hint 2: The differential equation involving a Laplacian can be most easily solved by Fourier transforms. See Appendix A.6 for some more help.

3. Give a definition of the correlation length from your answer (or the solution!) to the first part, and show that the correlation length ξ follows:

$$\xi \propto |T - T_c|^{-\nu}, \tag{7.79}$$

   where $\nu = 1/2$ for both $T > T_c$ and $T < T_c$.

4. From your answer to the previous part, show that there is a critical ratio:

$$\frac{\xi(T \to T_c^+)}{\xi(T \to T_c^-)} = \sqrt{2}. \tag{7.80}$$

5. A more subtle critical exponent governs the nature of the thermodynamic fluctuations at criticality ($T = T_c$). Away from criticality, as we've just seen, the correlations fall exponentially with distance. A key feature of a system *at* criticality is that it has no characteristic lengthscale—it is fractal. As a consequence, correlation functions can only fall as a power of distance:

$$\langle\phi(\vec{r})\phi(\vec{r}')\rangle \propto |\vec{r} - \vec{r}'|^{-(d-2)+\eta} \text{ at } T = T_c, \tag{7.81}$$

   where η is yet another critical exponent, which in the present case takes the value $\eta = 0$. Show that this is the case by evaluating this for $d = 3$.

W7.6   Susceptibility of the XY Model for $T < T_c$

Suppose we have a system of XY spins. There is an intruiging piece of physics about the way in which the spins respond to an externally applied field: below T_c, the susceptibility depends on the direction in which the field is applied! This worksheet gives you the chance to see why this is the case.

1. Let's start by considering the simpler case of an Ising order parameter, ϕ. Assume that everything is spatially uniform for now.

   (a) Suppose that $T > T_c$. What is ϕ if there are no applied fields? If we apply a small field of magnitude h, how does ϕ respond, in terms of χ?

   (b) Now consider $T < T_c$. The equilibrium value of $\phi = \bar{\phi}$ is nonzero. What is the new equilibrium value, in terms of χ, if we apply a small field h?

2. Let's see what is different for the case of a two-component XY order parameter, $\phi = \phi_1 \hat{\mathbf{e}}_1 + \phi_2 \hat{\mathbf{e}}_2$. Remember that the spin can point in any direction in an abstract two-dimensional plane.

   (a) Write down the expression for the energy in the presence of an externally applied field, $\mathbf{h} = h_1 \hat{\mathbf{e}}_1 + h_2 \hat{\mathbf{e}}_2$.

   (b) If $T > T_c$, do you expect the susceptibility to depend on the direction of the applied field? Write an expression for ϕ in the presence of a small field, $\mathbf{h}$, in terms of χ.

   (c) Now consider $T < T_c$. What is the equilibrium order parameter in terms of a magnitude $|\phi| = \bar{\phi}$ and an arbitrary angle $0 \le \theta < 2\pi$?

   (d) Without loss of generality, assume that $\theta = 0$ so that ϕ points in the $\hat{\mathbf{e}}_1$ direction. Let's apply a weak magnetic field—we can apply it in either the $\hat{\mathbf{e}}_1$ or the $\hat{\mathbf{e}}_2$ direction. Do you expect ϕ to respond the same in both cases? Why or why not? *Hint*: Are the $\hat{\mathbf{e}}_1$ and $\hat{\mathbf{e}}_2$ directions equivalent?

Let $\hat{\mathbf{e}}_\| = \cos(\theta)\hat{\mathbf{e}}_1 + \sin(\theta)\hat{\mathbf{e}}_2$ be a unit vector in the direction of the order parameter, $\phi = \bar{\phi}\hat{\mathbf{e}}_\|$. Anything *parallel* to $\hat{\mathbf{e}}_\|$ is called longitudinal, and anything *perpendicular* to it is transverse. Let $\chi_\|$ and $\chi_\perp$ denote the longitudinal and transverse susceptibility, respectively.

3. We can think of our magnetic field as having a longitudinal and a transverse component, $\mathbf{h} = h_\| \hat{\mathbf{e}}_\| + h_\perp \hat{\mathbf{e}}_\perp$ (where $\hat{\mathbf{e}}_\perp$ is an in-plane unit vector perpendicular to $\hat{\mathbf{e}}_\|$, i.e. $\vec{\mathbf{e}}_\perp = -\sin(\theta)\vec{\mathbf{e}}_1 + \cos(\theta)\vec{\mathbf{e}}_2$).

   (a) Write down the expression for ϕ in the presence of small $\mathbf{h}$, in terms of $\chi_\|$ and $\chi_\perp$.

   (b) What do you think is larger, $\chi_\|$ or $\chi_\perp$? *Hint:* How is the magnitude/direction of the order parameter affected by a longitudinal versus a transverse perturbation?

Now that we understand what longitudinal and transverse susceptibility mean, let's explicitly calculate $\chi_\perp(\vec{r})$ and $\chi_\|(\vec{r})$ as functions of position. For T slightly below T_c we can use a Ginzburg-Landau form of the free energy,

$$\mathcal{F}_{var}[\phi] = \int d\vec{r} \left[V_0 + \frac{\alpha}{2}\phi^2 + \frac{u}{4}\phi^4 + \frac{\kappa}{2}|\nabla\phi|^2 - \mathbf{h}\cdot\phi \right],$$

where $\phi^2 = \phi_1^2 + \phi_2^2$, $\phi^4 = (\phi^2)^2$, and $|\nabla\phi|^2 = \sum_\alpha |\nabla\phi_\alpha|^2$. (Remember $\alpha < 0$ since we're below T_c). If $\mathbf{h}(\vec{r}) = 0$, then $\phi(\vec{r})$ is spatially uniform with magnitude $\bar{\phi} = \sqrt{|\alpha|/u}$. Without loss of generality, let's assume the ordering is in the $\theta = 0$ direction, so $\bar{\phi} = \bar{\phi}\hat{\mathbf{e}}_1$.

4. If we apply a weak magnetic field, then the order parameter shouldn't change much, so we can use an ansatz,

$$\phi(\vec{r}) = \bar{\phi} + \delta\phi(\vec{r}),$$

where $\delta\phi = \delta\phi_1\hat{\mathbf{e}}_1 + \delta\phi_2\hat{\mathbf{e}}_2$ is the response to the perturbation. Which component describes the longitudinal response? The transverse response?

5. Write down expressions for ϕ^2, ϕ^4, and $|\nabla\phi|^2$, keeping terms up to quadratic order in $\delta\phi_1$ and $\delta\phi_2$.

6. Substitute these expressions into the free energy and show that

$$\mathcal{F}_{var}[\phi] = \mathcal{F}_{var}[\bar{\phi}] + \int d\vec{r} \left[|\alpha|(\delta\phi_1)^2 + \frac{\kappa}{2}|\nabla\delta\phi_1|^2 - h_1\delta\phi_1 \right]$$

$$+ \int d\vec{r} \left[\frac{\kappa}{2}|\nabla\delta\phi_2|^2 - h_2\delta\phi_2 \right].$$

7. What does the lack of a $(\delta\phi_2)^2$ term imply about the nature of the transverse susceptibility? Does this agree with your qualitative assessment earlier?

W7.7   A Domain Wall in an Ising System

So far, in a phase with broken Ising symmetry, we have assumed that one of the two states is chosen across the entire extent of the system. However, there can be situations where different regions of the system occupy different broken-symmetry states. These regions are known as domains, and the boundaries between them are known as domain walls.

1. Consider an Ising system for $T < T_c$ in a continuum limit where the state of the system is described by an order parameter field, ϕ. Suppose that
   - far to the left there is a slight negative magnetic field,
   - far to the right there is a slight positive magnetic field, and
   - around the origin there are no magnetic fields.

Do you expect ϕ to be positive or negative for $x \ll 0$? How about for $x \gg 0$? Sketch a graph of what you expect $\phi(x)$ to look like.

2. If T is not too far below T_c, then the free energy of the system has a Landau-Ginzberg form,

$$\mathcal{F}[\phi] = \int d\vec{r} \left[\frac{\alpha}{2} \phi^2 + \frac{u}{4} \phi^4 + \frac{\kappa}{2} \left(\frac{\partial \phi}{\partial x} \right)^2 \right], \tag{7.82}$$

where $\alpha = T - T_c < 0$. Let $\bar{\phi} = \sqrt{-\alpha/u}$. What is $\lim_{x \to \infty} \phi(x)$? How about $\lim_{x \to -\infty} \phi(x)$?

3. Suppose that our domain wall has a width W.

   (a) Label W in your sketch from part (1).

   (b) Give an argument for why W cannot be too small.

   (c) Give an argument for why W cannot be too large.

4. Let's come up with a rough estimate for W. In this "calculation" you are free to drop any constant factors; we just want to find out how W depends on the parameters α, u, and κ.

   (a) Sketch $\frac{\alpha}{2} \phi^2 + \frac{u}{4} \phi^4$. What is the height of the "free energy barrier" to cross from one minimum to the other?

   (b) The free energy cost of a domain wall is the difference in free energy between a state without a domain wall and a state with a domain wall. What is the cost that arises from the first two terms in Eq. 7.82, $\frac{\alpha}{2} \phi^2 + \frac{u}{4} \phi^4$?

      Hint: To estimate an integral, multiply a typical value of the integrand by the range of x where it is nonzero.

   (c) Now let's look at the last term. What is $\frac{\partial \phi}{\partial x}$ roughly?

   (d) Determine, up to a constant, the free energy cost of the domain wall arising from the last term in the free energy.

   (e) Estimate W by minimizing the sum of parts (b) and (d).

5. You can actually arrive at this same result from just dimensional analysis. Let E denote the dimension of free energy, M the dimension of ϕ, and L the dimension of length.

   (a) What are the dimensions of α, u, and κ?

   (b) What is the only combination that gives dimension of length?

8 *THE SELF-CONSISTENT GAUSSIAN APPROXIMATION

Dimension matters.

So far, while some of our discussion has been more general, our perspective has been rooted in a mean-field description of phases of matter. At this level of approximation the problem looks pretty similar independent of the details of the microscopic model (within certain broad limits) and even the spatial dimension. For instance, the critical exponents that we have found appear to be the same in all cases. In this chapter we will begin to refine our understanding of these problems using a slightly more sophisticated form of mean-field theory. Since the central feature of a mean-field analysis is to relate the unsolvable problem that we wish to understand to a solvable problem that in some way "best" represents the same physics, the first step in this process is to analyze a somewhat richer class of solvable problems.

Having familiarized ourselves with the sorts of mathematical manipulations that underlie our analysis, we will omit some of the intermediate algebraic steps in the remaining chapters—trusting that these can be reconstructed by the interested reader.

8.1 The Gaussian Model

The Gaussian model is a model in which the Hamiltonian is a quadratic function of the microscopic degrees of freedom, or in other words, a model in which the Boltzmann weight is Gaussian. To begin with, let us consider a lattice model in which we have a real field, $\phi(\vec{R})$, defined on each site. We will consider the case in which there is a global Ising symmetry such that the Hamiltonian is invariant if we replace all the fields by $\phi(\vec{R}) \to -\phi(\vec{R})$. We will also assume that there is translational symmetry in the sense that each lattice site is equivalent to every other lattice site, i.e., the Hamiltonian is invariant if we replace all fields by $\phi(\vec{R}) \to \phi(\vec{R} + \vec{e}_j)$, where $\vec{e}_j$ is any vector that connects a pair of nearest-neighbor sites. (For this to be an exact symmetry, we need to imagine periodic boundary conditions for our system.)

To be explicit, we can consider this model with short-range interactions

$$H = \frac{J}{2} \sum_{\langle \vec{R}, \vec{R}' \rangle} \left[\phi(\vec{R}) - \phi(\vec{R}') \right]^2 + \frac{\mu}{2} \sum_{\vec{R}} \phi^2(\vec{R}), \qquad (8.1)$$

where the sum over $\langle \vec{R}, \vec{R}' \rangle$ is over all pairs of nearest-neighbor sites. This Hamiltonian is bounded below provided that $\mu \geq 0$ and $2zJ \geq -\mu$, where z is the number of nearest neighbors of a single site. To simplify the discussion, we will further assume that $J > 0$.

Now the partition function can be defined in terms of an integral over all configurations of the fields:

$$Z = \int D\phi \, \exp[-\beta H], \qquad (8.2)$$

where

$$\int D\phi \equiv \prod_{\vec{R}} \int_{-\infty}^{\infty} d\phi(\vec{R}), \qquad (8.3)$$

and various correlations can be defined similarly, for instance,

$$G(\vec{R}, \vec{R}') \equiv \langle \phi(\vec{R})\phi(\vec{R}') \rangle = Z^{-1} \int D\phi \, \phi(\vec{R})\phi(\vec{R}') \, \exp[-\beta H]. \qquad (8.4)$$

This is a lot of Gaussian integrals. If there are V sites in our system (before we take the thermodynamic limit), then there are V Gaussian integrals to perform. (V is an integer, but we can think of it as the "volume" of our system.) To perform these integrals, we first make a change of variables into a Fourier transformed form, which takes advantage of the translation invariance:

$$\phi(\vec{R}) = V^{-1/2} \sum_{\vec{k}} e^{i\vec{k}\cdot\vec{R}} \phi_{\vec{k}}, \qquad (8.5)$$

where

$$\phi_{\vec{k}} = V^{-1/2} \sum_{\vec{R}} e^{-i\vec{k}\cdot\vec{R}} \phi(\vec{R}). \qquad (8.6)$$

Here $\phi_{\vec{k}}$ is the amplitude of the plane wave component with wavevector $\vec{k}$. In going between the "position" representation and the "wavevector" representation, we have made use of the fundamental theorem of Fourier transform:

$$\sum_{\vec{k}} e^{i\vec{k}\cdot\vec{R}} = V\delta_{\vec{R},\vec{0}}. \qquad (8.7)$$

There are a few details of this transformation that we need to get straight, as discussed explicitly in Appendix A.3. One point that is important is that the number of distinct $\vec{k}$ points must be the same as the number of $\vec{R}$ points—going to Fourier space cannot change the number of independent degrees of freedom. Consequently, the values of $|\vec{k}|$ over which we sum must be bounded to a range of $\vec{k}$ space called "the first Brillouin zone" (BZ).[1]

---

1. To see how this arises, note that it is the set of factors, $e^{i\vec{k}\cdot\vec{R}}$, rather than the vectors $\vec{k}$ themselves that enter the Fourier transform. Thus, for any $\vec{G}$ (known as a reciprocal lattice vector) such that $e^{i\vec{G}\cdot\vec{R}} = 1$

It still may seem that the Fourier description involves a doubling of the number of degrees of freedom since the amplitudes $\phi_{\vec{k}}$ are in general complex, i.e., it has a real and an imaginary part. However, the condition that $\phi(\vec{R})$ is real for all $\vec{R}$ is equivalent to the requirement that $\phi_{\vec{k}} = \phi_{-\vec{k}}^{\star}$. Having realized this, it is straightforward to show that

$$\int D\phi \equiv \prod_{\vec{k}}' \int_{-\infty}^{\infty} d\phi_{\vec{k}}' d\phi_{\vec{k}}'', \tag{8.9}$$

where $\prod_{\vec{k}}'$ means the product over all $\vec{k}$ in the first BZ with the identification $\vec{k} \leftrightarrow -\vec{k}$, and ϕ' and ϕ'' refer to the real and imaginary parts of ϕ.

The advantage of this choice of coordinates is that H now can be expressed as a sum of independent modes, without any cross-terms between modes of differing $\vec{k}$:

$$H = \frac{1}{2} \sum_{\vec{k}}' \left[Jg(\vec{k}) + \mu \right] \left| \phi_{\vec{k}} \right|^2. \tag{8.10}$$

Because there are no interactions in the Fourier basis, the various thermodynamic quantities can be found in a straightforward way:

$$Z = \prod_{\vec{k}} \left[\frac{T\pi}{Jg(\vec{k}) + \mu} \right]^{1/2} \tag{8.11}$$

with

$$g(\vec{k}) \equiv \sum_{\vec{R} = \text{n.n.} \vec{0}} \left[1 - e^{\vec{k} \cdot \vec{R}} \right], \tag{8.12}$$

and

$$G(\vec{R}, \vec{R}') = \frac{1}{V} \sum_{\vec{k}}' \frac{e^{i\vec{k} \cdot (\vec{R} - \vec{R}')}}{Jg(\vec{k}) + \mu} \tag{8.13}$$

Finally, taking the thermodynamic limit, we find

$$F = \frac{T}{2} \int \frac{d\vec{k}}{(2\pi)^d} \ln \left[\frac{Jg(\vec{k}) + \mu}{\pi T} \right]. \tag{8.14}$$

---

for all lattice vectors $\vec{R}$, it follows that $\vec{k}$ and $\vec{k} + \vec{G}$ are equivalent. For instance, on a cubic lattice with unit lattice constant, any vector of the form $\vec{G} = 2\pi n \hat{e}_j$ is a reciprocal lattice vector. The result is that, to avoid redundancy, we can restrict the sum over $\vec{k}$ to the points that are closer to the origin, $\vec{0}$, than to any non-zero $\vec{G}$. It is straightforward to prove that for a system with periodic boundary conditions, the number of $\vec{k}$ points in the first BZ is equal to the number of lattice points,

$$V \equiv \sum_{\vec{R}} 1 = \sum_{\vec{k} \in 1^{st} BZ} 1. \tag{8.8}$$

and

$$G(\vec{R}, \vec{R}') = T \int \frac{d\vec{k}}{(2\pi)^d} \frac{e^{i\vec{k}\cdot(\vec{R}-\vec{R}')}}{Jg(\vec{k}) + \mu} \equiv G(\vec{R} - \vec{R}'), \tag{8.15}$$

where the integral is over the first BZ and d is the number of spatial dimensions.

We can also ask what happens if we perturb the Gaussian model with a small external field, h,

$$H \to H - \sum_{\vec{R}} h(\vec{R})\phi(\vec{R}). \tag{8.16}$$

It is a feature of the Gaussian model that the response to such a field can be computed exactly. To see this, we first rewrite the Hamiltonian in terms of the Fourier transform of the symmetry-breaking field, such that

$$\sum_{\vec{R}} h(\vec{R})\phi(\vec{R}) = \sum_{\vec{k}} h_{\vec{k}}\phi_{-\vec{k}}. \tag{8.17}$$

Now we see that the integrals needed to perform various thermodynamic averages are simply Gaussian integrals with shifted origins. We can handle this by defining new integration variables, $\phi_{\vec{k}} = \bar{\phi}_{\vec{k}} + \delta\phi_{\vec{k}}$, where

$$\bar{\phi}_{\vec{k}} = [Jg(\vec{k}) + \mu]^{-1} h_{\vec{k}}, \tag{8.18}$$

and then expressing the various integrals over fields as integrals over $\delta\phi_{\vec{k}}$. The result is that the integrals over $\delta\phi_{\vec{k}}$ are the same as the previous integrals were for $h = 0$, and thus

$$F[h] = F[0] - \frac{1}{2} \sum_{\vec{k}} \frac{|h_{\vec{k}}|^2}{[Jg(\vec{k}) + \mu]}, \tag{8.19}$$

$$\langle \phi(\vec{R}) \rangle = \bar{\phi}(\vec{R}), \tag{8.20}$$

where $\bar{\phi}(\vec{R})$ is the Fourier transform of $\bar{\phi}_{\vec{k}}$, and

$$\langle \phi(\vec{R})\phi(\vec{R}') \rangle = \bar{\phi}(\vec{R})\bar{\phi}(\vec{R}') + G(\vec{R} - \vec{R}'). \tag{8.21}$$

Among other things, this means that the uniform susceptibility, defined from the relation $\langle \phi \rangle = \chi h$, is given as

$$\chi = 1/\mu. \tag{8.22}$$

Note that the susceptibility is always finite but diverges as $\mu \to 0$. This suggests that we will want to associate this limit with a system approaching a phase transition.

8.1.1 The Nature of the Solution

We have solved this problem. However, before we use this solution for all sorts of things, we need to understand what it means.

First, although the Gaussian model includes interactions between fields on neighboring sites, it still corresponds to a noninteracting problem: each of the plane wave components $\phi_{\vec{k}}$ independently contributes to the energy. Furthermore, the Hamiltonian is quadratic in the magnitude of each of the modes. Each mode can be treated as a simple harmonic oscillator. The "stiffness" of a mode with wave vector $\vec{k}$ is $Jg(\vec{k}) + \mu$. The long wavelength modes (small $|\vec{k}|$) are more easily excitable than the short-wavelength modes (large $|\vec{k}|$). This is due to the energetic penalty from the unequal magnitude of ϕ on neighboring sites. In the limit of very long wavelength excitations ($k \to 0$), there is still an energy per unit site of $\mu|\phi|^2/2$. In the limit $\mu \to 0$, the amplitude of these small-k excitations diverges.

It is already clear from Eq. 8.14 that this model has no phase transitions as a function of T; F is manifestly an analytic function of T so long as $T > 0$. Moreover, it is easy to see that the phase in question does not break any symmetries of the underlying problem—this is a model that has only a high-temperature (fully symmetric) phase. For instance, as shown in Eq. 8.15, the correlation function G is translationally invariant in that it depends only on the distance between sites, $\vec{R} - \vec{R}'$.

In general, G depends on details of the lattice structure as encoded in $g(\vec{k})$. But the long-distance behavior of these fluctuations is much less sensitive to details. If we consider the integral expression in Eq. 8.15 for large $\vec{R} - \vec{R}'$, it is clear that the contributions from everything other than small $\vec{k}$ are negligible due to the rapid oscillation of the exponential factor. Thus, to obtain an understanding of how correlations decay at long distances, we can use the Taylor series expansion of g at small $\vec{k}$. Since g is an even function of $\vec{k}$ and $g(\vec{0}) = 0$, we can approximate g as $g(\vec{k}) = k_a g^{a,b} k_b + \cdots$, where summation convention is assumed, $\ldots$ refers to terms of order k^4 and higher, and $g^{a,b}$ is a real, symmetric matrix, obtained from the second partial derivatives of $g(\vec{k})$ evaluated at $\vec{k} = \vec{0}$. It then follows that this matrix can be diagonalized by choosing an appropriate set of basis vectors (which in fact correspond to the principal axes of the lattice involved), in terms of which,

$$g(\vec{k}) = \sum_a g_a k_a^2 + \cdots . \tag{8.23}$$

We therefore conclude that for large $|\vec{R}|$,

$$G(\vec{R}) \sim \frac{T}{J} \int \frac{d\vec{k}}{(2\pi)^d} \frac{e^{i\vec{k}\cdot\vec{R}'}}{k^2 + \tilde{\mu}}, \tag{8.24}$$

where $\tilde{\mu} = \mu/J$ and $\vec{R}' \equiv \sum_a g_a^{-1/2} R_a \hat{e}_a$ is the displacement vector rescaled appropriately to remove the anisotropy from Eq. 8.23.

In other words, at long distances, the correlation function has a universal form independent of details of the lattice—it is the Fourier transform of a simple Lorentzian, which it to say it has the Ornstein-Zernicke form derived in Section 7.3.2. In fact, with a little thought, you can convince yourself that this result is extremely general—even if we had included more complicated interactions between further distant sites (so long as it were not too long-ranged)—we would get the same asymptotic form of this correlation function.

We also want to think about the short-distance behavior of correlations in the Gaussian model. Specifically, the mean squared fluctuation of ϕ, $\langle \phi^2 \rangle = G(\vec{0})$, is

$$G(\vec{0}) = \frac{T}{J} \int \frac{d\vec{k}}{(2\pi)^d} \frac{1}{g(\vec{k}) + \tilde{\mu}}. \tag{8.25}$$

This is a decreasing function of μ, which depends on details of the lattice structure through the $\vec{k}$ dependence of g. However, there is an interesting limit, which is the limit as $\mu \to 0$. From our expression for the correlation length ξ, we can see that in this limit the correlation length diverges. This can be thought of as some approximate version of what happens as we approach a critical point. For $\mu = 0$, the integrand in Eq. 8.25 diverges at $\vec{k} = \vec{0}$. However, the effect of this is very different depending on the dimension. For $d > 2$, this is an integrable divergence—the integral is finite despite the fact that the integrand diverges. On the other hand, for $d \leq 2$, the integral diverges. We will see that this difference has major implications concerning the difference between critical phenomena in low dimensions and high dimensions.

In preparation for the uses we will make of the Gaussian model, let us explore the asymptotic form of $G(\vec{0})$ as $\mu \to 0^+$ in different dimensions. To be on familiar territory, let us start by considering the case of $2 < d < 4$. It is simple algebra to write

$$\beta J G(\vec{0}) = A_d - B_d(\tilde{\mu}), \tag{8.26}$$

where

$$A_d = \int \frac{d\vec{k}}{(2\pi)^d} \frac{1}{g(\vec{k})} \tag{8.27}$$

$$B_d(\tilde{\mu}) \equiv \int \frac{d\vec{k}}{(2\pi)^d} \frac{\tilde{\mu}}{g(\vec{k})[g(\vec{k}) + \tilde{\mu}]}. \tag{8.28}$$

A_d is the mean squared value of ϕ at the critical point. It is finite because, although $1/g(\vec{k})$ diverges in proportion to k^{-2} as $k \to 0$, the reciprocal space volume near $k = 0$ is small in proportion to $k^{d-1} dk$, so the integral is convergent near small k. If we were to approximate g by a quadratic form and evaluate this integral over all k, the same magic that makes the integral converge for small k would make $\int d^d k / k^2$ diverge if we were to extend the integral to arbitrarily large values of k. We say that this integral is infra-red convergent (meaning that the contribution from the small k part of the integral is finite) and ultraviolet divergent (meaning that the answer would diverge were we not to restrict the range of integration at large k). Thus, A_d depends on the nature of the lattice and details of the interactions we include in our model, since it is dominated by large values of k.

The extra factor of $g(\vec{k})$ in the dominator means that B_d is more strongly dominated by the small k portions of the integral as $\tilde{\mu} \to 0$. In particular, so long as $d < 4$, we can replace $g(k)$ by its quadratic approximation and carry out the integration, since the integral is ultraviolet convergent. In this way we find that

$$B_d(\tilde{\mu}) \sim \tilde{\mu}^{(d-2)/2} \tilde{B}_d = \xi^{-(d-2)} \tilde{B}_d, \tag{8.29}$$

where $\tilde{B}_d$ is number given by the expression in Eq. 8.28 with $g(\vec{k}) = \sum_a g_a k_a^2$, and $\tilde{\mu} = 1$. In other words, the fluctuations of ϕ grow stronger in magnitude with decreasing $\tilde{\mu}$, or increasing ξ, but remain bounded at the critical point, where $\xi \to \infty$.

For $d \leq 2$, the situation is entirely different. The quantity A_d is infinite for all $d < 2$. In the Gaussian model, what that means is that as we approach closer and closer to the putative critical point at which the correlation length would diverge, the magnitude of the fluctuations of ϕ diverges. We might have imagined that the Gaussian model is the lowest order term in the expansion of the true Hamiltonian in powers of ϕ, and so is valid so long as the typical magnitude of ϕ is not too large. Thus, at the very least, this divergence of $\langle \phi^2 \rangle$ as $\tilde{\mu} \to 0$ should alert us to the fact that something must go very wrong with the Gaussian model near to criticality in $d \leq 2$.

$d = 2$ is not the only special dimension. So is $d = 4$. This we can tell by studying the expression for B_d. For $d < 4$ and small $\tilde{\mu}$, this integral is dominated by small values of $\vec{k}$; this is what allowed us to approximate g by a quadratic form and replace the integral over the first BZ by an integral over all $\vec{k}$. But for $d > 4$, this integral is ultraviolet divergent, but infrared convergent. We thus carry on the analogous sort of analysis as we did above:

$$B_d(\tilde{\mu}) = A_d^{(2)} - C_d(\tilde{\mu}), \tag{8.30}$$

where $A_d^{(2)}$ is given by an integral of the same form as Eq. 8.27 with $g(\vec{k}) \to [g(k)]^2$, and

$$C_d(\tilde{\mu}) \equiv \int \frac{d\vec{k}}{(2\pi)^d} \frac{\tilde{\mu}^2}{g^2(\vec{k})[g(\vec{k}) + \tilde{\mu}]}. \tag{8.31}$$

This, in turn, is ultraviolet convergent in $d < 6$, from which we conclude that for $4 < d < 6$,

$$C_d(\tilde{\mu}) \sim \tilde{\mu}^{(d-2)/2} \tilde{C}_d, \tag{8.32}$$

where $\tilde{C}_d$ is number given by the expression in Eq. 8.31 with $g(\vec{k}) = \sum_a g_a k_a^2$, and $\tilde{\mu} = 1$. In other words, for $4 < d < 6$,

$$\langle \phi^2 \rangle = A_d - A_d^{(2)} \tilde{\mu} + \tilde{C}_d \tilde{\mu}^{(d-2)/2} + \dots, \tag{8.33}$$

where ... is higher order terms in powers of $\tilde{\mu}$. Just to make certain the point is clear, since $1 < (d - 2)/2$ for $d > 4$, the third term (with its non-analytic $\tilde{\mu}$ dependence) is small compared to the second term, which is linear in $\tilde{\mu}$.

8.1.2 Higher Order Correlations and Wick's Theorem

There is another point that we will want to know about the Gaussian model. This concerns higher order correlation functions. It is a property of Gaussian integrals that all higher order correlation functions can be expressed in terms of the quadratic correlations. You will prove this in worksheet W8.1 (see also Appendix A.9). It is really a simple observation, but it is so important that it gets a fancy name—Wick's theorem. In the present context, Wick's theorem implies that

$$\langle \phi(\vec{R}_1)\phi(\vec{R}_2)\phi(\vec{R}_3)\phi(\vec{R}_4)\rangle = G(\vec{R}_1 - \vec{R}_2)G(\vec{R}_3 - \vec{R}_4)$$
$$+ G(\vec{R}_1 - \vec{R}_3)G(\vec{R}_2 - \vec{R}_4)$$
$$+ G(\vec{R}_1 - \vec{R}_4)G(\vec{R}_2 - \vec{R}_3).$$

To make sure that the meaning of this important expression is clear, it is a property of Gaussian integrals that the average of the product of four fields can be expressed in terms of the averages of pairs of fields, by summing over all possible ways to pair those fields up. For a single Gaussian integral with a single variable x, this gives the identity $\langle x^4 \rangle = 3\left[\langle x^2 \rangle\right]^2$.

8.2 Using the Gaussian Model in a Variational Approximation

Now let us consider a more complicated problem—one that has the possibility of having a phase transition. We consider a model, commonly referred to as ϕ^4 theory, that has a form analogous to that of the LG functional we have treated before, but now this serves as our Hamiltonian

$$H = \frac{J}{2}\sum_{\langle \vec{R},\vec{R}' \rangle}\left[\phi(\vec{R}) - \phi(\vec{R}')\right]^2 - \frac{|\alpha|}{2}\sum_{\vec{R}}\phi^2(\vec{R}) + \frac{u}{4}\sum_{\vec{R}}|\phi(\vec{R})|^4$$

$$= \frac{J}{2}\sum_{\langle \vec{R},\vec{R}' \rangle}\left[\phi(\vec{R}) - \phi(\vec{R}')\right]^2 + \frac{u}{4}\sum_{\vec{R}}\left[|\phi(\vec{R})|^2 - \phi_0^2\right]^2 - \sum_{\vec{R}}\epsilon_0, \qquad (8.34)$$

and the expressions for the partition functions and various correlation functions are given by the same integrals over ϕ as in Eqs. 8.2 and 8.4. Now, however, the Boltzmann weight is not a Gaussian, so we really cannot evaluate these integrals. Here,

$$\phi_0 = \sqrt{|\alpha|/u} \ \text{ and } \ \epsilon_0 = |\alpha|^2/4u. \qquad (8.35)$$

Note that so long as $u > 0$, the energy of this model is bounded below, regardless of the sign of α. For simplicity, we will restrict our attention to the problem with $J > 0$ (i.e., we will still consider a "ferromagnetic" problem), in which H has two minima corresponding to the $\vec{R}$ independent configurations $\phi(\vec{R}) = \phi_0$ and $\phi(\vec{R}) = -\phi_0$. If we were to treat this problem the way we treated the Landau-Ginzburg theory, by minimizing with respect to $\phi(\vec{R})$, we would conclude that the model corresponds to the low temperature broken-symmetry phase of some system. Indeed, at $T = 0$, the ground-state configuration is obtained in precisely this fashion. Moreover, at low enough T, even though we are instructed to integrate over all values of ϕ, we expect that all averages will be dominated by low-energy configurations in which for almost all $\vec{R}$, $\phi(\vec{R})$ is close to $\pm\phi_0$; so we can think of this as a fancy version of the Ising model. One goal of our analysis will be to identify a T_c at which, even though $\alpha > 0$, this model undergoes a transition from a low-temperature broken-symmetry phase for $T < T_c$ to a high-temperature symmetric phase for $T > T_c$.

8.2.1   The Symmetric Phase at $T > T_c$

What we are going to do to approximately solve this problem is to use the Gaussian model as a variational approximation. We will take the trial Hamiltonian, H_{tr}, to be the Guassian model in Eq. 8.1, where we will treat μ as a variational parameter. The variational free energy, as always, is

$$F_{var}(\mu) = F_{tr} + \langle [H - H_{tr}] \rangle_{tr} \tag{8.36}$$

$$= F_{tr} + \sum_{\vec{R}} \left[\frac{u}{4} \langle \phi^4(\vec{R}) \rangle_{tr} - \frac{|\alpha| + \mu}{2} \langle \phi^2(\vec{R}) \rangle_{tr} \right].$$

From Wick's theorem, it follows that $\langle \phi^4(\vec{R}) \rangle_{tr} = 3 \left[\langle \phi^2(\vec{R}) \rangle_{tr} \right]^2$ and from the derivative theorem that $dF_{tr}/d\mu = \frac{1}{2} \sum_{\vec{R}} \langle \phi^2(\vec{R}) \rangle_{tr}$. Hence,

$$\frac{dF_{var}}{d\mu} = \frac{1}{2} \frac{dG(\vec{0})}{d\mu} \left[3uG(\vec{0}) - |\alpha| - \mu \right], \tag{8.37}$$

where $G(\vec{R}) = \langle \phi(\vec{R})\phi(\vec{0}) \rangle_{tr}$ is the correlation function of the Gaussian model we have just finished discussing, which depends implicitly upon μ. Thus, minimizing the free energy with respect to μ leads to the self-consistent expression:

$$\mu = -|\alpha| + 3uG(\vec{0}; \mu). \tag{8.38}$$

Remember that the Gaussian model is only well defined when $\mu > 0$, and that it always describes a symmetric phase of the system. Thus, wherever we find a solution of the self-consistency equation with $\mu > 0$, we know that we are dealing with the high-T disordered phase. We will identify T_c as the temperature at which $\mu \to 0^+$. Clearly, the value of μ we obtain from the above is always larger than $-|\alpha|$. Although the ground-state configuration has a spontaneously broken symmetry, at large enough T we expect to find a solution with $\mu > 0$, corresponding to the disordered phase. If we follow the evolution of the system as a function of decreasing T for fixed $|\alpha|$, we expect μ to decrease monotonically until, at some point, we encounter a critical point at which $\mu \to 0^+$. We will not attempt to explore beyond this point, since we know that the Gaussian model is unphsyical for $\mu < 0$, but at least in this way we can hope to approximately identify the critical point of our interacting problem.

Here is where the first surprise arises. In $d \leq 2$, $G(\vec{0}; \mu)$ diverges as $\mu \to 0$. Thus, no matter how large $|\alpha|$ is, we will never find a solution to this self-consistency equation with $\mu = 0$ for any nonzero T. At the present level of approximation, we conclude that we will never have a transition for $d \leq 2$. (We know for certain that this conclusion is wrong for the Ising model, since we have already encountered Onsager's solution in $d = 2$. But please do not feel that this means this whole analysis is worthless—we will see that there is more here than meets the eye.)

Let us now address the most interesting range of dimensions, $2 < d < 4$. We have spent considerable time analyzing the small μ behavior of G and we know from this that

$$\mu = -m + \frac{3uT}{J}\left[A_d - \left(\frac{\mu}{J}\right)^{(d-2)/2} \tilde{B}_d + \cdots\right], \tag{8.39}$$

where ... is higher order terms in powers of μ/J. To get a feel for what this equation means, we identify T_c as the point at which this equation is solved for $\mu = 0$; this gives us the expression

$$T_c = \frac{mJ}{3uA_d} . \tag{8.40}$$

Next, let us determine the value of μ at a temperature T that is very slightly above T_c. Here μ will be small, and hence $(\mu/J)^{(d-2)/2} \gg (\mu/J)^1$, meaning that it is justified to neglect the term on the left-hand side of the Eq. 8.39 as well as ... on the right-hand side. Thus,

$$\left(\frac{\mu}{J}\right) = \mu_0 \left(\frac{T - T_c}{T_c}\right)^{2/(d-2)} + \cdots, \tag{8.41}$$

where ... refers to terms that are higher order in $(T - T_c)/T_c$, and

$$\mu_0 = \left(\frac{\alpha J}{3uT_c\tilde{B}_d}\right)^{2/(d-2)} . \tag{8.42}$$

From our understanding of the relation between μ and the correlation length in the Gaussian model, we immediately see that this implies $\xi \sim (T - T_c)^{-\nu}$ with our (first) dimension-dependent critical exponent,

$$\nu = 1/(d-2). \tag{8.43}$$

This gives $\nu = 1$ for $d = 3$ (which differs from any mean-field exponent we have derived until now), but it approaches the familiar value $\nu \to 1/2$ as $d \to 4$. Similarly, since the susceptibility $\chi = 1/\mu \sim (T - T_c)^{-\gamma}$, we obtain $\gamma = 2/(d-2)$ or $\gamma = 2$ in $d = 3$ and $\gamma = 1$ in $d = 4$.

There is a rather glaring omission in our discussion so far—we have been able to use this approach only to treat the symmetric phase at $T \geq T_c$. There is an obvious way to extend the approach to $T < T_c$—we could include an external field which breaks the symmetry as a variational parameter, $H_{tr} \to H_{tr} - h \sum_{\vec{R}} \phi_{\vec{R}}$, and then treat h as an additional variational parameter. However, this approach produces unphysical results in ways we will address in a bit. So, for now, we will continue to focus our attention only on the results in the high-temperature phase.

8.3 *Exact Results for $O(N)$ Model as $N \to \infty$

Before we assess the validity of the new mean-field treatment we have just discussed, we introduce a more general form of ϕ^4 theory. Now, we imagine that, rather than specifying the microstates of our system by giving the value of a scalar field on each lattice site, we consider an N-component vector $\phi(\vec{R})$ with components $\phi_a(\vec{R})$, with $a = 1, 2, \ldots, N$. (Note, $\vec{R}$ is a d-dimensional vector in real-space and ϕ is a N-component vector in another abstract field space, hence the different notation.) Whereas before we considered our model to have a global "Ising" symmetry, we now imagine that we have a global N-dimensional rotational symmetry. Specifically, we will consider only problems in which the Hamiltonian is invariant under a transformation in which we "rotate" all the fields, $\phi_a(\vec{R}) \to \mathcal{R}_{a,b}\phi_a(\vec{R})$, where $\mathcal{R}$ is any orthogonal $N \times N$ matrix. An orthogonal matrix is one in which $\mathcal{R}^T\mathcal{R} = 1$. Adopting the language of group theory, this is called an $O(N)$ symmetry (Section 5.4). Physically, one can think of this as an invariance under all rotations and reflections of ϕ.

In particular, we can define an $O(N)$ symmetric version of ϕ^4 theory with Hamiltonian

$$H = \frac{J}{2} \sum_{\langle \vec{R}, \vec{R}' \rangle} \left| \phi(\vec{R}) - \phi(\vec{R}') \right|^2 + \frac{u}{4N} \sum_{\vec{R}} \left[\left| \phi(\vec{R}) \right|^2 - N\phi_0^2 \right]^2, \tag{8.44}$$

where $|\phi|^2 \equiv \sum_a \phi_a^2$. For $N = 1$, this is just the model in Eq. 8.34, but for $N = 2$ and $N = 3$ it is, respectively, a version of the XY and Heisenberg models we have already mentioned. The factors of N have been added judiciously, so that each term in H is of order N—which will prove necessary when we come to take the large N limit. Following the same logic, we can treat this problem variationally by introducing an $O(N)$ Gaussian model as our trial Hamiltonian

$$H_{tr} = \frac{J}{2} \sum_{\langle \vec{R}, \vec{R}' \rangle} \left| \phi(\vec{R}) - \phi(\vec{R}') \right|^2 + \frac{\mu}{2} \sum_{\vec{R}} \left| \phi(\vec{R}) \right|^2. \tag{8.45}$$

The $O(N)$ symmetry of this model implies that $\langle \phi_a(\vec{R})\phi_b(\vec{0}) \rangle_{tr} = \delta_{ab}G(\vec{R})$, and from Wick's theorem it follows that $\langle \left| \phi(\vec{R}) \right|^4 \rangle_{tr} = N(N+2)G^2(\vec{0}; \mu)$. Thus, minimizing the variational free energy with respect to μ leads to the self-consistency condition (with $|\alpha| \equiv u|\phi_0|^2$)

$$\mu = -|\alpha| + \frac{(N+2)}{N} u \, G(\vec{0}). \qquad (8.46)$$

How lovely—we can solve this problem (to this level of approximation) without doing anything new. All we need to do is replace $u \to u(1 + 2/N)$ in all the calculations we did above. Indeed, just as d appeared simply as a parameter, so does N, so we can consider this problem for any real $N > -2$. (Believe it or not, it is sometimes useful to think of this problem extended to negative values of N, although we will not explore this in this book.)

But here is what is even better—the approximate solution we have been discussing becomes the exact solution of the problem in the limit $N \to \infty$! Proving this is beyond the scope of this book, but the intuition behind this statement is very simple.[2] In effect, what we have done is to treat the ϕ^4 term approximately. Let us look at the problem from the perspective of a single component, ϕ_1. The first two terms in H are already quadratic functions of ϕ_1. The other term has a quadratic and a quartic term, $a(\vec{R})\phi_1^2(\vec{R}) + (u/4N)\phi_1^4(\vec{R})$, where $a(\vec{R}) = (u/2N)\sum_{a=2}^{N} \phi_a^2(\vec{R})$. Of these, the second term vanishes in the limit $N \to \infty$ and so can be ignored. The first term at first seems dangerous, as it depends on the value of all the other components of the field. But if we think of each of the N components of the field as approximately independent random variables, then by the central limit theorem we know that in the large N limit, $a(\vec{R}) \to (u/2N)\sum_{a=2}^{N}\langle\phi_a^2(\vec{R})\rangle \to (u/2)G(\vec{0})$. This is just what our mean-field solution assumed.

While this sounds abstractly exciting, you may justifiably ask whether solving the properties of an $O(N)$ model in the $N \to \infty$ limit is of any use whatsoever. Surely no physical system has such a large symmetry. We have mostly focused on the Ising model, for which $N = 1$, and while 1 is not a small number it is surely not a large number. We have mentioned previously physical circumstances in which there is an $O(2)$ (as in a superfluid) or an $O(3)$ (as in a ferrofluid). But there is a question, even here, whether $N = 2$ or $N = 3$ is large enough to be well approximated by $N \to \infty$. So let us compare some of the present results with those obtained by the previous (Weiss) mean field theory, and with exact results we know from other methods.

Firstly, the Weiss mean-field theory leads to the prediction that there is a phase transition to an ordered phase at low T in any dimension for any symmetry of the fields involved. The large N approach implies that no transitions can occur in $d \leq 2$. As we have already seen, there can indeed be no broken-symmetry states (at nonzero T and in the absence of long-range interactions) in $d = 1$, while for $N = 1$ (the Ising case) a finite T transition to a ferromagnetic phase is possible in $d = 2$. However, as

---

2. See discussion of "spherical model" in Baxter's book, *Exactly Solved Models in Statistical Mechanics* (Academic Press, London, 1982).

we will discuss in Section 8.3.1, for systems with continuous symmetries (i.e., for $N \geq 2$), spontaneous symmetry breaking is impossible for $d \leq 2$. The case of $N = 2$ in $d = 2$ turns out to be extremely subtle and interesting—while there can be no ordered phase, there turns out to be a phase transition at a nonzero temperature to a phase that has correlations that fall as a power of distance, rather than the familiar (Ornstein-Zernicke) exponential decay. Needless to say, this behavior is not captured by the large N approach. However, on the basis of analysis that goes beyond the scope of the present book, it has been established that for all $N > 2$, not only is it true that there are no finite temperature phase transitions in $d = 2$, but many of the more detailed features of the large N results are accurately reproduced by the large N mean-field theory.

As far as critical exponents in $d > 2$ are concerned, one can compare the results we have obtained, $\nu = 1/(d-2)$, and $\gamma = 2/(d-2)$, with the best known results (Table 1.4) and with the results of Weiss mean-field theory, $\nu = 1/2$ and $\gamma = 1$. It is immediately clear that both mean-field theories and the exact results all agree on the values of these exponents for $d = 4$. (There are logarithmic corrections known to arise in $d = 4$ that should really be discussed separately, so strictly speaking the agreement between all the results applies only for $d > 4$.) However, in $d = 3$, the large N exponents are considerably closer to the exact results than those of Weiss mean-field theory—even for the case of $N = 1$. Moreover, taking a step back, even though the critical exponents are not obtained precisely, we have now obtained an insight—which is confirmed by more rigorous analysis—that the critical exponents are dimension-dependent for dimensions between the lower critical dimension, $d = 2$ (below which no ordered phase exists), and $d = 4$ (above which mean-field exponents are exact).

8.3.1   The Ordered Phase and the Mermin-Wagner Theorem

Let us now turn to the properties of the ordered phase at $T < T_c$. We will treat problems with different symmetries by considering the $O(N)$ ϕ^4 theory in Eq. 8.44. At $T = 0$, the ground-state properties are obtained simply by minimizing H with respect to ϕ. The result is the spatially uniform solution,

$$\phi(\vec{R}) = \sqrt{N}\phi_0\Omega, \tag{8.47}$$

where Ω is any N-component unit vector, and the ground-state energy is $E_0 = 0$. The arbitrariness of Ω encodes the ground-state degeneracy that is a property of systems with a low-T phase with a spontaneously broken symmetry. In the Ising case, $N = 1$, the two ground states correspond to $\Omega = 1$ and $\Omega = -1$. For the $N = 2$, the XY model, the possible ground states are specified by a single phase, $0 \leq \theta < 2\pi$, such that $\Omega = \langle \cos(\theta), \sin(\theta) \rangle$. For the $N = 3$, the Heisenberg case, Ω can be thought of as specifying a point on the unit 3-sphere. It is important to remember that Ω is a vector not in the d-dimensional space of the lattice but rather, in an internal "spin-space," which has dimension unrelated to d.

Still restricting our attention to $T = 0$, we can look at the effect of applying a small symmetry-breaking field, $H \to H - \sum_{\vec{R}} h \cdot \phi_{\vec{R}}$. Clearly, the ground state is now unique—it is the state in which Ω is aligned parallel to h. Indeed, if we define ω to be the angle between h and Ω, i.e., $h \cdot \Omega = |h| \cos(\omega)$, then we can easily see that the energy of each potential ground state is

$$E_0 = -V|h|\sqrt{N}\phi_0 \cos(\omega) + \mathcal{O}(h^2). \tag{8.48}$$

Because this energy scales with the volume of the system, no matter how small $|h|$ is or how minute the misalignment of the ground state, in the thermodynamic limit $V \to \infty$ the difference in energy between the true ground state ($\omega = 0$) and any other state ($\omega \neq 0$) diverges.

Now let us consider the problem at low but nonzero T. It seems very reasonable to assume that the only important configurations of the fields are "close" to the ground-state configuration. Without loss of generality, we can choose our small symmetry-breaking field, h, to be a vector of length h along the $\hat{e}_1$ direction, so that in the ground state, Ω is a unit vector in the same direction. We define the deviation from the ground-state configuration as $\delta\phi$ such that

$$\phi_a(\vec{R}) = \sqrt{N}\phi_0 + \delta\phi = \begin{cases} \bar{\phi} + \sigma(\vec{R}) & \text{for} \quad a = 1 \\ \pi_a & \text{for} \quad a > 1, \end{cases} \tag{8.49}$$

where $\sigma(\vec{R})$ encodes the longitudinal fluctuations (i.e., $\bar{\phi}$ is the uniform component of ϕ, defined so that $\sum_{\vec{R}} \sigma(\vec{R}) = 0$), and $\pi(\vec{R})$ is the transverse fluctuations, i.e., $\pi_1(\vec{R}) = 0$. Importantly, the longitudinal and transverse fluctuations enter the problem very differently. Specifically, if we expand the Hamiltonian to quadratic order in the amplitude of the fluctuations, we find

$$H = E_0(\bar{\phi}) + H_\parallel[\sigma] + H_\perp[\pi] + \cdots, \tag{8.50}$$

where σ is a scalar (one-component) field, π is an $(N-1)$-component vector field, $\ldots$ signifies higher order terms in powers of $\delta\phi$,

$$E_0(\bar{\phi}) = V\left[-h\bar{\phi} + \frac{\mu}{2}[\bar{\phi} - \sqrt{N}\phi_0]^2\right], \tag{8.51}$$

$$H_\parallel[\sigma] = \frac{J}{2} \sum_{\langle \vec{R}, \vec{R}' \rangle} \left[\sigma(\vec{R}) - \sigma(\vec{R}')\right]^2 + \frac{\mu}{2} \sum_{\vec{R}} \left[\sigma(\vec{R})\right]^2, \tag{8.52}$$

$$H_\perp[\pi] = \frac{J}{2} \sum_{\langle \vec{R}, \vec{R}' \rangle} \left|\pi(\vec{R}) - \pi(\vec{R}')\right|^2, \tag{8.53}$$

and μ, rather than being a variational parameter, is given by $\mu \equiv 2u\phi_0^2$.

Fortunately, we know how to treat this problem:

i) $\bar{\phi} - \sqrt{N}\phi_0 = h/\mu$ is simply the shift in the order parameter produced by the external field. It represents a single degree of freedom in a macroscopic system and so its fluctuations can be ignored. This correction vanishes as $h \to 0$ and so is often simply ignored.

ii) The fluctuations are governed by quadratic Hamiltonians, so we know all we need to know about how to compute their thermal fluctuations. Immediately, we see that the longitudinal fluctuations look just like the fluctuations of ϕ itself in the disordered state. Specifically, at $T \ll T_c$,

$$\langle \phi_1(\vec{R})\phi_1(\vec{R}') \rangle = \bar{\phi}^2 + G(\vec{R} - \vec{R}'; \mu) + \cdots. \tag{8.54}$$

The part of this correlation function that is independent of $\vec{R} - \vec{R}'$ is called the "disconnected" piece, and the remainder is the "connected" piece. Here ... refers to corrections due to the higher order terms in powers of $\delta\phi$ that we have neglected.

iii) For $N > 1$ we must also consider the transverse fluctuations. These are also governed by a Gaussian theory, but in this case it looks like a Guassian theory at criticality, $\mu = 0$. In other words, for any $a > 1$,

$$\langle \phi_a(\vec{R})\phi_b(\vec{R}') \rangle = \delta_{a,b}\, G(\vec{R} - \vec{R}'; \mu = 0) + \cdots. \tag{8.55}$$

For $d > 2$, this is all well and good. This gives us a sensible description of the correlations deep in the ordered phase at $T \ll T_c$. A general way to frame this result is that there exists a set of normal modes—known as "Goldstone modes"—that have an energy that vanishes at long-wavelength in proportion to $|\vec{k}|^2$. The vanishing of the energy as $\vec{k} \to \vec{0}$ simply reflects the fact that a small spatially uniform rotation of the order parameter orientation (keeping its magnitude fixed) converts one ground-state into another, and so costs no energy. Indeed, based on this intuition, it is possible to show that the expression for the transverse correlations in Eq. 8.55 is asymptotically exact at low T in the broken-symmetry phase.

However, for $d \leq 2$, we know that $G(\vec{0}, \mu) \to \infty$ as $\mu \to 0$. That is to say, the mean squared magnitude of the transverse fluctuations is infinite in $d \leq 2$. At the very least, this tells us that the Gaussian approximation we have adopted is suspect— we cannot expect to expand in powers of $\delta\phi$ under conditions in which $\langle \delta\phi^2 \rangle$ is infinite! This should remind you of our previous result that we were not able to approach the critical point from above in the Gaussian approximation unless $d > 2$. But the present failure is more serious than the former one. Here we have not made a variational ansatz—all we have assumed is that we are in a broken-symmetry state at low T where one would expect the fluctuations to be parametrically small. For $N = 1$, where the supposed broken symmetry is discrete, we have encountered no inconsistency. But for $N \geq 2$, where the putative broken symmetry is continuous, what we have uncovered is an inconsistency in our starting assumption: if we were to suppose the existence of a phase with a broken continuous symmetry in $d \leq 2$, we would find that the magnitude of the fluctuations about that state would be infinite at any nonzero T. What we have encountered is an approximate version of a general theorem (that can be proved rigorously) known as the Mermin-Wagner theorem: under very general circumstances, no phase with a spontaneously broken continuous symmetry is possible in $d \leq 2$.

8.4 Worksheets for Chapter 8

W8.1   Gaussian Integrals and Wick's Theorem (Isserlis' Theorem)

Here we will prove a few properties of Gaussian integrals, as well as Wick's (probability) theorem.

Review on Gaussian integrals

First we will review some basic facts about Gaussian integrals of the form:

$$\int_{-\infty}^{\infty} dx\, x^{2n} e^{-x^2/2\lambda}. \tag{8.56}$$

If you are already very familiar with this, you can skip to the next section.

1. A general (nonnormalized, zero-centered) Gaussian distribution is:

$$G(x; \lambda) = e^{-x^2/2\lambda}. \tag{8.57}$$

   Show that the definite integral

   $$I = \int_{-\infty}^{\infty} dx\, G(x; \lambda) = \sqrt{2\pi\lambda}. \tag{8.58}$$

   Hint: Use cylindrical coordinates $x^2 + y^2 = \rho^2$ to evaluate the quantity

   $$I^2 = \int_{-\infty}^{\infty} \int_{-\infty}^{\infty} dx\, dy\, G(x; \lambda) G(y; \lambda). \tag{8.59}$$

2. Since $G(x; \lambda)$ is even, all integrals of the form

   $$\int_{-\infty}^{\infty} dx\, x^k\, G(x; \lambda) \tag{8.60}$$

   evaluate to zero for odd powers k. Thus, only integrals of the form

   $$\int_{-\infty}^{\infty} dx\, x^{2n} e^{-x^2/2\lambda} \tag{8.61}$$

   are nonzero for integer $n \geq 0$. Show that:

   $$\int_{-\infty}^{\infty} dx\, x^{2n} e^{-x^2/2\lambda} = \lambda(2n-1) \int_{-\infty}^{\infty} dx\, x^{2n-2} e^{-x^2/2\lambda}. \tag{8.62}$$

   Hint: One way to do this is by integration by parts.

3. For our last integration exercise, prove the following integrals. *Hint:* Complete the square for the exponents, i.e., find c for $-x^2/2\lambda + bx = -(x/\sqrt{2\lambda} - c)^2 + c^2$.

(a) $Z = \displaystyle\int_{-\infty}^{\infty} dx\, e^{-x^2/2\lambda + bx} = e^{b^2\lambda/2}\sqrt{2\pi\lambda}$.

(b) $\displaystyle\int_{-\infty}^{\infty} dx\, x\, e^{-x^2/2\lambda + bx} = b\lambda Z$.

(c) $\displaystyle\int_{-\infty}^{\infty} dx\, x^2 e^{-x^2/2\lambda + bx} = (\lambda + b^2\lambda^2)Z$.

Since you now have all the tricks in the bag to do Gaussian integrals, we will give you the following expressions to save you from the tedium (but feel free to prove them!):

$$\int_{-\infty}^{\infty} dx\, x^3 e^{-x^2/2\lambda + bx} = (3b\lambda^2 + b^3\lambda^3)Z,$$

$$\int_{-\infty}^{\infty} dx\, x^4 e^{-x^2/2\lambda + bx} = (3\lambda^2 + 6b^2\lambda^3 + b^4\lambda^4)Z.$$

Wick's (probability) theorem

A multivariate (multidimensional) Gaussian distribution of N variables can be written as:

$$G(\{x\}; \{\lambda\}) = \prod_{j}^{N} \exp(-x_j^2/2\lambda_j),$$

where each x_j is an independent random variable. We can define a "partition function" or normalization factor as

$$Z = \int \prod_{j}^{N} \left[dx_j \exp(-x_j^2/2\lambda_j) \right].$$

Since the x_js are independent, this integral factorizes into:

$$Z = \prod_{j} \int dx_j \exp(-x_j^2/2\lambda_j) = \prod_{j} \sqrt{2\pi\lambda_j} = \prod_{j} Z_j. \qquad (8.63)$$

With this in place, we can define the expectation values of various quantities, in general of the form:

$$\langle A(\{x\}) \rangle = Z^{-1} \int A(\{x\}) \prod_{j} \left[dx_j \exp(-x_j^2/2\lambda_j) \right]. \qquad (8.64)$$

Let us now calculate some expectation values. You may want to consult results from the previous sections.

As a freebie, $\langle x_j \rangle = 0$ for any j, since the integrand is odd for that j.

(a) Show that $\left\langle \prod_j (x_j)^{n_j} \right\rangle = \prod_j \left\langle (x_j)^{n_j} \right\rangle$ for any nonnegative integer n_j.

(b) Show that $\langle x_j x_k \rangle = \delta_{jk} \lambda_j$, and thus $\left\langle (x_j)^2 \right\rangle = \lambda_j$.

(c) Show that, specifically, $\left\langle (x_k)^4 \right\rangle = 3 \left\langle (x_k)^2 \right\rangle^2$.

(d) Argue that

$$\langle x_p x_q x_r x_s \rangle = \langle x_p x_q \rangle \langle x_r x_s \rangle + \langle x_p x_r \rangle \langle x_q x_s \rangle + \langle x_p x_s \rangle \langle x_q x_r \rangle \tag{8.65}$$

is true.

Here have proved Wick's probability theorem up to the 4th-order expectation values, but this can be further generalized to higher (even) powers, which can be proven with combinatorics.[3]

Importantly, for the expectation value of a single variable to the power of $2n$:

$$\left\langle x^{2n} \right\rangle = (2n-1)!! \left\langle x^2 \right\rangle^n, \tag{8.66}$$

where $!!$ is the double factorial

$$(2n-1)!! = (2n-1)(2n-3) \cdots 3 \cdot 1. \tag{8.67}$$

As an example of the higher order general expression, the sixth-order one is:

$$\begin{aligned}
&\langle x_1 x_2 x_3 x_4 x_5 x_6 \rangle \\
&= \langle x_1 x_2 \rangle \langle x_3 x_4 \rangle \langle x_5 x_6 \rangle + \langle x_1 x_2 \rangle \langle x_3 x_5 \rangle \langle x_4 x_6 \rangle + \langle x_1 x_2 \rangle \langle x_3 x_6 \rangle \langle x_4 x_5 \rangle \\
&+ \langle x_1 x_3 \rangle \langle x_2 x_4 \rangle \langle x_5 x_6 \rangle + \langle x_1 x_3 \rangle \langle x_2 x_5 \rangle \langle x_4 x_6 \rangle + \langle x_1 x_3 \rangle \langle x_2 x_6 \rangle \langle x_4 x_5 \rangle \\
&+ \langle x_1 x_4 \rangle \langle x_2 x_3 \rangle \langle x_5 x_6 \rangle + \langle x_1 x_4 \rangle \langle x_2 x_5 \rangle \langle x_3 x_6 \rangle + \langle x_1 x_4 \rangle \langle x_2 x_6 \rangle \langle x_3 x_5 \rangle \\
&+ \langle x_1 x_5 \rangle \langle x_2 x_3 \rangle \langle x_4 x_6 \rangle + \langle x_1 x_5 \rangle \langle x_2 x_4 \rangle \langle x_3 x_6 \rangle + \langle x_1 x_5 \rangle \langle x_2 x_6 \rangle \langle x_3 x_4 \rangle \\
&+ \langle x_1 x_6 \rangle \langle x_2 x_3 \rangle \langle x_4 x_5 \rangle + \langle x_1 x_6 \rangle \langle x_2 x_4 \rangle \langle x_3 x_5 \rangle + \langle x_1 x_6 \rangle \langle x_2 x_5 \rangle \langle x_3 x_4 \rangle.
\end{aligned} \tag{8.68}$$

Wick's theorem for linear functions

We can generalize the above to a generalized linear function. Consider the same multivariate Gaussian distribution of N variables,

$$G(\{x\}; \{\lambda\}) = \prod_j^N \exp(-x_j^2 / 2\lambda_j),$$

3. See L. Isserlis, *Biometrika*, **12**, 134–139 (1918).

and a general linear function of these variables,

$$F_a = \sum_j f_j^a x_j,$$

where f_j^a are the coefficients for each of the terms linear in x_j.

We will prove a few identities of the expectation value of linear functions using Wick's probability theorem. Again, as a freebie, $\langle F_a \rangle = \sum_j f_j^a \langle x_j \rangle = 0$.

1. Show that:

$$\langle F_a F_b \rangle = \sum_j f_j^a f_j^b \lambda_j.$$

2. Show that:

$$\langle F_a F_b F_c \rangle = 0.$$

3. Show that:

$$\langle F_a F_b F_c F_d \rangle = \langle F_a F_b \rangle \langle F_c F_d \rangle + \langle F_a F_c \rangle \langle F_b F_d \rangle + \langle F_a F_d \rangle \langle F_b F_c \rangle.$$

Indeed, Wick's theorem for general linear functions generalizes the same way as the basic Wick's theorem.

Expectation of linear functions for Gaussian random variables

1. Consider a linear function $f = f_0 + f_k x_k$ for some specific value of k (no summations here). By completing the square as previously, show that $\langle e^f \rangle = \exp\{\langle f \rangle + (1/2)[\langle f^2 \rangle - \langle f \rangle^2]\}$. *Hint:* Calculate $\langle f \rangle$ and $\langle f^2 \rangle$ along the way.

2. Now let us consider a generalized linear function $f = f_0 + \sum_j f_j x_j$. Show that the expression $\langle e^f \rangle = \exp\{\langle f \rangle + (1/2)[\langle f^2 \rangle - \langle f \rangle^2]\}$ is still true for the generalized function.

W8.2   Mean-Field Approximations to the 0D ϕ^4 Theory

Let us consider the ϕ^4 theory, but with only a single site. This is similar to considering the statistical properties of the single site Ising model. Let us consider the Hamiltonian

$$H(\phi) = \frac{\tilde{\alpha}}{2}\phi^2 + \frac{u}{4}\phi^4, \tag{8.69}$$

so we can consider that ϕ is the order parameter. For this exercise, let us work in units where the temperature is always 1, such that $T = \beta = 1$. The partition function is then defined as

$$Z = \int_{-\infty}^{\infty} d\phi\, e^{-H(\phi)}, \tag{8.70}$$

and the free energy as

$$F = -\ln Z.$$

Since the Hamiltonian is even in ϕ, only expectation values of even powers of ϕ will be nonzero, defined as:

$$\langle \phi^{2n} \rangle = Z^{-1} \int_{-\infty}^{\infty} d\phi \, \phi^{2n} e^{-H(\phi)}. \tag{8.71}$$

1. We will begin with some rescaling of parameters. We will assume $u > 0$ to ensure the energy is bounded below. By using a variable substitution of the form $\theta = \phi u^{1/4}$, and an appropriate definition of α, show that:

   (a)
   $$F(\tilde{\alpha}, u) = \frac{\ln u}{4} + \mathcal{F}(\alpha),$$

   where

   $$\mathcal{F}(\alpha) = F(\tilde{\alpha} = \alpha, u = 1). \tag{8.72}$$

   (b)
   $$\langle \phi^{2n} \rangle = u^{-n/2} G_{2n}(\alpha),$$

   where

   $$G_{2n}(\alpha) = \langle \phi^{2n} \rangle_{\tilde{\alpha}=\alpha, u=1}. \tag{8.73}$$

   The point of this exercise is show that the physics is in fact entirely contained in the (rescaled) coefficient α, and therefore in the functions $\mathcal{F}(\alpha)$ and $G_{2n}(\alpha)$, which only depend on α.

2. While the functions $\mathcal{F}(\alpha)$ and $G_{2n}(\alpha)$ *can* be evaluated exactly in terms of special functions, this is not particularly illuminating. Here let us use a variational approximation and see how well we can do compared to the exact solution. Since we have shown that α is the physically important parameter, we will work with $\tilde{\alpha} = \alpha$ and $u = 1$ such that $F = \mathcal{F}$ and $G_{2n} = \langle \phi^{2n} \rangle$.
   Consider a trial Hamiltonian:

   $$H_{tr} = \frac{\mu}{2} \phi^2.$$

   (a) Evaluate the variational free energy $F_{var} = F_{tr} + \langle H - H_{tr} \rangle_{tr}$.

   (b) Minimize F_{var} to find the value of the variational parameter μ that best approximates the system. Keeping in mind that $\mu > 0$ due to the $\ln \mu$ term in the free energy, show that

   $$\mu = \frac{\alpha + \sqrt{\alpha^2 + 12}}{2}.$$

   (c) Evaluate $G_2(\alpha)$ in the trial ensemble. Substitute in the optimal μ from the last step and expand $G_2(\alpha)$ to three leading orders, for both small and large α. *Hint:* To expand about $\alpha \to \infty$, substitute in $\gamma = \alpha^{-1}$, expand about $\gamma = 0$, then reverse the substitution.

3. So far we have used a trial Hamiltonian that respected the symmetry of the system. However, in general we can have spontaneously broken symmetries in the trial Hamiltonian as well. Consider now the trial Hamiltonian:

$$H_{tr} = \frac{\mu}{2}\phi^2 - h\phi.$$

(a) Evaluate the variational free energy $F_{var} = F_{tr} - \langle H - H_{tr} \rangle_{tr}$.

(b) Minimize F_{var} to find values of the parameters μ and h that best approximate the system. Show that there are two solutions (remember $\mu > 0$ by necessity):

$$h = 0, \; \mu = \frac{\alpha + \sqrt{\alpha^2 + 12}}{2}$$

and

$$h = \sqrt{-\alpha\mu^2 - 3\mu}, \; \mu = -\alpha + \sqrt{\alpha^2 - 6}.$$

(c) Interpret these solutions. According to the mean-field approximation, is there a phase transition for some critical value of $\alpha = \alpha_c$? Should there be a phase transition? *Hint:* For your reference, here are the various free energies plotted as a function of α.

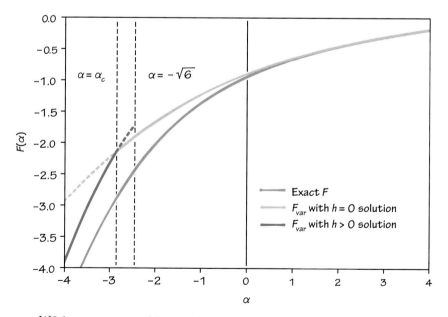

FIGURE W8.1. A comparison of the exact free energy and various approximations to it.

9 *QUENCHED DISORDER

Platonic perfection is elegant, but the imperfections of real-world materials host a beauty all their own.

9.1 Systems with Quenched Disorder

Quenched degrees of freedom, or quenched disorder, are degrees of freedom that are not in thermal equilibrium, but which are determined by a dynamical process that has become arrested at some stage and "frozen in." For instance, there are many defects found in all real "crystalline" materials, with concentrations that can depend on the thermal history of the material, including impurity atoms, vacancies, interstitials, dislocations, and grain boundaries. At temperatures significantly below the melting temperature, atomic rearrangements in a solid are so slow as to be negligible for many purposes. Therefore, one is stuck with whatever distribution of these defects happens to occur in the particular sample under study. Indeed, while liquids and gases are generally able to fully equilibrate on laboratory timescales, solids always have quenched disorder.

In this chapter, we generalize the statistical mechanics approach to treat systems with quenched disorder. This approach divides the degrees of freedom into two subsets, one of which achieves thermal equilibrium and the other of which is frozen in, like the crystal defects we have described. Formally, what this means is that the constrained equilibrium properties of the system are governed by a Hamiltonian, $H(C; s)$, which depends (as before) on the microstate s of the bulk of the degrees of freedom that are in thermal equilibrium, but also on the "configuration" C of the quenched degrees of freedom. The free energy of such a system is dependent on the particular configuration C that has frozen into a sample:

$$F[C] = -T \log \{Z[C]\} \quad \text{where} \quad Z[C] = \sum_s e^{-\beta H(C;s)}. \tag{9.1}$$

9.2 Configuration Averages

If we are to make progress on this problem, we need to understand the nature of the configuration-dependent Hamiltonian, $H(C; s)$, and the ensemble of configurations. On physical grounds, we expect the particular configuration of imperfections

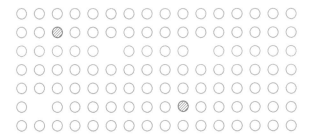

FIGURE 9.1. A schematic of quenched disorder in a crystalline system. A red circle represents an impurity atom, and a missing circle represents a vacancy.

in a given crystal to be chosen from a random ensemble with probability measure $P(C)$. For instance, if there is a small concentration of impurity atoms in an otherwise perfect crystalline system, one would imagine that the concentration of impurities might be fixed (as it reflects the purity of the constituents from which the crystal was synthesized), but exactly where in the crystal each impurity sits is a matter of random circumstance at the time the crystal grew. This is illustrated in Figure 9.1.

If we were to grow a large number of crystal samples in this manner, each would end up with a different configuration of impurities. The average free energy of all these samples is

$$\overline{F(C)} \equiv \int dC\, P(C) F(C), \tag{9.2}$$

where $\overline{(\dots)}$ signifies the configuration average and $\int dC$ means an integral (or sum) over all possible configurations of quenched disorder. The interpretation of the configuration average is quite different than that of the thermal average we have discussed so far. It is not immediately clear why we even care about the configuration average, rather than the $F(C)$ for the particular realization of quenched disorder in a sample we are studying. Under very general circumstances, however, there is no difference between $F(C)$ and $\overline{F(C)}$ in the thermodynamic limit!

This property is known as "self-averaging." Let us pause for a minute and discuss why we expect physical quantities to be self-averaging. In general, physics is local. What this means is that the local properties of a solid in one region are independent of those in another region that is far away compared to the correlation length ξ. Thus, if we divide a macroscopic system into a large number of subsystems of size large compared to ξ, but small compared to the size of the system, we can treat each region as essentially independent of the others. The free energy density in each region is thus dependent on the disorder configuration within that region and is independent of the configuration in neighboring regions. If we look at the average free energy density over N regions, its variance decreases as $1/N$—and so the mean squared deviation between $F(C)/V$ and $\overline{F(C)}/V$ vanishes as $V \to \infty$. Any property of the system that is local in the above sense will be self-averaging as a consequence of the central limit theorem.[1]

The fact that we can deal with configuration-averaged, rather than configuration-dependent, quantities is a great simplification—similar in nature to that involved in computing thermal averages in statistical mechanics instead of time averages. One obvious simplification is that averaged quantities reflect the symmetries of the *configuration distribution*, $P(C)$, rather than the much smaller set of symmetries manifested by a particular disorder configuration, C. For instance, if we were to compute the thermal average of a local density at some point in space, $\langle \rho(\vec{r}) \rangle$, its value will vary in a complicated way from one point in space to another—depending, for instance, on whether there happen to be more or less impurities nearby in configuration C. But if the ensemble of disorder configurations described by $P(C)$ is *on average* translationally invariant, the corresponding configuration-averaged quantity, $\overline{\langle \rho(\vec{r}) \rangle}$, will be independent of $\vec{r}$ (modulo the possibility of spontaneous translation symmetry breaking). Just as the ergodic theorem relates time-averaged quantities to thermal averages over an equilibrium ensemble, the property of self-averaging relates the spatial average of local quantities in a given configuration to the configuration average.

9.2.1   Example: Array of Harmonic Oscillators

The configuration average is what is call a "quenched" average, and it is different in important ways from a thermal average. To illustrate this, let us consider the simple model problem of an array of decoupled classical harmonic oscillators subjected to

---

1. In classical systems, except at a critical point, there is always a finite correlation length. Thus, as long as the system is large compared to the correlation length, self-averaging can be expected. Near a classical critical point, where the correlation length diverges, finite-size effects in the form of sample-to-sample variations in the properties of a finite-size system (configuration dependence of quantities) will persist to anomalously large system sizes. In quantum systems there is a strange sort of emergent lengthscale, ξ_Q, which characterizes the lengthscale over which quantum coherence is preserved. The ground-state, even of a complicated interacting system, is a single coherent state. It is a peculiar and unavoidable feature of quantum mechanics that the nature of the states depends not only on the Hamiltonian but also on the boundary conditions, i.e., quantum mechanics is generically nonlocal. Thus, under many circumstances, even when the system being studied is nowhere near any critical point, the quantum coherence length will tend to diverge, $\xi_Q \to \infty$ as $T \to 0$. If we are considering the ground state properties of a system, to properly approach the thermodynamic limit we must first take the system size to infinity, and only then take the $T \to 0$ limit. Conversely, there is a whole cornucopia of interesting phenomena associated with "mesoscopic" phenomena—i.e., phenomena associated with systems that are large compared to any microscopic length, but small compared to ξ_Q. For further discussion, see Akkerman and Montambaux, *Mesoscopic Physics of Electrons and Photons* (Cambridge Univ. Press, Cambridge, 2007) and Altshuler, Lee, and Webb, *Mesoscopic Phenomena in Solids*, Modern Problems in Condensed Matter Sciences **30**, ed. by Agronovich and Maradudin (North Holland, Amsterdam, 1991)

a linear force which varies randomly from site to site. The Hamiltonian is thus

$$H = \sum_j \left\{ \frac{K}{2} x_j^2 - x_j f_j \right\},$$ (9.3)

where the Gaussian variables, x_j, are assumed to be in equilibrium at temperature T, while f_j is a quenched random force. We we will further assume that the quenched variables on each site are independent random variables, governed by a probability density, $P(\{f\}) = \prod_j p(f_j)$, with mean zero, and variance σ:

$$\overline{f_j} = 0, \quad \overline{f_i f_j} = \delta_{i,j} \sigma^2.$$ (9.4)

For this simple problem, the configuration-dependent free energy is easily computed,

$$F(\{f\}) = F_0(K) - \sum_j \left(\frac{f_j^2}{2K} \right); \quad F_0(K) = N \frac{T}{2} \ln\left[\frac{2\pi T}{K} \right],$$ (9.5)

from which it follows at once that the configuration-averaged free energy density is

$$\frac{\overline{F}}{N} = \frac{F_0}{N} - \frac{\sigma^2}{2K}.$$ (9.6)

To illustrate how different this is from a thermal average, let us consider instead the case in which f_j represents forces that are part of the equilibrium description rather than being quenched. The new problem thus consists of two coupled harmonic oscillators,

$$H' = \sum_j \left\{ \frac{K}{2} x_j^2 - x_j f_j + \frac{T}{2\sigma^2} f_j^2 \right\},$$ (9.7)

where the parameters are chosen so that the thermal distribution of f_j in the absence of coupling to x_j (i.e., in a constrained ensemble with $x_j = 0$) is the same as that for the quenched version, $\langle f_j \rangle_0 = 0$ and $\langle f_i f_j \rangle_0 = \delta_{i,j} \sigma^2$. (Note that to make this problem look as similar as possible to the problem with quenched disorder, we have made the effective spring constant, T/σ^2, temperature-dependent in such a way that the variance of f_j is T-independent.) This problem is still simple enough that we can exactly compute the free energy by performing the appropriate Gaussian integrals for Z. The result is

$$\frac{F'}{N} = \frac{F_0(K)}{N} + \frac{F_0(T/\sigma^2)}{N} + \frac{T}{2} \ln\left[1 - \frac{\sigma^2}{KT} \right],$$ (9.8)

where the first two terms are the free energy of the decoupled problem and the last term is the result of the coupling between x_j and f_j.

Comparing this with $\overline{F}$ in Eq. 9.6, it is apparent that the change in the free energy produced by the interaction between the x_j and f_j is identical to leading-order in σ^2.

(To see, this, expand the logarithm in Eq. 9.8 to leading order in σ^2/KT.) However, for larger σ (or lower T), the two results are very different. Indeed, the system in which f_j is a thermal (rather than quenched) variable is thermodynamically unstable (reflected in the fact that the argument of the logarithm is negative) whenever $\sigma^2 > TK$, while no such instability occurs for the quenched system for finite σ^2. The big physical difference between the two cases is manifest here: in the quenched case, while the presence of the random forces f_j affects the thermodynamic state of the remaining degrees of freedom, the distribution of f_js is given once and for all, independently of T or any other parameters governing the thermal state of the system. In contrast, when all degrees of freedom can equilibrate there is a back action of the system on the "random" variables (sometimes referred to as "annealed disorder"), so that their distribution is T-dependent and more generally responds to the thermal state of the other degrees of freedom.

9.3 Gaussian Model in a Random Field

To get a better feeling for the effects of disorder, we turn to our new friend, the Gaussian model. As we have seen in the previous chapter, this can either be considered as a solvable model in its own right, or (with the parameters entering it determined from self-consistent mean-field equations) as a variational solution of a more general interacting model. We thus consider a system with a Boltzmann weight corresponding to the Hamiltonian

$$H[\{\phi\},\{h\}] = \frac{\mu}{2}\sum_j \phi_j^2 + \frac{J}{2}\sum_{<i,j>}[\phi_i - \phi_j]^2 - \sum_j h_j\phi_j, \qquad (9.9)$$

where, as before, ϕ_i represents a thermal variable at each site, while now h_j is a quenched random variable. As we shall see, we will not need to specify the precise ensemble—all that will be needed is the mean, which we will assume vanishes, $\overline{h_j} = 0$, and the variance, which we will assume is local, $\overline{h_i h_j} = \delta_{i,j}\sigma^2$. (As an exercise, the reader is encouraged to consider what differences would arise if somewhat different assumptions were made concerning the distribution, $P[\{h\}]$.)

The first step is to perform the many Gaussian integrals over the variables, ϕ_j, needed to compute the configuration-dependent free energy, $F[\{h\}]$. Fortunately, we have already done this in deriving Eq. 8.19, where we treated h_j as a set of externally applied fields. The result (undoing the Fourier transform used to derive an explicit expression) is

$$F[\{h\}] = F_0 - \frac{1}{2T}\sum_{i,j} h_i G(\vec{R}_i - \vec{R}_j)h_j, \qquad (9.10)$$

where F_0 is the free energy of the system in the absence of disorder and $G(\vec{R})$ is given in Eq. 8.15. From this, moreover, we can readily derive expressions for the thermal average quantities

$$\langle \phi_j \rangle = -\partial F / \partial h_j = T^{-1} \sum_j G(\vec{R}_i - \vec{R}_j) h_j \qquad (9.11)$$

and

$$\langle \phi_i \phi_j \rangle - \langle \phi_i \rangle \langle \phi_j \rangle = -T \partial^2 F / \partial h_i \partial h_j = G(\vec{R}_i - \vec{R}_j). \qquad (9.12)$$

The next step is to calculate the requisite configuration averages. This is straightforward. The result is

$$\frac{\overline{F}}{N} = \frac{F_0}{N} - \frac{\sigma^2}{2} \chi_0 , \qquad (9.13)$$

$\overline{\langle \phi_j \rangle} = 0$, and

$$\overline{\langle \phi_i \phi_j \rangle} = G(\vec{R}_i - \vec{R}_j) + \sigma^2 \chi^{(2)}(\vec{R}_i - \vec{R}_j), \qquad (9.14)$$

where the first term is the result in the absence of disorder, and in the second term

$$\chi^{(2)}(\vec{R}) \equiv \frac{1}{T^2} \sum_j G(\vec{R} - \vec{R}_j) G(\vec{R}_j) = \int \frac{d\vec{k}}{(2\pi)^d} \frac{e^{i\vec{k}\cdot\vec{R}}}{[\mu + Jg(\vec{k})]^2} . \qquad (9.15)$$

What does this result signify? Let us focus on the expression for $\overline{\langle \phi_i \phi_j \rangle}$. The first term is the familiar contribution from thermal fluctuations—for fixed parameters, it has a magnitude proportional to T. Moreover, as we discussed in Chapter 8, both the range of the correlations and their magnitude depend critically on the parameter μ/J; where the range of the correlations diverges as $\mu/J \to 0$. The second term has a very different character. It does not come from thermal fluctuations at all—for fixed parameters, it is T-independent, and it is a measure of the magnitude of the local values of the order parameter, $\overline{\langle \phi_i \rangle \langle \phi_j \rangle}$. Specifically, evaluated at $i = j$, it is the mean squared magnitude of the thermally averaged order parameter. Most strikingly, this term is even more singular as $\mu/J \to 0$; while G remains finite in this limit in any number of spatial dimensions $d > 2$, $\chi^{(2)}$ diverges for any $d \le 4$!

When we considered the Gaussian model as a variational approximation to a more complex model, we recognized $\mu \to 0$ as a treatment of the high-temperature (disordered) phase upon approach to the critical temperature, T_c. The fact that G diverges as $\mu \to 0$ reflects the fact that many types of order (not including Ising order) are forbidden for $d \le 2$. The above result, thus, suggests that there is a similar notion of a lower critical dimension in the presence of disorder, but now with a critical dimension of 4 rather than 2. Indeed, it was argued forcefully by a recent Nobel laureate, on the basis of sophisticated theoretical analysis which is, however, based on the simple considerations we have just discussed, that there is a general principle[2]

---

2. This is a technical subject for which no introductory review exists. The issues involved are still very much being investigated. The original proposition appeared in "Random Magnetic Fields, Supersymmetry, and Negative Dimensions," Parisi and Soulas, *Phys. Rev. Lett.* **43**, 744 (1979). A recent paper

of "dimensional reduction," i.e., that problems in the presence of random field disorder in dimension d exhibit the same sorts of critical properties as the same system without disorder in $d' = d - 2$. Let us explore this suggestion more carefully.

9.4   The Imry-Ma Argument

This issue was addressed first using a simple intuitive analysis by Imry and Ma, an analysis similar to the droplet argument we considered in Section 3.2.1. It turns out that the results of this analysis have been proven rigorously—something that is rare in physics.[3] To be specific, we will consider the case of an Ising ferromagnet, with nearest-neighbor exchange coupling J and with independently distributed random fields h_j with zero mean and variance σ^2. The question we wish to address is whether the ground state, in the limit $T \to 0$, is ferromagnetically ordered or if the random field necessarily destroys the ferromagnetic long-range order. If the ground state is not ordered, then it seems unlikely that there will be a broken-symmetry phase at any nonzero T as well. Conversely, if there is a broken-symmetry ground state, then the Peierls argument should still apply.

We thus start with a ground state in the absence of disorder, taking all the spins to be "up." We then look at a region of spins of linear dimension L, and ask how the energy of the state would change if we were to turn all these spins down. As with the droplet argument, there will be two terms. One comes from the surface of the region where there is an interface between the inside down spins and the outside up spins. This term goes as $L^{d-1}J$, the surface (hyper)area times the energy of a broken bond. The other term represents the change in energy from the random fields, which is just (twice) the sum of all the random fields in the region. By the central limit theorem, the typical size of this goes as the square root of the number of independent random variables, i.e., the (hyper)volume of the region, $\sigma\sqrt{L^d}$. Therefore, for large L,

$$\Delta E = L^{d-1}2J - A\sqrt{\sigma^2 L^d} = L^{d-1}J\left[1 - A(\sigma/J)L^{-(d-2)/2}\right], \tag{9.16}$$

where the precise value of A will depend on the particular region chosen. For half of all regions, A will be positive, and for others it will be negative, but it will only rarely have a magnitude that is very large or very small compared to 1.

Clearly, in any region in which A is negative, it costs energy to flip the spins. However, for regions in which A is positive, there will be an energy gain so long as $A(\sigma/J) > L^{(d-2)/2}$. In $d > 2$, this condition will never be satisfied for large L. Thus, while it may pay to flip the spins in the occasional small domain in which the random fields are particularly favorable to down spins, all such domains will be small. Moreover, in the limit of weak disorder ($\sigma/J \ll 1$), even such small domains will be

---

that discusses the current understanding of this topic is Tarjus, Balog, and Tissier, *Euro. Phys. Lett.* **103**, 61001 (2013).

3. J. Imbrie, *Commun. in Math Phys.* **98**, 145 (1985).

very rare. Thus, we conclude that the ferromagnetic state is robust to weak disorder in $d > 2$. On the other hand, for $d < 2$, even for weak disorder, if we take L to be sufficiently large it will always pay to flip the spins in any domain in which A is positive. We thus conclude that ferromagnetism is forbidden in the presence of disorder in $d < 2$. $d = 2$ is clearly the critical dimension—it requires a somewhat more sophisticated version of the Imry-Ma analysis to resolve the issue in this case. However, when this is done, one concludes that ferromagnetic order is also precluded in $d = 2$.

Thus, it is true that random field disorder precludes ferromagnetism in $d = 2$, although in the absence of disorder ferromagnetism survives to a finite critical temperature. However, contrary to the expectations of dimensional reduction, ferromagnetism is possible in $d = 3$!

9.5   Larkin-Ovchinikov-Lee-Rice Argument

We have already discovered that the lower critical dimension for breaking a discrete symmetry (allowed in $d > 1$) is smaller than that for a continuous symmetry ($d > 2$). It is thus reasonable to wonder whether the same may be true of the effects of disorder.

As a first pass at this problem, let us try to generalize the Imry-Ma argument to the case of a continuous symmetry, i.e., where our local order parameter is specified by a spin $\mathbf{S}_j$ of fixed magnitude that can point in an arbitrary direction in an N-dimensional order parameter space. Again, we start with a uniformly ferromagnetic ground-state, and consider the energy cost or gain of reorienting the spins in the interior of a region of linear dimension L. Now, however, rather than making a domain wall, we make a gradual twist of the spins over the entire length of the system. As we saw in our discussion of Landau-Ginzburg theory, the cost of such an order parameter texture will come from a contribution to the energy density proportional to $J|\vec{\nabla}\mathbf{S}|^2 \sim J/L^2$. Thus, integrating this over the domain, we find that

$$\Delta E = AL^{d-2}J - A'\sqrt{\sigma^2 L^d} = L^{d-2}J\left[A - A'(\sigma/J)L^{-(d-4)/2}\right], \qquad (9.17)$$

where A is a number of order 1 that depends on the precise way we twist the order parameter, and A' is, again, a random variable that depends on the particular configuration of random fields within the domain.

This line of argument leads us to the conclusion that no ferromagnetic state which breaks a continuous symmetry can arise in the presence of random field disorder in $d < 4$. In this case, $d = 4$ is the lower critical dimension, which again requires slightly more sophisticated analysis, from which it follows that random field disorder precludes the existence of a broken continuous symmetry in all $d \leq 4$.

In this case, the conclusion is congruent with what we had guessed on the basis of our analysis of the Gaussian model. Indeed, we can perform a more systematic analysis of this problem, which makes a more direct connection with the previous analysis.

Let us start from the problem with no disorder, and let us assume that we are at temperature $T < T_c$, where the system is an ordered ferromagnet, in which $\langle \mathbf{S} \rangle = m(T)\mathbf{\Omega}$, where $\mathbf{\Omega}$ is a unit vector in order parameter space. Since the ferromagnetic state spontaneously breaks the rotational symmetry of the spins, the direction of $\mathbf{\Omega}$ is arbitrary—presumably determined by the thermal history in which we cooled the system through T_c in the presence of a symmetry-breaking field along the $\mathbf{\Omega}$ axis, which we subsequently took to 0. Now, we wish to address the effect of weak disorder on the ordered state. If weak disorder has a small effect, we can safely conclude that the existence of the ordered state is robust—at least for weak enough disorder. Conversely, if we find that if even a small amount of disorder produces a disproportionately large distortion of the equilibrium state, we will be inclined to infer that it precludes a broken-symmetry phase.

Consider, therefore, the response of the system to a random magnetic field that is small in magnitude, $H \to H - \sum_j \mathbf{S}_j \cdot \mathbf{h_j}$. The result can be expressed to linear order in the random field as

$$\langle \mathbf{S}_j \rangle = m(T)\mathbf{\Omega} + \sum_i \left\{ \chi_{\parallel}(\vec{R}_{ij})\mathbf{\Omega}(\mathbf{\Omega} \cdot \mathbf{h}_i) + \chi_{\perp}(\vec{R}_{ij}) \left[\mathbf{h}_i - \mathbf{\Omega}(\mathbf{\Omega} \cdot \mathbf{h}_i) \right] \right\} + \dots, \quad (9.18)$$

where ... represents higher order terms in powers of h. Here χ is the linear response function, computed for the disorder-free system. It is related in the usual way to the equilibrium fluctuations of $\mathbf{S}_j$ by the fluctuation-dissipation theorem—$\chi = T^{-1}G$. However, since we are talking about the response of the system in a state which breaks the rotational symmetry of the order parameter space, the linear response is different when the applied field $\mathbf{h}_j$ is parallel to $\mathbf{\Omega}$—the longitudinal response—or perpendicular to it—the transverse response, as discussed in Section 7.4.

In terms of these responses, we can compute the mean squared magnitude of the order parameter:

$$\overline{\langle \vec{S}_j \rangle \langle \vec{S}_i \rangle} = M^2 + \frac{\sigma^2}{N} \left[\chi_{\parallel}^{(2)}(\vec{R}_{ij}) + (N-1)\chi_{\perp}^{(2)}(\vec{R}_{ij}) \right], \quad (9.19)$$

where

$$\chi_{\parallel}^{(2)}(\vec{R}_{ij}) = \sum_q \chi_{\parallel}(\vec{R}_{iq})\chi_{\parallel}(\vec{R}_{qj}) \quad (9.20)$$

and

$$\chi_{\perp}^{(2)}(\vec{R}_{ij}) = \sum_q \chi_{\perp}(\vec{R}_{iq})\chi_{\perp}(\vec{R}_{qj}). \quad (9.21)$$

For T near T_c, computing these response functions may be difficult. However, for the present purpose, we do not care about critical phenomena. Indeed, we have already shown in Section 8.3.1 that deep in the ordered phase, the transverse correlations are well approximated by those of the Gaussian model with $\mu = 0$ (reflecting the existence of Goldstone modes), while the longitudinal response can, to first approximation, be approximated by the Gaussian model with a nonzero μ that grows in proportion to $T_c - T$. In other words,

$$\chi_{\parallel}^{(2)}(\vec{R}_{ij}) \approx \int \frac{d\vec{k}}{(2\pi)^d} \frac{e^{i\vec{k}\cdot\vec{R}_{ij}}}{[\mu + Jg(\vec{k})]^2} \tag{9.22}$$

and

$$\chi_{\perp}^{(2)}(\vec{R}_{ij}) \approx \int \frac{d\vec{k}}{(2\pi)^d} \frac{e^{i\vec{k}\cdot\vec{R}_{ij}}}{[Jg(\vec{k})]^2}. \tag{9.23}$$

Note that $\chi_{\perp}^{(2)}(\vec{0})$ diverges in all $d \leq 4$. In other words, in the broken-symmetry state, even the smallest amount of quenched disorder produces an infinite change in the pattern of magnetization. This establishes that breaking of a continuous symmetry cannot occur in $d \leq 4$ (i.e., in any physically accessible dimension) in the presence of random field disorder.

One commonsense caveat is worth making at this point. Theorems are certainly very appealing in physics, and this result is more or less rigorous. But theorems can also be misleading. If a system has long-range order in the absence of disorder, then even if this description breaks down in real systems with disorder, it stands to reason that the system must have at least quite long finite range order if the disorder is weak. In other words, while the ferromagnetic correlations in $d \leq 4$ will likely fall exponentially with distance in materials with quenched disorder, that correlation length must diverge as $\sigma \to 0$. Thus, as a practical matter, in sufficiently clean systems it can often be reasonable to some level of accuracy to ignore the effects of disorder.

9.6   Other Effects of Disorder

What we have discussed in this chapter up to now is what is referred to as "random field" disorder—which is to say disorder that couples linearly to the relevant order parameter. What characterizes this sort of disorder is that it locally breaks the relevant symmetries—in the case of the Ising ferromagnet, it represents a local preference for one or the other direction of the magnetic order. Thus, in this case, the symmetry that is spontaneously broken in a ferromagnetic phase is not, strictly speaking, a symmetry at all—it is a statistical symmetry in the sense that it is a symmetry of the disorder ensemble, rather than of each realization of the disorder. When this flavor of quenched disorder is present, it typically has more significant qualitative effects on the nature of the phases than any other form of disorder. In this sense, the surprise may not be so much that the existence of a distinct broken-symmetry phase is precluded in low dimensions ($d \leq 2$ for the Ising case, $d \leq 4$ for a continuous symmetry); rather, it is remarkable that one can still define a broken-symmetry phase at all.

There is also the possibility of disorder that respects the order parameter symmetries. For example, one can consider the case of an Ising model,

$$H = -\sum_{<i,j>} J_{ij}\sigma_i\sigma_j, \tag{9.24}$$

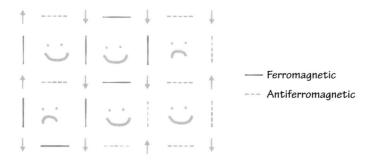

FIGURE 9.2. A schematic of a spin glass. The solid red and dashed blue lines represent ferromagnetic ($J_{ij} > 0$) and antiferromagnetic ($J_{ij} < 0$) interactions, respectively. If a plaquette has an odd number of antiferromagnetic interactions, it is frustrated, i.e., it is impossible to minimize all interactions simultaneously. Unfrustrated and frustrated plaquettes are depicted by smiley and frowny faces, respectively. Shown in gray is an example of an optimal spin configuration.

where the local exchange couplings, J_{ij}, depend on the disorder configuration in a way that varies from location to location in the system. Now $\{J_{ij}\}$ are quenched random variables. This problem still possesses a global "time-reversal symmetry," under $\sigma_j \to -\sigma_j \ \forall \ j$. Thus, in the high-temperature symmetric phase, it must generally be that $\langle \sigma_j \rangle = 0$; a low-temperature phase in which $\langle \sigma_j \rangle = m_j(T) \neq 0$ exists only if this symmetry is spontaneously broken. Note, however, that unlike in the disorder-free case, we expect that $m_j(T)$ varies from site to site, i.e., is j-dependent, since for a given realization of the quenched disorder the neighborhood of each site is somewhat different.

There are two extreme versions of random J disorder that are worth considering—although of course there can be intermediate cases:

i) If J_{ij} varies randomly in magnitude, but is everywhere positive, the ground-states at $T = 0$ are not in doubt; there are two such states, one with $\sigma_j = +1$ for all j, and the other its flipped-spin partner. This case is often referred to as "random T_c disorder." This name is somewhat misleading—generally, one finds that there is a unique T_c, such that for all j, $m_j = 0$ for $T > T_c$ and $m_j \neq 0$ for $T < T_c$. However, the same name carries the intuitive information that if one were to imagine the periodic extension of the local environment to the entire system, then one would have a locally defined mean-field T_c of the form $T_{cj} = \sum_{i = \text{n.n.} j} J_{ij}$. Random T_c disorder is relatively innocuous. In any case in which there is a broken-symmetry phase at low T in the absence of disorder, this phase will generally persist in the presence of disorder.[4] Indeed, in many cases, the only effect of the disorder is to shift the value of T_c and to slightly change the character of the broken-symmetry phase, although in some cases it can change the values of the critical exponents as well.

ii) If the J_{ij} is random in sign, the situation is considerably more complicated. As shown in Figure 9.2, this introduces "frustration" into the problem, in the sense that

---

4. There can be exceptions to this with certain extreme distributions of J, but these are somewhat special.

there is in general no configuration of the spins, σ_j, that simultaneously minimizes all the interactions in the Hamiltonian. This problem is referred to as the "spin-glass" problem—or, more specifically, this is the "Ising spin glass." It is an extremely rich and complicated problem. While much has been learned about it, there remain many mysteries.[5] Moreover, this turns out to be a paradigmatic representation of a host of "hard" problems that arise across disciplines of science and engineering. If we think of H as being not an energy, but some quantity that we wish to optimize, then finding the ground-state (or operationally as good, the low-T equilibrium state) of this problem may be a valuable goal. Such optimization problems arise in many practical contexts. Spin glasses are also studied as models of more general complex systems, including neural networks and social networks. A theory of spin-glasses whose applicability in physical dimensions, $d \leq 3$, remains highly controversial—but which turns out to be a right theory of many other problems with a large degree of frustration—was awarded a portion of the Nobel Prize in physics in 2021.

To give a taste of the richness of the problem, we end this section by briefly discussing the nature of the order parameter that characterizes "spin-glass order." As already mentioned, if we consider configuration-dependent quantities, then there is a clear way to distinguish the ordered from the symmetric phase; the magnetization $m_j(T)$ on any one site, j, vanishes for $T > T_{sg}$, and is nonzero for $T < T_{sg}$.[6] However, unlike in a ferromagnet, due to the frustration in the system, the sign of $m_j(T)$ varies from location to location in the system such that even for $T < T_{sg}$, the spatial average of $m_j(T)$, which is equal to its configuration average, vanishes:

$$\overline{m_j(T)} = N^{-1} \sum_j m_j(T) = 0. \tag{9.25}$$

Since this quantity cannot be used as an order parameter, we thus are driven to consider a somewhat more complex quantity—known as the Edwards-Anderson order parameter:

$$A_{ij}(T) \equiv \overline{\langle \sigma_i \rangle \langle \sigma_j \rangle}. \tag{9.26}$$

This quantity vanishes for all $T > T_{sg}$, since the thermal average preserves time-reversal symmetry. For $T < T_{sg}$, this quantity is nonzero. The site diagonal element, A_{jj}, represents the mean squared value of the local ordered moment—which as the average of a nonnegative quantity can vanish only if m_j itself vanishes identically (i.e., by symmetry). For large distance between the sites, R_{ij}, the relative sign of m_i and m_j becomes increasingly random, and thus $A_{ij} \to 0$ as $R_{ij} \to \infty$, even in the spin-glass phase.

---

5. See, for instance, Fischer and Hertz, *Spin Glasses* (Cambridge Univ. Press, Cambridge, 1991), or "Equilibrium behavior of the spin-glass ordered phase," Fisher and Huse, *Phys. Rev. B* **38**, 386 (1988).

6. There is a subtlety, even here; as discussed previously, a careful analysis of spontaneously broken symmetry involves applying a symmetry-breaking field, and then taking the limit as this field goes to zero after first taking the thermodynamic limit. However, for spin-glass order, the nature of the appropriate symmetry-breaking field is far from obvious.

10 *RENORMALIZATION GROUP PERSPECTIVE

Deep thoughts and much waving of hands as a parting gift to the student.

In this final chapter, we introduce the modern perspective on the theory of phase transitions based on an approach to statistical mechanics known as the renormalization group (RG). Unlike the straightforward calculations we have presented up until now, much of what we will cover in this chapter has a more qualitative flavor—sounding, perhaps, more like philosophy than science. One reason for this is that actual RG calculations tend to be more technical and complex than anything we have covered so far. But more generally, it is as a conceptual framework—more so than as a method for obtaining explicit results—that the RG is so powerful. Even where only skeletal results can be obtained, the RG provides a way of thinking about emergent physics that is comprehensive and intellectually satisfying[1].

To place the discussion in context, imagine an abstract space in which each point represents a possible Hamiltonian of a statistical mechanical system. The different directions in the space correspond to parameters such as the strength of certain interactions, the temperature, the chemical potential, and so forth. There are points in this space which represent materials at specified conditions—one for metallic Cu at room temperature, one for superfluid ^{4}He, one for $Nd_2Fe_{14}B$, the strongest commercially available ferromagnet, etc. There are also points in this space that correspond to simple models—the Ising ferromagnet on the cubic lattice, the ideal Bose and Fermi gases, the $O(N)$ ϕ^4 field theory, etc. Importantly, we will assume that points that lie close to one another represent similar systems. In particular, we picture that this space is divided up into simply connected subregions such that all points in a given subregion exhibit the same macroscopic phase of matter. The boundaries between these regions are "critical" hypersurfaces. A point on a critical hypersurface describes a system that is tuned to the phase transition between the two phases that lie on either side. (Places where hypersurfaces intersect correspond

---

1. There are numerous textbooks that cover this material, for example, *Scaling and Renormalization in Statistical Physics* by John Cardy (Cambridge University Press, Cambridge, UK, 1996) and E. Fradkin's, *Quantum Field Theory* (Princeton University Press, Princeton, 2021).

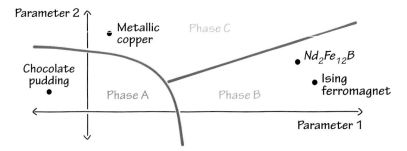

FIGURE 10.1. Schematic of an abstract space of Hamiltonians (see text).

to a system at a multicritical point.) A schematic two-dimensional version of this space is sketched in Figure 10.1.

This picture provides the basic justification for everything we have done. The fact that there is something generic to be understood about phases and phase transitions is based on the supposition that a system lying in a given region exhibits similar behaviors to all other such systems—especially long-distance equilibrium behaviors. On the empirical side, all ferromagnets have properties in common with each other, as do all superconductors and all liquids. But once we accept that this generality applies to the set of Hamiltonians realized in nature, there is no barrier to extending the notion to Hamiltonians that exist only as theoretical abstractions. If the Hamiltonian of a model lies in the same region as the Hamiltonian of a particular physical system, then the calculated properties of the model can be used to make qualitative statements about the nature of the phase of that physical system. Moreover, the closer the model and the physical Hamiltonian are to each other, the more similar we expect their behaviors to be.

To go further, we need to be a bit more explicit about what we mean by "similar behaviors." There are certain features that characterize a phase of matter. Any point inside a given region must share these features precisely with all other points. For instance, all points within a region must exhibit the same pattern of spontaneously broken symmetries and the same asymptotic functional form of connected correlation functions. Any crystalline system must have a finite shear modulus while any fluid system must be able to flow. There are additional features that are associated with the boundaries between regions; any system that is evolved along a trajectory from a point in one region to a point in a neighboring region (e.g., as a function of temperature or of some parameter in the Hamiltonian itself) must pass through the boundary at a unique point, i.e., must undergo a phase transition.

At the mean-field level, we have already begun to see that much about the character of a phase transition is independent of what trajectory we take through the critical surface; the nature of the thermodynamic singularities (e.g., the specific heat critical exponent, α), the existence of a diverging correlation length, and the divergence of a particular susceptibility are features that characterize all such trajectories. Moreover, as we have already mentioned, at an empirical level, a form of "universality" is observed in experiment—a set of critical exponents and amplitude ratios take on *exactly* the same values for a host of different physical systems (Section 1.3). Given that this is the case, there is every reason to expect the same degree of universality no matter if we are looking at a physical system or a model system. This gives us confidence that we should be able to compute these universal properties exactly, even if a model system differs in important details from any physical system, so long as both evolve through the same critical surface.

The universality of critical phenomena is not magical. In essence, it follows from the existence of a diverging correlation length. At a microscopic scale, each physical system—and each model problem—is unique with its own idiosyncrasies. But when we take a coarse-grained view and study behaviors on a larger lengthscale, these features tend to get averaged away. For example, looking at a fluid in a wineglass, we are not typically aware of the number of different chemical constituents and how they interact with each other;[2] macroscopic properties depend only on select essential features such as the density, viscosity, color, etc. Likewise, for most purposes, the properties of a copper wire at room temperature are not very sensitive to its chemical purity or its crystalline perfection. Many material properties such as these are averaged over macroscopic scales, so this sort of coarse-graining is always working to our advantage. But other features do depend on microscopic details. Crucially, however, the particular behaviors that are important near a continuous phase transition often concern properties averaged over lengthscales set by the correlation length. To the extent that this is the case, as the correlation length grows, the microscopic details become less and less important, and only the few features that survive this averaging remain. RG gives us a way to make this intuition precise.

10.1 Scaling Theory

We have already gotten used to the idea that the way a system behaves at and near criticality is characterized by a set of critical exponents. The existence of these exponents can be motivated by a set of heurristic arguments. Moreover, under certain assumptions, the critical exponents satisfy a set of "scaling relations" relating various exponents to one another. These scaling relations can be derived by assuming an asymptotic power-law behavior, and then demanding that certain thermodynamic quantities be consistent with one another.

Among the six exponents we have discussed, there are four scaling relations, derived below in Eqs. 10.4, 10.6, 10.9, and 10.12. This means that only two of them are truly independent.

---

2. Although if it is a glass of wine, a finely developed palate may still be able to detect small quantities of a particular chemical present in the wine from one vinyard in one especially blessed year.

A scaling relation involving η

The method for deriving scaling relations can be most easily seen by considering the exponent η. In mean-field theory we found that the order parameter correlation function at criticality falls as $\langle \phi(\vec{r})\phi(\vec{0})\rangle \sim |\vec{r}|^{-(d-2)}$. A special feature of a power law is that, unlike in the exponential function, there is no characteristic lengthscale associated with it. Indeed, a defining feature of a continuous critical point is that it is a point at which the correlation length diverges. Even if we do not expect the mean-field theory to give results that are correct in all details, we can still expect the exact solution to exhibit power-law correlations. Thus, in analogy with the exponents we have already defined, we can define another critical exponent, η, such that at criticality,

$$G(\vec{r}) \equiv \langle \phi(\vec{r})\phi(\vec{0})\rangle \sim A\ |\vec{r}|^{-(d-2+\eta)} \quad \text{for}\ \ T = T_c, \tag{10.1}$$

where A is a constant. (The exponent η is related to what is called the "anomalous dimension" of the order parameter, for reasons that we will touch on below.)

   Importantly, slightly away from criticality, we still expect the same power-law behavior to apply over distances $|\vec{r}| < \xi$, followed by exponential decrease for larger $|\vec{r}|$. (See discussion in Section 7.3.2.) Recalling the relation between susceptibility and equilibrium fluctuations, this implies that for T slightly above T_c, the susceptibility must go as

$$\chi = T^{-1} \int d\vec{r}\ \langle G(\vec{r})\rangle$$

$$\sim T^{-1}\ A\ \xi^{2-\eta}. \tag{10.2}$$

Next, we make the assumption that the divergence of the correlation length and the susceptibilies themselves are governed by power laws, $\chi \sim \chi_0\ t^{-\gamma}$ and $\xi \sim \xi_0\ t^{-\nu}$ (where $t \equiv (T - T_c)/T_c$). Substituting these into Eq. 10.2 yields

$$\chi_0 t^{-\gamma} \sim T^{-1}\ A\ (\xi_0 t^{-\nu})^{2-\eta}, \tag{10.3}$$

which can only be consistent if the power of t on both sides matches. Thus, we obtain our first scaling relation,

$$\gamma = \nu(2-\eta). \tag{10.4}$$

Observe that this relation is satisfied by the mean-field values, $\nu = 1/2$, $\eta = 0$, and $\gamma = 1$, in agreement with our assertion that the mean-field values are exact for $d > 4$. This scaling relation is also satisfied by the exact value of these exponents in $d < 4$ (e.g., extracted from numerical solutions of model problems) for various symmetry of order parameters, as can be checked from Table 1.4.

A scaling relation involving β

A similar scaling relation can be derived from another consistency condition, this time involving the behavior of $G(\vec{r})$ for T slightly below T_c. At lengthscales small compared to the correlation length, we expect $G(\vec{r})$ to behave as it does at criticality, while at large distances it should approach a finite value, m^2, to reflect the spontaneously broken symmetry:

$$G(\vec{r}) \sim A|r|^{-(d-2+\eta)}, \qquad |r| \ll \xi,$$

$$G(\vec{r}) \to m^2, \qquad\qquad |r| \to \infty \qquad \text{for } 0 \ll T < T_c. \qquad (10.5)$$

The lengthscale ξ is identified as the scale at which a crossover from "short-distance" critical behavior to "long-distance" broken symmetry occurs. Requiring that G be continuous at $|\vec{r}| \sim \xi$ implies that $m^2 \sim A\xi^{-d+2-\eta}$. However, m itself is governed by a critical exponent below criticality, $m \sim m_0|t|^\beta$. Matching exponents of t leads to the condition

$$2\beta = \nu(d - 2 + \eta)$$

$$= \nu d - \gamma, \qquad\qquad (10.6)$$

where in the second line we have used the first scaling relation, Eq. 10.4. Note that this relation is satisfied by the exact critical exponents listed in Table 1.4, but is satisfied by the mean-field exponents (with $\beta = 1/2$) only for $d = 4$. We have already noted that the mean-field exponents are exact in $d \geq 4$, so the fact that $d = 4$ is special from this perspective is not unexpected. (The breakdown of this scaling relation in $d > 4$ has an interesting origin, but one that is beyond the scope of the present book.)[3]

The nonanalytic part of the free energy

To obtain additional relations between critical exponents, we have to follow another, somewhat more subtle line of reasoning which leads to a deeper appreciation of the character of the critical state. Consider the free energy density, F, as a function of temperature and magnetic field, B. To focus on critical phenomena we consider two different sorts of contributions to F—a nonanalytic piece, F_{cr}, that depends on the degrees of freedom that are somehow essentially involved in the phase change, and an analytic piece that involves all the other microscopic degrees of freedom that evolve smoothly between the phases. This separation is intuitively reasonable even if there is ambiguity in making this separation in actual circumstances.

We will focus our attention on F_{cr}. Our first observation is that F_{cr} must take on a particular form because it has units of energy per unit volume. At temperatures near criticality, a relevant scale of energy is the (zero-field) T_c itself. For the noncritical degrees of freedom, the relevant volume scale is a microscopic scale of order a^d, where a is the lattice constant in the case of a crystal. However, the critical degrees

---

3. This is related to the concept of *dangerously irrelevant operators* in RG theory.

of freedom are correlated over much larger distance scales, as reflected in the fact that $\xi/a \gg 1$, and consequently the relevant volume is the correlation volume ξ^d. Therefore, by dimensional analysis, it follows that

$$F_{cr}(T, B) = T_c\, \xi^{-d}\, \tilde{F}_{cr}(t, b), \tag{10.7}$$

where $\tilde{F}_{cr}$ is a dimensionless function which depends on details of the system. For later convenience, we have expressed the arguments of $\tilde{F}_{cr}$ as the dimensionless measure of the temperature, t, and an appropriate dimensionless measure of the magnetic field strength, $b = (B/T_c)|t|^{-x_b}$. (The rationale for including the power of t in defining b and the appropriate value of x_b will be discussed below.) All we have done so far is to simply cast the original function in a somewhat different form by separating out the dimensionful and dimensionless parts, as can be done for many problems in physics.

A scaling relation involving α

Let us begin by approaching the critical point from above ($t > 0$) in the absence of a magnetic field ($b = 0$). The essential assumption is that the "scaling function" $\tilde{F}$ can be expressed as $\tilde{F}(t, 0) = \tilde{F}_0 + \tilde{F}_1(t)$, where $\tilde{F}_1(t) \sim t^{x_1}$ is "small" in the sense that it vanishes in proportion to a positive power of t (i.e., $x_1 > 0$). Given that the correlation length diverges upon approach to criticality as $\xi(t) \sim \xi_0\, t^{-\nu}$, we can take derivatives of F (Eq. 2.14) to derive the expression for the specific heat,

$$C = t^{-\alpha}\left[C_0 + \mathcal{O}(t^{x_1}) + \mathcal{O}(t)\right] + \cdots, \tag{10.8}$$

where $C_0 = (2 - \alpha)(1 - \alpha)\xi_0^{-d}\tilde{F}_0$, and the specific heat exponent is related to the correlation length exponent by the scaling relation

$$\alpha = 2 - \nu d, \tag{10.9}$$

where . . . represents the contribution from the noncritical part of F (which is thus a bounded, analytic function of t). Again, this scaling relation is satisfied by the exact critical exponents listed in Table 1.4, but is satisfied by the mean-field exponents only in $d = 4$.[4]

---

4. Because we are talking about a critical point, it is not justified to assume that $\tilde{F}(t, 0)$ is an analytic function of t at $t \to 0$. There will always be analytic pieces of $\tilde{F}$, and these will contribute to the $\mathcal{O}(t)$ corrections in Eq. 10.8. On the other hand, if $0 < x_1 < 1$, the corrections to the leading scaling behavior can vanish more slowly on approach to criticality than might naively have been expected. It is also important to note that the same sort of analysis can be applied for the approach to criticality from below, $t < 0$. Although F must be a continuous function of T, this does not preclude a discontinuity in $\tilde{F}(t, 0)$ at $t = 0$ (i.e., $\tilde{F}_0$ could be different as $t \to 0$ from above or below). As a result, there are generically distinct values of the "critical amplitude," C_0, upon approach to criticality from above and below.

A scaling relation involving δ

The B dependence of $\tilde{F}_{cr}$ is more complicated. In the presence of a fixed nonzero $B > 0$, there is no phase transition—the Ising symmetry is already (explicitly) broken—and therefore F must be an analytic function of T (and hence of t). For $t > 0$, $F(T, B)$ must be an analytic function of B for small B, which means that $F(T, B) = F(T, 0) - \frac{1}{2}\chi(T)B^2 + \mathcal{O}(B^4)$. Correspondingly, this means that $\tilde{F}(t, b) = \tilde{F}(t, 0) - \frac{1}{2}[T_c\chi\xi^d t^{2x_b}]\,b^2 + \mathcal{O}(b^4)$. Since $\chi \sim \chi_0 t^{-\gamma}$, it is clear that the B dependence of F becomes increasingly strong as $t \to 0$, meaning that the range of B over which this expansion is valid gets increasingly narrower. However, with the suitable choice of

$$x_b = \frac{1}{2}\left[\gamma + \nu d\right] \tag{10.10}$$

(at least to lowest order) the dependence of $\tilde{F}$ on b is uniform at small t, $\tilde{F}(t, b) = \tilde{F}(t, 0) - \frac{1}{2}\tilde{\chi}_0\,b^2 + \mathcal{O}(b^4)$ with $\tilde{\chi}_0 = T_c\chi_0\xi_0^d$. Put another way, in the critical regime, the field scale relative to which we measure the applied field B vanishes as $B^\star \sim T_c\,t^{x_b}$ upon approach to criticality, so that the scaled field, b, is the correct measure of its strength. The exponent x_b is called the scaling dimension of B—a concept to which we will return.

Precisely at $T = T_c$, the magnetization, $M(T_c, B) = -dF/dB \sim |B|^{1/\delta}$, has a nonanalytic B-dependence characterized by the critical exponent δ. Clearly, this means that $F(T_c, B) - F(T_c, 0) \sim -\frac{1}{2}\chi_1\,T_c^2|B/T_c|^{1+1/\delta}$. (The factors of T_c are included to make the units work out.) Even if the system is at a temperature ever so slightly larger than T_c, for $B > B^\star$ (i.e., for $b \gg 1$), the B dependence must be essentially indistinguishable from that at $T = T_c$. In other words,

$$\tilde{F}_{cr}(t, b) - \tilde{F}_{cr}(t, 0) \sim -\frac{1}{2}\begin{cases} \tilde{\chi}_0\,b^2 & \text{for} \quad |b| \ll 1 \\ \tilde{\chi}_1\,[t^y]\,|b|^{(\delta+1)/\delta} & \text{for} \quad |b| \gg 1, \end{cases} \tag{10.11}$$

with $y = x_b(\delta + 1)/\delta - \nu d$ and $\tilde{\chi}_1 = \chi_1 T_c\xi_0^d$. Since this function must cross over continuously from the small to the large b behavior at $b \sim 1$, independently of t, this scaling description can be consistent only if $y = 0$, i.e., if [5]

$$\gamma = \nu d\left(\frac{\delta - 1}{\delta + 1}\right). \tag{10.12}$$

10.1.1 Relevant and Irrelevant Perturbations

Now, finally, we have reached the conceptual climax of the analysis. In addition to the temperature and magnetic field, let us consider a family of distinct systems undergoing the same sort of phase transition. Each Hamiltonian is parameterized

---

5. We could have gone through the same line of argument as T approaches T_c from below, using the result that for $t < 0$ and small but nonzero B, $F(T, B) \sim F(0, 0) - m(T)|B| + \cdots$, which would have led to the same value of x_b, and the scaling relation $\beta = \nu d/(1 + \delta)$.

by a set of interactions, J_j, which we consider to be "perturbations" to the original model. We define a coordinate system in the space of Hamiltonians relative to a special point on the critical surface $T = T_c$, $B = 0$, and $J_j = 0$ $\forall$ j. We then generalize the above analysis to write the critical part of the free energy for any problem "close" to this reference problem as

$$F_{cr}(T, B, J_1, J_2, \ldots) = T_c \xi^{-d} \tilde{F}_{cr}(t, b, g_1, g_2, \ldots), \tag{10.13}$$

with

$$b = (B/T_c)|t|^{-x_b}, \quad \text{and} \quad g_j = (J_j/T_c)|t|^{-x_j}, \tag{10.14}$$

where the exponents x_j are such that $\tilde{F}_{cr}$ is a well-behaved function of its arguments. The exponents x_j are known as the "scaling dimensions" of the various interactions. The parameters J_j are the strengths of interactions in units of energy, while the rescaled versions g_j are dimensionless measures of the importance of those interactions.

Any interaction for which $x_j > 0$ is called a "relevant" interaction. For such an interaction, the dimensionless measure of its importance, g_j, increases as the critical point is approached, i.e., as $t \to 0$. This implies that such an interaction will somehow change the critical behavior. An example of a relevant perturbation is B for an Ising ferromagnet, since, as we have seen, $x_b > 0$; moreover, we know that there is no phase transition if $B \neq 0$.

Conversely, any interaction for which $x_j < 0$ is "irrelevant." In this case, the closer the system is to $T = T_c$, the smaller the dimensionless measure of the strength of the interaction g_j, and hence the less its effect on the thermodynamics.

If it is possible to write F_{cr} in this manner, there is a remarkable consequence: for any system which differs from the reference system only by a set of *irrelevant* perturbations, the importance of all perturbations g_j will decrease in magnitude as the critical point is approached, meaning that the differences between the reference system and the perturbed system becomes, increasingly small. Close enough to T_c,

$$F_{cr}(T, B, J_1, J_2, \ldots) \approx T_c \xi^{-d} \tilde{F}_{cr}(t, b, 0, 0, \ldots) \equiv T_c \xi^{-d} \tilde{F}_{cr}(t, b) . \tag{10.15}$$

Consequently, any system with nonzero values of J_j should exhibit the same critical behavior as a system with all $J_j = 0$—not only in the critical exponents but also in the form of the scaling function, $\tilde{F}_{cr}(t, b)$. These properties are universal in the sense that they are the same for all such systems.

The renormalization group approach—which we will describe qualitatively below—gives a constructive route to such a scaling description. Most importantly, it gives a precise criterion for identifying the optimal reference system in the above construction—this is defined as the "fixed point" of a certain abstract dynamics in the parameter space defined above. It gives us some way of determining which interactions are irrelevant and alerting us to the presence of possible "relevant" interactions (which could destablize the entire picture). In special cases, it even gives us a way to compute the critical exponents and universal pieces of the scaling functions, although in most situations this requires heavy-duty numerical analysis.

One peculiarity is that since the renormalization group is focused on determining universal properties, it is almost useless for computing non-universal properties. However, what is universal is not necessarily important, and vice versa. A clearly important example of a nonuniversal feature of a phase transition is the value of T_c; the absolute temperature at which a material transitions to a superconducting or ferromagnetic phase should have no effect on the universal properties of the phase transition. But it is of great practical importance whether a material has a T_c at room temperature or at 1K.

10.2   Renormalization Group–Overview

Often, to understand A you must know B; to appreciate B you must first understand A! Here A = the high level considerations of the present section, and B = the explicit examples in Secs. 10.3 and 10.4; cyclic reading is thus suggested.

The renormalization group refers to a transformation that turns one problem in statistical mechanics into another. Roughly speaking, the effect of the RG transformation is to "zoom out" and look at a system on longer lengthscales. Suppose we were looking at a sample of a ferromagnet under a microscope. An RG transformation corresponds to reducing the magnification of the microscope by a factor of $s > 1$. In this analogy, the resolution of the microscope is limited—meaning that at the highest level of magnification, the details cannot be resolved beyond a microscopic lengthscale, a. At an s-fold reduced level of magnification, we cannot see anything smaller than $a' = s \cdot a$. Crucially, the transformation preserves the large-scale structure of the system; "essential" information about the phase of matter is retained, but small-scale structure is averaged away.

Since critical phenomena involve patterns on the longest lengthscales, we are often interested in what happens after repeated RG transformations, i.e., as $s \to \infty$. Away from a critical point, the correlation length is finite, so it does not take much zooming out before all structured patterns are too small to see. Closer to a critical point, where the correlation length is longer, the patterns will persist for longer before they disappear. Exactly at a critical point, there is no characteristic length, so as the magnification is reduced, the patterns do not disappear; there is structure on all lengthscales. If the RG is properly implemented, a system at a critical point will not be changed by the transformation at all, i.e., it is a fixed point. This is why the RG is so useful in considering critical behavior.

The RG transformation can easily be incorporated into our picture of an abstract space of Hamiltonians: it transforms one system into another system, characterized by a new temperature and/or new coupling strengths, g_i. Thus, the RG transformation takes one point H into another point H'. Repeated transformations correspond to a series of points which trace out a trajectory in Hamiltonian space.

Typically, RG calculations are formulated with effective field theories, and the transformation is applied in a continuous manner. This is done by taking the scaling factor s to grow continuously with a "time" parameter t, such that after an RG time t, the system has undergone a net RG transformation with $s = e^t$. If we think of g_i as the coordinates of a point in Hamiltonian space, the dynamics of the RG flow can be determined from a set of differential equations for $\frac{dg_i}{dt}$. As we will see, the long RG time behavior of the flow is directly related to the long-distance behavior of thermodynamic correlations.

10.2.1   The Nature of the RG Transformation

There are two basic steps in the RG transformation. The first is referred to as "coarse-graining," corresponding to zooming out, as described above. The second is "rescaling," or changing the units with which we measure length and ϕ. The two steps are schematically illustrated in Figure 10.2.

It is important to recognize that this transformation is a calculational device, not a law of physics; as we shall elucidate below, the transformation must be carefully engineered so as to have the desired fixed point. To have a concrete model in mind, we will outline these steps starting with the ϕ^4 version of the Ising model (Eq. 8.34) with preferred values $\phi(\vec{R}) \sim \pm\phi_0$.

Coarse-graining

Coarse-graining involves averaging over fluctuations on short distances. Before coarse-graining, the thermodynamics is described in terms of the configurations of fields on each lattice site, with Boltzmann weights $P[\{\phi_j\}] \propto e^{-H/T}$. In the coarse-graining step we divide the lattice into smaller subregions and only consider the average spin within each subregion,

$$\tilde{\phi}_b = \frac{1}{n} \sum_{\substack{\text{sites } j \\ \text{in subregion } b}} \phi_j,$$

where n is the number of sites in a subregion. The "coarse-grained variable" $\tilde{\phi}_b$ is calculated by performing a partial thermal average over the subset of microstates consistent with a given $\tilde{\phi}_b$ (see Section 10.3). The size of each subregion, $a' = s \cdot a$, is larger than the microscopic lattice scale, a, but should still be small compared to any scale of macroscopic relevance. Crucially, the same long-wavelength thermodynamics can also be described in terms of these coarse-grained variables; thermal averages are governed by a new effective Boltzmann weight, $P[\{\tilde{\phi}_a\}] \propto e^{\tilde{H}/T}$, in terms of a new effective Hamiltonian $\tilde{H}$.

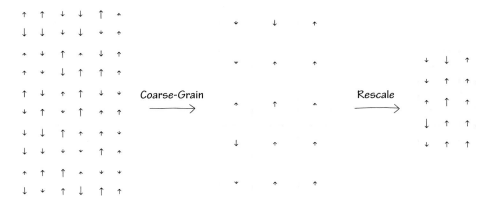

FIGURE 10.2. Schematic of the two basic steps in the RG transformation. The first step, coarse-graining, is to average out the short-distance degrees of freedom—in this case, by calculating the average within blocks of four spins. The second step, rescaling, is to change units of length and of ϕ in a way that the overall long-distance properties remain the same after the transformation. In this case, lengths have been multiplied by a factor of 1/2, and spins by a factor of $\sqrt{4} = 2$ (see Section 10.3).

Computing $\tilde{H}$ is the calculationally difficult part of an actual RG analysis. In general, even if H contains only nearest-neighbor terms, $\tilde{H}$ will include interactions between further neighbors and multi-spin interactions. Still, it is clear from the nature of the transformation that while $\tilde{H}$ may be complicated, it is at least short-ranged. Thus, if we start with a version of H which is sufficiently general to include any sort of interaction that could result from coarse-graining, then $\tilde{H}$ will be of the same form as H, just with different, "renormalized" coupling strengths.

Rescaling

There are still two issues after the coarse-graining step. Firstly, the lattice constants are clearly different in the two problems—we can take care of this by changing our unit of length, so that we report all lengths in the new problem in units of a'. Secondly, the coarse-grained variables $\tilde{\phi}$ will tend to have different scales than the microscopic variables ϕ. In the example we are considering, the microscopic field has a typical scale $|\phi| \sim \phi_0$. But since the coarse-grained variable $\tilde{\phi}$ is the value of ϕ averaged over a given region, it will have a different characteristic magnitude—as we shall see in Section 10.3. So we will want to rescale our effective fields suitably as well, $\phi' = Z\tilde{\phi}$. Expressing our coarse-grained Hamiltonian, $\tilde{H}$, in terms of these rescaled variables, we arrive at a new renormalized Hamitonian, H'.

It is preferable to choose units of the Hamiltonian so that we are dealing with dimensionless quantities. Since we are interested in the Boltzmann weight, we will focus on H/T and H'/T. Moreover, we will specify these in terms of a set of dimensionless measures of the interaction strength, $\{g_j\}$ and $\{g'_j\}$. For instance, in an Ising model, rather than specifying the strength of the interaction between nearest-neighbor spins, J, we will specify $g_1 = J/T$. In addition, we will include in the set of

gs all sorts of dimensionless interactions between further neighbors (although still of finite range) and multi-spin interactions.

The transformation

With these ingredients in place, we conclude that for a given system we can compute thermal averages using either the original or renormalized versions of the Hamiltonian, units of length, and field scales:

$$H(\{g\}) \longrightarrow H'(\{g'\}), \tag{10.16}$$

$$\vec{r} \longrightarrow (a/a')\,\vec{r},$$

$$\phi(\vec{r}) \longrightarrow Z_\phi(\{g\})\,\phi(\vec{r}),$$

where ϕ is the field we use to locally describe the state of the system. The outputs of the RG calculation are the renormalization factor Z_ϕ, and the relation between the initial and the renormalized couplings,

$$g'_j = g_j + \beta_j(\{g\}). \tag{10.17}$$

The key point is that the RG transformation maps the Hamiltonian onto itself with different parameters. We thus can iterate the transformation—the second step will proceed the same way as the first step. Indeed, as mentioned above, we can view Eq. 10.17 as a sort of dynamical equation for the parameters: we can define a discrete RG time, t, such that $a \to (a'/a)^t a$, in which case

$$g_j(t+1) - g_j(t) = \beta_j(\{g(t)\}). \tag{10.18}$$

This equation is then to be solved starting from the initial conditions, $g_j(0) = g_j$, i.e., the microscopic values of the various parameters. This can also be used to compute thermodynamic correlation functions; for example, defining $G(\vec{r}; \{g\}) \equiv \langle \phi(\vec{r})\phi(\vec{0}) \rangle$, we see that

$$G(\vec{r}; \{g(t+1)\}) = Z_\phi^2 G(\vec{r}a/a'; \{g(t)\}). \tag{10.19}$$

Eventually, because it is simpler, we will reformulate this problem in terms of an effective field theory, so that we can escape the discreteness imposed by a lattice and consider an infinitessimal version of the coarse-graining transformation, i.e., the case in which $a/a' = 1 - \epsilon$, with ϵ small. In this case, we can approximate the above difference equation by a 1st-order differential equation,

$$\frac{dg_j}{dt} = \beta_j(\{g\}). \tag{10.20}$$

Having identified this problem with a dynamical system, it is natural to integrate these equations to long times—which is to say to coarse-grain until we are left with a description at macroscopic lengthscales. Even without explicitly computing the "beta functions" β_j, we can guess quite a bit about the expected behavior of these equations. If we start at high temperatures, where the initial values of all the interactions are small, then we expect the correlations between neighboring regions to become increasingly weak the more we coarse-grain. In other words, starting with initial data corresponding to high temperatures (i.e., all $|g_j| \ll 1$), we expect that under RG all the couplings will become increasingly weak—$g_j(t) \to 0$ as $t \to \infty$. Conversely, if we start at low temperatures, where (at least) some particular interactions are large—corresponding to a phase with long-range order—then the more we coarse-grain, the less we will be sensitive to small fluctuations about the fully ordered state. In other words, the coarse-grained system will appear more and more like the $T \to 0$ ground state, and hence $|g_j(t)| \to \infty$.

But there are other behaviors that are possible for dynamical systems. Most importantly, under some circumstances there will exist (at least) one "fixed point" such that

$$\beta_j(\{g^*\}) = 0 \ \forall \ j. \tag{10.21}$$

A system with interactions tuned to a fixed point is scale invariant—the effective Hamiltonian that describes the thermal average looks the same at all lengthscales. This manifestly describes some sort of critical system. To see why this is the case, imagine computing the order parameter correlation function at the fixed point. From Eq. 10.19 it follows that

$$G(\vec{r}; \{g^*\}) = Z_\phi^2 G(\vec{r}a/a'; \{g^*\}). \tag{10.22}$$

It is simply a matter of substitution to verify that

$$G(\vec{r}; \{g^*\}) = G_0(\{g^*\}) \, |\vec{r}|^{-2x_\phi} \tag{10.23}$$

is a general solution to this iterated equation, where the exponent x_ϕ is determined from the relation

$$Z_\phi(\{g^*\}) = |a/a'|^{x_\phi}. \tag{10.24}$$

Comparing this result to the expression in Eq. 10.1, we see that the anomalous dimension is related to x_ϕ as $\eta = 2x_\phi - d + 2$; it is called "anomalous" since if one were to rescale the fields to leave the Gaussian model invariant (i.e. for the Gaussian fixed point defined below) we would find $x_\phi = d - 2$ and hence $\eta = 0$. The correlations computed for the fixed-point Hamiltonian are thus seen to be pure power laws—as expected at a critical point!

To get a handle on the sort of critical system it is that results in these power-law correlations, let us consider what happens if we start with a microscopic

Hamiltonian that is close to the fixed point Hamiltonian, i.e.,

$$g_j(0) = g_j^* + \delta g_j(0), \tag{10.25}$$

with $|\delta g_j|$ assumed small. We thus can expand our dynamic equation to linear order in δg_j, so that

$$\delta g_j(t+1) = \Gamma_{ij}\delta g_j(t) + \cdots, \tag{10.26}$$

where summation convention is assumed,

$$\Gamma_{ij} \equiv \left.\frac{d\beta_i(\{g\})}{dg_j}\right|_{\{g\}=\{g^*\}}, \tag{10.27}$$

and ... indicates higher order terms in powers of δg. Note that Γ is a property of the fixed point Hamiltonian. Its eigenvalues and eigenvectors determine the nature of the RG "flow" in Hamiltonian space in the neighborhood of the fixed point. Specifically, let

$$\Gamma_{ij}f_j^{(a)} = e^{\gamma_a}f_i^{(a)}, \tag{10.28}$$

and let us express δg in this basis as

$$\delta g_j(t) = \sum_a \delta\tilde{g}_a(t)f_j^{(a)}. \tag{10.29}$$

We can readily solve the resulting equations to find

$$\delta\tilde{g}_a(t) = e^{\gamma_a t}\delta\tilde{g}_a(0) + \cdots, \tag{10.30}$$

where as before ... represents higher order terms in δg.

Until now, we have not paid any attention to the particular way in which we characterized the interactions in the problem—for instance, g_1 might characterize the strength of the interaction between nearest-neighbor sites, and g_2 of that between second-nearest neighbors. Now, however, the RG treatment gives us reason to prefer a more natural basis for describing interactions—the eigenvectors of Γ. On physical grounds, we expect the eigenvalues γ_a to be real.[6] To draw a connection to our scaling analysis earlier, any interaction for which $\gamma_a < 0$ is "irrelevant," any for which $\gamma_a > 0$ is "relevant," and any for which $\gamma_a = 0$ is "marginal."

Suppose at a given fixed point, all perturbations were irrelevant, i.e., $\gamma_a < 0\ \forall\ a$. In this case, even if the microscopic system of interest is not precisely described by the fixed point Hamiltonian, the behavior of the system at large lengthscales resembles that of the fixed point. Specifically, if we consider a coarse-grained version of the original problem, then the longer the lengthscale probed, the more the properties of the system will resemble the properties of the fixed-point problem. In

---

6. While in certain strange limits it is possible to find problems in which complex eigenvalues arise, their physical interpretation is still unclear. Certainly, complex interactions, g_j, make no physical sense.

the parameter space we have visualized, the fixed point Hamiltonian corresponds to a specific point in the space; if all possible perturbations about this Hamiltonian are irrelevant in the technical sense just described, this implies that there must exist a region of nonzero extent surrounding the fixed point in which the renormalization group flows converge on the fixed point at large RG time; this region is known as the "basin of attraction" of the fixed point.

The one problem with this is that we do not expect to find such a fixed point. We have identified critical points with surfaces in our parameter space separating regions of different phases. For any point on such a surface (i.e., any system at its critical point) that means that there is one direction in parameter space—the direction that is locally perpendicular to the critical surface—which is clearly highly relevant in the colloquial sense of the word. For instance, if we hold everything but the temperature fixed, then at $T = T_c$ we expect critical behavior, but for T ever so slightly higher than T_c we are in the disordered phase, so at a long distance all correlations fall exponentially with distance, while for $T < T_c$, the long-distance behavior is characterized by a spontaneously broken symmetry. Clearly, for this sort of critical point we expect there to be at least one relevant perturbation.

Consider, therefore, the case of a fixed point with a single relevant perturbation, which we will take to be $\delta \tilde{g}_1$. Now, the basin of attraction of the fixed point constitutes a surface of co-dimension 1, known as the critical surface. To visualize this, consider a fixed point that lies somewhere on the surface separating phase A from phase B (Figure 10.3). At the fixed point, this surface is perpendicular to $\vec{f}^{(1)}$. Any microscopic system which lies within the critical surface will flow under the RG transformation ever closer to the fixed point. In other words, at long distances, the behavior of all systems in the basin of attraction of the fixed point will be identical—they will exhibit the same universal critical behavior. However, even if the microscopic system corresponds to a point close to the fixed point, if $\delta \tilde{g}_1(0)$ is nonzero, then at long RG time the flow will carry the system away from the fixed point. This is precisely the behavior we expect of a critical point—one parameter needs to be tuned in order to tune the system to criticality. An extension of this is a fixed point with two (or more) relevant directions—these fixed points correspond to multicritical points which can be accessed only by tuning two (or more) parameters to their critical values.

This is perhaps the most important result of the RG approach—it gives a clear and crisp understanding of how universality arises. It moreover gives us a more precise understanding of the origin of power-law correlations of systems at criticality as well as provides (in principle) a way to compute the critical exponents. It also allows us to gain a more general understanding of the exponents that govern the approach to criticality.

Suppose we return to the example of a fixed point with a single relevant direction, and ask what happens if we start with a system nearby but slightly displaced from the critical surface. Under RG flow, the strength of the relevant interaction, $\tilde{g}_1(t)$, grows. After a while, therefore, it will no longer be small. At that point we need to treat the

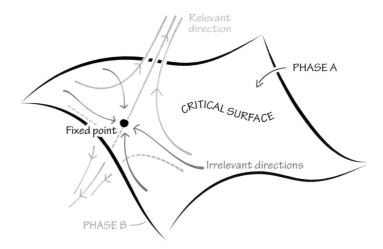

FIGURE 10.3. Schematic of a critical surface separating phase A from phase B. Any system lying exactly on the critical surface flows into the fixed point after repeated coarse-graining (red arrows), but if the relevant parameter is tuned away from its critical value, then the system will not lie on the surface and instead flow into either phase A or phase B at long distances (blue arrows).

problem in a different way—we can no longer expand the beta function about the fixed point. However, what we can do is to identify a characteristic lengthscale at which this happens, i.e., at which $\tilde{g}_1(t^*) \sim 1$:

$$t^* = |\gamma_1|^{-1} \ln|1/\delta\tilde{g}_1(0)| \quad \rightarrow \quad \xi = a(a'/a)^{t^*} = a|\delta\tilde{g}_1(0)|^{-\nu}, \tag{10.31}$$

where $|\delta\tilde{g}_1(0)|$ is the "distance" to the critical surface, and

$$\nu = [\gamma_1]^{-1} \ln[a'/a] \tag{10.32}$$

is the correlation length exponent. Note that ξ appears in the RG treatment as a crossover scale. For small $|\delta\tilde{g}_1(0)|$, the RG flows leave the system close to the critical surface—i.e., correlation functions look similar to those at criticality for lengthscales small compared to ξ. However, the flows carry the system far from the critical surface—i.e., the correlation functions are distinct from those at criticality—at distances larger than ξ. By itself, this analysis does not tell us what the different behavior at longer distances might be—that requires separate analysis. However, assuming we know the properties of the two phases on either side of the critical point, this should not be too mysterious.

10.3   RG for the Central Limit Theorem

The central limit theorem from statistics serves as an interesting application of the RG perspective. Recall the statement of the theorem: when a large number M of independent random variables are drawn from a given probability distribution, the distribution of the mean approaches a Gaussian distribution, with a width that decreases in proportion to $1/\sqrt{M}$ (Appendix A.7). Importantly, the ultimate distribution is Gaussian, *no matter the initial probability distribution* (as long as the mean and variance are both finite).

From the viewpoint of RG, the universal behavior arises because the Gaussian distribution is a fixed point of an appropriate RG operation in the space of probability distributions. In this example, the RG operation transforms one probability distribution into another by coarse-graining and rescaling, as described shortly. Upon iteratively applying this operation, any probability distribution (of finite mean and variance) approaches the Gaussian fixed point.

To make the connection between this problem and statistical mechanics, imagine we were studying the equilibrium properties of a system consisting of decoupled sites. (We could, for instance, be thinking about a high-temperature limit in which T is large compared to the interactions between neighboring sites.) At each site is a random variable ϕ_j, with the same probability distribution $p(\phi)$ on all sites. We are interested in the thermodynamic average of ϕ_j over all sites. We can define an RG operation which transforms this problem into a new, coarse-grained problem as follows. First we divide the lattice into blocks. Each block contains n sites. Next, we calculate the average field of all the sites within a block, and we call that the coarse-grained field,

$$\tilde{\phi}_b = \frac{1}{n} \sum_{\substack{\text{sites } j \\ \text{in block } b}} \phi_j, \tag{10.33}$$

This procedure is illustrated in Figure 10.2.

The probability distribution of the coarse-grained fields can be written as

$$P_n(\tilde{\phi}; p) = \int \prod_{j \in a} p(\phi_j) d\phi_j \, \delta\left[\tilde{\phi} - n^{-1} \sum_{j=1}^{n} \phi_j\right], \tag{10.34}$$

where, the integral runs over all configurations of the fields ϕ_j within block b, and the delta function enforces the condition that $\tilde{\phi}$ is the average value within the block.

Now we have a new probability distribution, P_n, of the average spin in a block. This is a coarse-grained version of the original problem. Importantly, the procedure can be iterated to construct blocks of blocks, and then blocks of blocks of blocks.

The other necessary ingredient of the RG transformation is the rescaling. When the average of n variables is taken, the width of the distribution shrinks by a factor of $\sqrt{n}$ (see Problem 4.2). However, if we wish the Gaussian distribution to be a fixed point, we must ensure that it transforms into itself; its width cannot shrink. To compensate for this, after the coarse-graining step, we must *rescale* the variable by a factor of $\sqrt{n}$. If this rescaling is not properly chosen, the distribution will not approach a useful fixed point upon repeated transformations.[7]

The two steps together, the coarse-graining and the rescaling, constitute a single iteration of the RG operation:

$$p(\phi) \rightarrow q(\phi') \equiv p'(\phi')n^{-1/2}P_n(n^{-1/2}\phi;p), \qquad (10.35)$$

where P_n is defined in Eq. 10.34 and ϕ' represents the *rescaled* version of the average spin within a block, $\phi' = \sqrt{n}\tilde{\phi}$. If $p(\phi)$ is a Gaussian with mean $\bar{\phi}$ and variance σ, then the result of this transformation, $q(\phi')$, is also a Gaussian, with

$$\bar{\phi} \rightarrow \sqrt{n}\,\bar{\phi}, \qquad (10.36)$$

$$\sigma \rightarrow \sigma.$$

In particular, a Gaussian distribution with zero mean ($\bar{\phi} = 0$) is a fixed point of this transformation. Moreover, the central limit theorem assures us that this is an attractive fixed point—that distributions within a certain "basin of attraction" will flow toward, rather than away from, this fixed point. To see this, in Problem 10.1 you will be asked to iterate Eq. 10.35 starting with a variety of different "microscopic" distributions, $p_0(\phi)$, to see that they all approach the Gaussian upon repeated RG transformations.

10.4   The RG in Slightly Less than 4D

Now let us consider actually carrying out an RG analysis of an interesting problem in statistical mechanics, and in particular a system at or near a critical point where correlations between neighboring sites are strong and complicated. On the face of it, carrying out such an explicit RG analysis may be no easier than solving the problem directly—i.e., beyond the reach of existing theory. However, in certain cases, it is possible to identify a small parameter—even if it seems artificial at first glance— that allows theoretical control of the approach. The two most significant of these are known as the ϵ expansion and the $1/N$ expansion. In the former, the small parameter is $\epsilon = 4 - d$, where d is the dimensionality of space. This approach is thus valid in the limit of dimension "close to 4." In the latter, large N refers to an order parameter with many equivalent components, as in the $O(N)$ theories we have already discussed.

It is not clear why we should consider these small parameters to be reason- able. Nonetheless, the ϵ-expansion and the $1/N$-expansion give qualitatively useful insight into the broader problem of general d and N. Astoundingly, for the case of a

---

7. To be concrete, suppose we were to begin with a distribution with finite variance, and we attempted to construct an RG with a rescaling $\phi \rightarrow \phi/n^{\alpha}$. If $\alpha > 0.5$, then the width of the distribution would diverge as $n \rightarrow \infty$; if $\alpha < 0.5$, it would go to zero. Only $\alpha = 0.5$ leads to a fixed point. The rescaling must be chosen carefully.

3D Ising model ($d = 3, N = 1$), the approaches can yield good quantitative results, even though, with $\epsilon = 1$ and $1/N = 1$, neither expansion parameter is obviously small.

Carrying out the analysis to derive the expansions is mathematically involved, so we will focus on general features of the result. The details of the calculations are worked out in great detail in Worksheet W10.2. The steps we go through are essentially the same as we sketched above, in Section 10.2; they are repeated here for pedagogic clarity. Students interested in the details are encouraged to read one of many advanced texts containing a more complete treatment.[8]

Formulation of the problem

To have a concrete model problem in mind, let us consider the $O(N)$ ϕ^4 theory as in Eq. 8.44. Since critical behavior involves lengthscales large compared to a lattice constant, it is more sensible to reformulate this problem as a continuum field theory. What this means is that instead of fields defined at lattice points, ϕ_j, the microstates are specified by the configurations of a field $\phi(\vec{r})$ at all points $\vec{r}$ in a d-dimensional space, as when we constructed the Ginzburg-Landau free energy. However, now, rather than just finding the field configurations that minimize the free energy, we have to account for all possible field configurations. Namely, we wish to compute thermodynamic averages by integrating over all $\phi(\vec{r})$ with a Boltzmann weight $\sim e^{-H/T}$, where H is the Hamiltonian functional, $H[\phi]$, i.e., the energy associated with each field configuration. The Hamiltonians that describe systems near their critical points are constrained by the same notions that arise in the Ginzburg-Landau theory.

Specifically, locality implies that H can be written as an integral over a Hamiltonian density

$$\frac{H[\phi]}{T} = \int d^d r \, \mathcal{H}(\phi, \vec{\nabla}\phi, \partial_a \partial_b \phi, \ldots), \tag{10.37}$$

where $\mathcal{H}$ is a function of the field and its derivatives at a point in space. Moreover, $\mathcal{H}$ must be an analytic function, so it can be expressed as a power series in its arguments. Under conditions in which the relevant values of ϕ are relatively small and slowly varying, we can keep terms to low order in a joint expansion in powers of ϕ and of gradients. Symmetry further limits the terms that arise; for instance, if the system has an Ising symmetry under $\phi \to -\phi$, then we know that $\mathcal{H}$ must be invariant under this transformation, in which case it can only have even powers of ϕ. Consequently, we are led to consider the field-theoretic version of the ϕ^4 theory that we

8. Chaikin and Lubensky, *Principles of Condensed Matter Physics* (Cambridge University Press, Cambridge, 2000) or Cardy, *Scaling and Renormalization in Statistical Physics* (Cambridge University Press, Cambridge, 1996).

have already treated variationally in Section 8.2;

$$\mathcal{H} = \frac{\alpha}{2}|\phi|^2 + \frac{\kappa}{2}|\nabla\phi|^2 + \frac{u}{4}|\phi|^4 + \frac{v}{6}|\phi|^6 + \cdots, \tag{10.38}$$

where the ... includes terms proportional to $|\nabla^2\phi|^2$, $|\phi(\nabla\phi)|^2$, and $|(\nabla\phi)^2|^2$, and other terms with high powers of $\phi(\vec{r})$ and its derivatives.

This is physically sensible, but with the understanding that variations in $\phi(x)$ on lengthscales shorter than a do not really mean anything. This is a central subtlety in the definition of any field theory that we have previously hidden by always defining models on a lattice. If we were to take the prescription literally, then in integrating over "all" field configurations, we would include possible structures that occur on arbitrarily small lengthscales. However, at small scales, there are all sorts of microscopic physical structures in a system that are not captured by $\phi(\vec{r})$. When we considered systems on a lattice, we chose the lattice constant, a, to represent the smallest physical scale—roughly corresponding to the distance between constituent atoms—for which our model is relevant. Now, in the field theory description, we likewise need to recognize that there is a shortest scale at which the field-theoretic description applies—which in analogy to a lattice constant we will also call a. This is the field-theoretic version of the coarse-graining scale that we have been discussing. In some way or other we will have to supplement the theory defined in Eq. 10.38 to ensure that fluctuations on shorter scales than a make no significant contribution to any interesting thermally averaged quantity.

To be explicit, let us recall the Gaussian model we analyzed in Section 8.1.1. In Eq. 8.25 we obtained an expression for the expectation value of $\langle|\phi|^2\rangle$. Since we carried out the calculations on a lattice, we obtained (as we should) a finite answer. But we also observed that this quantity is "ultraviolet divergent," meaning that if we were to keep the leading order in the gradient expansion, as in Eq. 10.38,[9] we would get a divergent result in any $d \geq 2$. Formally, this is the statement that $\int d^d k/[\mu + k^2]$ diverges for $d \geq 2$. A divergent $\langle|\phi|^2\rangle$ is obviously unacceptable—among other things, it is inconsistent with the notion that the magnitude of ϕ is typically small. The way to avoid the divergence is to introduce some upper cutoff to the integral, $k_{max} \sim 1/a$, corresponding to the shortest lengthscale a that the model is meant to describe. The field theory in Eq. 10.38 is not well defined until we specify a "regularization" scheme that suppresses the contribution of fluctuations with $|\vec{k}|a > 1$.

Outline of the RG transformation

Now, we are in a position to perform the RG transformation. The first thing we do is perform the average over fluctuations of the fields on scales between a and ae^t.

---

9. This is equivalent to approximating the lattice structure factor by its leading-order expression at long wavelengths, $g(\vec{k}) \sim |\vec{k}|^2$.

Then, we rescale length by a factor of $1/e^t$ such that the new short-distance cutoff scale is taken to be the unit of length. We also need to cleverly rescale the fields in such a way that their fluctuations get neither larger nor smaller as we average over short-distance fluctuations. The result is a new, renormalized field theory that describes the same long-distance physics without including explicitly the shortest distance fluctuations of the original theory.

In other words, we start with the system described in Eq. 10.38 with parameters $(\alpha, \kappa, u, v, \ldots)$, and short-distance cutoff scale a. Applying the RG to this system will transform it into a new system with a different set of parameters, $(\alpha', \kappa', u', v', \ldots')$, and short-distance cutoff scale $a' = ae^t$. The two systems have the same large-scale behavior, but the new system is zoomed out by a factor of e^t and coarse-grained accordingly. Both the old and the new system can be represented by points in an abstract space of Hamiltonians.

Flow equations and fixed points

To actually implement the RG, we need to write down a set of equations for $(\alpha', \kappa', u', v' \ldots')$ in terms of $(\alpha, \kappa, u, v, \ldots)$. This generally cannot be implemented exactly in any "interacting theory" that includes terms beyond those of order ϕ^2. However, if the interactions are weak, it can be carried out to leading order by a simple form of dimensional analysis. The results can then be improved by considering the effects of interactions perturbatively.

The dimensional analysis is straightforward. Firstly, we express all lengths in units of a, i.e., we transform to rescaled coordinates, $\tilde{\vec{r}} = \vec{r}/a$. Secondly, since at criticality the leading term in $\mathcal{H}$ is the gradient term, a natural choice of units for the fields is such that $\kappa \to 1$, i.e., we transform to rescaled field $\tilde{\phi} = \sqrt{\kappa a^{d-2}}\phi$. In these rescaled units, $\mathcal{H}$ appears with the dimensionless coupling constants, $\tilde{\alpha} = (\alpha/\kappa)a^2$, $\tilde{\kappa} = 1$, $\tilde{u} = (u/\kappa^2)a^{4-d}$, and $\tilde{v} = (v/\kappa^3)a^{6-2d}$. In the absence of fluctuations, the averaging over fluctuations on lengthscales between a and ae^t does not change the dimensionless coupling constants. Thus, to leading order, the transformation properties of the coupling constants are given by $\alpha \to \alpha e^{2t}$, $\kappa \to \kappa$, $u \to ue^{(4-d)t}$, and $v \to ve^{(6-2d)t}$. We could likewise easily obtain the leading order transformation properties of the other higher order terms in $\mathcal{H}$ we have left out.

Carrying out the perturbation theory is technical but standard. It is done very carefully and thoroughly in Worksheet W10.2. The resulting RG flow equations have the form of

$$\frac{d\kappa}{dt} = 0 \tag{10.39}$$

$$\frac{d\alpha}{dt} = 2\alpha + A_2 u + \cdots \tag{10.40}$$

$$\frac{du}{dt} = (4-d)u - A_4 u^2 + \cdots \tag{10.41}$$

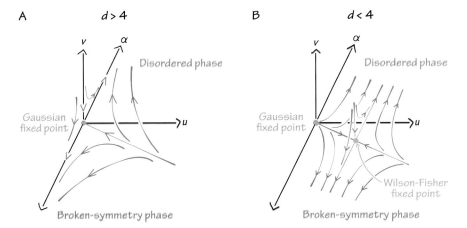

FIGURE 10.4. Fixed points of the RG transformation for an $O(N)$ model. (A) For $d > 4$, the Gaussian fixed point is a stable fixed point. (B) For $d < 4$, the Gaussian fixed point is unstable but the Wilson-Fisher fixed point is stable.

$$\frac{dv}{dt} = (6 - 2d)v - A_6 u^2 + \cdots \qquad (10.42)$$

$$\ldots, \qquad (10.43)$$

where ... signifies higher powers of u, v, α, and other coefficients we have not included. The coefficient of the first term reflects the dimensional analysis we have just reviewed, whereas the coefficients A_n (and the higher order terms) can only be calculated through explicitly implementing the perturbation theory.

This set of equations has a fixed point at $(0, \kappa, 0, 0, \ldots)$, where all the coefficients except κ are zero. This is known as the Gaussian fixed point, since it is the Gaussian theory evaluated at criticality. The fact that the factor of 2 in Eq. 10.40 is positive implies that α is a relevant perturbation. This is unsurprising, since we have come to identify α with $T - T_c$.

However, crucially, for dimension $d \approx 4$ the factor multiplying u in its flow equation—$(4 - d)$—is negative for $d > 4$ but positive for $d < 4$. This means that the $u = 0$ fixed point is stable (u is irrelevant) for $d > 4$ but is unstable for $d < 4$ (u is relevant). By contrast, v (along with all the higher order terms represented by ... in the flow equation) is irrelevant for $d \approx 4$. We thus have two different situations, depending on whether d is slightly greater or less than 4. This is depicted in Figure 10.4.

In the case with $d > 4$, the only relevant perturbation of the Gaussian fixed point is α. For $\alpha > 0$, the RG flows diverge from the critical surface into the disordered phase, while for $\alpha < 0$, the flows carry the system ever deeper into the low-T (broken-symmetry) phase. The $\alpha = 0$ situation corresponds to the system at $T = T_c$. Since α is the only relevant perturbation, *all other coefficients constants go to zero as $t \to \infty$.*

Herein lies the explanation for the remarkable fact that mean-field theory gives the correct description of critical behavior in all $d > 4$! Critical systems in $d > 4$ will flow into the Gaussian fixed point after following the RG flow for long times. The deviations away from mean-field theory are irrelevant; upon viewing the system at longer and longer lengthscales, the differences between the Gaussian fixed point and the system at hand vanish.

In contrast, in the $d < 4$ case, the Gaussian fixed point has *two* relevant perturbations—α and u—so it is not the fixed point which governs critical behavior. Even if u starts off with a small value, the action of the RG will eventually cause it to grow.[10] Rather, in $d < 4$, the critical behavior is governed by another fixed point of these equations, known as the Wilson-Fisher fixed point, with

$$\alpha^\star = -(A_2/2A_4)\epsilon + \cdots, \qquad (10.44)$$

$$u^\star = (1/A_4)\epsilon + \cdots, \qquad (10.45)$$

where $\epsilon \equiv 4 - d$ and ... signifies terms that are higher power in ϵ. Importantly, the Wilson-Fisher fixed point has only one relevant perturbation: $\alpha - \alpha^\star$. (This can be seen by expanding the flow equation to linear order in powers of $(\alpha - \alpha^\star)$ and $(u - u^\star)$.)

The ϵ expansion

In $d = 3$, ϵ has a value of 1, so there would seem no justification for neglecting all the higher order powers of ϵ. At this point, those who lack imagination might begin to despair. However, those who dare to study the abstract problem where $\epsilon \ll 1$ would realize that everything is under control. The Wilson-Fisher fixed point is "close" to the Gaussian fixed point, so the perturbation theory is justified and the results are self-consistently valid. We can calculate the critical exponent ν from the flow out of the fixed point and the exponent η from the rescaling of ϕ (which we have not discussed in any detail above).

As mentioned previously, the coefficients A_2, A_4, etc., can be computed with laborious perturbation theory. With time and hard work, intrepid physicists have carried out the perturbative analysis to ever higher orders. Now the RG flow equations have been studied up to fifth order in powers of ϵ. Results have been obtained not only for the Ising case, but for the $O(N)$ case and other possibilities.[11]

---

10. This is true unless we were to have a doubly fine-tuned scenario where both α and u are zero, as in our discussion of multicritical points. In this case we see that the Gaussian fixed point in $3 < d < 4$ corresponds to a multicritical point. Since v flows as $dv/dt = 2(3 - d)v + \cdots$, it is relevant for $d < 3$.

11. In fact, there is considerably more information available from this RG treatment. For instance, it is possible to compute the values of certain universal amplitude ratios of the sort we discussed in Section 4.4.2. Moreover, the exponents that govern the flow of the leading-order irrelevant

$$\nu = \frac{1}{2} + \frac{N+2}{4(N+8)}\epsilon + \frac{N+2}{8(N+8)^3}\left(N^2 + 23N + 60\right)\epsilon^2$$

$$+ \frac{N+2}{32(N+8)^5}\Big[2N^4 + 89N^3 + 1412N^2 + 5904N + 8640$$

$$- 192(0.60103)(5N+22)(N+8)\Big]\epsilon^3 + \mathcal{O}(\epsilon^4)$$

$$\eta = \frac{N+2}{2(N+8)^2}\epsilon^2 + \frac{N+2}{8(N+8)^4}\left(-N^2 + 56N + 272\right)\epsilon^3 + \mathcal{O}(\epsilon^4)$$

Table 10.1. Critical exponents of the $O(N)$ fixed point in the ϵ expansion

In Table 10.1, we show the values for the critical exponents ν and η calculated to third order in ϵ for arbitrary N. (The other critical exponents can be obtained through the scaling relations.) In the limit $N \to \infty$, the results line up with the self-consistent Gaussian approximation (Chapter 8). As it turns out, these exponents evaluated for $\epsilon = 1$ and $N = 1$ are quite reasonable for the 3D Ising model; the results are not so far from the best numerical simulations. But the more important point is that the RG gives us a clear way to understand why the critical exponents are universal, why they depend on d and N but on no other details, and why mean-field theory is so successful for $d \geq 4$. More generally, this approach provides a general framework for how to think about universality.

10.5   Beyond Broken Symmetry: Superconductors, Topological Order, Gauge Theories

In case you think you now know all there is to know, here are some final mind boggling observations . . .

Distinguishing phases by broken symmetries is extremely effective and conducive to a reductionist understanding of the topic. There are, however, circumstances in which this notion needs to be extended: there are distinct phases of matter that differ not by any pattern of broken symmetry but rather in a more subtle manner, associated with notions of gauge invariance and "toplogical order."[12]

---

interactions—those that flow most slowly toward the fixed point within the critical surface—can be computed and used to extract information on how measurable quantities begin to deviate from their universal values when measured at finite lengthscales, e.g., the correction to scaling exponent x_1 in Eq. 10.8.

12. For these advanced topics see *Quantum Field Theory of Many-Body Systems: From the Origin of Sound to an Origin of Light and Electrons*, Wen (Oxford University Press, Oxford, 2004) or *Field Theoretic Aspects of Condensed Matter Physics: An Overview*, by Fradkin in the Encyclopedia of Condensed Matter Physics 2e.

Guage invariance is familiar from the theory of electricity and magnetism. There, the gauge fields—the vector and scalar potentials $\vec{A}$ and A_0—are introduced as a computational device. But there is a redundancy in this description—the physical electric ($\vec{E}$) and magnetic ($\vec{B}$) fields are unchanged if we transform $A_\mu \rightarrow A_\mu + \partial_\mu \chi$, where χ is any function of space and time. Since these fields are computational devices, we expect that any physical quantity we might hope to describe should depend only on the values of $\vec{E}$ and $\vec{B}$ and so should be invariant under the above transformation—i.e., they should be "gauge invariant."

There is, in fact, a general theorem in statistical mechanics, known as Eliztur's theorem, that pertains to this. It states that in any problem in which the Boltzmann weight is invariant under a set of gauge transformations—which may be the familiar gauge transformation we have just discussed or some more exotic "emergent" gauge invariance—the thermodynamic average of any quantity must itself be gauge invariant. In other words, while spontaneous symmetry breaking is possible in the thermodynamic limit, as we have discussed at length, spontaneous breaking of gauge invariance is forbidden. In particular, any order parameter that describes a given phase of matter must itself be gauge invariant.

The most immediate problem this observation raises concerns the theory of superconductors. As we have mentioned, superfluids can be characterized as broken-symmetry states in which the symmetry involved is related to particle number conservation. This is pretty strange sounding, but possibly no worse than broken symmetries related to momentum conservation—i.e., crystals. Superconductors are much like superfluids, except in this case the particles involved (pairs of electrons) are charged, unlike the neutral particles in superfluids such as ^{4}He. In most discussions of superconductors, the distinction between superconductors and superfluids is not terribly important. Because the speed of light is so large, for many purposes fluctuations of the electromagnetic field can be neglected in the theory of the superconducting state. Thus, superconductors are often considered to be states of broken symmetry, just like superfluids.

However, since gauge invariance is related to charge conservation, the complete theory of superconductors should not be discussed in terms of broken symmetries, but rather in terms of a more exotic form of order. While this distinction can be swept under the rug in the case of superconductors, the same theoretical framework applies to many other situations. There is a profound analogy between superconductivity and the quantum field theory of the electro-weak interaction as well as with quantum chromodynamics, the theory of the strong interaction. However, in this context, the fluctuations of the gauge fields involved are not an afterthought, but a central feature of the problem. In this analogy, phase transitions that occurred in the early universe are analogous to a transition to a superconducting state. The spectrum of massive and massless particles (e.g., the Higgs particle and the photon) that make up our current environment can be traced to the nature of these transitions.

One of the most successful ways to classify states that goes beyond broken symmetries is known as topological order. The notion is abstract and sometimes hard to relate directly to measurable properties. The key idea is to consider the properties of a phase of matter that depend only on the large scale geometry of space—for instance, on a curved space, such as the surface of a sphere, or a space with higher genus, such as on the surface of a torus. We then notice that while many thermodynamic properties of the system depend on details, there are a few quantities that are topological in the sense that they take on discrete values that do not change as finite scale structures of the Hamiltonian or the embedding space are changed. One example is that in certain topologically ordered quantum phases of matter, there is a ground-state degeneracy, $n(g)$, that in the thermodynamic limit depends only on the genus, g, of the space. Since $n(g)$ must be an integer, it cannot change continuously, so a phase of matter can be associated with a set of integers, $n(g)$. More generally, topological properties are always discrete and so are ideal for use in classifying distinct phases of matter. Superconductivity (as well as the fractional quantum Hall liquids and other still less familiar "spin liquid" phases of matter) can all be well characterized as states with topological order.

10.6   Problems

10.1. *Central limit theorem and universality.* In this problem you will investigate how various probability distributions approach the Gaussian distribution as $n \to \infty$. The setup is that n random variables ϕ_j are each independently drawn from a distribution $p(\phi)$; we are interested in their average $\Phi \equiv n^{-1} \sum_{j=1}^{n} \phi_j$. The distribution of Φ is given in Eq. 10.34.

Consider the following four probability distributions:

$$\text{(Ising)} \quad p(\phi) = \begin{cases} 1/2 & \text{if } \phi = 1, \\ 1/2 & \text{if } \phi = -1, \\ 0 & \text{otherwise.} \end{cases} \tag{10.46}$$

$$\text{(Uniform)} \quad p(\phi) = \begin{cases} 1/2 & \text{if } -1 < \phi < 1, \\ 0 & \text{otherwise.} \end{cases} \tag{10.47}$$

$$p(\phi) = \frac{2a^3}{\pi} \frac{1}{(\phi^2 + a^2)^2}. \tag{10.48}$$

$$\text{(Lorentzian)} \quad p(\phi) = \frac{a}{\pi} \frac{1}{\phi^2 + a^2}. \tag{10.49}$$

For each of the four cases, plot the distribution of ϕ, and plot the distribution of Φ for a few small values of n. Explore what happens as n gets large. How does Φ have to be rescaled for the distribution of Φ to converge to a fixed distribution as $n \to \infty$?

The last case, the Lorentzian, is different from the others because its variance is infinite. The central limit theorem does not hold in this case. See if you can find the limiting distribution and its scaling with n!

Hint: There are a few ways to plot the distribution of Φ. The most straightforward way is to do a "Monte Carlo" simulation with your computer by drawing n samples of ϕ and taking the average. If you do this many times, you can plot a histogram of Φ. Alternatively, you can try to do the plot analytically by applying an inverse Fourier Transform to Eq. 10.34. You may find it helpful to know that in each case:

$$\tilde{p}(k) = \cos(k), \tag{10.50}$$

$$\tilde{p}(k) = \sin(k)/k,$$

$$\tilde{p}(k) = [1 + |k|a]\ \exp[-|k|a],$$

$$\tilde{p}(k) = \exp[-|k|a].$$

10.7 Worksheets for Chapter 10

W10.1 Renormalization Group of the 1D Ising Chain

As a concrete and tractable example, let us calculate the critical properties of the 1D Ising model using the method of renormalization group.

Recall that the Hamiltonian of the 1D Ising model is:

$$H = -J\sum_j \sigma_j\sigma_{j+1} - h\sum_j \sigma_j + \sum_j \varepsilon_0.$$

Here we have included an additive constant term in the energy of ε_0 per site, which does not modify the physics. We will see why we added this term shortly. H can be rewritten in terms of a function $K(\sigma_j, \sigma_{j+1})$:

$$H = \sum_j^N K(\sigma_j, \sigma_{j+1}),$$

$$K(\sigma_a, \sigma_b) = -J\sigma_a\sigma_b - \frac{h}{2}(\sigma_a + \sigma_b) + \varepsilon_0.$$

We will work in units where $T = \beta = 1$, so the partition function can then be written as the trace of a product over transfer matrices:

$$Z = \mathrm{Tr}\left[t^N\right],$$

$$t = e^{-\varepsilon_0}\begin{bmatrix} e^{J+h} & e^{-J} \\ e^{-J} & e^{J-h} \end{bmatrix}.$$

The idea is to coarse-grain every two sites into one, renormalize the coupling constants while doubling the number of sites, and repeat the procedure to see which parameters are relevant and irrelevant.

1. We start by decimating the original chain into blocks of two spins each. Show that this amounts to calculating $Z = \text{Tr}\left[(t^2)^{N/2}\right]$, and calculate $\tilde{t} = t^2$. This corresponds to the transfer matrix between block spins.

2. We now want to map a new chain of N block spins onto the original chain, thus giving us a new partition function $\tilde{Z} = \text{Tr}\left[\tilde{t}^N\right]$ where each transfer matrix has the form:
$$\tilde{t} = e^{-\tilde{\varepsilon}_0} \begin{bmatrix} e^{\tilde{J}+\tilde{h}} & e^{-\tilde{J}} \\ e^{-\tilde{J}} & e^{\tilde{J}-\tilde{h}} \end{bmatrix}.$$

Since $\tilde{t} = t^2$, show that:

$$\tilde{J} = J + \frac{1}{4}\ln\left[\frac{\left(e^{-4J} + e^{-2h}\right)\left(e^{-4J} + e^{2h}\right)}{4\cosh^2(h)}\right], \tag{10.51}$$

$$\tilde{h} = h + \frac{1}{2}\ln\left[\frac{\cosh(h+2J)}{\cosh(h-2J)}\right], \tag{10.52}$$

$$\tilde{\varepsilon}_0 = 2\varepsilon_0 - \frac{1}{4}\ln\left[8\cosh^2(h)\left(\cosh(2h) + \cosh(4J)\right)\right]. \tag{10.53}$$

3. There are of course many ways to simplify the equations, but the reason we wrote out Eqs. 10.51 to 10.53 the way we did is so that now we can express them as beta functions:

$$\tilde{J} - J = \beta_J(J, h),$$

$$\tilde{h} - h = \beta_h(J, h),$$

$$\tilde{\varepsilon}_0 - 2\varepsilon_0 = f(J, h).$$

We note here that the ε_0 is not a physically significant parameter, but naturally arises as a part of the RG iteration process, so we use the label $f(J, h)$ instead of calling it a beta function.

A fixed point occurs when beta functions are equal to zero. Show that the fixed point in $J - h$ parameter space is $(J, h) = (0, 0)$, and determine the "relevance" of J and h.

We have been doing our calculations with $T = \beta = 1$, which means that the parameter J is indeed actually J/T. The $J = 0$ fixed point, interpreted as temperature, means that $T \to \infty$ is really the fixed point, and therefore no finite temperature phase transition occurs.

There is in fact another fixed point at $J \to \infty$ (check that there is!) we neglected to analyze. This corresponds to the "ordered phase" of the 1D Ising model at $T = 0$, but as we have discovered that J flows toward 0 for any finite J, the $T = 0$ fixed point is unstable and does not indicate a true phase transition.

4. Finally, let us briefly take a look at the correlation length ξ. From the above analysis, we can see that, if we start with some coupling constant J_0, then we can write down a recurrence relation for the coupling constant after n RG iterations:

$$J_n = J_{n-1} + \beta_J(J_{n-1}, h = 0) = J_{n-1} + \frac{1}{2}\ln\left[\frac{1 + e^{-4J_{n-1}}}{2}\right].$$

(a) Assume that we are far away from the fixed point, so J_0 is large and therefore e^{-4J} is small. Show that:

$$J_n = J_0 - \frac{n}{2}\ln(2) + \mathcal{O}\left(e^{-4J_{n-1}}\right). \tag{10.54}$$

Give an approximate limit on n for this approximation to hold.

(b) We are now going to push the limit of this approximation, that is, use a value of n as large as possible such that Eq. 10.54 will break down for any larger n. Let us call this value n', and the corresponding value $J_{n'} = a \sim \mathcal{O}(1)$, indicating that it is on the order of unity. Justify this limit on the approximation.

(c) The correlation length ξ is defined through the two-site correlation function:

$$\langle \sigma_0 \sigma_L \rangle_J = e^{-L/\xi(J)},$$

where L is the distance between the two sites. Physically, the correlation length should depend on the coupling strength J, and for $J \sim \mathcal{O}(1)$, let $\xi(J) \equiv \xi_1$.

Consider now the separation distance $L = l2^{n'}$ (the same n' from part 4b), where l is the distance between each site. Since each RG iteration coarse-grains each two adjacent spins into one, the correlation of two sites at distance L with coupling J_0 corresponds to the correlation of two sites at distance l after n' RG iterations with coupling $J_{n'}$. In equation, this means:

$$\langle \sigma_0 \sigma_L \rangle_{J_0} = \langle \sigma_0 \sigma_l \rangle_{J_{n'}}. \tag{10.55}$$

Using the approximations

$$J_{n'} = a \sim \mathcal{O}(1),$$

$$\xi(a) = \xi_1 \sim l,$$

show that $\xi(J_0) = \left[\xi_1 e^{-2a}\right] e^{2J_0}$.

(d) Recall that the exact solution for the 1D Ising model gives us the correlation length:

$$\xi_{exact} = \left| \ln \left[\tanh(\beta J) \right] \right|^{-1}. \tag{10.56}$$

Assuming that $\beta J \gg 1$, expand ξ_{exact} to leading order in $e^{\beta J}$. Does it match up with our RG calculation?

With knowledge of only the leading term in the beta function, we have successfully computed the correct J_0 dependence of the correlation length. There is a number of order 1 in the prefactor—$[\xi_a e^{-2a}]$—that we cannot compute from simple RG analysis. This requires a more complete solution. But the important (exponential) temperature dependence of the correlation length is seen directly from knowledge of the leading-order behavior of the beta function. (Recall, with the temperature included, J_0 really means J_0/T.)

W10.2   Perturbative Renormalization Group of the ϕ^4 Theory in d Dimensions

In this exercise, we will carry out the renormalization group calculation using perturbation expansion of the ϕ^4 Hamiltonian described in the chapter. Before any calculations, let us detail the setup of the problem first.

Setting up the system

Consider a Hamiltonian density for a translationally invariant system with a global $\theta \to -\theta$ symmetry:

$$\mathcal{H}[\theta(\vec{r})] = \mathcal{H}_0 + \mathcal{H}_1,$$

$$\mathcal{H}_0 = \frac{\kappa}{2} |\nabla \theta|^2,$$

$$\mathcal{H}_1 = V(\theta)$$

$$= V_0 + \frac{V_2}{2} \theta^2 + \frac{V_4}{4} \theta^4 + \frac{V_6}{6} \theta^6 + \cdots,$$

where $\theta = \theta(\vec{r})$ is a position-dependent real scalar field in d dimensions. The two terms of the Hamiltonian density are $\mathcal{H}_0$, which is the noninteracting part, and $\mathcal{H}_1 = V$, which we call the interacting part. If $V = 0$, then this is a noninteracting system corresponding to the Gaussian model, which we have solved exactly before. Here we will see what happens when we turn on a small V, which we will write as an expansion in terms even in θ (due to the global symmetry), and develop the RG equations perturbatively.

The (total) Hamiltonian is:

$$H = \int_\Omega d^d \vec{r} \left(\mathcal{H}_0 + \mathcal{H}_1 \right) = H_0 + H_1. \tag{10.57}$$

We will consider the system in d-dimensional space with periodic boundary conditions over length L, such that the domain[13] of the volume integral is $\Omega = L^d$.

We will define a quantity S (sometimes called the "action") as:

$$S = \frac{1}{T} \left(H_0 + H_1 \right) = S_0 + S_1. \tag{10.58}$$

The Boltzmann factor is then defined in terms of the action:

$$e^{-S} = e^{-H_0/T} e^{-H_1/T}. \tag{10.59}$$

The partition function is consequently:

$$Z = \int D\theta \, e^{-S}$$
$$= \int D\theta \, \exp\left[-\frac{1}{T} \int_\Omega d^d \vec{r} \left(\frac{\kappa}{2} |\nabla\theta|^2 + V(\theta) \right) \right], \tag{10.60}$$

where $\int D\theta$ indicates the functional integral over all configurations of θ over Ω.

As the last (but critical) step of our setup, let us consider some smallest length-scale l below which $\theta(\vec{r})$ cannot meaningfully vary (so l is akin to the lattice constant of a microscopic model). This allows us to consider the field $\theta(\vec{r})$ as a sum of its Fourier components $\theta_{\vec{k}}$:

$$\theta(\vec{r}) = L^{-d/2} \sum_{|\vec{k}|<\Lambda} e^{i\vec{k}\cdot\vec{r}} \theta_{\vec{k}}, \tag{10.61}$$

where $\Lambda \propto l^{-1}$ is the (spatial) frequency cutoff imposed due to the microscopic lengthscale, and is for the time being an arbitrary value. This allows us to define the functional integral as:[14]

$$\int D\theta = \prod_{|\vec{k}|<\Lambda} \int d\theta_{\vec{k}} \tag{10.62}$$

---

13. We will play a bit fast and loose here, with Ω indicating both the domain and its volume.

14. The logic is that the functional integral, after defining a cutoff lengthscale l, is

$$\int D\theta = \prod_{l \leq |\vec{r}_n|}^{\Omega} \int d\theta(\vec{r}_n)$$

for all "lattice points" $\vec{r}_n$ separated by l inside Ω. Since the Jacobian of the Fourier transform is 1, we can rewrite this integral in Fourier space. This is performed more explicitly in the chapter on the self-consistent Gaussian approximation.

The notation here really means, written out for the components $\vec{k} = k_j\hat{k}$,

$$\prod_{|\vec{k}|<\Lambda} = \prod_{0 \le k_1} \prod_{0 \le k_2}^{|\vec{k}|<\Lambda} \cdots \prod_{0 \le k_d}, \tag{10.63}$$

where $k_j = \frac{2\pi n_j}{L}$ for integer n_j, such that all the $\vec{k}$ points are in the first "hyperoctant."

Since in general Fourier decomposition gives complex coefficients, the components $\theta_{\vec{k}}$ are really

$$\theta_{\vec{k}} = \theta_{\vec{k}}^{Re} + i\theta_{\vec{k}}^{Im}, \tag{10.64}$$

and the integration means:

$$\int d\theta_{\vec{k}} = \int d\theta_{\vec{k}}^{Re} \int d\theta_{\vec{k}}^{Im}. \tag{10.65}$$

Overview of RG procedure

In the following diagram, we will take a bird's-eye view of a single RG step we will carry out:

$$F(\kappa, V_0, V_2, V_4, \dots; \Lambda) \xrightarrow[\text{Averaging over } \tilde{\Lambda} \le |\vec{k}|<\Lambda]{\text{Coarse-graining}} \tilde{F}(\kappa, \tilde{V}_0, \tilde{V}_2, \tilde{V}_4, \dots; \tilde{\Lambda})$$

$$\text{Rescaling} \downarrow {\scriptstyle \kappa \to 1, \Lambda \to 1} \qquad\qquad\qquad\qquad \text{Rescaling} \downarrow {\scriptstyle \kappa \to 1, \tilde{\Lambda} \to 1}$$

$$F'(\kappa = 1, E_0, \alpha, u, \dots; \Lambda = 1) \xrightarrow[{\scriptstyle \alpha \to \tilde{\alpha},\, u \to \tilde{u},\, \cdots}]{\text{RG flow}} \tilde{F}'(\kappa = 1, \tilde{E}_0, \tilde{\alpha}, \tilde{u}, \dots; \tilde{\Lambda} = 1)$$

The initial physical system is represented by the free energy F, with parameters κ, $V_0, \dots$, and a lengthscale cutoff of Λ^{-1}. There are two operations we can perform on this system.

- We can nondimensionalize the parameters, such that the physics is rewritten in terms of dimensionless quantities in units of κ and Λ. This is represented by the free energy F' and dimensionless parameters $E_0, \alpha, \dots$. The lengthscale Λ^{-1} and Gaussian coupling strength κ are normalized to 1. In some sense, this procedure yields the fundamental description of the system for given physical parameters.
- We can perform a coarse-graining procedure, where we average away a subset of short lengthscale physics. The resulting system, represented by $\tilde{F}$ and parameters $\tilde{V}_0, \dots$, will reflect the remaining longer lengthscale physics. The cutoff lengthscale will be updated as well to a new value $\tilde{\Lambda}^{-1}$ to reflect the coarse-graining.

To quantify how coarse-graining affected the physics, we will need to nondimensionalize the new system $\tilde{F}$ as well and rewrite it in terms of dimensionless quantities in units of κ and $\tilde{\Lambda}$. This is represented by the free energy $\tilde{F}'$ and new dimensionless parameters $\tilde{E}_0$, $\tilde{\alpha}$, $\ldots$. Once again the new lengthscale $\tilde{\Lambda}^{-1}$ and Gaussian coupling κ are normalized to 1.

Now that parameters in F' and $\tilde{F}'$ are on equal footing, we can examine how coarse-graining affected the physics, and devise a description for the "RG flow" of the physical parameters. Since now we can consider the coarse-grained system $\tilde{F}$ as the new initial system, this is an iterative process. If coarse-graining is performed with infinitesimal steps (which it is), then the flow of parameters can be formulated as differential equations.

Nondimensionalization via rescaling

RG calculation involves iteratively coarse-graining and rescaling parameters, and we will start with the rescaling step. We will first rescale lengths using the natural lengthscale of the system, Λ^{-1} (remember Λ has units of inverse length),

$$\vec{x} = \Lambda \vec{r} \quad \text{and} \quad \vec{q} = \frac{\vec{k}}{\Lambda}, \tag{10.66}$$

which preserves the dot product $\vec{k} \cdot \vec{r} = \vec{q} \cdot \vec{x}$. This rescales the volume as

$$\Omega' = \Omega \Lambda^d$$

and (very importantly) the limits of the Fourier frequencies as

$$|\vec{k}| < \Lambda \rightarrow |\vec{q}| < 1.$$

1. We will also need to rescale the field θ. First, the energy of the system is measured in units of temperature T, and we would like to express all coupling constants in units of T. Since we are considering $\mathcal{H}_1 = V$ as a perturbation to $\mathcal{H}_0 = \kappa |\nabla \theta|^2 / 2$, we will rescale parameters such that $\kappa/T \rightarrow 1$. We will define a new field,

$$\phi(\vec{x}) = z\theta(\vec{r}), \tag{10.67}$$

with z being the scaling factor for the field, such that after rescaling the noninteracting action is

$$S_0[\theta] = \frac{1}{T} \int_\Omega d^d\vec{r}\, \frac{\kappa}{2} |\nabla_{\vec{r}}\, \theta(\vec{r})|^2 \overset{\text{def.}}{\equiv} S_0'[\phi] = \frac{1}{2} \int_{\Omega'} d^d\vec{x}\, |\nabla_{\vec{x}}\, \phi(\vec{x})|^2, \tag{10.68}$$

so that the action is a dimensionless quantity. Show that

$$z = \sqrt{\frac{\kappa}{T}}\, \Lambda^{(2-d)/2}. \tag{10.69}$$

2. Now let us turn our attention to the interaction term $\mathcal{H}_1 = V(\theta)$. We want to rescale V such that it is dimensionless. Considering that it originally has units

of energy per volume, we would like to define a new interaction potential,

$$V'[\phi(\vec{x})] = \frac{V[\theta(\vec{r})]}{T\Lambda^d}, \tag{10.70}$$

such that the new potential V' is dimensionless and a function of the new field ϕ. We will assign it an expansion in terms even in ϕ,

$$V'(\phi) = E_0 + \frac{\alpha}{2}\phi^2 + \frac{u}{4}\phi^4 + \frac{u_6}{6}\phi^6 + \cdots, \tag{10.71}$$

which satisfies

$$S_1[\theta] = \frac{1}{T}\int_\Omega d^d\vec{r}\, V[\theta(\vec{r})] \stackrel{\text{def.}}{=} S_1'[\phi] = \int_{\Omega'} d^d\vec{x}\, V'[\phi(\vec{x})], \tag{10.72}$$

such that the action is again a dimensionless quantity. Show that:

$$E_0 = \frac{V_0}{T\Lambda^d}, \qquad \alpha = \frac{V_2}{\kappa\Lambda^2}, \qquad u = \frac{TV_4}{\kappa^2}\Lambda^{d-4}, \qquad u_6 = \frac{T^2 V_6}{\kappa^3}\Lambda^{2d-6}.$$

By nondimensionalizing the parameters and rescaling the frequency cutoff, we can now compare (apples to apples!) the parameters before and after the coarse-graining step, which we will now work on.

Coarse-graining with momentum shell RG

The method we will use to coarse-grain the system is called "momentum shell RG." In terms of the Fourier components $\theta_{\vec{k}}$, larger $\vec{k}$ corresponds to higher spatial frequency, or shorter lengthscale, features. The idea is to average out the physics over an "outer shell" in $\vec{k}$ space, and see how the rescaled parameters (E_0, α, u, u_6, ...) change.

Let us introduce a frequency $\tilde{\Lambda}$ that serves as the cutoff for "fast modes," such that:

$$\theta(\vec{r}) = \theta_s(\vec{r}) + \theta_f(\vec{r}), \tag{10.73}$$

where

$$\theta_s(\vec{r}) = L^{-d/2}\sum_{|\vec{k}|<\tilde{\Lambda}} e^{i\vec{k}\cdot\vec{r}}\theta_{\vec{k}},$$

$$\theta_f(\vec{r}) = L^{-d/2}\sum_{\tilde{\Lambda}\leq|\vec{k}|<\Lambda} e^{i\vec{k}\cdot\vec{r}}\theta_{\vec{k}},$$

so θ_f contains the configurations of θ with short-range "fast" features, and θ_s contains the remaining long wavelength "slow" features. Our goal in this section is to integrate away the fast features to obtain a new effective interaction potential:

$$V(\theta) \xrightarrow{\text{RG}} \tilde{V}(\theta_s).$$

1. Let us first examine the Boltzmann factor of the noninteracting Hamiltonian, e^{-S_0}, in a bit more detail. Show that:

$$S_0[\theta] = \frac{\kappa}{2T} \sum_{|\vec{k}|<\Lambda} |\vec{k}|^2 \left|\theta_{\vec{k}}\right|^2. \tag{10.74}$$

Hint 1: Remember that $\nabla e^{i\vec{k}\cdot\vec{r}} = i\vec{k}e^{i\vec{k}\cdot\vec{r}}$.
Hint 2: Remember that:

$$\int_L dx \, \exp\left(i\frac{2\pi n}{L}x\right) = \begin{cases} L & \text{for } n=0 \\ 0 & \text{for integer } n \neq 0. \end{cases} \tag{10.75}$$

Hint 3: Since θ is a real-valued function, by definition of Fourier transforms, complex conjugation gives you:

$$\theta_{\vec{k}}^* = \theta_{-\vec{k}}.$$

2. The previous exercise shows that S_0 can be written as a sum of terms each with independent wave vector $\vec{k}$, so we can indeed separate it into two terms just like we did $\theta = \theta_s + \theta_f$, as

$$S_0[\theta] = S_0[\theta_s] + S_0[\theta_f], \tag{10.76}$$

where

$$S_0[\theta_s] = \frac{\kappa}{2T} \sum_{|\vec{k}|<\tilde{\Lambda}} |\vec{k}|^2 \left|\theta_{\vec{k}}\right|^2,$$

$$S_0[\theta_f] = \frac{\kappa}{2T} \sum_{\tilde{\Lambda}\leq|\vec{k}|<\Lambda} |\vec{k}|^2 \left|\theta_{\vec{k}}\right|^2,$$

such that the functional integral can be rewritten as

$$\int D\theta = \int D\theta_s D\theta_f. \tag{10.77}$$

Now, let us formally define what it means to integrate away the short wavelength features. The statistical properties of the original system are determined by the partition function

$$Z = \int D\theta \, \exp\left[-S_0[\theta_s] - S_0[\theta_f] - \frac{1}{T}\int_\Omega d^d\vec{r} \, V(\theta)\right]. \tag{10.78}$$

If we want to average over just the fast components, we need to define an appropriate partition function,

$$Z_f = \int D\theta_f \, e^{-S_0[\theta_f]},$$

such that thermal average of any operator over the short wavelength physics is defined as

$$\left\langle \hat{A}[\theta_s, \theta_f] \right\rangle_f = Z_f^{-1} \int D\theta_f \, \hat{A}[\theta_s, \theta_f] e^{-S_0[\theta_f]}. \tag{10.79}$$

With this, we would like to define a new effective action $\tilde{S}[\theta_s]$ depending only on the slow components of the field, such that the new Boltzmann factor is defined by:

$$e^{-\tilde{S}[\theta_s]} \stackrel{\text{def.}}{\equiv} \frac{e^{-S_0[\theta_s]}}{Z_f} \int D\theta_f \, \exp\left[-S_0[\theta_f] - \frac{1}{T} \int_\Omega d^d\vec{r} \, V(\theta_s + \theta_f)\right] \tag{10.80}$$

$$= e^{-S_0[\theta_s]} \left\langle \exp\left[-\frac{1}{T} \int_\Omega d^d\vec{r} \, V(\theta_s + \theta_f)\right]\right\rangle_f. \tag{10.81}$$

Show that this definition lets us naturally calculate the expectation value of operators depending only on θ_s as

$$\left\langle \hat{O}[\theta_s] \right\rangle = Z^{-1} \int D\theta \, \hat{O}[\theta_s] e^{-S} = \tilde{Z}^{-1} \int D\theta_s \, \hat{O}[\theta_s] e^{-\tilde{S}[\theta_s]}, \tag{10.82}$$

where:

$$\tilde{Z} = \int D\theta_s \, e^{-\tilde{S}[\theta_s]},$$

and thus describes the coarse-grained system with high frequency components averaged away.

3. The formal manipulations from the last part shows us that the physics of the long wavelength components can indeed be described by the new action $\tilde{S}[\theta_s]$ defined by:

$$e^{-\tilde{S}[\theta_s]} = e^{-S_0[\theta_s]} \left\langle \exp\left[-\frac{1}{T} \int_\Omega d^d\vec{r} \, V(\theta_s + \theta_f)\right]\right\rangle_f. \tag{10.83}$$

So far, this is an exact expression that we cannot evaluate exactly in general. We would like to approximate it perturbatively in powers of V. Notice that by taking the logarithm we get:

$$\tilde{S}[\theta_s] = S_0[\theta_s] - \ln\left\{\left\langle \exp\left[-\frac{1}{T} \int_\Omega d^d\vec{r} \, V(\theta_s + \theta_f)\right]\right\rangle_f\right\}. \tag{10.84}$$

An important identity can help us evaluate the second term:

$$\ln\left[\langle e^{\lambda x}\rangle\right] = \lambda\langle x\rangle + \frac{\lambda^2}{2}\left[\langle x^2\rangle - \langle x\rangle^2\right]$$

$$+ \frac{\lambda^3}{6}\left[\langle x^3\rangle - 3\langle x^2\rangle\langle x\rangle + 2\langle x\rangle^3\right] + \cdots. \tag{10.85}$$

This is called the cumulant expansion, which works for the random variable x with any distribution.[15] To first order in V, the logarithm terms gives:

$$\ln\left\{\left\langle\exp\left[-\frac{1}{T}\int_\Omega d^d\vec{r}\, V(\theta_s + \theta_f)\right]\right\rangle_f\right\} = -\frac{1}{T}\int_\Omega d^d\vec{r}\,\langle V(\theta_s + \theta_f)\rangle_f + \cdots. \tag{10.86}$$

Show that:

$$\langle V(\theta_s + \theta_f)\rangle_f = V_0 + V_s(\theta_s) + \langle V_f(\theta_f)\rangle_f \tag{10.87}$$

$$+ \frac{3V_4}{2}\theta_s^2\langle\theta_f^2\rangle_f + \frac{V_6}{2}\left[5\theta_s^4\langle\theta_f^2\rangle_f + 15\theta_s^2\langle\theta_f^2\rangle_f^2\right] + \cdots,$$

where:

$$V_s(\theta_s) = \frac{V_2}{2}\theta_s^2 + \frac{V_4}{4}\theta_s^4 + \frac{V_6}{6}\theta_s^6 + \cdots,$$

$$V_f(\theta_f) = \frac{V_2}{2}\theta_f^2 + \frac{V_4}{4}\theta_f^4 + \frac{V_6}{6}\theta_f^6 + \cdots.$$

Hint 1: Look at the definitions of $\langle\hat{O}\rangle_f$ and $S_0[\theta_f]$ again. Is there anything familiar (perhaps from the Gaussian integrals we considered before)?
Hint 2: You might recall that the expectation value of linear functions F in Gaussian distributions is:

$$\langle F^{2n}\rangle = (2n-1)(2n-3)\cdots(1)\langle F^2\rangle.$$

4. The above exercise tells us that we only have to calculate $\langle\theta_f^2\rangle_f$, and all other terms can be obtained from it. Let us calculate this quantity.

---

15. This expansion can be proved by noting

$$\langle e^{\lambda x}\rangle = \left\langle 1 + \lambda x + \frac{\lambda^2}{2}x^2 + \frac{\lambda^3}{6}x^3 + \cdots\right\rangle$$

$$= 1 + \lambda\langle x\rangle + \frac{\lambda^2}{2}\langle x^2\rangle + \frac{\lambda^3}{6}\langle x^3\rangle + \cdots$$

and that

$$\ln(1+y) = y - \frac{y^2}{2} + \frac{y^3}{3} + \cdots.$$

We recover the cumulant expansion by collecting terms in powers of λ.

(a) Since θ_f is real, $\theta_f = \theta_f^*$:

$$\theta_f^2 = \theta_f \theta_f^* = L^{-d} \sum_{\tilde{\Lambda} \leq |\vec{k}| < \Lambda} \sum_{\tilde{\Lambda} \leq |\vec{k}'| < \Lambda} e^{i(\vec{k}-\vec{k}')\cdot\vec{r}} \theta_{\vec{k}} \theta_{\vec{k}'}^*.$$

Use this expression to show that

$$\left\langle \theta_f^2 \right\rangle_f = L^{-d} \sum_{\tilde{\Lambda} \leq |\vec{k}| < \Lambda}^{\text{Shell}} \frac{T}{\kappa |\vec{k}|^2}, \tag{10.88}$$

where now the summation is over all hyperoctants (i.e., components of $\vec{k}$ can now be negative), and is thus a "shell" in d-dimensional $\vec{k}$ space.

(b) Use the continuum approximation

$$\left(\frac{2\pi}{L}\right)^d \sum_{\vec{k}} \to \int d^d \vec{k},$$

to show that

$$\left\langle \theta_f^2 \right\rangle_f = \frac{A_d}{(2\pi)^d} \frac{T}{\kappa} \frac{\Lambda^{d-2}}{(d-2)} \left[1 - \left(\frac{\tilde{\Lambda}}{\Lambda}\right)^{d-2} \right], \tag{10.89}$$

where A_d is the surface area of a d-dimensional hypersphere.[16]

(c) We have now performed the bulk of the coarse-graining step, which is to integrate over $\tilde{\Lambda} \to \Lambda$ in $\vec{k}$ space. Since eventually we want to consider this as a continuous process in some RG time t, let us choose:

$$\tilde{\Lambda} = e^{-\delta t} \Lambda,$$

where δt is a small increment in RG time t. Show that:

$$\left\langle \theta_f^2 \right\rangle_f = \frac{X_d \, \delta t}{z^2} + \mathcal{O}(\delta t^2), \tag{10.91}$$

where $X_d = A_d / (2\pi)^d$.

---

16. This is the factor that comes in when doing volume integrals with spherical symmetry. In 2D, $A_2 = 2\pi$ since

$$\int d^2 \vec{r} = \int 2\pi\rho \, d\rho,$$

and in 3D, $A_3 = 4\pi$ since

$$\int d^3 \vec{r} = \int 4\pi r^2 \, dr. \tag{10.90}$$

In general,

$$A_d = \frac{2\pi^{d/2}}{\Gamma(d/2)},$$

and it is a finite small positive number for all d.

Constructing the RG differential equations

We now have all the ingredients ready, or mise en place if you will, to assemble the differential equations that describe the RG flow. The original physical system is described by the action:

$$S[\theta] = T^{-1} \int_\Omega d^d \vec{r} \left[\frac{\kappa}{2} |\nabla \theta|^2 + V_0 + \frac{V_2}{2} \theta^2 + \frac{V_4}{4} \theta^4 + \frac{V_6}{6} \theta^6 + \cdots \right]. \tag{10.92}$$

After coarse-graining away the fast components θ_f, we are left with an effective action as a functional of the slow components θ_s (see Eqs. 10.84–10.87), which we would like to write in the form:

$$\tilde{S}[\theta_s] = T^{-1} \int_\Omega d^d \vec{r} \left[\frac{\kappa}{2} |\nabla \theta_s|^2 + \tilde{V}(\theta_s) \right], \tag{10.93}$$

where

$$\tilde{V}(\theta_s) = \tilde{V}_0 + \frac{\tilde{V}_2}{2} \theta_s^2 + \frac{\tilde{V}_4}{4} \theta_s^4 + \frac{\tilde{V}_6}{6} \theta_s^6 + \cdots. \tag{10.94}$$

We will first find the new coefficients $\tilde{V}_{2n}$ in terms of the original coefficients and parameters, then see how their rescaled versions ($\tilde{E}_0$, $\tilde{\alpha}$, $\tilde{u}$, $\cdots$) changed compared to the original system and derive the flow equations.

1. From Eq. 10.87, show that:

$$\tilde{V}_0 = V_0 + \frac{V_2}{2} \left\langle \theta_f^2 \right\rangle_f + \frac{3V_4}{4} \left\langle \theta_f^2 \right\rangle_f^2 + \frac{5V_6}{2} \left\langle \theta_f^2 \right\rangle_f^3 + \cdots,$$

$$\tilde{V}_2 = V_2 + 3V_4 \left\langle \theta_f^2 \right\rangle_f + 15 V_6 \left\langle \theta_f^2 \right\rangle_f^2 + \cdots,$$

$$\tilde{V}_4 = V_4 + 10 V_6 \left\langle \theta_f^2 \right\rangle_f + \cdots,$$

$$\tilde{V}_6 = V_6 + \cdots.$$

2. Since we have integrated away all features for $\tilde{\Lambda} \le |\vec{k}| < \Lambda$, the cutoff is now $\tilde{\Lambda}$, which will be the new lengthscale for rescaling the parameters. Following the exact same procedure as previously, the whole suite of rescaled parameters is:

$$\tilde{x} = \tilde{\Lambda} \vec{r}, \qquad \tilde{z} = \sqrt{\frac{\kappa}{T}} \tilde{\Lambda}^{(2-d)/2}, \qquad \tilde{\phi}(\tilde{x}) = \tilde{z} \theta_s(\vec{r}), \qquad \tilde{V}'(\tilde{\phi}) = \frac{\tilde{V}(\theta_s)}{T \tilde{\Lambda}^d},$$

$$\tilde{V}'(\tilde{\phi}) = \tilde{E}_0 + \frac{\tilde{\alpha}}{2} \tilde{\phi}^2 + \frac{\tilde{u}}{4} \tilde{\phi}^4 + \frac{\tilde{u}_6}{6} \tilde{\phi}^6 + \cdots,$$

$$\tilde{E}_0 = \frac{\tilde{V}_0}{T \tilde{\Lambda}^d}, \qquad \tilde{\alpha} = \frac{\tilde{V}_2}{\kappa \tilde{\Lambda}^2}, \qquad \tilde{u} = \frac{T \tilde{V}_4}{\kappa^2} \tilde{\Lambda}^{d-4}, \qquad \tilde{u}_6 = \frac{T^2 \tilde{V}_6}{\kappa^3} \tilde{\Lambda}^{2d-6}.$$

Recall that to view coarse-graining as a continuous process in some RG time t, we have defined $\tilde{\Lambda} = e^{-\delta t}\Lambda$. We can then consider the rescaled parameters to be continuous functions of t in the following sense (take $\tilde{u}$ and u, for example):

$$u_\Lambda = u(t),$$

$$\tilde{u}_{\tilde{\Lambda}} = u(t + \delta t)$$

$$= u(t) + \frac{du}{dt}\bigg|_t \delta t + \mathcal{O}(\delta t^2).$$

On the other hand, $\tilde{u}$ is a function of δt and can be written as:

$$\tilde{u}(\delta t) = \tilde{u}(0) + \frac{d\tilde{u}}{d(\delta t)}\bigg|_{\delta t \to 0} \delta t + \mathcal{O}(\delta t^2).$$

Since $\tilde{u}(0) = u_\Lambda = u(t)$, if we consider an infinitesimal δt, we can construct a differential equation for $u(t)$ by

$$\frac{du}{dt} = \frac{d\tilde{u}}{d(\delta t)}\bigg|_{\delta t \to 0}, \tag{10.95}$$

and the construction in Eq. B.10.61 applies for all of the rescaled parameters. Use this to show that:

$$\frac{dE_0}{dt} = E_0 d + \frac{X_d}{2}\alpha,$$

$$\frac{d\alpha}{dt} = 2\alpha + 3X_d u,$$

$$\frac{du}{dt} = (d-4)u + 10X_d u_6,$$

$$\frac{du_6}{dt} = (2d-6)u_6 + \mathcal{O}(u_8).$$

where u_8 is the coefficient for 8th-order term in the expansion for V'.

Hint 1: Since this amounts to finding the 1st-order expansion in δt, $e^{\lambda \delta t} = 1 + \lambda \delta t$ is a perfectly fine expansion to use everywhere. You also only need to keep terms to order δt elsewhere.

Hint 2: Look back at expressions for $E_0, \alpha, \ldots$, etc. to simplify your results.

APPENDIX A

Appendices

A.1 The Qubit and Pauli Matrices

The Hamiltonian of an arbitrary 2-state quantum system (qubit) can be represented by a 2×2 Hermitian[1] matrix H, which can be written as a linear combination of Pauli matrices with real coefficients as

$$H = \epsilon_0 I + b_x \tau_x + b_y \tau_y + b_z \tau_z, \tag{A.1}$$

where

$$I = \begin{bmatrix} 1 & 0 \\ 0 & 1 \end{bmatrix}, \qquad \tau_x = \begin{bmatrix} 0 & 1 \\ 1 & 0 \end{bmatrix},$$

$$\tau_y = \begin{bmatrix} 0 & -i \\ i & 0 \end{bmatrix}, \qquad \tau_z = \begin{bmatrix} 1 & 0 \\ 0 & -1 \end{bmatrix}. \tag{A.2}$$

So for a general 2×2 matrix of the form $\begin{bmatrix} a & b \\ c & d \end{bmatrix}$, we can find the coefficients for its Pauli decomposition with:[2]

$$\epsilon_0 = \frac{a+d}{2}, \qquad b_x = \frac{b+c}{2},$$

$$b_y = i\frac{b-c}{2}, \qquad b_z = \frac{a-d}{2}. \tag{A.3}$$

With these coefficients, the entries of H are

$$H = \begin{bmatrix} \epsilon_0 + b_z & b_x - ib_y \\ b_x + ib_y & \epsilon_0 - b_z \end{bmatrix}, \tag{A.4}$$

and the eigenvalues are

---

1. A Hermitian matrix is one that which is equal to its conjugate transpose, $M = M^\dagger$.
2. This decomposition works in general, regardless of Hermitianness.

$$\lambda_\pm = \epsilon_0 \pm \sqrt{b_x^2 + b_y^2 + b_z^2} = \epsilon_0 \pm b, \tag{A.5}$$

where we may interpret b as the magnitude of the vector $\vec{b} = (b_x, b_y, b_z)$. The normalized eigenvectors are:

$$|\psi_+\rangle = \frac{1}{a}\begin{pmatrix} b_z + b \\ b_x + ib_y \end{pmatrix}, \qquad |\psi_-\rangle = \frac{1}{a}\begin{pmatrix} ib_y - b_x \\ b_z + b \end{pmatrix}, \qquad a = \sqrt{2b(b_z + b)}, \tag{A.6}$$

such that

$$H|\psi_\pm\rangle = \lambda_\pm |\psi_\pm\rangle. \tag{A.7}$$

Conversely, any 2×2 Hermitian matrix—for instance, the transfer matrix we encountered in the exact solution of the 1D Ising model—can be interpreted as the Hamiltonian of a qubit. Indeed, to get physical intuition into what the above results mean, consider H to be the Hamiltonian of a spin-half particle in a magnetic field. The eigenvalues are thus the energy levels. The terms in Eq. A.1 then correspond to a state-independent constant energy, ϵ_0, and a magnetic field of $\vec{b} = (b_x, b_y, b_z)$; the energy levels are $\lambda_\pm = \epsilon_0 \pm b$, where the splitting is proportional to the magnetic field strength. The energy eigenstates are thus states in which the component of spin along the direction of the magnetic field is sharply defined—i.e., the spin points in a direction parallel or antiparallel to the magnetic field.

Using Eqs. (A.3), we see that the transfer matrix in Section 3.4 may be decomposed as

$$t = \frac{e^{\beta(J+h)} + e^{\beta(J-h)}}{2}I + \frac{e^{-\beta J} + e^{-\beta J}}{2}\tau_x + \frac{e^{\beta(J+h)} - e^{\beta(J-h)}}{2}\tau_z \tag{A.8}$$

$$= e^{\beta J}\cosh(\beta h)I + e^{-\beta J}\tau_x + e^{\beta J}\sinh(\beta h)\tau_z, \tag{A.9}$$

allowing its eigenvalues to be read off as

$$\lambda_\pm = e^{\beta J}\cosh(\beta h) \pm \sqrt{e^{2\beta J}\sinh^2(\beta h) + e^{-2\beta J}}. \tag{A.10}$$

Using the analogy to the quantum mechanics of a spin-half particle in a magnetic field, it is possible to evaluate $\langle \psi_+|\tau_z|\psi_+\rangle$ without explicitly finding the eigenvectors. In this analogy, $\langle \psi_+|\tau_z|\psi_+\rangle$ is the expectation of the z component of spin when the particle is in a state with the spin vector parallel to the magnetic field. In other words, it is the projection onto the z-axis of a unit vector in the direction of the effective $\vec{b}$:

$$\langle \psi_+|\tau_z|\psi_+\rangle = \frac{b_z}{\sqrt{b_z^2 + b_x^2}} = \frac{e^{\beta J}\sinh(\beta h)}{\sqrt{e^{2\beta J}\sinh^2(\beta h) + e^{-2\beta J}}}.$$

A.2 Low-Temperature Expansion of the Ising Model

We define $y = \exp[-2J/T]$ and $x = \exp[-2h/T]$, and write the free energy of a finite-size system with N sites—which is a function of J, h, and T—as

$$F(J, h, T) = E_0 + TNf(x, y), \tag{A.11}$$

where E_0 is the ground-state energy,

$$E_0/N = -\frac{z}{2}J - h, \tag{A.12}$$

where z is the number of nearest neighbors (i.e., 4 for the square lattice). Moreover, from this we can compute the magnetization as

$$m = -\frac{1}{N}\frac{\partial F}{\partial h} = 1 + 2x\frac{\partial f}{\partial x}. \tag{A.13}$$

To begin, let us consider h finite so that at low T both x and y are small. Eventually, because f is independent of N as $N \to \infty$, our first goal will be to obtain expressions for $f(x, y)$ expressed as a power series in powers of x:

$$f(x, y) = \sum_{n=1}^{\infty} x^n f_n(y). \tag{A.14}$$

However, as a first step we will consider obtaining such an expansion for the partition function itself,

$$Z = e^{-E_0/T}\left[\sum_{n=0}^{\infty} x^n Z_n(y)\right], \tag{A.15}$$

even though Z is not a well-behaved quantity as $N \to \infty$ since it depends exponentially on N. Here $x^n Z_n$ is the contribution to the partition function (with the ground-state energy subtracted) from all configurations with n overturned spins, and of course $Z_0 = 1$. We can compute the expansion for f from the expansion for Z with the result that

$$Nf_1(y) = Z_1, \tag{A.16}$$

$$Nf_2(y) = Z_2 - \frac{Z_1^2}{2}, \tag{A.17}$$

$$Nf_3(y) = Z_3 - Z_1 Z_2 + \frac{Z_1^3}{3}, \tag{A.18}$$

$$Nf_4(y) = Z_4 - \frac{Z_2^2 + 2Z_1 Z_3}{2} + Z_1^2 Z_2 - \frac{Z_1^4}{4}, \tag{A.19}$$

and so on.

Now we compute the first few Z_ns. Z_1 is easy: the cost to flip a single spin is $2zJ$ and the number of configurations with exactly one flipped spin is N, so

$$Z_1 = Ny^z. \tag{A.20}$$

Computing Z_2 is a bit more complicated. The energy to flip two spins that are not nearest neighbors of each other is $4zJ$ and the number of ways to do this is $N(N-z-1)/2$. The number of ways to flip two spins that are nearest neighbors of each other is $Nz/2$ and the energy of this is $4zJ - 4J$. Thus,

$$Z_2 = \frac{N(N-z-1)}{2}y^{2z} + \frac{Nz}{2}y^{2z-2} = \frac{N(N-1)}{2}y^{2z} + \frac{Nz}{2}\left[y^{2z-2} - y^{2z}\right]. \tag{A.21}$$

Computing Z_3 is still more involved. The total number of ways to flip 3 spins is $N(N-1)(N-2)/6$. However, not all of these configurations have the same energy. In most of them, none of the three spins is next to another, so the energy is $6zJ$. We will take this as a first approximation but then correct for the configurations that have a different energy. The second expression in the equation for Z_2 can be interpreted as doing this. Specifically, one special case is configurations in which of the three spin-flips, two are nearest neighbors and one is not a nearest neighbor of either of the other two. There are $Nz(N-2z)/2$ such configurations and they have energy $6zJ - 4J$. And then there are those configurations in which one spin is a nearest neighbor of the two other flipped spins, but neither of these other flipped spins is a nearest neighbor of the other. For a lattice—such as the square or cubic lattices—in which there are no elementary triangles, these are all the cases with three neighboring spin-flips. The number of such configurations is $Nz(z-1)/6$ and the energy is $6zJ - 8J$. Putting all this together we get

$$Z_3 = \frac{N(N-1)(N-2)}{6}y^{3z} + \frac{Nz(N-2z)}{2}\left[y^{3z-2} - y^{3z}\right]$$
$$+ \frac{Nz(z-1)}{2}\left[y^{3z-4} - y^{3z}\right]. \tag{A.22}$$

We see two things here. One is that this is a really terrible expansion for a large system. In the large N limit, $Z_n \sim N^n$, and so the expansion in Eq. A.15 is useless unless $x \ll N^{-1}$. The second is that while we started this exercise as an expansion in powers of x, it is also organized in powers of y in the sense that for small y, the leading-order contribution to $Z_1 \sim y^z$, and the leading-order contribution to $Z_2 \sim y^z y^{z-2}$, which for $z > 2$ (i.e., in dimension greater than 1) is a higher power than for Z_1, and $Z_3 \sim y^z y^{2z-4}$, which is still smaller. So we will not, ultimately, have to rely on small x to control our expansion, even though we used it as an intermediate step.

However, things look better when we reanalyze the same information in terms of an expansion for f. The first two terms in this expansion can be expressed in the same form for any lattice structure:

$$f_1 = y^z,\tag{A.23}$$

$$f_2 = \frac{y^{2z-2}}{2}\left[z - (z+1)y^2\right].\tag{A.24}$$

The higher order terms depend on the lattice topology. For lattices with no triangular plaquettes (such as the square, hexagonal, and hypercubic lattices),

$$f_3 = y^{3z-4}\left[\frac{z(z-1)}{2} - z^2 y^2 + \left(\frac{1}{3} + \frac{z(z-1)}{2}\right)y^4\right].\tag{A.25}$$

With increasing order, the topology of the lattice (in addition to the number of nearest neighbors) becomes increasingly important. On a square lattice,

$$f_4 = y^8\left[1 + \mathcal{O}(y^2)\right].\tag{A.26}$$

The cool thing to notice is that all the terms of order N^2 in Z_2 cancel against the corresponding terms in $-Z_1^2/2$ and that all the order N^3 and N^2 terms in Z_3 cancel against the corresponding terms in $-Z_1 Z_2 + Z_1^3/3$, and so on. Thus, as expected, the expansion for f does not have any of the problems that the expansion for Z has—all terms to all orders are independent of N. Moreover, we can see that each of these terms is a series in y, but higher order terms always start at higher orders in y. Thus, we can reorganize this into an expansion in powers of y—i.e., we can obtain the low-temperature expansion even in the limit $h \to 0$ ($x = 1$).

Specifically, taking the terms we have already derived up through f_4 and rearranging them into a series in powers of y, we obtain expressions for all contributions up through order y^8, i.e., the first term that we would need further work to compute is of order y^{10}. Similarly, to compute the spontaneous magnetization, m, in the limit $h \to 0$ we express it as a derivative of f as in Eq. A.13, and evaluate it at $h \to 0$. The result (for the square lattice) is readily seen to be

$$m = 1 - 2y^4 - 8y^6 - 34y^8 + \mathcal{O}(y^{10}).\tag{A.27}$$

You can find places[3] where this series is computed to order y^{76}!

A.3   The Fourier Series and Transforms

We have used properties of Fourier series and Fourier transforms at various places throughout this book. Here we summarize some of the key results we have used. To begin with, let $f(\vec{r})$ be any periodic function of a d-dimensional vector, $\vec{r}$, i.e., it satisfies the periodicity condition $f(\vec{r}) = f(\vec{r} + L\hat{e}_a)$, where $\hat{e}_a$ is a unit vector in the a

---

3. See I. G. Enting et al., *J. Phys. A: Math. Gen.* **27**, 6987 (1994).

direction, $\hat{e}_a \cdot \hat{e}_b = \delta_{a,b}$, and L is the linear dimension of the system. Subject to certain mild conditions, any periodic function $f(\vec{r})$ can be represented as a Fourier series:

$$f(\vec{r}) = \sum_{\vec{k}} e^{i\vec{k}\cdot\vec{r}} F(\vec{k}), \qquad (A.28)$$

where $\vec{k} = (2\pi)/L \sum_{a=1}^{d} m_a \hat{e}_a$ is also a d-dimensional vector and m_a are integers. Moreover,

$$F(\vec{k}) = L^{-d} \int d^d\vec{r}\, e^{-i\vec{k}\cdot\vec{r}} f(\vec{r}), \qquad (A.29)$$

where the spatial integral extends over the entire volume and

$$\int d^d\vec{r} \equiv \prod_{a=1}^{d} \int_0^L dr_a = L^d. \qquad (A.30)$$

$F(\vec{k})$ is the Fourier transform of $f(\vec{r})$.

Since we are typically interested in taking the thermodynamic limit, we should view the periodic boundary conditions as being a convenient computational device. In particular, we will be interested in the above expressions in the limit that $L \to \infty$. In this limit, the discrete allowed values of $\vec{k} = (2\pi/L)\vec{n}$ become more and more closely spaced, so it is eventually possible to replace sums over them with integrals. The way this limit is approached can be seen using the Poisson summation formula (which we quote without derivation):

$$\sum_{\vec{n}} g(2\pi\vec{n}/L) = \sum_{\vec{m}} \int d\vec{q}\, e^{2\pi i \vec{q}\cdot\vec{m}} g(2\pi q/L) \qquad (A.31)$$

$$= L^d \left\{ \int \frac{d\vec{k}}{(2\pi)^d} g(\vec{k}) + \text{error} \right\}, \qquad (A.32)$$

where $g(\vec{k})$ is any "reasonable" function, $\vec{n}$ and $\vec{m}$ are vectors with integer components, and

$$\text{error} = \sum_{\vec{m}\neq\vec{0}} \int \frac{d\vec{k}}{(2\pi)^d} e^{iL\vec{k}\cdot\vec{m}} g(\vec{k}). \qquad (A.33)$$

Since the integrand in the contribution from nonzero $\vec{m}$ (which we have identified as the "error") is an increasingly rapidly oscillating function of the length, L, this term typically decreases exponentially with increasing L and hence is negligible for large systems. As a consequence, with a different normalization convention, it is possible to express the same relations between functions and their Fourier transforms in the thermodynamic limit as

$$\tilde{F}(\vec{k}) = \int d^d\vec{r}\, e^{-i\vec{k}\cdot\vec{r}} f(\vec{r}), \qquad (A.34)$$

in which case

$$f(\vec{r}) = \int \frac{d\vec{k}}{(2\pi)^d} \, e^{i\vec{k}\cdot\vec{r}} \, \tilde{F}(\vec{k}), \tag{A.35}$$

where $\tilde{F}(\vec{k}) = \lim_{L\to\infty} L^d F(\vec{k})$.

We have often considered the situation in which a quantity of interest, $f(\vec{R})$, is defined on a regular lattice, specified by discrete lattice points $\vec{R}$. Still we arrange to have periodic boundary conditions, so that $f(\vec{R}) = f(\vec{R} + L\hat{e}_a)$. Again, we can express such functions in a Fourier series such that

$$f(\vec{R}) = \sum_{\vec{k}} \frac{e^{i\vec{k}\cdot\vec{R}}}{L^{d/2}} F(\vec{k}), \tag{A.36}$$

where now, however, instead of being expressed as an integral over space,

$$F(\vec{k}) = \sum_{\vec{R}} \frac{e^{-i\vec{k}\cdot\vec{R}}}{L^{d/2}} f(\vec{R}). \tag{A.37}$$

Here, because both space and Fourier space are discrete, it is conventional to use yet another normalization convention; if we measure L in units of the lattice constant, then L^d is the number of sites in the system, so the sums are normalized by factors of the square root of this number.

There is one other important difference between the situations in which functions are defined on a lattice of points and in continuum space. Our original function, $f(\vec{R})$, is now entirely specified by L^d numbers—the sums over $\vec{R}$ being restricted to those values. Correspondingly, although the allowed values of $\vec{k}$ are still of the form $\vec{k} = (2\pi/L)\vec{n}$ with $\vec{n}$ an integer valued vector, it cannot be true that the sum extends over all (infinite) possible values of $\vec{n}$. Indeed, for any given periodic lattice, there exists an infinite set of vectors—known as the reciprocal lattice vectors $\vec{G}$— such that

$$\exp[i\vec{R}\cdot\vec{G}] = 1 \ \ \forall \ \vec{R}. \tag{A.38}$$

For instance, if $\vec{R} = \vec{m}$ forms a hypercubic lattice, then any $\vec{G} = (2\pi)\vec{n}$ is a reciprocal lattice vector that satisfies Eq. A.38. In other words, it follows from Eq. A.37 that for any $\vec{k}$, $F(\vec{k}) = F(\vec{k} + \vec{G})$. We are thus forced to restrict the sum in Eq. A.36 to "distinct" values of $\vec{k}$. The way we do this is by defining what is known as a Brillouin zone (BZ): the BZ associated with each reciprocal lattice vector $\vec{G}$ is the region of $\vec{k}$ space that is closer to $\vec{G}$ than to any other reciprocal lattice vector. The first BZ is then the set of points $\vec{k}$ that are closer to the origin than to any nonzero $\vec{G}$. (One needs to be careful in treating points on the "surface" of a BZ—that is to say points that are equidistant from two distinct reciprocal lattice vectors—but we will not discuss this further here.) We will always interpret

$$\sum_{\vec{k}} \equiv \sum_{\vec{k}\in 1^{\text{st}}\text{ BZ}}. \tag{A.39}$$

Importantly, there are always exactly the same number of points in the first BZ as there are lattice points in the orignal problem, so that

$$\sum_{\vec{R}} 1 = \sum_{\vec{k}} 1 = L^d. \tag{A.40}$$

A.4   The Calculus of Variations

In Section 7.3 we transitioned from treating a system with physical properties, such as the magnetization m_j, defined on a discrete lattice, to describing the system in terms of fields, such as $\phi(x)$, defined on a continuous space of points, x. This brings with it the necessity to generalize various familiar notions from the study of functions of many variables, such as the mean-field free energy $F_{mf}(\{m_j\})$, to the study of "functionals," which is to say functions of functions, such as the Landau-Ginzberg free energy $F_{LG}[\phi]$. Furthermore, where in the discrete case we identified optimal mean-field textures by searching for conditions in which all the partial derivatives vanish $\partial F_{mf}/\partial m_j = 0$, when treating a field theory we seek the analogous conditions for the functional derivative, $\delta F_{LG}/\delta\phi = 0$.

This situation is analogous to the least-action principle of classical mechanics. We briefly summarize it here for the purpose of analogy. The goal of classical mechanics is to determine the trajectory $\bar{y}(t)$ traveled by a particle. The action, $S[y]$, is a functional which associates a real number with any possible (continuous) trajectory, $y(t)$, with fixed starting and ending points, $y_0 = y(t_0)$ and $y_f = y(t_f)$. The trajectory with the smallest action is the one that the particle actually takes. The way to find this trajectory is to compute the change in S produced by a tiny change in the trajectory $\delta y(t)$, i.e., $\delta S[y] = S[y(t) + \delta y(t)] - S[y(t)]$. The trajectory which minimizes the action is the one whose variation is stationary no matter how the path is altered, i.e., where $\delta S = 0$ to first order in δy for all possible $\delta y(t)$ subject to the condition that $\delta y(t_0) = \delta y(t_f) = 0$. The result is a differential equation for the trajectory $\bar{y}(t)$—the Euler-Lagrange equation.

In the Landau-Ginzburg theory we are faced with an analogous problem. We wish to find the field configurations $\phi(x)$—the "order parameter textures"—that minimize the free energy functional, $F[\phi(x)]$. This requires computing the functional derivative of F and finding its zeros.

To be explicit, we consider the 1D ϕ^4 theory for which

$$F[\phi(x)] = \int dx \left\{ -h(x)\phi(x) + \frac{\alpha}{2}\phi(x)^2 + \frac{u}{4}\phi(x)^4 + \frac{\kappa}{2}\phi'(x)^2 \right\}. \tag{A.41}$$

The functional derivative is the 1st-order change in F if we change the argument $\phi(x) \to \phi(x) + \delta\phi(x)$. (The curly delta symbol is a notational reminder that we are now dealing with deviations in functions rather than deviations in numbers).

Ignoring terms of order $(\delta\phi(x))^2$ or higher, we find

$$F[\phi(x) + \delta\phi(x)] = \int dx\Big\{ -h(x)\left[\phi(x) + \delta\phi(x)\right]$$

$$+ \frac{\alpha}{2}\left[\phi(x)^2 + 2\phi(x)\,\delta\phi(x)\right]$$

$$+ \frac{u}{4}\left[\phi(x)^4 + 4\phi(x)^3\,\delta\phi(x)\right]$$

$$+ \frac{\kappa}{2}\left[\phi'(x)^2 + 2\phi'(x)\,\delta\phi'(x)\right] + \dots \Big\}. \qquad (A.42)$$

To deal with the term proportional to $\delta\phi'(x)$, we integrate by parts:

$$\int dx\phi'(x)\,\delta\phi'(x) = -\phi''(x)\,\delta\phi(x) + \underline{(\phi'(x)\,\delta\phi(x))}\big|_{-\infty}^{+\infty}, \qquad (A.43)$$

where the boundary term vanishes. The integration by parts moves a derivative from the $\delta\phi$ onto the ϕ' and picks up a minus sign, leaving us with

$$\delta F = F[\phi(x) + \delta\phi(x)] - F[\phi(x)]$$

$$= \int dx \left\{ -h(x) + \alpha\phi(x) + u\phi(x)^3 - \kappa\phi''(x) \right\} \delta\phi(x). \qquad (A.44)$$

The function in curly brackets is thus the functional derivative of F, represented as $\delta F/\delta\phi$. The only way for this to vanish for any possible deviation $\delta\phi(x)$ is if it vanishes for all x. Hence the field configuration $\phi(x)$ which minimizes the free energy must satisfy the differential equation

$$\delta F/\delta\phi(x) = \alpha\phi(x) + u\phi(x)^3 - \kappa\phi''(x) - h(x) = 0. \qquad (A.45)$$

We have thus reduced the problem of finding the equilibrium configurations—the textures $\phi(x)$ that minimize the free energy $F[\phi(x)]$—to solving a differential equation.

A.5 The Continuum Limit

Now we will take the continuum limit mathematically and convert the free energy from the function of many variables, $F[m_j]$, into the functional of the field, $F[\phi(\vec{r})]$. For simplicity, we begin in one dimension, where the free energy looks like

$$F[m_j] = \sum_{j=1}^{N}\left[V(m_j) + \frac{J}{2}(m_{j+1} - m_j)^2 \right], \qquad (A.46)$$

where sites are labeled as $j = 1, 2, \dots, N$ and ϕ_j is the value of the order parameter at the jth site. Here $V(\phi_j)$, given in Eq. 4.30, is the "one-site" free energy.

In the discrete world, the order parameter ϕ_j takes on values for *discrete* values of $j = 1, 2, \ldots, N$, one value for each lattice site. In the continuum world, the order parameter $\phi(x)$ takes on values for *continuous* values of x, which can be any real number on the number line. Notice the differing notation: in the discrete case ϕ_j, the label for the site j is a *subscript*, while in the continuous case $\phi(x)$, the label for the position x is an *argument of a function*.

Since we want to turn the discrete case into the continuous case, it is helpful to first write down the discrete ϕ_j in a functional form as well. The position of the jth lattice site is $x_j = ja$, where a is the lattice spacing. The discrete order parameter field can be written as $\phi(x_j) := \phi_j$, where the function is defined only at positions x that are multiples of the lattice constant a. Then

$$F[\phi(x)] = \sum_j \left[V(\phi(x_j)) + \frac{J}{2}[\phi(x_{j+1}) - \phi(x_j)]^2 \right]. \tag{A.47}$$

In the limiting case as $a \to 0$, the order parameter field $\phi(x_j)$ becomes $\phi(x)$ and the sum becomes an integral. Recall that an integral is defined as

$$\sum_j f(x_j)\Delta x \longrightarrow \int f(x)\,dx, \tag{A.48}$$

where the spacing Δx corresponds to the lattice constant a of our problem. Correspondingly the first term of Eq. A.47 becomes

$$\sum_j V(\phi(x_j)) \longrightarrow \frac{1}{a} \int V(\phi(x))\,dx. \tag{A.49}$$

The second term of Eq. A.47 is related to the difference in ϕ between neighboring sites. In the continuum limit it becomes a spatial derivative. Applying the finite difference approximation

$$\frac{f(x_j + \Delta x) - f(x_j)}{\Delta x} \longrightarrow f'(x_j), \tag{A.50}$$

we find that the second term becomes

$$\frac{J}{2}[\phi(x_{j+1}) - \phi(x_j)]^2 \longrightarrow \frac{Ja^2}{2}[\phi'(x_j)]^2 \tag{A.51}$$

in the continuum limit. The overall free energy thus becomes

$$F = \int_0^L \mathcal{F}(x)\,dx, \tag{A.52}$$

where

$$\mathcal{F}(x) = \tilde{V}(\phi(x)) + \frac{\kappa}{2}(\phi'(x))^2,$$

$$\tilde{V}(x) = \frac{1}{a}V(x),$$

$$\kappa = Ja. \tag{A.53}$$

Here the free energy density $\mathcal{F}(x)$ may be interpreted as the local contribution at x to the total free energy. It includes contributions both from a "Landau" term, which depends only on the value of the field there, as well as a "gradient" term, which depends on the spatial variations in the field. Notably, the free energy density does not depend explicitly on the position x.

The gradient term is the key difference compared to the Landau theory of Chapter 6. It penalizes spatial variation in the order parameter. In higher dimensions, it generalizes to

$$F = \int_V \left[\tilde{V}(\phi(\vec{r})) + \frac{\kappa}{2}(\vec{\nabla}\phi) \cdot (\vec{\nabla}\phi) \right] d^d r. \tag{A.54}$$

A.6 Fourier Transform of a Lorentzian: The Ornstein-Zernike Law

We have seen that, under multiple circumstances, the continuum expression for the Fourier transform of various correlation functions is a Lorentzian—that is to say,

$$G(\vec{k}) = G_0 \left[\frac{1}{|\vec{k}|^2 + \kappa^2} \right], \tag{A.55}$$

where, in d dimensions, $|\vec{k}|^2 = \sum_{j=1}^{d} k_j^2 = k^2$. Here $G_0 \kappa^{-2} = G(\vec{k} = \vec{0})$ is typically an interesting susceptibility and κ is the inverse of an appropriately defined correlation length. Specifically, we are often interested in evaluating the behavior at long distances, $|\vec{r}| \to \infty$, of the inverse Fourier transform

$$\tilde{G}(\vec{r}) = G_0 \int \frac{d^d \vec{k}}{(2\pi)^d} \frac{e^{i\vec{k}\cdot\vec{r}}}{k^2 + \kappa^2}. \tag{A.56}$$

Getting to this form may involve scaling distances appropriately in different directions, so that the $\vec{k}$ dependence is isotropic. If the original problem was defined on a lattice, then getting to this form also involved neglecting terms higher than quadratic in powers of $\vec{k}$. In general, however, this can always be justified for large enough $|\vec{r}|$, where it can be shown that the integral is dominated by small $|\vec{k}|$ in any case. So while this is a very particular integral, it is one that describes generic correlations in many circumstances.

There are a few things we can tell about this integral even before performing it. The first is that since $G(\vec{k})$ depends only on $|k|$, and not on the direction of $\vec{k}$, the same must be true of the $\vec{r}$ dependence of $\tilde{G}$, i.e., $\tilde{G}(\vec{r}) = \tilde{G}(|\vec{r}|)$ (where we have used the same symbol for the two functions in the interest of notational parsimony). This can be seen by using our freedom to define a coordinate system such that, for any given $\vec{r}$, we can chose the x-axis of $\vec{k}$ to lie along the direction of $\vec{r}$. We can, moreover, rescale the integration variable $\vec{k} \to \vec{k}/\kappa$, from which we can infer that the κ dependence of the result can be expressed as

$$\tilde{G}(\vec{r};\kappa) = \kappa^d \tilde{G}(\kappa r;1). \tag{A.57}$$

We will discuss the result first for the important special cases of $d = 1$ and $d = 3$, and then will analyze it in a manner that applies to all values of d.

A.6.1 $d = 1$

In this case, we have a single integral to perform:

$$\tilde{G}(|r|) = G_0 \int_{-\infty}^{\infty} \frac{dk}{2\pi} \frac{e^{ik|r|}}{\kappa^2 + k^2}. \tag{A.58}$$

This is a familiar integral. There are many ways to do it (including looking it up in a table of integrals). The best way, for those familiar with the theory of functions of a complex variable, is to treat it as a contour integral closed in the upper half plane (since $|r| \ge 0$). A slightly devious method is to work backward from the fact that a Lorenzian is the Fourier transform of an exponential, i.e.,

$$\int_{-\infty}^{\infty} dr\, e^{-\kappa|r|}\, e^{-ikr} = \frac{2\kappa}{k^2 + \kappa^2}. \tag{A.59}$$

Since $2\kappa/[k^2 + \kappa^2]$ is the Fourier transform of $e^{-\kappa|r|}$, it follows that the latter function must be the inverse Fourier transform of the former. Thus

$$\tilde{G}(|r|) = \frac{G_0}{2\kappa} e^{-\kappa|r|}. \tag{A.60}$$

A.6.2 $d = 3$

In $d = 3$, we can use spherical coordinates to simplify the integral, where we chose the x-axis above such that $\vec{r} \cdot \vec{k} = |r|k\cos(\theta)$:

$$\tilde{G}(r) = \frac{G_0}{(2\pi)^3} \int_0^{\infty} k^2 dk \int_0^{\pi} \sin\theta\, d\theta \int_0^{2\pi} d\varphi \frac{e^{ik|\vec{r}|\cos\theta'}}{\kappa^2 + k^2}. \tag{A.61}$$

The integral over φ gives a factor of 2π. The integral over θ is performed by making the substitution $u = \cos\theta \implies du = -\sin\theta\, d\theta$:

$$\int_0^{\pi} \sin\theta\, d\theta\, e^{ikr\cos\theta} = -\int_1^{-1} du\, e^{ikru} = \frac{2\sin(kr)}{kr}. \tag{A.62}$$

The final integral over k can be evaluated using a few manipulations. Firstly

$$\tilde{G}(r) = \frac{G_0}{(2\pi)^2 r} \int_0^\infty k\,dk \frac{2\sin(kr)}{q^2 + k^2} \tag{A.63}$$

$$= -\frac{G_0}{(2\pi)^2 r} \frac{d}{dr} \int_0^\infty dk \frac{2\cos(kr)}{\kappa^2 + k^2}. \tag{A.64}$$

Given that the integrand is an even function of k, we can extend the integral from $-\infty$ to ∞ and drop the factor of 2. The same symmetry implies that we can use the identity $e^{ikr} = \cos(kr) + i\sin(kr)$ to replace the cosine by the exponential—i.e., the sin integral vanishes by symmetry. Thus

$$\tilde{G}(r) = -\frac{G_0}{(2\pi)^2 r} \frac{d}{dr} \int_{-\infty}^\infty dk \frac{e^{ikr}}{\kappa^2 + k^2}. \tag{A.65}$$

This is the same integral we already performed in the 1D case. As a result,

$$\tilde{G}(r) = \frac{G_0}{4\pi r} e^{-\kappa r}. \tag{A.66}$$

A.6.3   Arbitrary d

To perform this analysis for arbitrary d, we will make use of a trick. Since $\kappa^2 + k^2 \geq 0$:

$$\frac{1}{q^2 + k^2} = \int_0^\infty d\lambda\, e^{-\lambda(q^2 + k^2)}. \tag{A.67}$$

Since $k^2 = \sum_j k_j^2$ and $\vec{k} \cdot \vec{r} = \sum_j k_j r_j$, we can write $e^{i\vec{k}\cdot\vec{r} - \lambda k^2} = \prod_j e^{ik_j r_j - \lambda k_j^2}$, and hence write the integral Eq. A.56 as

$$\tilde{G}(\vec{r}) = G_0 \int_0^\infty d\lambda\, e^{-\lambda\kappa^2} \prod_j \int_{-\infty}^\infty \frac{dk_j}{(2\pi)} e^{-\lambda k_j^2 + ik_j r_j}. \tag{A.68}$$

Using

$$\int_{-\infty}^\infty dx\, e^{-ax^2 + ibx} = \sqrt{\frac{\pi}{a}} e^{-\frac{b^2}{4a}}, \tag{A.69}$$

the integrals over k_j can be evaluated to give

$$\tilde{G}(\vec{r}) = G_0 \int_0^\infty d\lambda\, e^{-\lambda\kappa^2} \prod_j^d \left[\sqrt{\frac{1}{4\pi\lambda}} e^{-\frac{r_j^2}{4\lambda}} \right] \tag{A.70}$$

$$= \frac{G_0}{(4\pi)^{d/2}} \int_0^\infty \frac{d\lambda}{\lambda^{d/2}} e^{-\lambda\kappa^2 - \frac{r^2}{4\lambda}}.$$

At this stage, we have transformed the problem in any d into a single integral. This integral is not trivial. However, for what it is worth, it can be expressed in terms of K_ν, which is known as the modified Bessel function of the second kind, which has (as can be checked in a book of special functions) an integral representation

$$K_\nu(x) = (1/2)(x/2)^\nu \int_0^\infty dt\; t^{-\nu-1}\; e^{-t-x^2/4t}. \tag{A.71}$$

In other words, by changing the integration variables according to $t = \lambda\kappa^2$, we find that the general expression for the inverse Fourier transform of a Lorentzian in d dimensions is

$$\tilde{G}(\vec{r}) = \frac{G_0\kappa^{d-2}}{(2\pi)^{d/2}} \left(\frac{1}{\kappa r}\right)^{(d-2)/2} K_\nu(\kappa r). \tag{A.72}$$

From this expression we can find the critical exponent $\nu = (d-2)/2$ (Section 7.3.2). To make sure we have not made any mistakes, we can consult the same table of integrals to find that

$$K_{-1/2}(x) = K_{1/2}(x) = \sqrt{\frac{\pi}{2x}}e^{-x}, \tag{A.73}$$

which reproduces the previously obtained results for $d = 1$ and $d = 3$.

While it is always aesthetically pleasing to have exact analytic results for expressions, the expression in terms of Bessel functions is not entirely transparent. Moreover, we are really primarily interested in understanding the behavior of these functions in the limit of large r, which is to say when $\kappa r \gg 1$. This we can derive from the large x asymptotic form of Eq. A.71.

We will do this using what is known as the "method of steepest descents," which can be found in many textbooks on math methods. We define the function

$$f(t) = x^2/4t + t + (\nu+1)\ln[t], \tag{A.74}$$

in terms of which

$$K_\nu(x) = \frac{1}{2}\left(\frac{x}{2}\right)^\nu \int_0^\infty dt\; \exp[-f(t)]. \tag{A.75}$$

We can find the point at which the integrand is biggest, which is equivalent to the point at which $f(t)$ is smallest, i.e., at $t = t^\star$, where

$$t^\star = (1/2)\left[\sqrt{x^2 + (\nu+1)^2} - (\nu+1)\right] = (x/2)\left[1 + (\nu+1)x^{-1} + \dots\right], \tag{A.76}$$

where in the second expression we have expanded the result in powers of x^{-1} (assumed small). We are guessing that for large x, the integral will be dominated by the range of t near $t^\star$, where we can approximate f by the first terms in its Taylor series:

$$f(t) = f(t^\star) + \frac{(t-t^\star)^2}{2}f''(t^\star) + \dots, \tag{A.77}$$

where

$$f''(t^\star) = \frac{2}{t^\star}\left[1 + \frac{(\nu+1)}{2t^\star}\right] = \frac{4}{x}[1 + \dots]. \tag{A.78}$$

We now check that our guess is right. If we keep only the first two terms in the expansion of f, and if we do not worry about the limits of integration (i.e., we integrate over all t), then we are left with a Gaussian integral that we can evaluate explicitly:

$$K_\nu(x) = \frac{1}{2}\left(\frac{x}{2}\right)^\nu \sqrt{\frac{2\pi}{f''(t^\star)}}\, e^{-f'(t^\star)}[1 + \dots], \tag{A.79}$$

where ... signifies all errors we have made as a result of these approximations. We can estimate this by noting that the range of $|t - t'|$ that makes significant contributions to the integral is at most a few times $1/\sqrt{f''(t^\star)} \sim \sqrt{x}$—for example, the mean square value of $|t - t'|$ is precisely $1/f''(t^\star)$. For large x, this is a small fraction of $t' \sim x \gg \sqrt{x}$. It is possible to show that the expression in Eq. A.79 is the first term in an asymptotic series for the Bessel function, and that all the terms in ... are small in proportion to increasingly large powers of $1/x$. Finally, taking the large x expressions for $f(t^\star)$ and $f''(t^\star)$ gives

$$K_\nu(x) = \sqrt{\frac{\pi}{2x}}e^{-x}[1 + \dots]. \tag{A.80}$$

Putting this all together, we obtain an asymptotic form of the Ornstein-Zernike law in arbitrary dimensions:

$$\tilde{G}(\vec{r}) \sim \frac{G_0 \kappa^{d-2}}{2(2\pi)^{(d-1)/2}}\left(\frac{1}{\kappa r}\right)^{(d-1)/2} e^{-\kappa r}. \tag{A.81}$$

A.6.4   At Criticality: Ornstein-Zernike as $\kappa \to 0$.

With κ having the interpretation of an inverse correlation length, the limit $\kappa \to 0$ corresponds to a critical point, where the correlation length diverges. (Since $G(r)$ is proportional to the inverse Fourier transform of $1/k^2$, this problem may also be recognizable as the d-dimensional Coulomb or Newtonian gravity potential laws.)

The results here can be inferred from the above in the limit $\kappa \to 0$, but since the $K_\nu(x)$ Bessel function is singular in the $x \to 0$ limit, doing it this way requires taking the limit carefully. Instead, it is easier to evaluate the integral in Eq. A.70 explicitly for $\kappa = 0$:

$$\tilde{G}(\vec{r}) = \frac{G_0}{(4\pi)^{d/2}}\int_0^\infty \frac{d\lambda}{\lambda^{d/2}} e^{-\frac{r^2}{4\lambda}}. \tag{A.82}$$

To evaluate this integral, we make the substitution $t = r^2/4\lambda$, in terms of which the integral runs from $-\infty$ to 0. Inverting the order of limits and collecting terms, the result is

$$\tilde{G}(\vec{r}) = \frac{G_0}{(4\pi)^{d/2}} \left(\frac{r^2}{4}\right)^{1-d/2} \int_0^\infty dt\ t^{d/2-2} e^{-t}$$

$$= \frac{G_0}{(4\pi)^{d/2}} \Gamma\left(\frac{d}{2}-1\right) \left(\frac{2}{r}\right)^{d-2},\tag{A.83}$$

where Γ is the gamma function. Just to make sure we have made no mistake, note that in $d = 3$, using the fact that $\Gamma(1/2) = \sqrt{\pi}$, we can confirm that we get the familiar form of the 3D Coulomb potential, $\tilde{G}(\vec{r}) = (G_0/4\pi)r^{-1}$.

A.7 Review of Probability

Probability distributions are familiar to us from elementary statistical mechanics. Here, we review them briefly.

Let $\{s\}$ be an underlying set of possibilities. In any given event, the probability of s is $P(s)$. Let each s have a property $X(s)$. The probability of observing X is then given by

$$P(X) = \sum_s P(s) \text{ for } X(s) = X,\tag{A.84}$$

that is, $P(X)$ is the fraction of events yielding that value of X. If X is a continuous variable, the definition is slightly modified to include events whose X falls within a small range as

$$\rho(X)dX = \sum_s P(s) \text{ for } X < X(s) < X + dX.\tag{A.85}$$

The average value of X is

$$\langle X \rangle = \int X\rho(X)dX,\tag{A.86}$$

which is the average value of X where each possible value of X is weighted by its probability.

The variance of a distribution is the typical size of the deviation from the average value, $X - \langle X \rangle$. It is given by the expression

$$\left\langle (X - \langle X \rangle)^2 \right\rangle = \langle X^2 \rangle - \langle X \rangle^2.\tag{A.87}$$

Multiple variables

There may be multiple observable variables X and Y. When there are multiple variables, one can define the joint distribution

$$\rho(X, Y)\ dX\ dY,\tag{A.88}$$

describing the probability of combinations of X and Y.

The conditional probability

$$P(X|Y) = \frac{P(X,Y)}{P(Y)} \tag{A.89}$$

is the probability of X occurring *given* that Y occurs. In other words, $P(X)$ counts the fraction of X occurrences out of *all* events, but $P(X|Y)$ counts the fraction only out of the events where Y occurs.

Independence

Sometimes two random variables are related to each other and other times they are not. This is captured by the notion of independence. We say that two random variables X and Y are independent if

$$P(X) = P(X|Y), \tag{A.90}$$

that is, if the distribution over X is exactly the same no matter the value of Y. Stated another way, if X and Y are independent, then knowing the value of Y gives no information about the value of X.

The definition of independence may also be rewritten as

$$P(X,Y) = P(X)P(Y), \tag{A.91}$$

or, in terms of probability distributions, as

$$\rho(X,Y)\, dX\, dY = \rho(X)\rho(Y)\, dX\, dY. \tag{A.92}$$

The central limit theorem

Because the central limit theorem is so foundational, it is worth studying it with all its bells and whistles—what the precise conditions are and how the limit is approached. Here, we derive it more casually, but still with some reference to the necessary assumptions.

Suppose there is a large number N of equally distributed random variables, x_j, so that

$$P(\{x\}) = \prod_{j=1}^{N} p(x_j). \tag{A.93}$$

We are interested in the average value of these quantities,

$$X \equiv N^{-1} \sum_{j=1}^{N} x_j, \tag{A.94}$$

whose distribution is

$$P_N(X) = \int \prod_{j=1}^{N} dx_j\, p(x_j)\delta(X - N^{-1}\sum_{j=1}^{N} x_j). \tag{A.95}$$

We can use the representation of the δ function as a Fourier transform

$$\delta(X) = (2\pi)^{-1} \int_{-\infty}^{\infty} dk\, \exp(iXk) \tag{A.96}$$

to re-express the above as

$$P_N(X) = (2\pi)^{-1} \int_{-\infty}^{\infty} dk\, e^{ikX} \left[\int dx p(x)e^{-ikx/N}\right]^N$$

$$= (2\pi)^{-1} \int_{-\infty}^{\infty} dk\, e^{ikX} \left[p_{k/N}\right]^N, \tag{A.97}$$

where p_k is the Fourier transform of $p(x)$.

The key observation now is that since N is large, the integral is increasingly dominated by the regions of k in which p_k is largest—the larger N is, the more true this is.

The first thing to note is that because $p(x) \geq 0$ for all x, it follows that p_k has its absolute maximum value at $k = 0$. To see this, note that

$$|p_k| = \left|\int dx p(x)e^{-ikx}\right| \leq \int dx \left|p(x)e^{-ikx}\right| \tag{A.98}$$

$$= \int dx p(x) = p_0 = 1,$$

where the final equality follows from the normalization of the probability. We thus know that for large N, the integral will be dominated by small values of k/N. It is also useful to note that because $p(x)$ is real, i.e., $p(x) = [p(x)]^*$, it follows that $p_k = p^*_{-k}$.

Building on these general properties, it is then useful (where possible) to express p_k in terms of a power series in k as

$$p_k = \exp[iak - bk^2 + ick^3 + \ldots], \tag{A.99}$$

where the normalization condition precludes a k^0 term, the behavior under complex conjugation implies that a, b, and c are real, and the condition that the maximum of $|p_k|$ is at $k = 0$ implies that $b \geq 0$. The $\ldots$ in the expansion refers to higher order terms in powers of k. These constants are related to various moments of the distribution,

$$\bar{x} \equiv \int dx p(x)x = i\left.\frac{dp_k}{dk}\right|_{k=0} = a, \tag{A.100}$$

$$\overline{x^2} \equiv \int dx p(x)x^2 = -\left.\frac{d^2 p_k}{dk^2}\right|_{k=0} = a^2 + 2b^2,$$

from which we identify $a = \bar{x}$ and $b = \sigma^2/2$. (see Problem 2.3) (As an aside, this is clearly possible only under the assumption that the distribution, $p(x)$, falls at least fast enough at large x that its second moment is defined. Indeed, the central limit theorem does not apply to a Lorenzian distribution, $p(x) = (\Delta/\pi)[(x-\bar{x})^2 + \Delta^2]^{-1}$.)[4]

If we now use this expression for p_k, we find that

$$P_N(X) = (2\pi)^{-1} \int_{-\infty}^{\infty} dk\, e^{ik(X-\bar{x}) - \sigma^2 k^2/2N} \left\{ 1 + icN^{-2}k^3 + \ldots \right\} \qquad \text{(A.101)}$$

$$= \frac{1}{\sigma\sqrt{\pi}} \exp\left[-N(X-\bar{x})^2/(2\sigma^2)\right] \left\{ 1 + cN(X-\bar{x})^3 + \ldots \right\}.$$

This is the central limit theorem. Up to corrections that are increasingly small as $N \to \infty$, starting with any (reasonable) probability distribution function $p(x)$ leads to a Gaussian distribution whose properties are entirely determined by the first and second moments of $p(x)$. Moreover, the width of the Gaussian shrinks in proportion to $1/\sqrt{N}$.

One aspect of the above derivation may not be entirely transparent. The first correction term is proportional to cN, and hence it is not apparent that it is negligible in the large N limit. This reveals a subtlety of the central limit theorem: clearly, from the leading term we can see that there is an extremely small probability of finding a situation in which $|X - \bar{x}| \gg \sigma/\sqrt{N}$. For this range of values, indeed the first correction term is small, in proportion to $N^{-1/2}$. However, even for large N, in situations in which we are interested in the *tails* of the distribution where $P_N(X)$ is extremely small, corrections to the central limit theorem can sometimes be of concern. This is why it is called the "central" limit theorem!

A.8   Deriving the Variational Principle for Noncommuting Hamiltonians

In Section 4.2, in deriving the variational principle, we assumed that the Hamiltonian H and the trial Hamiltonian, H_{tr}, were commuting operators. Specifically, according to the principles of quantum mechanics, in any given basis of states, H and H_{tr} are Hermitian matrices. If they are commuting matrices, we can choose a basis in which both are diagonal—that is to say, we assumed that each microstate s (chosen to be an eigenstate of H) has energies $E(s)$ and $E_{tr}(s)$. In the general quantum mechanical problem, this will not be the case. However, the same variational result can be derived with only a little more effort, as we show here.

---

4. Since the Fourier transform of a Lorentzian is exactly given as $p_k = e^{-ik\bar{x} - |k/\Delta|}$, the integral in Eq. A.97 can also be evaluated with the result that, for a Lorenzian distribution, $P_N(x) = p(x)$.

It is always the case that

$$Z = \text{Tr}\left[e^{-\beta H}\right] = \exp[-\beta F]. \tag{A.102}$$

Moreover, we can always express H as a sum of two operators as

$$H = H_{tr} + \tilde{H}, \tag{A.103}$$

although possibly H and H_{tr} may not commute (which implies $[H, \tilde{H}] \neq 0$ and $[H_{tr}, \tilde{H}] \neq 0$).

Define the imaginary time ($\tau = it$) propagators (which are Hermitian operators in Hilbert space),

$$U(\tau) = e^{-\tau H} \quad \text{and} \quad U_{tr} = e^{-\tau H_{tr}}, \tag{A.104}$$

such that these operators satisfy a Euclidean (no is) version of the Schrodinger equation (note that we have set $\hbar = 1$):

$$\frac{dU(\tau)}{d\tau} = -HU(\tau) \quad \text{and} \quad \frac{dU_{tr}(\tau)}{d\tau} = -H_{tr}U_{tr}(\tau), \tag{A.105}$$

where, since $[U, H] = [U_{tr}, H_{tr}] = 0$, the order of the operators is arbitrary. Note that if we make the identification $\tau = \beta = T^{-1}$, then $Z = \text{Tr}\left[U(\beta)\right]$.

Here we will borrow some techniques from time-dependent perturbation theory.[5] We can treat H_{tr} as the unperturbed Hamiltonian and $\tilde{H}$ as the perturbation to define the imaginary time "interaction representation" of the perturbation

$$\tilde{H}_I(\tau) = e^{\tau H_{tr}} \tilde{H} e^{-\tau H_{tr}}. \tag{A.106}$$

This leads to the differential equation for the imaginary time propagator in the interaction picture $U_I(\tau)$:

$$\frac{dU_I(\tau)}{d\tau} = -\tilde{H}_I(\tau) U_I(\tau) \tag{A.107}$$

subject to the initial condition that $U_I(0) = 1$. The solution to this differential equation is

$$U_I(\tau) = T_\tau\left[\exp\left\{-\int_0^\tau d\tau' \tilde{H}_I(\tau')\right\}\right] \tag{A.108}$$

$$= 1 + \sum_{n=1}^{\infty} \frac{(-1)^n}{n!} \int_0^\tau d\tau_1 \int_0^\tau d\tau_2 \cdots \int_0^\tau d\tau_n T_\tau\left[\tilde{H}_I(\tau_1)\tilde{H}_I(\tau_2)\cdots\tilde{H}_I(\tau_n)\right]$$

$$= 1 - \int_0^\tau d\tau' \tilde{H}_I(\tau') + \int_0^\tau d\tau' \int_0^{\tau'} d\tau'' \tilde{H}_I(\tau')\tilde{H}_I(\tau'') + \ldots,$$

---

5. For a good reference on some of the results quoted here, see, e.g., Chapter 4 of Peskin and Schroeder, *An Introduction to Quantum Field Theory* (CRC Press, Boca Raton, 1995).

where T_τ is the imaginary time-ordered product, i.e., it specifies that the products of operators should be ordered such that factors at the earlier imaginary times are to the right of those at later imaginary times. The first line of Eq. A.108 is the ordered exponential, which can be expressed as the Taylor series in the second line, the meaning of which is illustrated to second order in the series expansion in the third line.

Now we can show that, consistent with Eq. A.105, the propagator $U(\tau)$ can be expressed as

$$U(\tau) = U_{tr}(\tau)U_I(\tau) \tag{A.109}$$

by taking the τ derivative,

$$\frac{dU(\tau)}{d\tau} = \frac{dU_{tr}}{d\tau}U_I(\tau) + U_{tr}(\tau)\frac{dU_I(\tau)}{d\tau}$$

$$= -(H_{tr} + \tilde{H})U_{tr}U_I(\tau) = -HU(\tau), \tag{A.110}$$

thus showing that Eq. A.109 is true. By taking the trace of Eq. A.109 it follows that (with a change of variable $\tau \to \beta$)

$$Z = Z_{tr}\left\langle T_\beta\left[\exp\left\{-\int_0^\beta d\beta' \tilde{H}_I(\beta')\right\}\right]\right\rangle_{tr}, \tag{A.111}$$

where the expectation value is with respect to the trial ensemble:

$$\langle O \rangle_{tr} = Z_{tr}^{-1}\mathrm{Tr}\left[e^{-\beta H_{tr}}O\right]. \tag{A.112}$$

Since the exponential is again a convex operator, we can now use our favorite Jensen's inequality to see that

$$Z \geq Z_{tr}\exp\left\{-\int_0^\beta d\beta' \langle \tilde{H}_I(\beta')\rangle_{tr}\right\}; \tag{A.113}$$

we are able to bring the expectation value all the way inside the integral since it is a linear operator. We got rid of the time-ordered product by noting that

$$\langle \tilde{H}_I(\tau)\rangle_{tr} = \frac{1}{Z_{tr}}\mathrm{Tr}\left[e^{-\beta H_{tr}}e^{\beta H_{tr}}\tilde{H}e^{-\beta H_{tr}}\right] \tag{A.114}$$

$$= \frac{1}{Z_{tr}}\mathrm{Tr}\left[e^{-\beta H_{tr}}\tilde{H}\right] = \langle \tilde{H}\rangle_{tr} \tag{A.115}$$

is independent of τ.

Thus:

$$-\ln Z \le -\ln Z_{tr} + \int_0^\beta \langle \tilde{H} \rangle_{tr}, \tag{A.116}$$

$$\beta F \le \beta F_{tr} + \beta \langle \tilde{H} \rangle_{tr}, \tag{A.117}$$

$$F \le F_{tr} + \langle H - H_{tr} \rangle_{tr}. \tag{A.118}$$

This is the generalized variational principle—valid even for noncommuting operators.

A.9 Properties of Gaussian Distributions and Wick's Theorem

A Gaussian distribution is entirely specified by its mean $\bar{x}$ and its second moment $\overline{x^2}$:

$$P(x) = \left[\sigma\sqrt{\pi}\right]^{-1} \exp[-(x-\bar{x})^2/2\sigma^2], \tag{A.119}$$

where

$$\sigma^2 \equiv \langle [x-\bar{x}]^2 \rangle = \langle x^2 \rangle - \bar{x}^2. \tag{A.120}$$

Thus, if we know we have a Gaussian distribution, and we know its first two moments, we can compute (by doing simple integrals) any further moments. For instance, it is easy to see that

$$\langle [x-\bar{x}]^3 \rangle = 0 \tag{A.121}$$

and that

$$\langle [x-\bar{x}]^4 \rangle = 3\sigma^4. \tag{A.122}$$

Now consider a Gaussian distribution of N independent variables:

$$P(\{x\}) = \prod_{j=1}^N \left[\sigma_j\sqrt{\pi}\right]^{-1} \exp[-(x_j-\bar{x}_j)^2/2\sigma_j^2], \tag{A.123}$$

for which $\bar{x}_j$ is the mean of x_j and

$$\langle x_i x_j \rangle = \bar{x}_i \bar{x}_j + \delta_{i,j}\, \sigma_j^2. \tag{A.124}$$

Now consider the averages involving arbitrary N-component vectors:

$$\mathbf{v}^{(a)} = \sum_{j=1}^N v_j^{(a)} x_j. \tag{A.125}$$

Looking at the lowest moments of this, it is easy to see that

$$\langle \mathbf{v}^{(a)} \rangle = \sum_{j=1}^N v_j^{(a)} \bar{x}_j, \tag{A.126}$$

and that defining $\delta \mathbf{v}^a \equiv \mathbf{v}^{(a)} - \langle \mathbf{v}^{(a)} \rangle$,

$$\langle \delta \mathbf{v}^{(a)} \, \delta \mathbf{v}^{(b)} \rangle = \sum_{j=1}^{N} v_j^{(a)} v_j^{(b)} \, \sigma_j^2. \tag{A.127}$$

So far, this is straightforward and not very surprising. However, because a Gaussian distribution is entirely determined by its first and second moments, knowing these two moments should enable us to compute all higher moments, just as in the case of a single Gaussian variable. Indeed, the generalization of Eq. A.121 follows by symmetry, i.e., the third (and all odd) moments vanish, e.g.,

$$\langle \delta \mathbf{v}^{(a)} \, \delta \mathbf{v}^{(b)} \, \delta \mathbf{v}^{(c)} \rangle = 0. \tag{A.128}$$

What requires a bit more work to see is that the fourth moment can be written as a sum of products of second moments,

$$\langle \delta \mathbf{v}^{(a)} \, \delta \mathbf{v}^{(b)} \, \delta \mathbf{v}^{(c)} \, \delta \mathbf{v}^{(d)} \rangle = \langle \delta \mathbf{v}^{(a)} \, \delta \mathbf{v}^{(b)} \rangle \langle \delta \mathbf{v}^{(c)} \, \delta \mathbf{v}^{(d)} \rangle \tag{A.129}$$
$$+ \langle \delta \mathbf{v}^{(a)} \, \delta \mathbf{v}^{(c)} \rangle \langle \delta \mathbf{v}^{(b)} \, \delta \mathbf{v}^{(d)} \rangle + \langle \delta \mathbf{v}^{(a)} \, \delta \mathbf{v}^{(d)} \rangle \langle \delta \mathbf{v}^{(b)} \, \delta \mathbf{v}^{(c)} \rangle,$$

i.e., the sum of the three ways the fourth moment can be related to products of the second moments. The factor 3 in Eq. A.122, which we originally derived from doing an integral, from this perspective is simply the combinatoric factor expressing the number of ways 4 xs can be broken into two pairs.

Indeed, a similar relation applies to all higher moments. There are 15 ways to pair up 6 xs, and indeed you can check that for the single Gaussian,

$$\langle [x - \bar{x}]^6 \rangle = 15 \left[\langle [x - \bar{x}]^2 \rangle \right]^3. \tag{A.130}$$

In the context of statistical mechanics, this property of Gaussian integrals is known as Wick's theorem. It is the statement that in any ensemble in which the Hamiltonian is a quadratic function of some appropriate fields, all correlation functions can be expressed directly in terms of the average fields and the second moments. Thus, for instance, if we have a field $\phi(\vec{r})$ that is defined on points in space $\vec{r}$, and if we further are told that the Hamiltonian is a quadratic function of the fields—or equivalently that the probability of a given field configuration is determined by a Gaussian Boltzmann probability distribution function—then, defining $\delta\phi(\vec{r}) \equiv \phi(\vec{r}) - \langle \phi(\vec{r}) \rangle$, it follows from the properties of Gaussian integrals that

$$\langle \delta\phi(\vec{r}_1)\delta\phi(\vec{r}_2)\delta\phi(\vec{r}_3) \rangle = 0, \tag{A.131}$$

$$\langle \delta\phi(\vec{r}_1)\delta\phi(\vec{r}_2)\delta\phi(\vec{r}_3)\delta\phi(\vec{r}_4) \rangle = \langle \delta\phi(\vec{r}_1)\delta\phi(\vec{r}_2) \rangle \langle \delta\phi(\vec{r}_3)\delta\phi(\vec{r}_4) \rangle \tag{A.132}$$

$$+ \langle \delta\phi(\vec{r}_1)\delta\phi(\vec{r}_3) \rangle \langle \delta\phi(\vec{r}_2)\delta\phi(\vec{r}_4) \rangle + \langle \delta\phi(\vec{r}_1)\delta\phi(\vec{r}_4) \rangle \langle \delta\phi(\vec{r}_2)\delta\phi(\vec{r}_3) \rangle,$$

and so on for higher moments. Obviously, this is a good thing—it greatly reduces the number of calculations we need to do.

The derivation presented here—which is sufficient for present purposes—is a classical version of Wick's theorem. In a general quantum context, the fields in question generally do not not commute, so the order of the operators is important. Wick's theorem can be readily generalized to this case, but this is beyond the scope of the present treatment.

A.10   Rudiments of Group Theory

A group is comprised of a set of elements $G = \{g_1, g_2, g_3 \ldots\}$, and a binary operation $\circ$, called group multiplication, that combines two elements together to form another element $g_i \circ g_j = g_k$. To qualify as a group, the set and group law together $(G, \circ)$ must satisfy the following requirements (called group axioms):

1. **Closure**
   For all g_i and g_j that are in the set G, the operation $g_i \circ g_j = g_k$ yields g_k that is also in the set G.

2. **Associativity**
   For all g_i, g_j, and g_k that are in the set G, the associative law

   $$(g_i \circ g_j) \circ g_k = g_i \circ (g_j \circ g_k)$$

   is satisfied.

3. **Existence of identity element**
   There exists an element e in G such that for all g_i in G, $e \circ g_i = g_i = g_i \circ e$ is satisfied. e is called the *identity element*.

4. **Existence of inverse elements**
   For every element g_i in G, there exists an element g_i^{-1} also in G such that $g_i \circ g_i^{-1} = e = g_i^{-1} \circ g_i$ is satisfied.

Note that, the group is $(G, \circ)$, where both the underlying set G and the group law $\circ$ are considered together. As a matter of convenience, however, we will often just call this group G. Note also that the commutative law is not a requirement— in fact, groups in which multiplication is commutative are a special class of groups called "Abelian." In addition, there's an uniqueness theorem associated with each of axioms 3 and 4, which you will prove as an exercise to get yourself thinking about groups!

An example: The Z_4 group

To make all the previous statements concrete, consider a simple group, the Z_4 group. It can be represented as:

$$(\{1, i, -1, -i\}, \times). \tag{A.133}$$

So there are four elements, ± 1 and $\pm i$, and the group law is multiplication.

To verify that this is indeed a group as defined above, we can check the criteria one by one:

1. **Closure:** It is easy to see that multiplying any of the complex numbers ± 1 or $\pm i$ by any other produces another number in this set.
2. **Associativity:** Complex multiplication is manifestly associative.
3. **Existence of identity element:** Any element multiplied by 1 equals itself.
4. **Existience of inverse elements:** The inverse of 1 is 1, the inverse of -1 is -1, and i and $-i$ are inverses of each other.

To encapsulate the structure of a group, we often use a "multiplication table." Some of the above properties are more apparent when viewing the group this way:

Table A.1. Multiplication table for Z_4 represented using complex numbers.

$\times$	1	i	-1	$-i$
1	1	i	-1	$-i$
i	i	-1	$-i$	1
-1	-1	$-i$	1	i
$-i$	$-i$	1	i	-1

Any row or column of the multiplication table of a group consists of a permuted list of all the group elements in the underlying set,[6] which demonstrates closure, existence of identity, and existence of inverses.

Group as an abstraction

Before we move on, let us reinforce the idea that groups are fundamentally abstract. We can do away entirely with the reliance on complex numbers, and still talk about the Z_4 group. Consider the set of elements

$$(\{e, g_1, g_2, g_3\}, \circ) \tag{A.134}$$

with the group multiplication table:

Table A.2. Multiplication table for Z_4.

$\circ$	e	g_1	g_2	g_3
e	e	g_1	g_2	g_3
g_1	g_1	g_2	g_3	e
g_2	g_2	g_3	e	g_1
g_3	g_3	e	g_1	g_2

---

6. This is called the *rearrangement theorem*. Try proving it!

You may notice that this is the exact same table as Table A.1 if we make the identifications:

$$e \to 1,$$

$$g_1 \to i,$$

$$g_2 \to -1,$$

$$g_3 \to -i,$$

$$\circ \to \times.$$

In fact, when we speak of the Z_4 group, we are really talking about the contents of Table A.2, where the group elements, the group law, and the result of each binary combination using the group law are all abstract and untethered to any concrete mathematical object. We can always find a "representation" for the elements and group law as in Table A.1 if we wish to make it concrete. The abstraction allows us to use the idea of groups on different types of mathematical objects, underlying the power of group theory.

In summary, a group captures the "nice" abstract binary relations between elements of a "nice" set. In particular, the niceness of groups lends itself perfectly to the study of symmetries—the symmetries of an object form a group! As an illustrative example, we next consider the equilateral triangle to see how symmetry and groups are tied together.

A.10.1   Symmetries as a Group: An Example in a Triangle

In this section, we hope to illustrate one important point: the symmetry operations of an object form a group, or, in other words, a group encapsulates all the symmetries of an object. The equilateral triangle will be our prototypical object with symmetry.

Symmetries of an equilateral triangle

How many transformations can we do to an equilateral triangle such that it looks identical afterward? At a glance there seem to be five:

1. Rotation by $120°$.
2. Rotation by $-120°$.
3. Reflection across the line (3).
4. Reflection across the line (4).
5. Reflection across the line (5).

But there is another one; let's call it the zeroth operation:

0. *Do* nothing.

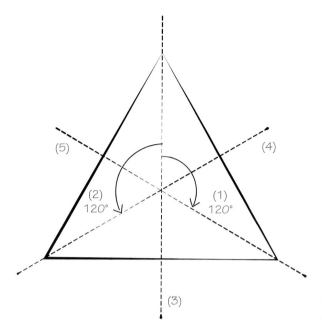

FIGURE A.1. Symmetries of an equilateral triangle.

While it may seem silly as of this moment, it is of paramount importance we are allowed to *do* nothing. It is also important that for any object, we always have this unique operation that leaves it invariant! Now that we have all six operations, let us give them some names:

Operation	Name
Do nothing	$g_0 \equiv e$
Rotation 120°	g_1
Rotation −120°	g_2
Reflection (3)	g_3
Reflection (4)	g_4
Reflection (5)	g_5

A multiplication table of symmetries

If you are flipping and rotating an equilateral triangle (either physically or mentally), you will notice that the result of several sequential symmetry operations is equivalent to another distinct operation, e.g., rotating by 120° twice is the same as rotating by −120°, or reflecting across (3) then reflecting across (4) is the same as rotating by 120°. We can enumerate all of the possibilities in the form of a multiplication table:

Table A.3. Multiplication table for $g_i \circ g_j$

g_i \ g_j	e	g_1	g_2	g_3	g_4	g_5
e	e	g_1	g_2	g_3	g_4	g_5
g_1	g_1	g_2	e	g_5	g_3	g_4
g_2	g_2	e	g_1	g_4	g_5	g_3
g_3	g_3	g_4	g_5	e	g_1	g_2
g_4	g_4	g_5	g_3	g_2	e	g_1
g_5	g_5	g_3	g_4	g_1	g_2	e

This is a nice table. If we ever want to find out the equivalent operation of a long chain of operations, e.g., "flip across (3), then rotate clockwise 120°, then flip across (5), then rotate counterclockwise 120°," we can just consult the table for the results of $g_3 \circ g_1 \circ g_5 \circ g_2$ (which will end up being e, the *do nothing* operation). Notice that we would not have been able to construct a multiplication table at all—let alone interpret it as a group multiplication table—without including the identity element e—*do nothing*.

Symmetry operations form a group

Hopefully we've planted enough hints in the last few sections so this is not a surprise: the set of all symmetry operations of an object form a group, where binary sequential application of the symmetry operations[7] is the group law. As an immediate example, look at the multiplication table A.3 again, and check that all four group axioms are satisfied.

But it is somewhat tedious to generate the multiplication table of every object's symmetry operations and then verify that the group axioms are satisfied. We can in fact have confidence they are, as long as we have enumerated all the symmetry operations:

1. **Closure**
   Since symmetry operations are transformations that leave the object invariant, we know that the composition of any two symmetry operations must also be a symmetry operation.
2. **Associativity**
   This one is a little tricky, so we won't dwell on it too much. In essence, the symmetry operations are mappings (from one configuration to another),

---

7. This means combining two symmetry operations into one equivalent operation.

and binary composition of mappings is associative. Therefore the symmetry operations must also be associative.

3. **Existence of identity element**

Here we see the importance of including the *do nothing* operation, which serves as the identity element. Intuitively, the identity operation is one that does not change the original object at all. In terms of being a member of the group, it means the identity operation does not change any other operation when composed with it.

4. **Existence of inverse**

As any symmetry operation is a mapping from one configuration to another, for each mapping there exists an inverse mapping that returns the object to its original configuration.

Let us reiterate the fact here that it is the *symmetry operations* that form a group, while the object itself is *invariant* under these operations. The group that is formed by the symmetry operations is called the *symmetry group* of the object. For example, the symmetry group of the equilateral triangle (Table A.3) is called D_3, which stands for *dihedral group of order 3*. Another example is the Z_4 group we have looked at (Table A.2), which stands for *cyclic group of order 4*. Since symmetry groups are abstractions of the symmetries of an object, one symmetry group can be used to describe the symmetries of many, sometimes very dissimilar, objects.

A.11   Validity of Mean-Field Theory in an Ising Ferromagnet with Long-Range Interactions

To more explicitly explore the issues concerning the accuracy of mean-field theory discussed in Section 4.5, we consider a modified version of the Ising ferromagnet,

$$H = -\sum_{ij} J(\vec{r}_i - \vec{r}_j)\sigma_i\sigma_j, \tag{A.135}$$

where $J(\vec{r}) \geq 0 \ \forall \ \vec{r}$ represents the interaction strength between two spins separated by a displacement $\vec{r}$. This is still a ferromagnet in the sense that the lowest energy microstates are the two states in which all the spins point in the same direction. The model we have studied until now has $J(\vec{r}) = J$ for $\vec{r}$ equal to any vector connecting two nearest-neighbor sites and $J(\vec{r}) = 0$ otherwise. But we could consider a case in which the range of interactions allowed by $J(\vec{r})$ is longer. We thus define an "effective" number z^{eff} and a typical magnitude of the interaction J as the implicit solutions to the equations

$$z^{\mathrm{eff}} J \equiv \sum_{\vec{r}} J(\vec{r}) \text{ and } z^{\mathrm{eff}} J^2 \equiv \sum_{\vec{r}} \left[J(\vec{r})\right]^2. \tag{A.136}$$

Our first undertaking will be to estimate the nature of corrections to mean-field theory, and to show that they tend to be relatively unimportant in the limit of large z^{eff}.

Let us first look at the effective field felt by the spin at site j:

$$b_j^{\text{eff}} = h + \sum_{k \neq j} J(\vec{r}_j - \vec{r}_k)\sigma_k. \tag{A.137}$$

The subtlety here is that this is not a number, but rather a function of the spins on all the z^{eff} sites that interact with site j. The Weiss mean-field theory we have developed is equivalent to replacing this field by its mean value

$$b = \langle b_j^{\text{eff}} \rangle = z^{\text{eff}} J m, \quad \text{where} \quad m = \langle \sigma_k \rangle, \tag{A.138}$$

and then computing m, self-consistently, as the magnetization of a system consisting of a single spin in the presence of the field b. Indeed, the mean-field solution of this more general problem involves an identical analysis as the original problem, with the single difference that in all expressions we make the substitution $zJ \to z^{\text{eff}} J$.

To estimate what sorts of errors this approximation entails, we compute the magnitude of the fluctuations of b_j^{eff} about its mean value, approximating the spin-correlations

$$[\Delta b]^2 \equiv \langle [b_j^{\text{eff}} - b]^2 \rangle \tag{A.139}$$

$$= \sum_{k,k'} J(\vec{r}_j - \vec{r}_k) J(\vec{r}_j - \vec{r}_{k'}) \langle (\sigma_k - m)(\sigma_{k'} - m) \rangle.$$

We can estimate the size of this quantity in the same self-consistent approximation we used to compute the mean by evaluating the spin-correlator in the trial ensemble (Eq. 4.19) with the result:

$$[\Delta b]^2 \approx z^{\text{eff}} J^2 \left(1 - m^2 \right). \tag{A.140}$$

We would expect the effects of the fluctuations about the mean to be negligible so long as $|\Delta b|/T \ll 1$. Let us therefore consider this ratio at various points in the phase diagram as a metric for the validity of mean-field theory. The mean-field transition temperature is $T_c = z^{\text{eff}} J$. At $T = T_c$, $m = 0$ and hence $\Delta b/T \approx (z^{\text{eff}})^{-1/2} \ll 1$. This is encouraging, since it is for T near T_c that we expect the most serious problems with mean-field theory. Obviously, at temperatures $T \gg T_c$, $\Delta b/T \approx (z^{\text{eff}})^{-1/2}(T_c/T)$ is even smaller. Moreover, since $(1 - m^2)/T^2 \to 0$ as $T \to 0$, this estimate suggests that mean-field theory should also be reliable at temperatures much below T_c as well. The result suggests that mean-field theory is exact in the limit $z^{\text{eff}} \to \infty$.

To the extent that this argument is valid, it implies that in cases in which each spin interacts ferromagnetically with a sufficiently large number of neighboring spins, mean-field theory should be arbitrarily accurate. This turns out to be only partially true, but it certainly is a part of the reason that mean-field theory proves to be successful in many cases.

There is one loophole in this line of reasoning—near T_c, important correlations develop between neighboring spins. In going from Eq. A.139 to Eq. A.140 we neglected the possibility of long-range correlations. Thus, while the arguments we have gone through are indeed valid well above and well below T_c, their validity upon close approach to T_c is a more subtle issue. More formally, there is an order of limits problem: mean-field theory becomes arbitrarily accurate for any fixed nonzero value of $T - T_c$ as $z^{\text{eff}} \to \infty$. However, for any fixed large value of z^{eff}, there can occur arbitrarily large failures of the mean-field analysis as $T \to T_c$.

A.12   *More about the Broken Symmetry in Superfluids

The nature of the broken symmetry in superfluids cannot be understood without treating it as a problem in quantum mechanics. Hence, this discussion has been relegated to an appendix, to be studied by students only once they have some familiarity with the subject.

To fully appreciate the nature of a spontaneously broken symmetry, we need to identify the sort of externally applied field h that would explicitly break the requisite symmetry. Then, we can follow the canonical prescription of taking the limit $h \to 0$ only after we have taken the thermodynamic limit. However, the symmetry involved is rather strange—to break it, we will need to add to the Hamiltonian an operator that violates particle number conservation.

In all cases considered in elementary quantum mechanics, the particle number is conserved, and hence there exists a complete set of states that are simultaneously eigenstates of the particle number operator, $\hat{N}$, and the Hamiltonian, H:

$$H|N,\alpha\rangle = E(N,\alpha)|N,\alpha\rangle \;\; ; \;\; \hat{N}|N,\alpha\rangle = N|N,\alpha\rangle. \tag{A.141}$$

Here N is the number of particles and α labels the individual N-particle eigenstates. In looking at the problem in this way, we have expanded the Hilbert space to a space with an arbitrary number of particles. Any state in this larger Hilbert space can now be expressed as

$$|\psi\rangle = \sum_{N,\alpha} \psi(N,\alpha)|N,\alpha\rangle, \tag{A.142}$$

where $\psi(N,\alpha)$ is a complex amplitude. However, so long as we consider only operators corresponding to observables that conserve particle number—that is to say, operators that act within a subspace with fixed N—the expectation value is invariant under the transformation

$$|\psi\rangle \to \sum_{N,\alpha} \psi(N,\alpha)e^{i\theta(N)}|N,\alpha\rangle, \tag{A.143}$$

where $\theta(N)$ is an arbitrary N-dependent phase. This is the symmetry that is spontaneously broken in a superfluid.

To simplify the discussion, we will focus on the nature of the broken symmetry in the limit $T \to 0$. In this case, what we care about are the ground states, labeled by $\alpha = \alpha_0$ for each N, so we introduce the simplified notation $|N, \alpha_0\rangle \equiv |N\rangle$ and $E(N, \alpha_0) \equiv E_0(N)$. For a system of N noninteracting bosons, each N particle energy eigenstate can be expressed as a direct product of single particle eigenstates. Moreover, without loss of generality, since the zero of energy is arbitrary, we can set $E_0(1) = 0$. (Correspondingly, so long as the single particle ground state is unique, $E(1, \alpha) > 0$ for all $\alpha \neq \alpha_0$.) However, since the energy of noninteracting particles is additive, it directly follows from this that $E_0(N) = N E_0(1) = 0$ for all N. Consequently, the state

$$|\Psi\rangle = \sum_N \Psi(N)|N\rangle \tag{A.144}$$

is also a zero energy eigenstate of H, i.e., a ground state, albeit one with an indeterminate number of particles.

In a more general interacting problem this is not precisely the case, but it reasonable to expect (and possible to prove) that the ground-state energy of weakly interacting bosons is approximately independent of N for a range of N that corresponds to the same overall density in the thermodynamic limit. Specifically, defining $\bar{N} \equiv nV$, where V is the volume and n is the mean density, the ground-state energy is roughly independent of N for $|N - \bar{N}|^2 \lesssim \bar{N}$. Thus, even if $\Psi(N)$ is nonzero for a range of N, the resulting state $|\Psi\rangle$ is to good approximation an eigenstate of H.

We now consider the problem in the presence of a symmetry-breaking term,

$$H \to H' = H - h^\star \hat{O}^\dagger - h\hat{O}, \tag{A.145}$$

where $\hat{O}$ is an operator that adds one particle to the system. The exact choice of this operator does not matter, but to be explicit, we will assume

$$\langle M|\hat{O}|N\rangle = N|o|\,\delta_{M,N+1}, \tag{A.146}$$

where $|o|$ (as the notation suggests) is a positive real number, and the factor of N ensures that the operator is properly extensive in the thermodynamic limit. Now, since

$$\langle\Psi|H'|\Psi\rangle = -|o| \sum_N N\left[h^\star \Psi^\star(N+1)\Psi(N) + c.c.\right], \tag{A.147}$$

the energy depends on the relative phase of the different parts of the wavefunction. (Here, c.c. means the complex conjugate.) Specifically, if we take $\Psi(N) = |\Psi(N)|e^{-iN\theta}$, $h = |h|^{i\chi}$, then

$$\langle\Psi|H'|\Psi\rangle \approx -2V|h|\,|\Phi|\cos[\theta - \chi] = -\int d\vec{r}\,\left[h^\star \Phi + c.c.\right], \tag{A.148}$$

where Φ is the intensive quantity

$$\Phi \equiv V^{-1}\langle \hat{O} \rangle = |o|\, n\, e^{i\theta}. \tag{A.149}$$

In deriving this we have assumed that $|\Psi(N)|$ is sufficiently slowly varying that we can approximate $|\Psi(N)| \approx |\Psi(N+1)|$, while it is sufficiently strongly peaked near $N = \bar{N}$ that we can approximate the explicit factor of N in the sum by $\bar{N}$. Moreover, we make use of the normalization of the wavefunction, $\sum_N |\Psi(N)|^2 = 1$. In the final expression, we have used the identity $\int d\vec{r}\, V^{-1} = 1$.

Whereas in the absence of the symmetry-breaking field, the energy of this state is independent of θ, in the presence of nonzero h, the energy is minimized when $\theta = \chi$. Moreover, the energy that favors this particular choice of phase is extensive, so it is significant so long as $|h| \gg [|o|nV]^{-1}$, i.e., provided we take $V \to \infty$ before $|h| \to 0$. In this limit, $\Phi \to |o|e^{i\chi}$ as $|h| \to 0$ in a manner such that $h = |h|e^{i\chi}$. In analogy to the way we determined the magnetization density $\vec{m}$ in a Heisenberg ferromagnet, Φ remains nonzero as $h \to 0$, and has a phase that is determined by the manner in which this limit is approached. The XY symmetry of the superfluid order parameter derives directly from number conservation, in the same way as the Ising symmetry of m comes directly from time-reversal symmetry.

APPENDIX B

Worksheet Solutions

B.1 Chapter 1 Worksheet Solutions

B.1.1 Critical Exponents

If we plot certain properties of materials as a function of a tuning parameter close to a phase transition, the functional dependence can often be described by some power law. Here let us gain some familiarity with the shapes of various power laws, and identify the correct power laws from real experimental data.

1. Sketch the following power laws:
   (a) $m \propto (T_C - T)^{\beta}$, where $0 < \beta < 1$.
   (b) $c \propto |T - T_C|^{-\alpha}$, where $0 < \alpha < 1$.

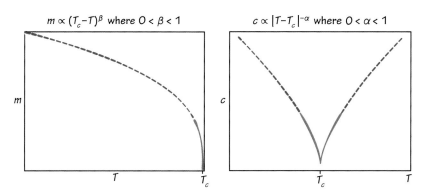

The blue part of the sketch is close to the critical point, and would correctly approximate the real behavior. Away from the critical point (in dashed red) power laws are not guaranteed to describe real behaviors.

   When you have access to graphing tools, try plotting out these power laws and observe how they change as you vary α and β!

2. A phase transition is considered "continuous" if the first derivative of the free energy is continuous across the phase boundary. Are the above transitions continuous given that:
   (a) m is a first derivative of the free energy with respect to an applied field, h?
   Since m is continuous across T_C and is a first derivative of the free energy, this indicates a continuous phase transition.

(b) c is a second derivative of the free energy with respect to temperature T?

The antiderivative of c w.r.t. T is $\propto |T - T_C|^{-\alpha+1}$, which is continuously across T_C. This indicates a continuous phase transition.

3. A critical exponent and the associated power law describe the behavior of various physical quantities *close to* the critical point. Let us see what this means in practice.

Suppose we have a magnet close to $T = T_c$ whose normalized magnetization as a function of the temperature behaves as:

$$m(T) = \sqrt{1 - (T/T_c)^2} \quad \text{for} \quad T < T_c. \tag{B.1.1}$$

(a) What is the meaning of the temperature scale T_c?

The critical temperature is the temperature at which entropic effects start dominating over consideration of minimizing energy. Typically (but not always), T_c of a phase transition corresponds roughly to the energy scale of the interactions responsible for that phase transition.

(b) Close to the phase boundary $(0 < 1 - |T/T_c| \ll 1)$, the normalized magnetization can be written as:

$$m(T) = a(T_c - T)^\beta. \tag{B.1.2}$$

Find the values of the constant of proportionality a and the critical exponent β such that Eq. B.1.2 correctly describes the behavior of Eq. B.1.1 close to the phase boundary (see Figure B.1.1).

Hint: Find the quantity that is small close to T_c, and expand Eq. B.1.1 w.r.t. that quantity.

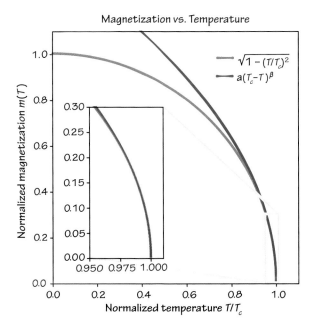

FIGURE B.1.1. Full functional dependence vs. power-law description near phase boundary.

Close to T_c, $T/T_c \approx 1$, so the quantity $1 - T/T_c$ is small. Let's define:

$$\eta = 1 - T/T_c. \tag{B.1.3}$$

Then we can rewrite Eq. B.1.1 as:

$$m(T) = \sqrt{(1 - T/T_c)(1 + T/T_c)}$$

$$= \sqrt{\eta(2 - \eta)}$$

$$= \sqrt{2\eta}\sqrt{1 - \frac{\eta}{2}}$$

$$\approx \sqrt{2\eta}\left(1 - \frac{\eta}{4} + \dots\right).$$

Hence, the leading-order term is $\sqrt{2\eta} = \sqrt{2}(1 - h_x)^{1/2}$. Comparing with Eq. B.1.2, we find that $\beta = 1/2$ and $a = \sqrt{2}/T_c^\beta$.

With the correct values for a and β, the power-law description very accurately describes the behavior near the phase boundary, but deviates quickly away from it. Moving forward, we must remember that many descriptions of critical phenomena are (approximately) correct only near phase boundaries!

4. Below is a plot of the superconducting gap energy of Tantalum close to the critical temperature as a function of the rescaled temperature $t = T/T_c$

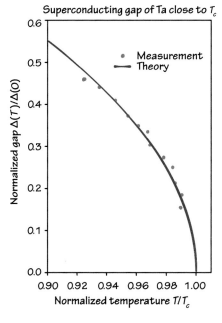

FIGURE B.1.2. Superconducting gap of Tantalum close T_c.

T/T_c	$\Delta(T)/\Delta(0)$
0.9904	0.1833
0.9895	0.1522
0.9858	0.2112
0.9843	0.2485
0.9781	0.2718
0.9689	0.3013
0.9674	0.3339
0.9612	0.3478
0.9542	0.3711
0.9458	0.4084
0.9349	0.4394
0.9249	0.4580

TABLE B.1. Tantalum gap energy temperature dependence

(a) Judging from the plot, what is the form of the normalized gap $\Delta(T)/\Delta(0)$ close to $t = 1$?

The curve appears to have a vertical tangent close to the critical temperature. The power law that has this form is $\propto (1 - T/T_C)^\beta$, with $0 < \beta < 1$

(b) Plotting the data on log-log axes, estimate the critical exponent, as well as the constant of proportionality.

If close to the phase transition the power law behaves as:

$$\frac{\Delta(T)}{\Delta(0)} = a\left(1 - \frac{T}{T_c}\right)^\beta, \tag{B.1.4}$$

then

$$\ln\frac{\Delta(T)}{\Delta(0)} = \ln a + \beta \ln\left(1 - \frac{T}{T_c}\right). \tag{B.1.5}$$

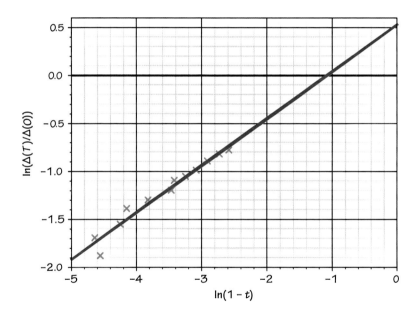

FIGURE B.1.3. Log-log plot of Tantalum superconducting energy gap.

Hence the slope is β and the y-intercept is $\ln(a)$. Estimating from the plot, $\beta \approx 0.49$ and $a \approx e^{0.52} = 1.68$.

Is the theory a good description close to criticality? Close to T_c the theoretical value is $\Delta(T)/\Delta(0) = 1.74\sqrt{1 - T/T_c}$.

In this case, the data and theory agree quite well.

B.2   Chapter 2 Worksheet Solutions

B.2.1   Calculating Thermal Averages

While we assume some of these exercises are just a review of prerequisites, it is never a bad idea to refresh our memory on these things.

Free energy and the partition function

The definition of the (Helmholtz) free energy is

$$F \equiv E - TS, \tag{B.2.1}$$

and that of the partition function is

$$Z = \sum_s e^{-\beta E_s}, \tag{B.2.2}$$

where E_s is the energy of state s, and $\beta = 1/T$, such that the probability of state s occurring is:

$$P(s) = \frac{e^{-\beta E_s}}{Z}. \tag{B.2.3}$$

1. The energy term E in the free energy is the expectation value of E_s. Find an expression for $E \equiv \langle E_s \rangle$ in terms of the partition function Z and inverse temperature β.

   We know that

   $$E \equiv \langle E_s \rangle \tag{B.2.4}$$

   $$= \frac{1}{Z} \sum_s E_s e^{-\beta E_s}. \tag{B.2.5}$$

   On the other hand, since taking a derivative is a linear operation,

   $$\frac{\partial Z}{\partial \beta} = - \sum_s E_s e^{-\beta E_s}, \tag{B.2.6}$$

   and, conveniently,

   $$\frac{\partial \ln Z}{\partial Z} = \frac{1}{Z}. \tag{B.2.7}$$

   So combining the above with the chain rule, we find:

   $$E = \frac{1}{Z} \sum_s E_s e^{-\beta E_s} \tag{B.2.8}$$

   $$= \frac{\partial \ln Z}{\partial Z} \frac{-\partial Z}{\partial \beta} \tag{B.2.9}$$

   $$= -\frac{\partial \ln Z}{\partial \beta}. \tag{B.2.10}$$

2. We assume the energy of each microstate is a function of generalized coordinates q_j, such that $E_s = E_s(q_j)$. Associated with each is a generalized force $f_{sj} = -\frac{\partial E_s}{\partial q_j}$. Find an expression for $f_j \equiv \langle f_{sj} \rangle$ in terms of Z, β, and q_j.

Similarly to the procedure above, first let's write down the definition of f_j:

$$f_j = \langle f_{sj} \rangle = \frac{1}{Z} \sum_s f_{sj} e^{-\beta E_s}. \tag{B.2.11}$$

We see that

$$\frac{\partial Z}{\partial q_j} = -\beta \sum_s \frac{\partial E_s}{\partial q_j} e^{-\beta E_s} \tag{B.2.12}$$

$$= \beta \sum_s f_{sj} e^{-\beta E_s}. \tag{B.2.13}$$

Then using the chain rule, we find:

$$f_j = \frac{1}{\beta} \frac{\partial \ln Z}{\partial Z} \frac{\partial Z}{\partial q_j} \tag{B.2.14}$$

$$= \frac{1}{\beta} \frac{\partial \ln Z}{\partial q_j}. \tag{B.2.15}$$

3. Since the partition function should depend on both temperature and the generalized coordinates, we can write it as $Z = Z(\beta, q_j)$. Start with the expression for the total differential of $\ln Z$,

$$d \ln Z = \frac{\partial \ln Z}{\partial \beta} d\beta + \frac{\partial \ln Z}{\partial q_j} dq_j, \tag{B.2.16}$$

and prove that $F = -T \ln Z$ follows from the definition Eq. B.2.1.
You may find the first law of thermodynamics useful:

$$dE = TdS - f_j dq_j. \tag{B.2.17}$$

Substituting in the results from the previous parts into $d \ln Z$, we see that

$$d \ln Z = -Ed\beta + \beta f_j dq_j. \tag{B.2.18}$$

Since $d(\beta E) = Ed\beta + \beta dE$, we have:

$$d \ln Z = -d(\beta E) + \beta dE + \beta f_j dq_j. \tag{B.2.19}$$

Substituting in the first law of thermodynamics:

$$d \ln Z = -d(\beta E) + \beta(TdS - f_j dq_j) + \beta f_j dq_j \tag{B.2.20}$$

$$= d(S - \beta E) \tag{B.2.21}$$

$$\Rightarrow \ln Z = S - \beta E \tag{B.2.22}$$

$$-T \ln Z = E - TS \tag{B.2.23}$$

$$\equiv F \tag{B.2.24}$$

The derivative theorem

We make use of this throughout the book, so understand it well!

4. Given that the Hamiltonian of a system is

$$H(s) = E(s) + \lambda A(s), \tag{B.2.25}$$

show that

$$\langle A \rangle = \frac{\partial F}{\partial \lambda}. \tag{B.2.26}$$

Hint: Use the definition of the partition function and $F = -T \ln Z$.

By definition

$$\langle A \rangle \equiv \frac{1}{Z} \sum_s A(s) e^{-\beta H(s)}. \tag{B.2.27}$$

Let's write out explicitly:

$$\frac{\partial F}{\partial \lambda} = \frac{-\partial T \ln Z}{\partial \lambda} \tag{B.2.28}$$

$$= -\frac{T}{Z} \frac{\partial Z}{\partial \lambda} \tag{B.2.29}$$

$$= -\frac{T}{Z} \sum_s -\beta A(s) e^{-\beta H(s)} \tag{B.2.30}$$

$$= \frac{1}{Z} \sum_s A(s) e^{-\beta H(s)} \equiv \langle A \rangle. \tag{B.2.31}$$

Thermal averages: Part two

Using Eq. 2.3, show that

5.

$$\left\langle [M_a - \langle M_a \rangle]^2 \right\rangle = \left\langle M_a^2 \right\rangle - \langle M_a \rangle^2. \tag{B.2.32}$$

6.

$$\langle [M_a - \langle M_a \rangle] [M_b - \langle M_b \rangle] \rangle \qquad \text{(B.2.33)}$$

$$= \langle M_a M_b \rangle - \langle M_a \rangle \langle M_b \rangle. \qquad \text{(B.2.34)}$$

Here we are trying to illustrate some basic properties of how random variables behave inside averaging operators. Let's reproduce Eq. 2.3 here:

$$\langle A \rangle = \sum_s P(s) A(s). \qquad \text{(B.2.35)}$$

Recall that

$$\sum_s P(s) = 1 \qquad \text{(B.2.36)}$$

and

$$\langle M_a \rangle = \sum_s P(s) M_a(s). \qquad \text{(B.2.37)}$$

Remember that $\langle M_a \rangle$ is just a number, so

$$\langle \langle M_a \rangle \rangle = \sum_s P(s) \langle M_a \rangle \qquad \text{(B.2.38)}$$

$$= \langle M_a \rangle \sum_s P(s) \qquad \text{(B.2.39)}$$

$$= \langle M_a \rangle. \qquad \text{(B.2.40)}$$

Back to the question at hand, we have

$$\langle [M_a - \langle M_a \rangle]^2 \rangle = \langle M_a^2 - 2 M_a \langle M_a \rangle + \langle M_a \rangle^2 \rangle \qquad \text{(B.2.41)}$$

$$= \sum_s P(s) \left[M_a^2 - 2 M_a \langle M_a \rangle + \langle M_a \rangle^2 \right] \qquad \text{(B.2.42)}$$

$$= \langle M_a^2 \rangle - 2 \langle M_a \rangle^2 + \langle M_a \rangle^2 \qquad \text{(B.2.43)}$$

$$= \langle M_a^2 \rangle - \langle M_a \rangle^2. \qquad \text{(B.2.44)}$$

$$\langle [M_a - \langle M_a \rangle] [M_b - \langle M_b \rangle] \rangle \qquad \text{(B.2.45)}$$

$$= \langle M_a M_b - M_a \langle M_b \rangle - \langle M_a \rangle M_b + \langle M_a \rangle \langle M_b \rangle \rangle \qquad \text{(B.2.46)}$$

$$= \langle M_a M_b \rangle - 2 \langle M_a \rangle \langle M_b \rangle + \langle M_a \rangle \langle M_b \rangle \qquad \text{(B.2.47)}$$

$$= \langle M_a M_b \rangle - \langle M_a \rangle \langle M_b \rangle. \qquad \text{(B.2.48)}$$

Fluctuation and susceptibility

7. Show that

$$\langle M_a M_b \rangle = T^2 \frac{1}{Z} \frac{\partial^2 Z}{\partial B_a \partial B_b}. \tag{B.2.49}$$

For a magnetic system, the Hamiltonian has the form:

$$H = E - \vec{M} \cdot \vec{B}. \tag{B.2.50}$$

So first of all:

$$\frac{\partial H}{\partial B_\alpha} = -M_\alpha. \tag{B.2.51}$$

And using the derivative theorem, we know that:

$$\frac{\partial F}{\partial B_\alpha} = -\langle M_\alpha \rangle. \tag{B.2.52}$$

By definition:

$$\langle M_a M_b \rangle \equiv \frac{1}{Z} \sum M_a M_b e^{-\beta H}. \tag{B.2.53}$$

Calculate the derivatives:

$$\frac{1}{Z} \frac{\partial^2 Z}{\partial B_a \partial B_b} = \frac{1}{Z} \sum \beta^2 M_a M_b e^{-\beta H} \tag{B.2.54}$$

$$= \beta^2 \langle M_a M_b \rangle \tag{B.2.55}$$

$$\Rightarrow \langle M_a M_b \rangle = T^2 \frac{1}{Z} \frac{\partial^2 Z}{\partial B_a \partial B_b}. \tag{B.2.56}$$

8. Given that

$$\chi_{a,b} = -\frac{1}{V} \left(\frac{\partial^2 F}{\partial B_a \partial B_b} \right), \tag{B.2.57}$$

use the product rule to show that:

$$\chi_{a,a} = \frac{1}{TV} \Delta_{M_a}^2, \tag{B.2.58}$$

where the fluctuation of M_a is

$$\Delta_{M_a}^2 = \langle M_a M_a \rangle - \langle M_a \rangle \langle M_a \rangle. \tag{B.2.59}$$

Let us just calculate the quantity $\chi_{a,a}$:

$$\chi_{a,a} = \frac{T}{V} \frac{\partial^2 \ln Z}{\partial B_a^2} \tag{B.2.60}$$

$$= \frac{T}{V} \left[\frac{\partial}{\partial B_a} \left(\frac{1}{Z} \frac{\partial Z}{\partial B_a} \right) \right] \qquad (B.2.61)$$

$$= \frac{T}{V} \left[-\frac{1}{Z^2} \left(\frac{\partial Z}{\partial B_a} \right)^2 + \frac{1}{Z} \frac{\partial^2 Z}{\partial B_a^2} \right] \qquad (B.2.62)$$

$$= \frac{T}{V} \left[-\frac{1}{T^2} \left(\frac{\partial(-T \ln Z)}{\partial B_a} \right)^2 + \frac{1}{T^2} \langle M_a^2 \rangle \right] \qquad (B.2.63)$$

$$= \frac{1}{TV} \left(\langle M_a^2 \rangle - \langle M_a \rangle^2 \right) \qquad (B.2.64)$$

$$= \frac{1}{TV} \Delta_{M_a}^2. \qquad (B.2.65)$$

B.2.2   Identifying Symmetries

The Ising ferromagnet is an example of a system with an ordered state that breaks a discrete symmetry. In the fully ordered ($T = 0$) state, there are two symmetry-related ground states, one with all spins up and the other with all spins down. In other systems, the symmetry broken can be a continuous symmetry. Below, we enumerate some common symmetry classes.

Symmetry	Form of the order parameter
Ising	$\sigma = \{\pm 1\}$
XY [O(2)]	$\vec{s} = (\cos\theta, \sin\theta); 0 \le \theta < 2\pi$
Heisenberg [O(3)]	$\vec{s} = (\cos\phi\sin\theta, \sin\phi\sin\theta, \cos\theta); |\vec{s}|^2 = 1$
q-state clock model	$\vec{s} = (\cos\theta, \sin\theta); e^{iq\theta} = 1$

For the following physical situations, indicate the model that fits their symmetry (breaking). We are introducing you to the language of symmetries, so don't worry if not every word makes sense yet.

1. Ferromagnetism of spins subject to an easy-axis magnetic anisotropy (i.e., spins prefer to point along a single axis).

   Since the spins can only take on one of two quantized values, this is Ising symmetry.

2. Ferromagnetism of spins subject to an easy-plane magnetic anisotropy (i.e., spins prefer to lie in a 2D plane).

   Since the spins can point in any direction on a circle, this is XY/O(2) symmetry.

3. Ferromagnetism of spins without any magnetic anisotropy.

Without any magnetic anisotropy, spins in our universe can point in any direction on a sphere; therefore this is Heisenberg/O(3) symmetry

4. Picking a preferred bond orientation on a honeycomb lattice.

All the bonds on a honeycomb lattice lie in one of three directions; therefore picking a preferred bond direction has 3-state clock symmetry. Note that the Ising symmetry is in fact identical to the 2-state clock symmetry.

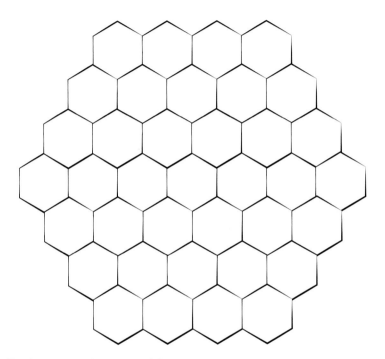

FIGURE B.2.1. Example of a honeycomb lattice.

5. Distortion of a tetragonal lattice (which has two equivalent axes).

Since the distortion can occur in one of two directions, this is again the Ising symmetry.

6. A superfluid transition, where the order parameter is a "condensate wave-function," specified by a complex number. As is true of all wavefunctions in quantum mechanics, the wavefunction can be multiplied by an overall phase without changing the expectation value of any operator.

While this example is somewhat more abstract than previous ones, notice that the order parameter can have any phase, which is a quantity between 0 to 2π (or interpreted as a point on a circle), so the symmetry is again XY/O(2).

B.3   Chapter 3 Worksheet Solutions

B.3.1   The Ising Hamiltonian–A Primer

In Section 3.1, we learned that in the absence of a magnetic field, the Hamiltonian of the Ising model is

$$H = -J_1 \sum_{\text{n.n.}} \sigma_i \sigma_j. \tag{B.3.1}$$

Before moving on to more complicated lattices and an infinite number of sites (i.e., thermodynamic limit $N \to \infty$), let us practice with a 4-site model. We will start simple, with just the nearest-neighbor (n.n.) interactions, J_1. With the above definition, $J_1 > 0$ is a ferromagnetic interaction, whereas $J_1 < 0$ is antiferromagnetic.

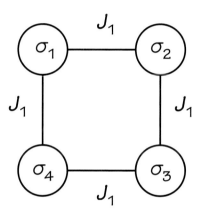

FIGURE B.3.1. Ising model with 4 sites coupled by nearest-neighbor interaction, J_1.

1. In a table, enumerate all of the possible spin configurations, and write down the corresponding energy of each spin configuration. You can reduce the size of the table by treating pairs of "flipped" states as equivalent, i.e., $\begin{smallmatrix} \uparrow\uparrow \\ \uparrow\uparrow \end{smallmatrix}$ is equivalent to $\begin{smallmatrix} \downarrow\downarrow \\ \downarrow\downarrow \end{smallmatrix}$.

Configuration(s)	Energy	Boltzmann factor
$\begin{smallmatrix} \uparrow \quad \uparrow \\ \uparrow \quad \uparrow \end{smallmatrix}$	$-4J_1$	$e^{4\beta J_1}$
$\begin{smallmatrix} \uparrow \quad \uparrow \\ \downarrow \quad \uparrow \end{smallmatrix} \; \begin{smallmatrix} \uparrow \quad \uparrow \\ \uparrow \quad \downarrow \end{smallmatrix} \; \begin{smallmatrix} \uparrow \quad \downarrow \\ \uparrow \quad \downarrow \end{smallmatrix}$ $\begin{smallmatrix} \uparrow \quad \downarrow \\ \uparrow \quad \uparrow \end{smallmatrix} \; \begin{smallmatrix} \downarrow \quad \uparrow \\ \uparrow \quad \uparrow \end{smallmatrix} \; \begin{smallmatrix} \uparrow \quad \uparrow \\ \downarrow \quad \downarrow \end{smallmatrix}$	0	1
$\begin{smallmatrix} \uparrow \quad \downarrow \\ \downarrow \quad \uparrow \end{smallmatrix}$	$4J_1$	$e^{-4\beta J_1}$

2. What are the corresponding Boltzmann factors of each configuration? Assuming that $J_1 > 0$, what is the ground-state $(T \rightarrow 0)$ configuration? What about for $J_1 < 0$?

> See the previous table for the Boltzmann factors. For $T \rightarrow 0$ $(\beta \rightarrow \infty)$, the Boltzmann factor (i.e., unnormalized probability) for one of the states dominates exponentially, depending on the sign of J_1. for $J_1 > 0$, the ferromagnetic state ($\uparrow\uparrow \atop \uparrow\uparrow$) is the ground-state; for $J_1 < 0$, the antiferromagnetic state $\uparrow\downarrow \atop \downarrow\uparrow$ is the ground-state; if $J_1 = 0$, there are no interactions between spins, and all states are equally likely to occur.
>
> Note that we did not enumerate the "flipped," a.k.a. globally spin-flipped, states. Therefore, at $T \rightarrow 0$ there are in fact twice the number of degenerate ground-states.

B.3.2 More on the 4-Site Problem

Let us extend the previous question and examine its behavior with next-nearest-neighbor (n.n.n.) (see Figure B.3.2) interactions, J_2:

$$H = -J_1 \sum_{\text{n.n.}} \sigma_i \sigma_j - J_2 \sum_{\text{n.n.n.}} \sigma_i \sigma_j. \tag{B.3.2}$$

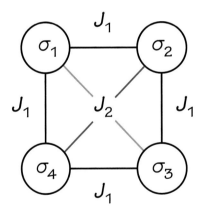

FIGURE B.3.2. Ising model with 4 sites coupled by nearest-neighbor interaction J_1, as well as next-nearest-neighbor interaction, J_2.

1. In a table, again enumerate the possible spin configurations, as well as the total energy of each configuration, treating pairs of flipped states as equivalent.

Configuration(s)	Energy	Boltzmann factor
↑ ↑ / ↑ ↑	$-4J_1 - 2J_2$	$e^{\beta(4J_1 + 2J_2)}$
↑ ↑ / ↓ ↑ ↑ ↑ / ↑ ↓ ↑ ↓ / ↑ ↑ ↓ ↑ / ↑ ↑	0	1
↑ ↑ / ↓ ↓ ↑ ↓ / ↑ ↓	$2J_2$	$e^{-2\beta J_2}$
↑ ↓ / ↓ ↑	$4J_1 - 2J_2$	$e^{\beta(-4J_1 + 2J_2)}$

2. Since we now have two parameters to tune, our phase diagram no longer lives on just a line, as before. Describe the nature of the ground state(s) in each region of the phase diagram shown in Figure W3.3, as well as on the phase boundaries.

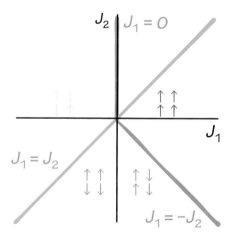

The phase diagram is separated into three regions by the thick blue lines. In the upper right region, the ground-state is the ferromagnetic state (shown in pink) with all spins aligned. The ground-state in the upper left region is one type of antiferromagnetism (shown in light blue). The lower region has a different type of antiferromagnetism (shown in darker blue) as its ground states.

On the boundaries (thick blue lines), the ground states from neighboring regions of the phase diagram have the same energy, and therefore are all possible degenerate ground-states.

Note that, as previously, the globally spin-flipped states are also possible.

3. If the four sites are arranged as a tetrahedron such that the interactions for any given pair of spins are identical, where does this situation arise in the phase diagram in Figure W3.3?

> If the pairwise interactions are identical, it implies that $J_1 = J_2$, which is shown in the previous figure as the translucent green line. In the ferromagnetic region (quadrant I), the ground state is again the ferromagnetic state (in pink). In the antiferromagnetic region (quadrant III), all types of antiferromagnetic states (both light and darker blue) have the same energy—there are many degenerate groundstates. This is a very simple model of geometric frustration.

B.3.3   The Droplet Argument in 2D

We start with the 2D Ising system in its ground state (all aligned) on an infinite square lattice. Without loss of generality, let us assume all the spins are pointing up. The Hamiltonian is

$$H = -J \sum_{\langle ij \rangle} \sigma_i \sigma_j. \tag{B.3.3}$$

1. What is the change in energy from a single pair of spins that is switched from ↑↑ to ↑↓?

> For a single pair of spins:
>
> $$E(\uparrow\uparrow) = -J$$
> $$E(\uparrow\downarrow) = J$$
> $$\Delta E = E(\uparrow\downarrow) - E(\uparrow\uparrow)$$
> $$= 2J. \tag{B.3.4}$$

2. Sketch out all the droplet configurations that have domain walls of the following lengths. A droplet is a group of connected spins, and the domain wall length L is the total length of the boundary enclosing the droplet.

> The solutions are graphically enumerated below. The blue dots on the solid black grid represent flipped spins, the pink dashed lines represent flipped bonds, the solid light blue lines represent the domain walls.
>
> (a) $L = 4$:
>
>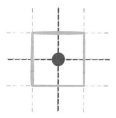

(b) $L = 6$:

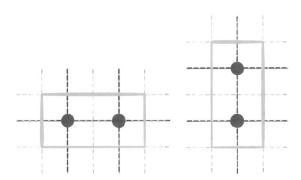

(c) $L = 8$:

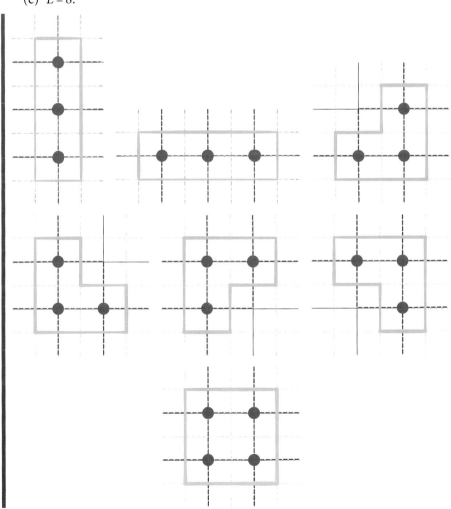

3. Assume that inside each droplet the spins are all pointing down, and outside the spins are all pointing up. What is the overall energy change ΔE from the ground state for each of the above cases? What is the general expression for $\Delta E(L)$?

> Every segment of the domain wall corresponds to a flipped bond, which we have determined to have energy $2J$. Therefore, the droplets have, in general, energy
>
> $$\Delta E(L) = 2JL. \tag{B.3.5}$$

4. For each of the prescribed wall lengths, how many different configurations $\mathcal{N}(L)$ were possible? If we assume there are on average α choices when forming each segment of the wall such that $\mathcal{N}(L) = \alpha(L)^L$, what is $\alpha(L)$ for each L?

> Since $\mathcal{N}(L) = \alpha(L)^L$, $\alpha(L) = \mathcal{N}(L)^{1/L}$, which can be evaluated numerically.
>
> | L | $\mathcal{N}(L)$ | $\alpha(L)$ |
> |---|---|---|
> | 4 | 1 | 1 |
> | 6 | 2 | 1.122 |
> | 8 | 7 | 1.275 |
> | 10 | 28 | 1.395 |

We get the above results by requiring the walls to form a closed, non-self-intersecting polygon. Without these constraints, a contiguous wall of length L will have 3^L ways of forming. From the above exercise, we should gain some intuition for the argument presented in the main text: for each closed wall of length L, we have at least one way and at most 3^L ways to construct it, i.e.,

$$1 \le \mathcal{N}(L) = \alpha(L)^L < 3^L. \tag{B.3.6}$$

By performing this algorithmically on a computer[1], the current best estimate of the asymptotic value is $\lim_{L \to \infty} \alpha(L) \approx 2.638$. You can get a sense of this from the plot in Figure B.3.3.

---

1. Jensen and Guttman, *J. Phys. A: Math. Gen.* **32**, 4867 (1999), and Clisby and Jensen, *J. Phys. A: Math. Theor.* **45**, 115202 (2012).

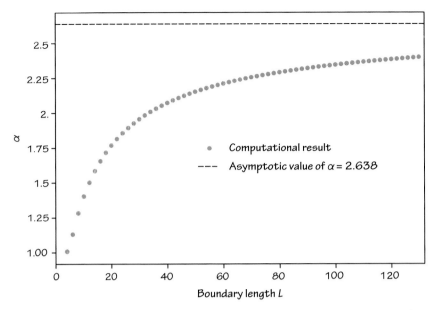

FIGURE B.3.3. Value of α evaluated computationally up to $L = 130$, and the asymptotic value.

5. We can assign a quantity

$$\Delta F(L) = \Delta E(L) - T\Delta S(L) \tag{B.3.7}$$

for the droplet states of length L, where $\Delta S(L) = \log \mathcal{N}(L)$ represents the change in entropy associated with these states. What is the physical meaning $\Delta F(L)$? *Hint:* The "partition function of droplets" can be written as:

$$Z = \sum_L \mathcal{N}(L)e^{-\beta\Delta E(L)}. \tag{B.3.8}$$

Starting with the definition of the partition function:

$$Z = \sum_L \mathcal{N}(L)e^{-\beta\Delta E(L)} \tag{B.3.9}$$

$$= \sum_L e^{\log \mathcal{N}(L)}e^{-\beta\Delta E(L)} \tag{B.3.10}$$

$$= \sum_L e^{-\beta[\Delta E(L)-\Delta S(L)/\beta]} \tag{B.3.11}$$

$$= \sum_L e^{-\beta\Delta F(L)}. \tag{B.3.12}$$

So we see that $\Delta F(L)$ tells us about the Boltzmann weight of a collection of droplet states of length L.

6. The partition function stops being analytic at a certain temperature T_c, indicating a phase transition. You can make the approximation that $\alpha(L) = \alpha$, i.e., a constant. Find T_c. *Hint:* The geometric series converges iff $|r| < 1$:

$$\sum_{k=0}^{\infty} ar^k = \frac{a}{1-r}. \tag{B.3.13}$$

Explicitly writing out the quantity we just defined:

$$\Delta F(L) = \Delta E(L) - T\Delta S(L) \tag{B.3.14}$$

$$= 2JL - TL\log\alpha. \tag{B.3.15}$$

Substituting into the partition function:

$$Z = \sum_L e^{-\beta(2JL - TL\log\alpha)} \tag{B.3.16}$$

$$= \sum_L e^{L(\log\alpha - 2J/T)}. \tag{B.3.17}$$

Here, we see that the common ratio is $r = \exp\left(\log\alpha - 2J/T\right)$. At T_c, the series diverges, so $|r| = 1$, implying:

$$T_c = \frac{2J}{\log\alpha}. \tag{B.3.18}$$

Above this temperature, it becomes entropically favorable for arbitrarily large droplets to form, therefore prohibiting any ordered phase.

B.3.4 Ising Model in 0D

The zero-dimensional Ising model is just an isolated spin. Since there are no interactions, there is only one term in the Hamiltonian:

$$H = -h\sigma. \tag{B.3.19}$$

Although this model won't necessarily teach us much about collective behavior, it is a nice example to see the machinery of statistical mechanics in action!

1. Given that the two possible states are $\sigma = \pm 1$, write down the partition function $Z(\beta, h) = \sum_\sigma e^{-\beta H(\sigma)}$.

$$Z = \sum_\sigma e^{\beta H(\sigma)}$$

$$= e^{\beta h} + e^{-\beta h} \tag{B.3.20}$$

$$= 2\cosh(\beta h).$$

2. Calculate the magnetization $m = \langle \sigma \rangle$:

   (a) Directly from the definition:

   $$\langle \sigma \rangle = \frac{1}{Z} \sum_\sigma \sigma e^{-\beta H(\sigma)}. \tag{B.3.21}$$

   $$\langle \sigma \rangle = \frac{1}{Z} \sum_\sigma \sigma e^{-\beta H(\sigma)}$$
   $$= \frac{(+1)e^{\beta h} + (-1)e^{-\beta h}}{2\cosh(\beta h)} \tag{B.3.22}$$
   $$= \frac{2\sinh(\beta h)}{2\cosh(\beta h)}$$
   $$= \tanh(\beta h).$$

   (b) From the free energy calculated via the partition function. Recall the derivative theorem from Section 2.1.3.

   Since $H = -h\sigma$, the derivative theorem tells us:

   $$\langle \sigma \rangle = -\frac{\partial F}{\partial h}. \tag{B.3.23}$$

   Thus, to calculate the magnetization:

   $$F = -T \log Z$$
   $$\langle \sigma \rangle = -\frac{\partial F}{\partial h} = T \frac{\partial \log Z}{\partial Z} \frac{\partial Z}{\partial h}$$
   $$= \frac{T}{Z} \frac{\partial [2\cosh(\beta h)]}{\partial h} \tag{B.3.24}$$
   $$= \beta T \frac{2\sinh(\beta h)}{2\cosh(\beta h)}$$
   $$= \tanh(\beta h).$$

3. Calculate the susceptibility $\chi = \frac{\partial m}{\partial h}$.

   One more derivative to calculate:

   $$m = \langle \sigma \rangle = \tanh(\beta h)$$
   $$\chi = \frac{\partial m}{\partial h} \tag{B.3.25}$$
   $$= \beta \operatorname{sech}^2(\beta h).$$

B.3.5   Ising Model in 1D–The Combinatorics Approach

For solving the Ising model in 1D, if the external field is $h = 0$, then we can employ a combinatorics method to calculate various quantities.[2] In this case, we have, again, the Hamiltonian:

$$H = -J \sum_j \sigma_j \sigma_{j+1}. \tag{B.3.26}$$

For our purposes, the spins occupy a 1D chain with N sites with periodic boundary conditions (i.e., $\sigma_{N+1} = \sigma_1$). Since the model has only nearest neighbor interactions, it is convenient to picture the ring equivalently as an ordered set of N bonds, on each of which the spins are either parallel ($E = -J$) or anti-parallel ($E = J$).

1. How many possible ways can k anti-parallel bonds be arranged on N total bonds? Try sketching out a few cases for small N.

    This is a classic scenario from combinatorics, where we have to choose k out of N objects. The answer is the binomial coefficient:

    $$\binom{N}{k} = \frac{N!}{k!(N-k)!}. \tag{B.3.27}$$

2. Can k be even/odd with periodic boundary conditions? Does this depend on the size N of the chain?

    Picture a single domain wall between σ_i and σ_{i+1}. This implies all spins up to σ_i must point in the same direction, and all spins from σ_{i+1} onward must point in the opposite direction. However, since the system is periodic, this poses a contradiction. Only even numbers of domain walls are allowed to exist. This is true regardless of the size or parity of the system.

3. Knowing that the ground-state (all bonds parallel) energy is $E_G = -NJ$, we can enumerate states based on their excitation energy E_k. How does E_k depend on the number (k) of anti-parallel bonds?

    Each domain wall consists of changing a pair of parallel spins to a pair of antiparallel spins, which has energy $2J$. Therefore, $E_k = 2Jk$.

4. Combining the results from the previous parts, express the partition function as a sum of the number of anti-parallel bonds.

    $$Z = 2e^{N\beta J} \sum_{k \text{ even}} \binom{N}{k} e^{-2\beta J k}, \tag{B.3.28}$$

---

2. This is mechanically easier than the general purpose solution via transfer matrices, which is the subject of the next exercise.

where we sum over all even k due to the constraint placed by the periodic boundary conditions. The overall factor $e^{N\beta J}$ accounts for the ground-state energy, and the factor of 2 accounts for the global spin-flip degeneracy.

5. Evaluate the sum to get an expression for the partition function, and then the free energy $F = -T \log Z$ in the thermodynamic limit $N \to \infty$. Is the expression analytic for all T? What does this say about the existence of a phase transition? *Hint:* The binomial theorem might come in handy here:

$$(a \pm b)^N = \sum_k^N \binom{N}{k}(\pm 1)^k a^{N-k} b^k, \tag{B.3.29}$$

where $\binom{N}{k} = \frac{N!}{k!(N-k)!}$ is the binomial coefficient.

The binomial theorem gives us a way to calculate the sum of only the odd or even terms. Consider:

$$(a+b)^N + (a-b)^N = \sum_k^N \binom{N}{k}(1+(-1)^k)a^{N-k}b^k \tag{B.3.30}$$

$$= 2 \sum_{k\,\text{even}}^N \binom{N}{k}a^{N-k}b^k.$$

We can now identify from our partition function, $a = 1$ and $b = e^{-2\beta J}$, thus:

$$Z = 2e^{N\beta J}\frac{(1+e^{-2\beta J})^N + (1-e^{-2\beta J})^N}{2} \tag{B.3.31}$$

$$= e^{N\beta J}\left[(1+e^{-2\beta J})^N + (1-e^{-2\beta J})^N\right].$$

In the thermodynamic limit $N \to \infty$, the first term dominates exponentially, so we can write:

$$\lim_{N\to\infty} Z = \left[e^{\beta J}(1+e^{-2\beta J})\right]^N \tag{B.3.32}$$

$$= \left[2\cosh(\beta J)\right]^N.$$

Thus the free energy is:

$$F = -T \log Z$$

$$= -NT \log\left[2\cosh(\beta J)\right]. \tag{B.3.33}$$

We can see that the free energy is properly extensive, and analytic down to $T \to 0$. This implies there are no phase transitions at finite temperatures.

B.3.6   Ising Model in 1D–The Transfer Matrix Approach

Let us now solve the 1D Ising model in full generality, external field h included. The Hamiltonian and some quantities of interest are:

$$H = -J \sum_j \sigma_j \sigma_{j+1} - h \sum_j \sigma_j, \tag{B.3.34}$$

$$Z = \sum_{\{\sigma\}} \exp\left[-\beta H(\{\sigma\})\right], \tag{B.3.35}$$

$$\langle \sigma_i \rangle = Z^{-1} \sum_{\{\sigma\}} \sigma_i \exp\left[-\beta H(\{\sigma\})\right], \tag{B.3.36}$$

$$\langle \sigma_i \sigma_j \rangle = Z^{-1} \sum_{\{\sigma\}} \sigma_i \sigma_j \exp\left[-\beta H(\{\sigma\})\right]. \tag{B.3.37}$$

We will once again assume a 1D chain with N sites and periodic boundary conditions $\sigma_{N+1} = \sigma_1$.

Reformulation of the problem

1. Taking inspiration from the previous approach where we considered the bonds between spins, find a function $K(\sigma_a, \sigma_b)$ such that we can recast the Hamiltonian in the form

$$H = \sum_j^N K(\sigma_j, \sigma_{j+1}). \tag{B.3.38}$$

It should satisfy $K(\sigma_a, \sigma_b) = K(\sigma_b, \sigma_a)$ to be interpreted as the energy of a bond between two spins.[3]

If we explictly write out all the terms in the Hamiltonian, we have:

$$H = -J(\sigma_1\sigma_2 + \sigma_2\sigma_3 + \cdots + \sigma_1\sigma_N) - h(\sigma_1 + \sigma_2 + \cdots + \sigma_N). \tag{B.3.39}$$

We have the freedom to write this in the form of

$$H = -J(\sigma_1\sigma_2 + \sigma_2\sigma_3 + \cdots + \sigma_1\sigma_N) - h\left[(a+b)\sigma_1 + (a+b)\sigma_2 + \cdots + (a+b)\sigma_N\right] \tag{B.3.40}$$

as long as $a + b = 1$. Now we can group the terms in a suggestive manner:

$$\begin{aligned} H = &-J\sigma_1\sigma_2 - h(a\sigma_1 + b\sigma_2) \\ &-J\sigma_2\sigma_3 - h(a\sigma_2 + b\sigma_3) \\ &-\cdots \\ &-J\sigma_N\sigma_1 - h(a\sigma_N + b\sigma_1). \end{aligned} \tag{B.3.41}$$

---

3. Mathematically, the choice of $K(\sigma_a, \sigma_b)$ is not unique, but the criterion we require allows for a physical interpretation, and more convenient algebra.

So to satisfy form of (B.3.38), we can use:

$$K(\sigma_a, \sigma_b) = -J\sigma_a\sigma_b - h(a\sigma_a + b\sigma_b). \tag{B.3.42}$$

As alluded to in the footnotes, any choice of $a + b = 1$ is faithful to the original Hamiltonian. However, if we choose $a = b = 1/2$, we can make the function symmetric. This allows us to interpret the function as the energy of the bond between a pair of spins, in the sense that given a magnetic field h, exchanging the two spins should yield the same energy regardless of their relative orientation. Furthermore, having symmetry in functions often leads to simpler algebra. Therefore, we arrive at:

$$K(\sigma_a, \sigma_b) = K(\sigma_b, \sigma_a) = -J\sigma_a\sigma_b - \frac{h}{2}(\sigma_a + \sigma_b). \tag{B.3.43}$$

2. The partition function, written out explicitly, is

$$Z = \sum_{\sigma_1 = \pm 1} \sum_{\sigma_2 = \pm 1} \cdots \sum_{\sigma_N = \pm 1} e^{-\beta K(\sigma_1, \sigma_2)} e^{-\beta K(\sigma_2, \sigma_3)} \cdots e^{-\beta K(\sigma_N, \sigma_1)}. \tag{B.3.44}$$

Recall that matrix operations can be expressed as sums:

$$(AB)_{ik} = \sum_j A_{ij}B_{jk}, \tag{B.3.45}$$

$$\mathrm{Tr}\,[A] = \sum_j A_{jj}. \tag{B.3.46}$$

Reinterpret the partition function as matrix multiplications such that

$$Z = \mathrm{Tr}\left[t^N\right]. \tag{B.3.47}$$

What is the matrix t? *Hint:* Start by interpreting the terms as matrix elements of t:

$$e^{-\beta K(\sigma_a, \sigma_b)} = \langle \sigma_a | t | \sigma_b \rangle = t_{\sigma_a \sigma_b}. \tag{B.3.48}$$

As the hint suggested, if we interpret each of the factors in the partition function as a matrix element, we get:

$$Z = \sum_{\sigma_1 = \pm 1} \sum_{\sigma_2 = \pm 1} \cdots \sum_{\sigma_N = \pm 1} \langle \sigma_1 | t | \sigma_2 \rangle \langle \sigma_2 | t | \sigma_3 \rangle \cdots \langle \sigma_N | t | \sigma_1 \rangle \tag{B.3.49}$$

$$= \sum_{\sigma_1 = \pm 1} \sum_{\sigma_2 = \pm 1} \cdots \sum_{\sigma_N = \pm 1} t_{\sigma_1 \sigma_2} t_{\sigma_2 \sigma_3} \cdots t_{\sigma_N \sigma_1}. \tag{B.3.50}$$

Here the two expressions are equivalent. The first line uses Dirac's notation, whereas the second line uses index notation. Both are presented here and below.

In Dirac notation, the identity matrix can be expressed as $I = \sum_i |\psi_i\rangle \langle \psi_i|$, where $|\psi_i\rangle$ is a set of orthonormal basis vectors of the relevant vector space. Here, each spin's vector space is spanned by $|\pm 1\rangle$, so one expression of the identity matrix is:

$$I = \sum_{\sigma_i = \pm 1} |\sigma_i\rangle \langle \sigma_i|. \tag{B.3.51}$$

Taking Eq. B.3.49, if we sum over spins σ_2 through σ_N, then the summations turn each $|\sigma_i\rangle \langle \sigma_i|$ into an identity matrix, leading to:

$$Z = \sum_{\sigma_1} \langle \sigma_1 | t^N | \sigma_1 \rangle$$
$$= \text{Tr}\left[t^N\right]. \tag{B.3.52}$$

Equivalently, using index notation, we see that each summation over σ_2 through σ_N in Eq. B.3.50 is a matrix multiplication, yielding:

$$Z = \sum_{\sigma_1} \left(t^N\right)_{\sigma_1 \sigma_1}$$
$$= \text{Tr}\left[t^N\right]. \tag{B.3.53}$$

To find the matrix t, we evaluate the quantity $K(\sigma_a, \sigma_b)$ for different values of σ_a and σ_b:

$$K(+1, +1) = -J - h,$$
$$K(+1, -1) = K(-1, +1) = J, \tag{B.3.54}$$
$$K(-1, -1) = -J + h.$$

Substituting in these values to $e^{-\beta K(\sigma_a, \sigma_b)}$, and interpreting the values as matrix elements, we get:

$$\langle +1 | t | +1 \rangle = t_{+1, +1} = e^{\beta(J+h)},$$
$$\langle +1 | t | -1 \rangle = t_{+1, -1} = e^{-\beta J},$$
$$\langle -1 | t | +1 \rangle = t_{-1, +1} = e^{-\beta J}, \tag{B.3.55}$$
$$\langle -1 | t | -1 \rangle = t_{-1, -1} = e^{\beta(J-h)}.$$

So the matrix is:

$$t = \begin{bmatrix} e^{\beta(J+h)} & e^{-\beta J} \\ e^{-\beta J} & e^{\beta(J-h)} \end{bmatrix}. \tag{B.3.56}$$

3. Using the cyclic property of the trace,[4] show that, for any values of i and j:

(a) The expectation value of any spin, which we will define as the magnetization m, is

$$\langle \sigma_i \rangle = \langle \sigma_j \rangle = Z^{-1} \text{Tr} \left[\tau_z t^N \right] \equiv m, \qquad \text{(B.3.57)}$$

where $\tau_z = \begin{bmatrix} 1 & 0 \\ 0 & -1 \end{bmatrix}$ is the standard Pauli-z matrix.

We find $\langle \sigma_i \rangle$ the standard way, by calculating the partition function normalized expectation value. To incorporate this into the Dirac formalism, recognize that:

$$\tau_z = \sum_{\sigma_i} |\sigma_i\rangle \, \sigma_i \, \langle \sigma_i| . \qquad \text{(B.3.58)}$$

Writing out $\langle \sigma_i \rangle$ in full, we see:

$$Z \langle \sigma_i \rangle = \sum_{\{\sigma_j\}} \sigma_i \langle \sigma_1 | t | \sigma_2 \rangle \cdots \langle \sigma_N | t | \sigma_1 \rangle =$$

$$\sum_{\{\sigma_j\}} \langle \sigma_1 | t | \sigma_2 \rangle \cdots \langle \sigma_{i-1} | t | \sigma_i \rangle \, \sigma_i \, \langle \sigma_i | t | \sigma_i + 1 \rangle \cdots \langle \sigma_N | t | \sigma_1 \rangle . \qquad \text{(B.3.59)}$$

We we're free to reorder the terms in the summation, because they are just numbers (i.e., individual matrix elements). The summation over σ_i results in the τ_z operator, where every other summation becomes the identity matrix, as previously, giving us:

$$Z \langle \sigma_i \rangle = \sum_{\sigma_1} \langle \sigma_1 | t^{i-1} \tau_z t^{N-i+1} | \sigma_1 \rangle$$

$$= \text{Tr} \left[t^{i-1} \tau_z t^{N-i+1} \right] . \qquad \text{(B.3.60)}$$

The index notation approach unravels similarly, using the fact that:

$$\left(\tau_z \right)_{\sigma_i \sigma_i'} = \sum_{\sigma_i'} \sigma_i \delta_{\sigma_i \sigma_i'} . \qquad \text{(B.3.61)}$$

And we can always harmlessly introduce the Kronecker delta into an existing matrix multiplication,

$$\sum_j A_{ij} B_{jk} = \sum_j \sum_{j'} A_{ij} \delta_{jj'} B_{j'k}, \qquad \text{(B.3.62)}$$

where we created a dummy index j' to make this work. Thus, we arrive at

---

4. Which is $\text{Tr}[AB] = \text{Tr}[BA]$. Combining this with the associative property of matrices $(AB)C = A(BC)$, the cyclic property can be extended to arbitrary number of matrices multiplied together, e.g. $\text{Tr}[ABCD] = \text{Tr}[BCDA] = \text{Tr}[CDAB] = \text{Tr}[DABC]$.

$$Z\langle\sigma_i\rangle = \sum_{\{\sigma_j\}} \sigma_i t_{\sigma_1\sigma_2}\cdots t_{\sigma_N\sigma_1}$$

$$= \sum_{\{\sigma_j\}} \sum_{\sigma_i'} t_{\sigma_1\sigma_2}\cdots t_{\sigma_{i-1}\sigma_i}\sigma_i\delta_{\sigma_i\sigma_i'}t_{\sigma_i'\sigma_{i+1}}\cdots t_{\sigma_N\sigma_1}$$

$$= \sum_{\{\sigma_j\}} \sum_{\sigma_i'} t_{\sigma_1\sigma_2}\cdots t_{\sigma_{i-1}\sigma_i}\left(\tau_z\right)_{\sigma_i\sigma_i'}t_{\sigma_i'\sigma_{i+1}}\cdots t_{\sigma_N\sigma_1} \qquad (B.3.63)$$

$$= \sum_{\sigma_1} \left(t^{i-1}\tau_z t^{N-i+1}\right)_{\sigma_1\sigma_1}$$

$$= \mathrm{Tr}\left[t^{i-1}\tau_z t^{N-i+1}\right],$$

which is the same as the Dirac notation approach. Index notation might be slightly less elegant in this case, but it nonetheless works.

Now we can finally apply the cyclic property of the trace by bringing the first t^{i-1} to the end, giving us:

$$Z\langle\sigma_i\rangle = \mathrm{Tr}\left[t^{i-1}\tau_z t^{N-i+1}\right]$$

$$= \mathrm{Tr}\left[\tau_z t^{N-i+1}t^{i-1}\right] \qquad (B.3.64)$$

$$\Rightarrow \langle\sigma_i\rangle = Z^{-1}\mathrm{Tr}\left[\tau_z t^N\right].$$

This step also shows that the specific value of i does not matter; $\langle\sigma_i\rangle = \langle\sigma_j\rangle$ for any value of i and j.

(b) The expected two-site correlation is

$$\langle\sigma_i\sigma_j\rangle = Z^{-1}\mathrm{Tr}\left[\tau_z t^{j-i}\tau_z t^{N-j+i}\right]. \qquad (B.3.65)$$

The solution to this is almost identical to that of the previous part, so we will skip the exposition and write out only the important steps. Without loss of generality we assume $j > i$. In Dirac notation:

$$Z\langle\sigma_i\sigma_j\rangle = \sum_{\{\sigma_j\}} \sigma_i\sigma_j \langle\sigma_1|t|\sigma_2\rangle\cdots\langle\sigma_N|t|\sigma_1\rangle$$

$$= \sum_{\{\sigma_j\}} \langle\sigma_1|t|\sigma_2\rangle\cdots\langle\sigma_{i-1}|t|\sigma_i\rangle\,\sigma_i\,\langle\sigma_i|t|\sigma_i+1\rangle$$

$$\cdots\langle\sigma_{j-1}|t|\sigma_j\rangle\,\sigma_j\,\langle\sigma_j|t|\sigma_j+1\rangle\cdots\langle\sigma_N|t|\sigma_1\rangle \qquad (B.3.66)$$

$$= \sum_{\sigma_1} \langle\sigma_1|t^{i-1}\tau_z t^{j-i}\tau_z t^{N-j+1}|\sigma_1\rangle$$

$$= \mathrm{Tr}\left[t^{i-1}\tau_z t^{j-i}\tau_z t^{N-j+1}\right]$$

$$\Rightarrow \langle\sigma_i\sigma_j\rangle = Z^{-1}\mathrm{Tr}\left[\tau_z t^{j-i}\tau_z t^{N-j+1}\right].$$

And in index notation:

$$
\begin{aligned}
Z\langle \sigma_i \sigma_j \rangle &= \sum_{\{\sigma_j\}} \sigma_i \sigma_j t_{\sigma_1 \sigma_2} \cdots t_{\sigma_N \sigma_1} \\
&= \sum_{\{\sigma_j\}} \sum_{\sigma_i'} \sum_{\sigma_j'} t_{\sigma_1 \sigma_2} \cdots t_{\sigma_{i-1} \sigma_i} \sigma_i \delta_{\sigma_i \sigma_i'} t_{\sigma_i' \sigma_{i+1}} \\
&\qquad \cdots t_{\sigma_{j-1} \sigma_j} \sigma_j \delta_{\sigma_j \sigma_j'} t_{\sigma_j' \sigma_{j+1}} \cdots t_{\sigma_N \sigma_1} \\
&= \sum_{\sigma_1} \left(t^{i-1} \tau_z t^{j-i} \tau_z t^{N-j+1} \right)_{\sigma_1, \sigma_1} \\
&= \mathrm{Tr} \left[t^{i-1} \tau_z t^{j-i} \tau_z t^{N-j+1} \right] \\
\Rightarrow \langle \sigma_i \sigma_j \rangle &= Z^{-1} \mathrm{Tr} \left[\tau_z t^{j-i} \tau_z t^{N-j+1} \right].
\end{aligned}
\tag{B.3.67}
$$

Calculations

Having reformulated the problem as matrix operations, we will perform the calculations. Let us use the notation:

$$
t | \psi_\pm \rangle = \lambda_\pm | \psi_\pm \rangle.
\tag{B.3.68}
$$

So $\lambda_\pm$ are the two eigenvalues of t, and $|\psi_\pm\rangle$ are the two corresponding eigenvectors.

1. In its eigenbasis, $t' = UtU^{-1}$ is diagonal (by construction) with the form

$$
t' = \begin{bmatrix} \lambda_+ & 0 \\ 0 & \lambda_- \end{bmatrix}
\tag{B.3.69}
$$

Express the partition function in terms of $\lambda_\pm$.

$$
\begin{aligned}
Z &= \mathrm{Tr} \left[t^N \right] \\
&= \mathrm{Tr} \left[(U^{-1} t' U)^N \right] \\
&= \mathrm{Tr} \left[U^{-1} (t')^N U \right] = \mathrm{Tr} \left[(t')^N \right],
\end{aligned}
\tag{B.3.70}
$$

where the last step again uses the cyclic property of the trace. This manipulation explicitly shows that the trace is invariant under basis transformations.

Since t' is a diagonal matrix, we can raise it to integer powers directly:

$$
(t')^N = \begin{bmatrix} \lambda_+^N & 0 \\ 0 & \lambda_-^N \end{bmatrix}.
\tag{B.3.71}
$$

So the partition function is:

$$
Z = \lambda_+^N + \lambda_-^N.
\tag{B.3.72}
$$

2. The trace can be computed in any basis, so we could write

$$\text{Tr}\,[A] = \langle \psi_+ | A | \psi_+ \rangle + \langle \psi_- | A | \psi_- \rangle. \tag{B.3.73}$$

Express the following quantities in terms of $\lambda_\pm$, $|\psi_\pm\rangle$, and τ_z:

(a) $\langle \sigma_i \rangle$.

Starting with the expression for $\langle \sigma_i \rangle$, but using the new basis in $|\psi_\pm\rangle$:

$$\begin{aligned} Z\langle \sigma_i \rangle &= \text{Tr}\left[\tau_z t^N\right] \\ &= \langle \psi_+ | \tau_z t^N | \psi_+ \rangle + \langle \psi_- | \tau_z t^N | \psi_- \rangle. \end{aligned} \tag{B.3.74}$$

Since this is the eigenbasis of t, we have:

$$t^N |\psi_\pm\rangle = \lambda_\pm^N |\psi_\pm\rangle. \tag{B.3.75}$$

Thus we get:

$$\langle \sigma_i \rangle = Z^{-1}\left(\lambda_+^N \langle \psi_+ | \tau_z | \psi_+ \rangle + \lambda_-^N \langle \psi_- | \tau_z | \psi_- \rangle\right). \tag{B.3.76}$$

(b) $\langle \sigma_i \sigma_j \rangle$.

Hint: The identity matrix can be expressed as:

$$I = |\psi_+\rangle \langle \psi_+| + |\psi_-\rangle \langle \psi_-|. \tag{B.3.77}$$

Using the expression for $\langle \sigma_i \sigma_j \rangle$ and again using the new eigenbasis $|\psi_\pm\rangle$:

$$\begin{aligned} Z\langle \sigma_i \sigma_j \rangle &= \text{Tr}\left[\tau_z t^{j-i} \tau_z t^{N-j+1}\right] \\ &= \langle \psi_+ | \tau_z t^{j-i} \tau_z t^{N-j+1} | \psi_+ \rangle + \langle \psi_- | \tau_z t^{j-i} \tau_z t^{N-j+1} | \psi_- \rangle. \end{aligned} \tag{B.3.78}$$

Let us focus on the first term, as there is a little trick involved, which is inserting the identity between the first and second τ_z:

$$\begin{aligned} & \langle \psi_+ | \tau_z t^{j-i} \tau_z t^{N-j+1} | \psi_+ \rangle \\ &= \langle \psi_+ | \tau_z t^{j-i} \left(|\psi_+\rangle \langle \psi_+| + |\psi_-\rangle \langle \psi_-|\right) \tau_z t^{N-j+1} | \psi_+ \rangle \\ &= \langle \psi_+ | \tau_z \lambda_+^{j-i} | \psi_+ \rangle \langle \psi_+ | \tau_z \lambda_+^{N-j+i} | \psi_+ \rangle \\ &\quad + \langle \psi_+ | \tau_z \lambda_-^{j-i} | \psi_- \rangle \langle \psi_- | \tau_z \lambda_+^{N-j+i} | \psi_+ \rangle \\ &= \lambda_+^N \left[|\langle \psi_+ | \tau_z | \psi_+ \rangle|^2 + \left(\frac{\lambda_-}{\lambda_+}\right)^{j-i} |\langle \psi_+ | \tau_z | \psi_- \rangle|^2\right]. \end{aligned} \tag{B.3.79}$$

Again, we used the fact that this is the eigenbasis of t. Using the same logic for the second term, we get:

$$\langle \psi_- | \tau_z t^{j-i} \tau_z t^{N-j+1} | \psi_- \rangle$$
$$= \lambda_-^N \left[|\langle \psi_- | \tau_z | \psi_- \rangle|^2 + \left(\frac{\lambda_+}{\lambda_-} \right)^{j-i} |\langle \psi_- | \tau_z | \psi_+ \rangle|^2 \right]. \tag{B.3.80}$$

Thus the final answer is:

$$\langle \sigma_i \sigma_j \rangle = Z^{-1} \left\{ \lambda_+^N \left[|\langle \psi_+ | \tau_z | \psi_+ \rangle|^2 + \left(\frac{\lambda_-}{\lambda_+} \right)^{j-i} |\langle \psi_+ | \tau_z | \psi_- \rangle|^2 \right] \right.$$
$$\left. + \lambda_-^N \left[|\langle \psi_- | \tau_z | \psi_- \rangle|^2 + \left(\frac{\lambda_+}{\lambda_-} \right)^{j-i} |\langle \psi_- | \tau_z | \psi_+ \rangle|^2 \right] \right\}. \tag{B.3.81}$$

3. Now, to do some heavy lifting, find the eigenvalues $\lambda_\pm$ and the corresponding normalized eigenvectors $|\psi_\pm\rangle$. *Hint:* If you need a refresher on how to handle 2×2 matrices, Appendix A.1 is a good place to be.

Briefly reiterating Appendix A.1, the transfer matrix can be written as a sum of Pauli matrices,

$$t = e^{\beta J} \cosh(\beta h) I + e^{-\beta J} \tau_x + e^{\beta J} \sinh(\beta h) \tau_z, \tag{B.3.82}$$

where we can identify

$$\epsilon_0 = e^{\beta J} \cosh(\beta h),$$
$$b_x = e^{-\beta J},$$
$$b_z = e^{\beta J} \sinh(\beta h).$$

So the eigenvalues are:

$$\lambda_\pm = e^{\beta J} \cosh(\beta h) \pm \sqrt{e^{-2\beta J} + e^{2\beta J} \sinh^2(\beta h)}. \tag{B.3.83}$$

Defining $b = \sqrt{b_x^2 + b_z^2}$, the normalized eigenvectors are:

$$|\psi_+\rangle = \frac{1}{\sqrt{2b}} \begin{pmatrix} \sqrt{b + b_z} \\ \sqrt{b - b_z} \end{pmatrix} \quad |\psi_-\rangle = \frac{1}{\sqrt{2b}} \begin{pmatrix} -\sqrt{b - b_z} \\ \sqrt{b + b_z} \end{pmatrix}. \tag{B.3.84}$$

This is a short algebraic manipulation from the form presented in the appendix.

4. Let us put it all together. Calculate explicitly, in the thermodynamic limit[5] $N \to \infty$:

---

5. Eliminate terms that are infinitely smaller than other terms, but retain the N dependence in the terms you keep.

(a) The partition function Z. Does this agree with the answer derived using combinatorics for $h = 0$?

Since $\lambda_+ > \lambda_-$, when we take the limit $N \to \infty$, all λ_+^N terms will be exponentially larger than λ_-^N terms. Therefore, in the partition function, and all subsequent expectation values, we can discard the λ_-^N terms as small in the thermodynamic limit. Thus:

$$\lim_{N \to \infty} Z = \lambda_+^N$$

$$= \left(e^{\beta J} \cosh(\beta h) + \sqrt{e^{-2\beta J} + e^{2\beta J} \sinh^2(\beta h)} \right)^N. \tag{B.3.85}$$

Without an external field, so $h = 0$, this reduces to

$$Z = \left(e^{\beta J} + e^{-\beta J} \right)^N = \left[2 \cosh(\beta J) \right]^N, \tag{B.3.86}$$

which does match the previous combinatorics calculation!

(b) The free energy $F = -T \log Z$.

From here on we will omit the limit symbol, since $N \to \infty$ is implied. The free energy is:

$$F = -T \log Z$$

$$= -TN \log \lambda_+ \tag{B.3.87}$$

$$= -TN \log \left(e^{\beta J} \cosh(\beta h) + \sqrt{e^{-2\beta J} + e^{2\beta J} \sinh^2(\beta h)} \right).$$

(c) The magnetization from the free energy:

$$m = -\frac{1}{N} \frac{\partial F}{\partial h}. \tag{B.3.88}$$

By generous application of the chain rule, we get:

$$\frac{\partial \log \lambda_+}{\partial h} = \frac{1}{\lambda_+} \frac{\partial \lambda_+}{\partial h}$$

$$= \beta \frac{e^{\beta J} \sinh(\beta h) + \frac{e^{2\beta J} \cosh(\beta h) \sinh(\beta h)}{\sqrt{e^{-2\beta J} + e^{2\beta J} \sinh^2(\beta h)}}}{e^{\beta J} \cosh(\beta h) + \sqrt{e^{-2\beta J} + e^{2\beta J} \sinh^2(\beta h)}} \tag{B.3.89}$$

$$= \beta \frac{e^{\beta J} \sinh(\beta h)}{\sqrt{e^{-2\beta J} + e^{2\beta J} \sinh^2(\beta h)}}.$$

So for the magnetization, we have:

$$m = -\frac{1}{N}\frac{\partial(-TN\log\lambda_+)}{\partial h}$$

$$= T\frac{\partial\log\lambda_+}{\partial h} \tag{B.3.90}$$

$$= \frac{e^{\beta J}\sinh(\beta h)}{\sqrt{e^{-2\beta J} + e^{2\beta J}\sinh^2(\beta h)}}.$$

(d) The magnetization as the expectation value:

$$m = \langle\sigma_i\rangle. \tag{B.3.91}$$

Does this agree with the previous part?

In the thermodynamic limit, the magnetization is:

$$\frac{\lambda_+^N}{Z}\langle\psi_+|\tau_z|\psi_+\rangle. \tag{B.3.92}$$

The scaling factor is 1, as $Z = \lambda_+^N$ in the thermodynamic limit as well. Using the eigenvectors we have calculated earlier,

$$m = \frac{1}{2b}\begin{pmatrix}\sqrt{b+b_z} & \sqrt{b-b_z}\end{pmatrix}\begin{bmatrix}1 & 0\\0 & -1\end{bmatrix}\begin{pmatrix}\sqrt{b+b_z}\\\sqrt{b-b_z}\end{pmatrix}$$

$$= \frac{1}{2b}[b+b_z - (b-b_z)] = \frac{b_z}{b} \tag{B.3.93}$$

$$= \frac{e^{\beta J}\sinh(\beta h)}{\sqrt{e^{-2\beta J} + e^{2\beta J}\sinh^2(\beta h)}},$$

which does agree with the previous part. As noted in Appendix A.1, we can also obtain this answer by considering t as the Hamiltonian of 2-state spin in an effective magnetic field!

(e) The expected two-site correlation:

$$\langle\sigma_i\sigma_j\rangle. \tag{B.3.94}$$

In the thermodynamic limit, the terms with prefactor λ_-^N drop out. We have also just calculated the magnetization to be $m = \langle\psi_+|\tau_z|\psi_+\rangle$, so the only term we need to find is:

$$\langle\psi_+|\tau_z|\psi_-\rangle = \frac{1}{2b}\begin{pmatrix}\sqrt{b+b_z} & \sqrt{b-b_z}\end{pmatrix}\begin{bmatrix}1 & 0\\0 & -1\end{bmatrix}\begin{pmatrix}-\sqrt{b-b_z}\\\sqrt{b+b_z}\end{pmatrix}$$

$$= -\frac{1}{b}\sqrt{b^2 - b_z^2} = -\frac{b_x}{b} \tag{B.3.95}$$

$$= -\frac{e^{-\beta J}}{\sqrt{e^{-2\beta J} + e^{2\beta J} \sinh^2(\beta h)}}.$$

Again, $\lambda_+^N/Z = 1$ in the thermodynamic limit, and we have:

$$\langle \sigma_i \sigma_j \rangle = m^2 + \frac{e^{-2\beta J}}{e^{-2\beta J} + e^{2\beta J} \sinh^2(\beta h)} \left(\frac{\lambda_-}{\lambda_+} \right)^{j-i}. \qquad (B.3.96)$$

(f) The connected correlation function, which is defined as

$$G(i,j) = \langle \sigma_i \sigma_j \rangle - \langle \sigma_i \rangle \langle \sigma_j \rangle, \qquad (B.3.97)$$

and recast the expression into the form:

$$G(i,j) \propto e^{-\frac{|i-j|}{\xi}}. \qquad (B.3.98)$$

What is the expression for the correlation length ξ?

Since $\langle \sigma_i \rangle \langle \sigma_j \rangle = m^2$, the connected correlation function is:

$$G(i,j) = \frac{e^{-2\beta J}}{e^{-2\beta J} + e^{2\beta J} \sinh^2(\beta h)} \left(\frac{\lambda_-}{\lambda_+} \right)^{j-i}. \qquad (B.3.99)$$

We can define

$$g_0(h) = \frac{e^{-2\beta J}}{e^{-2\beta J} + e^{2\beta J} \sinh^2(\beta h)} \qquad (B.3.100)$$

and rewrite $G(i,j)$ as

$$G(i,j) = g_0(h)e^{\log\left(\frac{\lambda_-}{\lambda_+}\right)(j-i)}. \qquad (B.3.101)$$

Since we know that $\lambda_- < \lambda_+$, and we assumed (without loss of generality) that $j > i$, the exponential decays with increasing separation between sites i and j. We can then cast this expression in terms of a correlation length ξ, ensuring that the relative signs of the quantities are correct:

$$G(i,j) = g_0(h)e^{-|i-j|/\xi}, \qquad (B.3.102)$$

where the correlation length is

$$\xi = \left| \log\left(\frac{\lambda_-}{\lambda_+} \right) \right|^{-1}. \qquad (B.3.103)$$

The absolute value ensures that the correlation length is a positive quantity.

(g) In the case that $h = 0$, how does the correlation length behave as a function of temperature?

Without an external field, the correlation length becomes:

$$\xi = \left| \log \left(\frac{e^{\beta J} - e^{-\beta J}}{e^{\beta J} + e^{-\beta J}} \right) \right|^{-1}$$

$$= \left| \log \left[\tanh(\beta J) \right] \right|^{-1}.$$

(B.3.104)

At higher temperatures, $\beta \to 0$, and the correlation length is small. The correlation length diverges at low temperatures as $\beta \to \infty$. However, it remains finite at any nonzero temperature, again showing that in the 1D Ising model there are no phase transitions at finite temperature.

B.3.7 Low-Temperature Expansion in 2D[6]

We can further understand the effect of dimensionality on the critical behavior of the Ising model by performing a quantitative expansion at low temperatures. While the text discusses this in general for any dimensionality d, here we will work out the details for $d = 2$ on a square lattice with N total sites. The Hamiltonian is

$$H = -J \sum_{\langle ij \rangle} \sigma_i \sigma_j,$$

(B.3.105)

and we will assume ferromagnetic interactions ($J > 0$).

The partition function

As a reminder, the partition function is

$$Z = \sum_{\{\sigma\}} e^{-\beta H(\{\sigma\})} = e^{4N\beta J} \sum_{E_i} g(E_i) e^{-\beta E_i},$$

(B.3.106)

where the first form is expressed as a sum over microstates, but we can equivalently express the partition function as an infinite series in which the states with excitation energy E_i are weighted by their multiplicities $g(E_i)$. As before, our starting point is the ground state, with all spins pointing up. The overall factor of $e^{4N\beta J}$ comes from accounting for the energy of the ground state, $E_G = -4NJ$, so we can use the notation E_i to indicate the excitation energy relative to the ground state.

1. The lowest energy state is the ground state, with $E_0 = 0$ and therefore Boltzmann factor of $e^0 = 1$. The first excited state is a single spin-flip, with $E_1 = 8J$. What are the next **two** possible excitation energies E_2 and E_3, and what are the

---

6. This is a tricky exercise; feel free to try your best on each step, then check your answer for correctness and understanding before moving on to the next step!

types of spin-flips associated with them? *Hint:* Excited spins are not necessarily adjacent to each other.

The next higher energy excitation is flipping two adjacent spins, which causes six bonds to flip, giving $E_2 = 6 \cdot 2J = 12J$. The next higher excitation is flipping three adjacent spins, or flipping four adjacent spins in a square. This will flip eight bonds, giving $E_3 = 8 \cdot 2J = 16J$. This excitation energy is also reachable by flipping two disconnected spins. Since each individual spin-flip causes four flipped bonds, this also gives $E_3 = 2 \cdot 4 \cdot 2J = 16J$.

2. Now we will need to count the multiplicity of states associated with each excitation energy. The ground state is unique, so $g(0) = 1$. The first excited state is a single spin, with N possible sites for this excitation, so $g(8J) = N$. Find $g(E_2)$ and $g(E_3)$. *Hint:* You may find it helpful to look at how we enumerated the droplet shapes in the previous exercise.

Refer to the solutions to the droplet exercise above, specifically enumerating the possible droplets of fixed domain wall lengths.

For states with energy E_2, there are two orientations for two spins adjacent to each other. Each configuration may be placed on N different sites, so $g(E_2) = 2N$.

For states with energy E_3, there are six distinct configurations for three connected spin-flips, and one configuration for four spin-flips with this energy. Each of the configurations may be placed on N different sites, so we have multiplicity $7N$ from the connected spin-flips.

For two disconnected spin-flips with energy E_3, we can count the ways to arrange them as such: the first spin may be placed on any of the N sites, whereas the second spin must be placed anywhere except on or adjacent to the first spin to give the correct energy E_3, so on $N - 5$ possible sites. Multiplying these two together will double count the number of distinct states, since the spin-flips are indistinguishable. Therefore, the multiplicity from the disconnected spin-flips is $N(N-5)/2$. Combining the contributions, $g(E_3) = 7N + N(N-5)/2 = N^2/2 + 9N/2$.

3. Now that we have all the ingredients, write down the next two orders in the expansion for Z:

$$Ze^{-4N\beta J} = 1 + Ne^{-8\beta J} + 2Ne^{-12\beta J} + \left(\frac{N^2}{2} + \frac{9}{2}N\right)e^{-16\beta J} + \mathcal{O}\left(e^{-\beta E_4}\right). \quad \text{(B.3.107)}$$

The magnetization

Since the lattice is translationally invariant, calculating the magnetization is the same as calculating the expectation value of a spin on any specific site, which we will call σ_1. So:

$$m = \langle \sigma_1 \rangle = \frac{1}{Z} \sum_{\{\sigma\}} \sigma_1 e^{-\beta H(\{\sigma\})} \tag{B.3.108}$$

$$= \frac{e^{4N\beta J}}{Z} \sum_{E_i} [g_+(E_i) - g_-(E_i)] e^{-\beta E_i}, \tag{B.3.109}$$

where the first line is the definition of the expectation value, and the second line is the form it takes as an infinite series. For each possible excitation energy E_i, there are $g_+(E_i)$ configurations in which $\sigma_1 = +1$, and $g_-(E_i)$ configurations in which $\sigma_1 = -1$, with $g_+(E_i) + g_-(E_i) = g(E_i)$.

1. The unique ground-state has all spins pointing up, so $g_+(0) = 1$ and $g_-(0) = 0$. The first excited state has exactly one configuration where σ_1 is flipped, so $g_-(E_1) = 1$ and $g_+(E_1) = N - 1$. Find $g_-(E_2)$ and $g_-(E_3)$. *Hint:* Again, you might want to revisit the droplet enumeration we did before. Think about how many unique ways in which you can place the droplets on the lattice such that they contain a specific site.

   For all of the connected spin-flips, each droplet configuration can be placed in unique ways equal to the number of spin-flips it contains. That is:
       For E_2 states, the two distinct droplet shapes can each be placed in two ways to contain a specific spin σ_1, so $g_-(E_2) = 2 \cdot 2 = 4$.
       For E_3 states with connected spin-flips, the six droplets with three spin-flips can each be placed in three ways, and the four spin-flip droplet can be placed in four ways. So from the connected spin-flips, the multiplicity is $6 \cdot 3 + 4 = 22$.
       For the disconnected spin-flips, if one flip is on the site σ_1 by requirement, then there are $N - 5$ ways to place the other spin-flip. Combining the contributions, $g_-(E_3) = 22 + (N - 5) = N + 17$.

2. From the definitions above, we see $g_+(E_i) - g_-(E_i) = g(E_i) - 2g_-(E_i)$. Use your results from previous parts to find the next two orders:

   Using answers from previous parts, substituting in values for $g(E_i)$ and $g_-(E_i)$, we find

   $$m = \frac{e^{4N\beta J}}{Z}\left[1 + (N-2)e^{-8\beta J} + (2N-8)e^{-12\beta J} \right.$$
   $$\left. + \left(\frac{N^2}{2} + \frac{5}{2}N - 34 \right) e^{-16\beta J} + \mathcal{O}\left(e^{-\beta E_4} \right) \right]. \tag{B.3.110}$$

3. To find the final answer for the magnetization, we will now incorporate the $1/Z$ normalization. Use the expression for Z found previously, and fully expand the expression for m up to order $\mathcal{O}\left(e^{-\beta E_3}\right)$. *Hint:* Multiply out the series completely, keeping terms up to the required order. You may find the following useful:

$$\frac{1}{1+x} = 1 - x + x^2 + \mathcal{O}\left(x^3\right). \tag{B.3.111}$$

When the dust settles, you will find that we are left with an expansion with no terms proportional to any power of N. Since m is an intensive quantity, the expansion (rightfully) reflects this property. Remarkably, as the expansion is carried out to higher orders, no further powers of N will appear!

Using the expression we just found in the previous part, and first cancelling out the overall factor of $e^{4N\beta J}$ from the partition function, we get (using the notation $e^{2\beta J} = u$):

$$m = \frac{1 + (N-2)u^{-4} + (2N-8)u^{-6} + \left(\frac{N^2}{2} + \frac{5}{2}N - 34\right)u^{-8} + \cdots}{1 + Nu^{-4} + 2Nu^{-6} + \left(\frac{N^2}{2} + \frac{9}{2}N\right)u^{-8} + \cdots}. \tag{B.3.112}$$

Using the expansion suggested in the hint, keeping terms to order u^{-8}, the expression becomes:

$$m = \left[1 + (N-2)u^{-4} + (2N-8)u^{-6} + \left(\frac{N^2}{2} + \frac{5}{2}N - 34\right)u^{-8} + \cdots\right]$$
$$\times \left[1 - Nu^{-4} - 2Nu^{-6} + \left(\frac{N^2}{2} - \frac{9}{2}N\right)u^{-8} + \cdots\right]. \tag{B.3.113}$$

Multiplying out the terms, and keeping terms up to u^{-8}, we finally get:

$$m = 1 - 2u^{-4} - 8u^{-6} - 34u^{-8} - \cdots. \tag{B.3.114}$$

And as advertised, m is a properly intensive quantity with no dependence on powers of N!

4. *The exact solution by Onsager and Yang[7] for the 2D Ising model gives the order parameter to be:

$$m = \left[1 - \frac{1}{\sinh^4(2\beta J)}\right]^{1/8}. \tag{B.3.115}$$

Does it agree with the expansion we just found?

---

7. L. Onsager, *Phys. Rev.* **65**, 117 (1944) and C. N. Yang, *Phys. Rev.*, **85**, 808 (1952)

Let us expand this expression in the low-temperature limit $\beta J \to \infty$. In order to make the low-temperature expansion, we make the same substitution $u = e^{2\beta J} = \tilde{u}^{-1}$ and Taylor expand about $\tilde{u} = 0$:

$$m = \left[1 - \frac{16}{(\tilde{u}^{-1} - \tilde{u})^4} \right]^{1/8}$$

$$= 1 - 2\tilde{u}^4 - 8\tilde{u}^6 - 34\tilde{u}^8 - 152\tilde{u}^{10} - \cdots$$

$$= 1 - 2u^{-4} - 8u^{-6} - 34u^{-8} - 152u^{-10} - \cdots.$$

And indeed, it agrees exactly with the expansion we have found order by order (up to the 8th-order term we have found). How cool is that!

The free energy, entropy, and heat capacity

Having found the expansion for the partition function, we can also find the free energy, entropy, and heat capacity. For all calculations below, keep terms up to order $\mathcal{O}\left(e^{-\beta E_3}\right)$.

1. Starting with the expansion for Z, find the expansion for the free energy $F = -T \log Z$. *Hint:* You may find the following useful:

$$\log(1 + x) = x - \frac{x^2}{2} + \mathcal{O}\left(x^3\right). \qquad \text{(B.3.116)}$$

Since the partition function has this form, we can simply plug in the expansion for Z and keep terms to order u^{-8}:

$$F = -T \log Z$$

$$= -T \left[Nu^{-4} + 2Nu^{-6} + \left(\frac{N^2}{2} + \frac{9}{2}N \right) u^{-8} - \frac{N^2}{2} u^{-8} + \cdots \right] \qquad \text{(B.3.117)}$$

$$= -NT \left(u^{-4} + 2u^{-6} + \frac{9}{2} u^{-8} + \cdots \right).$$

This calculation echoes the point from the text that it is $\log Z$ for which the expansion is meaningful.

2. Calculate the entropy $S = -\partial F/\partial T$.

Since we have been using powers of u, let us first calculate its temperature derivative:

$$\frac{\partial u^{-n}}{\partial T} = \frac{\partial}{\partial T} e^{-\frac{2nJ}{T}}$$

$$= -2nJe^{-\frac{2nJ}{T}}\frac{\partial T^{-1}}{\partial T} \tag{B.3.118}$$

$$= \frac{2nJ}{T^2}u^{-n}.$$

So, using the above expression for F and taking the T derivative (remember to use the product rule!):

$$S = -\frac{\partial F}{\partial T}$$

$$= N\left(u^{-4} + 2u^{-6} + \frac{9}{2}u^{-8} + \cdots\right) \tag{B.3.119}$$

$$+ \frac{2JN}{T}\left(4u^{-4} + 12u^{-6} + 36u^{-8} + \cdots\right).$$

3. Calculate the heat capacity per site:

$$c = \frac{C}{N} = \frac{T}{N}\frac{\partial S}{\partial T}. \tag{B.3.120}$$

Since this is also an intensive quantity, the correct answer should not have any terms proportional to any power of N.

Taking derivatives again, we find:

$$\frac{C}{T} = \frac{\partial S}{\partial T}$$

$$= \frac{2JN}{T^2}\left(4u^{-4} + 12u^{-6} + 36u^{-8} + \cdots\right)$$

$$- \frac{2JN}{T^2}\left(4u^{-4} + 12u^{-6} + 36u^{-8} + \cdots\right) \tag{B.3.121}$$

$$+ \frac{4J^2N}{T^3}\left(16u^{-4} + 72u^{-6} + 288u^{-8} + \cdots\right).$$

Many terms cancel nicely, leaving us with:

$$c = \frac{C}{N} = \frac{4J^2}{T^2}\left(16u^{u-4} + 72u^{-6} + 288u^{-8} + \cdots\right). \tag{B.3.122}$$

Since we performed the expansions correctly, the expression is properly intensive.

B.4   Chapter 4 Worksheet Solutions

B.4.1   Weiss Mean-Field Theory

Let us start with the now (hopefully) familiar Hamiltonian of the Ising model:

$$H = -J \sum_{\text{n.n.}} \sigma_i \sigma_j - h \sum_j \sigma_j. \tag{B.4.1}$$

We assume the spins exist on a N-site lattice with translational symmetry, and the spins interact only with their z nearest neighbors. An uniform external field h is also allowed. In this exercise we derive one of the simplest mean-field models.

1. Begin by considering the value of each spin in relation to the expectation value of the spins $\langle \sigma_i \rangle = m$, such that each individual spin can be written as:

$$\sigma_i = \langle \sigma_i \rangle + \delta \sigma_i$$

   where $\delta \sigma_i$ is the fluctuation of the spin relative to the average. Use this change of variables to rewrite the interaction term in the Hamiltonian such that fluctuations exist only as a 2nd-order term.

   Rewriting the interaction term (remember that $\langle \sigma_i \rangle = \langle \sigma_j \rangle = m$ due to translational symmetry):

$$\begin{aligned} \sigma_i \sigma_j &= (m + \delta \sigma_i)(m + \delta \sigma_j) \\ &= m^2 + m(\delta \sigma_i + \delta \sigma_j) + \delta \sigma_i \delta \sigma_j \\ &= -m^2 + m(\sigma_i + \sigma_j) + \delta \sigma_i \delta \sigma_j, \end{aligned} \tag{B.4.2}$$

   where in the last step we used $\delta \sigma_i = \sigma_i - m$.

2. Let us also make the nearest-neighbor sum more tractable by showing that:

$$\sum_{\text{n.n.}} = \frac{1}{2} \sum_i \sum_{j \text{ n.n. of } i}.$$

   We can ensure summation over the nearest-neighbor pairs by enumerating first each spin site i, and then each of its neighboring sites j. This gives us:

$$\sum_{\text{n.n}} \propto \sum_i \sum_{j \text{ n.n. of } i}.$$

   However, this leads to double counting of all the pairs, since if site $p - q$ is an n.n. pair, it will be counted first with q as a neighbor to p, then later with p as a neighbor to q. Thus, the correct expression is:

$$\sum_{\text{n.n}} = \frac{1}{2} \sum_i \sum_{j \text{ n.n. of } i}.$$

3. So far everything is exact, only with the interaction re-expressed, including a 2nd-order fluctuation term $\delta\sigma_i\delta\sigma_j$. The mean-field approximation entails ignoring second (and higher) order fluctuations,[8] such that $\delta\sigma_i\delta\sigma_j = 0$. Using this approximation and the rewritten form for nearest-neighbor summation, express the Hamiltonian as a single sum over all sites. *Hint:* Keep in mind that each site has z nearest neighbors.

After the mean-field approximation, the Hamiltonian looks like:

$$H = -\frac{J}{2} \sum_i \sum_{j \text{ n.n. of } i} \left[-m^2 + m(\sigma_i + \sigma_j) \right] - h \sum_i \sigma_i.$$

For the inner summation, since we are centered on σ_i, it will be counted z times. Each of its neighbors will be counted once. However, since each spin has z neighbors, this "counting once" will occur z times. Thus, once we consider the entire lattice, the summation will look like:

$$H = -\frac{J}{2} \sum_i (-zm^2 + 2zm\sigma_i) - h \sum_i \sigma_i,$$

with each spin counted z times as the central spin and z more times as neighbor to other spins. Consolidating the terms under one summation, we get:

$$H = \sum_i \left[\frac{zJm^2}{2} - (zJm + h)\sigma_i \right].$$

4. By now we have approximated the original interacting problem with N identical noninteracting effective problems. Solving the problem now amounts to solving any single copy. Note that we did not completely throw away the interaction—the thermodynamics of the interacting system is carried implicitly by the m and m^2 terms.

   Using the new Hamiltonian, calculate the partition function and the free energy, and obtain an expression for the average magnetization $m = \langle \sigma_i \rangle$. *Hint:* Note that the conjugate variable of σ is $-h$ in the Hamiltonian.

Starting with the "single spin" Hamiltonian:

$$H = \frac{zJm^2}{2} - (zJm + h)\sigma.$$

Following the same method as the 0D Ising model, partition function is then

$$Z_s = 2 \exp\left(-\frac{zJm^2}{2T}\right) \cosh\left(\frac{zJm + h}{T}\right),$$

---

8. The reason that this is justified (to an approximate extent) is that, while the term $\delta\sigma_i\delta\sigma_j$ itself may be large compared to m, the average of this term over neighboring pairs is much smaller.

and the free energy is

$$F_s = \frac{zJm^2}{2} - T\ln\left[2\cosh\left(\frac{zJm+h}{T}\right)\right].$$

Applying the "derivative theorem," we obtain:

$$\langle\sigma\rangle = -\frac{\partial F_s}{\partial h}$$

$$\implies m = \tanh\left(\frac{zJm+h}{T}\right).$$

5. *Often we may write the quantity $zJ = J_{\text{eff}}$ to indicate that this is the effective interaction strength after the mean-field procedure. What would happen to the mean-field Hamiltonian if we considered interactions beyond nearest neighbor, i.e., z_1 nearest-neighbor spins with interaction J_1, z_2 next-nearest-neighbor spins with interaction J_2, and so on? Can we still express the problem as N noninteracting spins?

The same procedure would apply, but now with a new summation for each type of interaction. However, by ignoring coupled fluctuations at any distance, the interactions all simplify to the same form,

$$H = \sum_i\left[\frac{(z_1J_1 + z_2J_2 + \cdots)m^2}{2} - \left[(z_1J_1 + z_2J_2 + \cdots)m + h\right]\sigma_i\right],$$

which we can now define $z_1J_1 + z_2J_2 + \cdots = J_{\text{eff}}$ to recover the same form as before:

$$H = \sum_i\left[\frac{J_{\text{eff}}m^2}{2} - (J_{\text{eff}}m + h)\sigma_i\right].$$

The expression for the magnetization is a transcendental self-consistent equation. Even though we cannot solve for the magnetization explicitly, this does give a tractable description of its behavior, which we will analyze in greater detail in following exercises.

B.4.2   Variational Principle

Here we prove a number of the assertions involved in our discussion of the variational principle in statistical mechanics.

1. Prove that $\langle e^{-\lambda\phi}\rangle \geq e^{-\lambda\langle\phi\rangle}$ for a random variable ϕ. This is a special case of Jensen's inequality relevant for statistical mechanics. *Hint:* Note that $1 = e^{\lambda\langle\phi\rangle - \lambda\langle\phi\rangle}$, and $e^x \geq 1 + x$ for all real x.

Start with the expression on the left-hand side without the expectation value,

$$e^{-\lambda\phi} = e^{-\lambda\phi}e^{\lambda\langle\phi\rangle - \lambda\langle\phi\rangle}$$

$$= e^{-\lambda\langle\phi\rangle}e^{-\lambda(\langle\phi\rangle-\phi)},$$

using the property of the exponential function from the hint:

$$e^{-\lambda\phi} \geq e^{-\lambda\langle\phi\rangle}\left[1 - \lambda(\langle\phi\rangle - \phi)\right].$$

Now we can take the expectation value of both sides. Note that $\langle\phi\rangle$ is not affected by this operation:

$$\left\langle e^{-\lambda\phi}\right\rangle \geq e^{-\lambda\langle\phi\rangle}\left\langle 1 - \lambda(\langle\phi\rangle - \phi)\right\rangle$$

$$\left\langle e^{-\lambda\phi}\right\rangle \geq e^{-\lambda\langle\phi\rangle}.$$

2. Now consider a Hamiltonian H and its partition function $Z = \text{Tr}[e^{-\beta H}]$, as well as a trial Hamiltonian H_{tr} and its partition function $Z_{tr} = \text{Tr}\left[e^{-\beta H_{tr}}\right]$, assuming that $[H, H_{tr}] = 0$. Use a similar approach to show that its Helmholtz free energy satisfies

$$-T\log Z = F \leq F_{tr} + \langle H - H_{tr}\rangle_{tr},$$

Where $\langle\rangle_{tr}$ denotes an expectation value calculated using Boltzmann weights determined by the trial Hamiltonian. *Hint:* Start with the expression of the partition function Z, and note that the identity $\mathbb{1} = e^{\beta H_{tr}}e^{-\beta H_{tr}}$.

Start by inserting the identity into the expression for the partition function:

$$Z = \text{Tr}\left[e^{-\beta H}e^{\beta H_{tr}}e^{-\beta H_{tr}}\right]$$

$$= \frac{Z_{tr}}{Z_{tr}}\text{Tr}\left[e^{-\beta H_{tr}}e^{-\beta H}e^{\beta H_{tr}}\right]$$

$$= Z_{tr}\text{Tr}\left[\frac{e^{-\beta H_{tr}}}{Z_{tr}}e^{-\beta(H-H_{tr})}\right]$$

$$= Z_{tr}\left\langle e^{-\beta(H-H_{tr})}\right\rangle_{tr}.$$

Here we have invoked Eq. 2.6 to obtain the expectation value in the trial ensemble. Now we can use the inequality we just proved:

$$Z \geq Z_{tr}e^{-\beta\langle H-H_{tr}\rangle_{tr}},$$

$$\log Z \geq \log Z_{tr} - \beta\langle H - H_{tr}\rangle_{tr},$$

$$-T\log Z \leq -T\log Z_{tr} + \langle H - H_{tr}\rangle_{tr},$$

$$F \leq F_{tr} + \langle H - H_{tr} \rangle_{tr} ,$$

where the right-hand side is what we call the variational free energy F_{var}. While the form of the variational inequality is the same for noncommuting H and H_{tr}, the proof is slightly more involved (see Appendix A.8).

3. The usefulness of the variational principle lies with the parameters $\{b_i\}$ in the trial Hamiltonian $H_{tr}(\{b_i\})$. Describe a general approach to optimize the parameters so that F_{var} provides the best possible approximation to F.

Since the variational inequality tells us that F_{var} is bound below by the real free energy F, the parameters $\{b_i\}$ that give the minimum F_{var} should in general give the best approximation. Thus, the approach is to solve the set of equations for $\{b_i\}$,

$$\frac{\partial F_{var}}{\partial b_i} = 0 \text{ for all } i,$$

with the usual caveat when minimizing multivariable functions, checking that the solution is not that of a local maximum or saddle point.

B.4.3 *Variational Mean-Field Theory–General Features

You can do this exercise before or after tackling the mean-field theory solution of the Ising model. In either case, this somewhat abstract exercise is meant to help you better understand the variational mean-field reasoning, without its being tied down to a specific physical model.

Consider a Hamiltonian of the form:

$$H = H_0 - \sum_j h_j \sigma_j. \tag{B.4.3}$$

This is a very general form. The second term is a bilinear coupling of the system, here denoted by a spin σ_j, to some external field h_j applied at site j. Everything else, including interaction terms, are relegated to the first term H_0, with the only requirement that it has no dependence on $\{h\} = \{h_j$ for all $j\}$ (this notation refers to the entire set of h_j). The free energy due to this Hamiltonian will be a function of the set of fields $\{h\}$ and the temperature T. From the derivative theorem, we know

$$\frac{\partial F(\{h\}, T)}{\partial h_k} = - \langle \sigma_k \rangle (\{h\}, T), \tag{B.4.4}$$

and higher order correlation terms can be calculated by taking successively higher order derivatives, which we will look at in greater detail in Chapter 7.

The quantity $\langle \sigma_k \rangle$ is the order parameter, which reflects symmetry-broken states of the system, and is therefore usually the first quantity we are interested in. However, the functional dependence of $F(\{h\}, T)$ is often difficult to

334 APPENDIX B. WORKSHEET SOLUTIONS

calculate analytically or even numerically. This is where the variational method comes in.

The mean-field trial Hamiltonian

Let us consider a trial Hamiltonian that we can solve of the form:

$$H_{tr} = H_{tr}^0 - \sum_j b_j \sigma_j, \tag{B.4.5}$$

where $\{b\}$ are "variational parameters" to be determined. While H_{tr} looks very similar to the original Hamiltonian at first glance, we emphasize two things: first, the term $-\sum_j b_j \sigma_j$ is what leads to the commonly named "mean field," since b_j takes on the role of "field" conjugate to σ_j (parallel to b_j in H); second, we should *be able to solve* the thermodynamics (free energy, expectation values, ...) of H_{tr} in its entirety, so H_{tr}^0 is shorthand for terms in H_{tr} that are not directly responsible for the "mean fieldness," but are within our ability to calculate nonetheless. H_{tr}^0 does not contain any $\{b\}$ dependence.[9]

We can use H_{tr} to calculate F_{tr}, which also follows the derivative theorem and gives us:

$$\frac{\partial F_{tr}}{\partial b_k} = -\langle \sigma_k \rangle_{tr}. \tag{B.4.6}$$

Here we use $\langle \ \rangle_{tr}$ to mean taking the expectation value in the trial ensemble. Note that $\langle \sigma_k \rangle_{tr}$ is a function of $\{b\}$ and T. Let us give it a new name (and just to make the $\{b\}$ dependence clear):

$$\langle \sigma_k \rangle_{tr} \equiv m_k(\{b\}, T) = m_k(b_j, T). \tag{B.4.7}$$

Here and onward, whenever variables with subscripts appear as arguments to functions, the set symbol $\{\}$ is implied.

1. Show that, given the form of Eqs. B.4.3 and B.4.5, the variational free energy can always be written in the form:

$$F_{var} = F_{tr} + F_{var}^0 - \sum_j (h_j - b_j) m_j, \tag{B.4.8}$$

   where

$$F_{tr} = F_{tr}(b_j, T) \tag{B.4.9}$$

   has no explicit dependence on $\{h\}$ and

$$F_{var}^0 = F_{var}^0(m_j, T) = \langle H_0 - H_{tr}^0 \rangle_{tr} \tag{B.4.10}$$

---

9. Though it may contain other variational parameters, for the sake of brevity, here we assume it does not, but the result here generalizes to where additional parameters exist.

has no explicit dependence on $\{h\}$, and depends only on $\{b\}$ implicitly through $\{m\}$.

Let us be explicit about dependence on variables every step of the way:

$$H_{tr} = H_{tr}(b_j),$$

$$Z_{tr} = Z_{tr}(b_j, T)$$

$$\implies F_{tr} = F_{tr}(b_j, T).$$

So F_{tr} has no h_j or m_j dependence.

$$\langle H - H_{tr}\rangle_{tr} = \langle H_0 - H_{tr}^0\rangle_{tr} - \left\langle \sum_j (h_j - b_j)\sigma_j \right\rangle_{tr}$$

$$= F_{var}^0\left[\langle\sigma_j\rangle_{tr}, T\right] - \sum_j (h_j - b_j)\langle\sigma_j\rangle_{tr}$$

$$= F_{var}^0(m_j, T) - \sum_j (h_j - b_j)m_j,$$

where $m_j = m_j(b_k, T)$. F_{var}^0 is a function of $\langle\sigma_j\rangle_{tr}$ because in taking the expectation value in the trial ensemble, any σ operator present will turn into $\langle\sigma\rangle_{tr}$. However, we have explicitly kept the $\{h\}$ and $\{b\}$ dependence as a separate term.

2. We have argued previously that the best approximation to the original free energy F is obtained by solving for

$$\frac{\partial F_{var}}{\partial b_k} = 0. \tag{B.4.11}$$

Show that, given Eq. B.4.8, this leads to the self-consistency condition:

$$\sum_j \left(\frac{\partial F_{var}^0}{\partial m_j} - h_j + b_j\right) \frac{\partial m_j}{\partial b_k} = 0. \tag{B.4.12}$$

Hint 1: If a function depends on a variable implicitly, partial differentiation requires the chain rule:

$$\frac{\partial f\left[g_\alpha(s_\beta)\right]}{\partial s_k} = \sum_\gamma \frac{\partial f}{\partial g_\gamma} \frac{\partial g_\gamma}{\partial s_k}. \tag{B.4.13}$$

Hint 2: $\{h\}$ are independent variables with $\partial h_j/\partial h_k = \delta_{jk}$. Derivatives of other variables with respect to h_j are 0.

The next few calculations are partial derivative soups, so buckle up!

$$\frac{\partial F_{var}}{\partial b_k} = \frac{\partial F_{tr}}{\partial b_k} + \frac{\partial F_{var}^0}{\partial b_k} - \sum_j \left[\frac{\partial (h_j - b_j)}{\partial b_k} m_j + (h_j - b_j) \frac{\partial m_j}{\partial b_k} \right]$$

$$\text{Chain rule} \rightarrow = -m_k + \sum_j \frac{\partial F_{var}^0}{\partial m_j} \frac{\partial m_j}{\partial b_k} - \sum_j \left[0 - \delta_{jk} m_j + (h_j - b_j) \frac{\partial m_j}{\partial b_k} \right]$$

$$= -m_k + m_k + \sum_j \left(\frac{\partial F_{var}^0}{\partial m_j} - h_j + b_j \right) \frac{\partial m_j}{\partial b_k}.$$

The chain rule was used since F_{var}^0 is still implicitly a function of $\{b\}$ through $\{m\}$. Setting the derivative to zero, we get:

$$\frac{\partial F_{var}}{\partial b_k} = 0 \implies \sum_j \left(\frac{\partial F_{var}^0}{\partial m_j} - h_j + b_j \right) \frac{\partial m_j}{\partial b_k} = 0.$$

Since $\partial m_j / \partial b_k \neq 0$ in general, the condition becomes, for each j:

$$\frac{\partial F_{var}^0}{\partial m_j} - h_j + b_j = 0. \tag{B.4.14}$$

3. The condition Eq. B.4.14 means that formally there is a set of solutions $\{b_j = b_j(h_j, T)\}$ that minimizes F_{var}, so we can define an optimized variational free energy F_{mf} as:

$$F_{mf}(h_j, T) \equiv F_{var}(b_j, h_j, T)\big|_{\frac{\partial F_{var}}{\partial b_j} = 0}, \tag{B.4.15}$$

where F_{mf} is a function of only $\{h\}$ and T. Show that, subject to Eq. B.4.14:

$$\frac{\partial F_{mf}}{\partial h_k} = -m_k(h_j, T), \tag{B.4.16}$$

where

$$m_k(h_j, T) = m_k(b_j, T)\big|_{\frac{\partial F_{var}}{\partial b_j} = 0}.$$

We will begin by treating F_{mf} just like F_{var}, take the derivative with respect to h_k, and substitute in Eq. B.4.14 when appropriate. Remember now that $\{b\}$ are no longer free parameters, but constrained by Eq. B.4.14 to be functions of $\{h\}$ and T.

$$\frac{\partial F_{var}}{\partial h_k} = \frac{\partial F_{tr}}{\partial h_k} + \frac{\partial F_{var}^0}{\partial h_k} - \sum_j \left[\frac{\partial (h_j - b_j)}{\partial h_k} m_j + (h_j - b_j) \frac{\partial m_j}{\partial h_k} \right]$$

$$\text{Chain rule} \to = \sum_j \left[\frac{\partial F_{tr}}{\partial b_j} \frac{\partial b_j}{\partial h_k} + \frac{\partial F^0_{var}}{\partial m_j} \frac{\partial m_j}{\partial h_k} - \delta_{jk} m_j + \frac{\partial b_j}{\partial h_k} m_j - (h_j - b_j) \frac{\partial m_j}{\partial h_k} \right]$$

$$= \sum_j \left[(-m_j + m_j) \frac{\partial b_j}{\partial h_k} + \left(\frac{\partial F^0_{var}}{\partial m_j} - h_j + b_j \right) \frac{\partial m_j}{\partial h_k} \right] - m_k(b_j, T),$$

$$\frac{\partial F_{var}}{\partial b_j} = 0 \implies = -m_k(h_j, T),$$

where in the last step we apply Eq. (B.4.14) and see the second term in parentheses satisfy the self-consistency equation, and it is therefore 0.

A small subtlety here is that when we applied the chain rule, we picked the most convenient/natural "intermediate" variables, as we have mentioned before.

Approximating the order parameter

For Hamiltonians of the form of Eq. B.4.3, the order parameter is defined as

$$\frac{\partial F(h_j, T)}{\partial h_k} = - \langle \sigma_k \rangle (h_j, T). \tag{B.4.17}$$

We can therefore obtain an approximate order parameter $\langle \tilde{\sigma}_k \rangle$ from any approximation to the free energy $\tilde{F}$ by:

$$\frac{\partial \tilde{F}}{\partial h_k} = - \langle \tilde{\sigma}_k \rangle. \tag{B.4.18}$$

Note that this procedure is general and does not depend on how we get the approximate free energy $\tilde{F}$.

4. Using the variational procedure with a trial Hamiltonian of the form Eq. B.4.5, we find the variational approximation $\tilde{F} = F_{mf}$. Show that the approximate order parameter is:

$$\langle \tilde{\sigma}_k \rangle = \langle \sigma_k \rangle_{tr} \Big|_{\frac{\partial F_{var}}{\partial b_j} = 0} = m_k(h_j, T). \tag{B.4.19}$$

Since our optimal approximate free energy is F_{mf}, the approximate order parameter is:

$$\langle \tilde{\sigma}_k \rangle = - \frac{\partial F_{mf}}{\partial h_k} = m_k(h_j, T). \tag{B.4.20}$$

That is, the approximate order parameter *is* the expectation value in the trial ensemble $\langle \sigma_k \rangle_{tr} = m_k(h_j, T)$, evaluated at the self-consistence conditions!

This result is not apparent a priori, and is a feature of the form of Eqs. B.4.3 and B.4.5. Fortunately, physical systems frequently have Hamiltonians of the form of

Eq. B.4.3, which makes picking Eq. B.4.5 as the trial Hamiltonian a useful general method of approximation.

This particular variational approximation is called the "mean-field theory." We have previously mentioned that the parameter b_j takes on the role of field, and is what we solve for in the self-consistency equation Eq. B.4.14. Its conjugate variable (in the thermodynamic sense) $\langle \sigma_j \rangle_{tr} = m_j$ turns out to be the order parameter in this approximation. Therefore, the physical interpretation of the parameter b_j is indeed a thermodynamically averaged field, or "mean field," experienced by σ_j.

Equivalence of b_j and m_j as variational parameters

Eqs. B.4.6 and B.4.7 define a new set of functions $\{m\}$ by

$$m_k(b_j, T) \equiv -\frac{\partial F_{tr}}{\partial b_k}, \tag{B.4.21}$$

meaning that we have the explicit functional dependence of $\{m\}$ on $\{b\}$. Formally, this implies that we can invert the relationship and write:

$$b_k = b_k(m_j, T). \tag{B.4.22}$$

We can then substitute this into the expression for the variational free energy,

$$F_{var}\left[b_k(m_j, T), h_k, T\right] = F_{var}(m_k, h_k, T), \tag{B.4.23}$$

meaning that we replace all dependence on $\{b_k\}$ explicitly with $\{m_k\}$.

While we can define an arbitrary set of functions that gives a correspondence:

$$g_k(b_j, T) \iff b_k(g_j, T),$$

only the set of functions $\{m\}$ defined by Eq. B.4.21 satisfies the following specific condition that we will now prove.

5. Show that

$$\frac{\partial F_{var}}{\partial m_k} = 0 \tag{B.4.24}$$

leads to the same self-consistency equations as Eq. B.4.14.

We proceed as before, but now with knowledge of the formal inverse relations that exist:

$$m_k(b_j, T) \iff b_k(m_j, T).$$

Taking the derivative:

$$\frac{\partial F_{var}}{\partial m_k} = \frac{\partial F_{tr}}{\partial m_k} + \frac{\partial F_{var}^0}{\partial m_k} - \sum_j \left[\frac{\partial(h_j - b_j)}{\partial m_k} m_j + (h_j - b_j)\frac{\partial m_j}{\partial m_k} \right]$$

$$= \sum_j \frac{\partial F_{tr}}{\partial b_j}\frac{\partial b_j}{\partial m_k} + \frac{\partial F^0_{var}}{\partial m_k} - \sum_j \left[-m_j \frac{\partial b_j}{\partial m_k} + (h_j - b_j)\delta_{jk} \right]$$

$$= \sum_j (-m_j + m_j)\frac{\partial b_j}{\partial m_k} + \frac{\partial F^0_{var}}{\partial m_k} - h_k + b_k.$$

Setting it to zero, we get the same self-consistency equations:

$$\frac{\partial F_{var}}{\partial m_k} = 0 \implies \frac{\partial F^0_{var}}{\partial m_k} - h_k + b_k = 0. \tag{B.4.25}$$

Physical interpretation of $F_{var}(m_k, h_k, T)$

Proving that Eq. B.4.24 leads to the same self-consistency equations tells us that $\{m\}$, the variational order parameters, can be used as the variational parameters in F_{var} instead of $\{b\}$.

While mathematically equivalent to eq. B.4.11, minimizing F_{var} with respect to $\{m\}$ gives a more transparent physical interpretation: the variational free energy is minimized with respect to the *order parameter* of the system. This also gives us an intuitive phenomenological understanding[10] of phase transitions, which we will explore in Chapters 6 and 7. While historically the name "mean-field theory" stems from the $\{b\}$ formulation of the variational theory, it is applicable also to this phenomenological variant, owing to the fact that formulating F_{var} in $\{b\}$ and $\{m\}$ are mathematically equivalent.

B.4.4   Mean-Field Solution of the Ising Model

You can do this exercise before or after tackling the general features of mean-field theory. In either case, this concrete exercise gives you a physically meaningful example of variational mean-field theory in practice.

Consider the Hamiltonian of the Ising ferromagnet with field h_j applied at each site σ_j:

$$H = -J \sum_{\langle i,j \rangle} \sigma_i \sigma_j - \sum_j h_j \sigma_j. \tag{B.4.26}$$

The derivative theorem tells us that the (sitewise) magnetization is:

$$\langle \sigma_j \rangle = -\frac{\partial F}{\partial h_j}. \tag{B.4.27}$$

---

10. Meaning that it is based on observed phenomena pertaining to the system, such as the order parameter and other macroscopic observables.

Let us employ the variational method to approximate the behavior of the system, using the trial Hamiltonian:

$$H_{tr} = -\sum_j b_j \sigma_j, \tag{B.4.28}$$

where $\{b_j\}$ are the variational parameters we will use to minimize F_{var}. We know from the derivative theorem that:

$$-\frac{\partial F_{tr}}{\partial b_j} = \langle \sigma_j \rangle_{tr} \equiv m_j, \tag{B.4.29}$$

where the last step is a definition for m_j.

Calculating averages with respect to the trial ensemble

1. As a small first step, calculate Z_{tr} from H_{tr} for only two spins. Simplify your answer to a single product.

   If there are only two spins, there are total of four configurations: $++$, $+-$, $-+$, and $--$. By definition:

   $$Z_{tr} = \sum_{++,+-,-+,--} e^{\beta b_1 \sigma_1} e^{\beta b_2 \sigma_2}$$

   $$= e^{\beta b_1} e^{\beta b_2} + e^{\beta b_1} e^{-\beta b_2} + e^{-\beta b_1} e^{\beta b_2} + e^{-\beta b_1} e^{-\beta b_2}.$$

   But notice that we can factor this into:

   $$Z_{tr} = \left(e^{\beta b_1} + e^{-\beta b_1} \right)\left(e^{\beta b_2} + e^{-\beta b_2} \right)$$

   $$= 2\cosh \beta b_1 \cdot 2\cosh \beta b_2.$$

2. Now consider an arbitrary number of spins. From H_{tr}, calculate Z_{tr}, then F_{tr}, $\langle \sigma_i \rangle_{tr}$, and $\langle \sigma_i \sigma_j \rangle_{tr}$. Show that $\langle \sigma_i \rangle_{tr} = \tanh \left(\beta b_i \right)$ and $\langle \sigma_i \sigma_j \rangle_{tr} = \langle \sigma_i \rangle_{tr} \langle \sigma_j \rangle_{tr}$ for $i \neq j$. *Hint:* The σs are independent in H_{tr}, so you can manipulate the sums and products easily. The previous part should come in handy.

   First let's calculate Z_{tr}:

   $$Z_{tr} = \sum_{\{\sigma\}} e^{-\beta H_{tr}} = \sum_{\{\sigma\}} e^{\beta \sum_j b_j \sigma_j} \tag{B.4.30}$$

   $$= \sum_{\{\sigma\}} \prod_j e^{\beta b_j \sigma_j}. \tag{B.4.31}$$

   Because the terms in the product are independent of each other, we can write Z_{tr} instead as a product of the sums of the terms. The previous part should be enough to convince you of this:

$$Z_{tr} = \prod_j \sum_{\sigma_j = \pm 1} e^{\beta b_j \sigma_j} \tag{B.4.32}$$

$$= \prod_j 2 \cosh(\beta b_j). \tag{B.4.33}$$

From Z_{tr}, we can calculate F_{tr}:

$$F_{tr} = -\frac{1}{\beta} \ln(Z_{tr}) \tag{B.4.34}$$

$$= -\frac{1}{\beta} \sum_j \ln \left[2 \cosh(\beta b_j) \right]. \tag{B.4.35}$$

For $\langle \sigma_i \rangle_{tr}$, we can calculate it the long way, as we did for Z_{tr}:

$$Z_{tr} \langle \sigma_i \rangle_{tr} = \sum_{\{\sigma\}} \sigma_i e^{-\beta H_{tr}} \tag{B.4.36}$$

$$= \sum_{\{\sigma\}} \sigma_i \prod_j e^{\beta b_j \sigma_j} \tag{B.4.37}$$

We can still appropriately reorder the sums and products; we just need to give special treatment to the term involving σ_i:

$$Z_{tr} \langle \sigma_i \rangle_{tr} = \left(\prod_{j \neq i} \sum_{\sigma_j = \pm 1} e^{\beta b_j \sigma_j} \right) \left(\sum_{\sigma_i = \pm 1} \sigma_i e^{\beta b_i \sigma_i} \right) \tag{B.4.38}$$

$$= \left(\prod_{j \neq i} 2 \cosh(\beta b_j) \right) 2 \sinh(\beta b_i), \tag{B.4.39}$$

$$\langle \sigma_i \rangle_{tr} = \tanh(\beta b_i). \tag{B.4.40}$$

Alternatively, knowing that each spin is independent in the trial Hamiltonian, we know that the thermal expectation of any spin can be calculated disregarding the rest of the spins, so:

$$\langle \sigma_i \rangle_{tr} = \frac{\sum_{\sigma_i = \pm 1} \sigma_i e^{\beta b_i \sigma_i}}{\sum_{\sigma_i = \pm 1} e^{\beta b_i \sigma_i}} \tag{B.4.41}$$

$$= \tanh(\beta b_i). \tag{B.4.42}$$

But of course the most convenient way is to make use of the derivative theorem:

$$\langle \sigma_i \rangle_{tr} = -\frac{\partial F_{tr}}{\partial b_i}. \tag{B.4.43}$$

For $\langle\sigma_i\sigma_j\rangle_{tr}$, we can once again calculate it the long way with all the spins, following steps almost identical to those above:

$$Z_{tr}\langle\sigma_i\sigma_j\rangle_{tr} = \sum_{\{\sigma\}}\sigma_i\sigma_j e^{-\beta H_{tr}}$$

$$= \sum_{\{\sigma\}}\sigma_i\sigma_j \prod_k e^{\beta b_k \sigma_k}$$

$$= \left(\prod_{k\neq i,j}\sum_{\sigma_k=\pm1} e^{\beta b_k \sigma_k}\right)\left(\sum_{\sigma_i=\pm1}\sigma_i e^{\beta b_i \sigma_i}\right)\left(\sum_{\sigma_j=\pm1}\sigma_j e^{\beta b_j \sigma_j}\right)$$

$$= \left(\prod_{k\neq i,j}2\cosh(\beta b_k)\right)4\sinh(\beta b_i)\sinh(\beta b_j),$$

$$\langle\sigma_i\sigma_j\rangle_{tr} = \tanh(\beta b_i)\tanh(\beta b_j).$$

But knowing each spin is independent, we can invoke the fact that the expectation value of the product of two independent random variables equals the product of their individual expectation values; thus:

$$\langle\sigma_i\sigma_j\rangle_{tr} = \langle\sigma_i\rangle_{tr}\langle\sigma_j\rangle_{tr} \tag{B.4.44}$$

$$= \tanh(\beta b_i)\tanh(\beta b_j). \tag{B.4.45}$$

3. Using these results, calculate $F_{var} = F_{tr} + \langle H - H_{tr}\rangle_{tr}$. It is not exactly an elegant expression, but will be useful soon. *Hint:* The notation $\langle\ \rangle_{tr}$ means, calculate expectation values with the trial ensemble, which means we can apply results of expectations we have already calculated above.

$$F_{var} = F_{tr} + \left\langle-J\sum_{<i,j>}\sigma_i\sigma_j - \sum_j(h_j - b_j)\sigma_j\right\rangle_{tr}$$

$$= F_{tr} - J\sum_{<i,j>}\langle\sigma_i\rangle_{tr}\langle\sigma_j\rangle_{tr} - \sum_j(h_j - b_j)\langle\sigma_j\rangle_{tr}$$

$$= -\frac{1}{\beta}\sum_j\ln\left[2\cosh(\beta b_j)\right] - J\sum_{<i,j>}\tanh(\beta b_i)\tanh(\beta b_j)$$

$$- \sum_j(h_j - b_j)\tanh(\beta b_j).$$

Derive the mean-field equations

To minimize the variational free energy F_{var}, we compute the derivative of F_{var} with respect to each b_k (we will use a different dummy index to not interfere with the existing dummy indices), and then set this derivative equal to zero:

$$\frac{\partial F_{var}}{\partial b_k} = 0. \tag{B.4.46}$$

1. Show that this gives the result:

$$b_j = h_j + J \sum_{i \text{ n.n. of } j} \tanh(\beta b_i). \tag{B.4.47}$$

We can use the above result for F_{var} and calculate derivatives w.r.t. b_j explicitly. To make life easier, we can leave the terms in the form of partial derivatives at first. Note that $\frac{\partial \langle \sigma_i \rangle_{tr}}{\partial b_j} = 0$ for $i \neq j$:

$$F_{var} = F_{tr} - J \sum_{\langle i,j \rangle}^{n.n.} \langle \sigma_i \rangle_{tr} \langle \sigma_j \rangle_{tr} - \sum_j (h_j - b_j) \langle \sigma_j \rangle_{tr},$$

$$\frac{\partial F_{var}}{\partial b_k} = -\langle \sigma_k \rangle_{tr} - J \sum_{\langle i,j \rangle} \left(\langle \sigma_i \rangle_{tr} \frac{\partial \langle \sigma_j \rangle_{tr}}{\partial b_k} + \frac{\partial \langle \sigma_i \rangle_{tr}}{\partial b_k} \langle \sigma_j \rangle_{tr} \right)$$

$$- \sum_j \left[-\delta_{jk} \langle \sigma_j \rangle_{tr} + (h_j - b_j) \frac{\partial \langle \sigma_j \rangle_{tr}}{\partial b_k} \right].$$

The second term requires some counting gymnastics, but you should be able to convince yourself with small test cases in 1-D/2-D that:

$$-J \sum_{\langle i,j \rangle} \left(\langle \sigma_i \rangle_{tr} \frac{\partial \langle \sigma_j \rangle_{tr}}{\partial b_k} + \frac{\partial \langle \sigma_i \rangle_{tr}}{\partial b_k} \langle \sigma_j \rangle_{tr} \right) = -J \sum_j \frac{\partial \langle \sigma_j \rangle_{tr}}{\partial b_k} \sum_{i \text{ n.n. of } j} \langle \sigma_i \rangle_{tr}$$
$$\tag{B.4.48}$$

The summation in the third term produces an overall $+\langle \sigma_k \rangle_{tr}$ term, which cancels out the first term, leaving us with:

$$\frac{\partial F_{var}}{\partial b_k} = \sum_j \frac{\partial \langle \sigma_j \rangle_{tr}}{\partial b_k} \left[-(h_j - b_j) - J \sum_{i \text{ n.n. of } j} \langle \sigma_i \rangle_{tr} \right].$$

Setting $\frac{\partial F_{var}}{\partial b_k} = 0$, we get:

$$b_j = h + J \sum_{i \text{ n.n. of } j} \langle \sigma_i \rangle_{tr} \tag{B.4.49}$$

$$= h + J \sum_{i \text{ n.n. of } j} \tanh(\beta b_i), \tag{B.4.50}$$

since $\partial \langle \sigma_j \rangle_{tr} / \partial b_k \neq 0$ in general.

2. Now, show that:

$$\frac{\partial F_{var}}{\partial h_k} = -m_k. \tag{B.4.51}$$

Hint: We know $\partial F_{tr}/\partial b_k = -\langle \sigma_k \rangle_{tr}$; what's $\partial F_{tr}/\partial h_k$? Use the chain rule for partial differentiation.

Since F_{tr} is implicitly a function of $\{h\}$ through $\{b\}$, we need to use the chain rule when calculating $\partial F_{tr}/\partial h_k$:

$$\frac{\partial F_{tr}}{\partial h_k} = \sum_j \frac{\partial F_{tr}}{\partial b_j} \frac{\partial b_j}{\partial h_k} \tag{B.4.52}$$

$$= -\sum_j \frac{\partial b_j}{\partial h_k} \langle \sigma_j \rangle_{tr}. \tag{B.4.53}$$

The other terms are also functions of $\{h\}$, so

$$\frac{\partial}{\partial h_k} \left(-J \sum_{\langle i,j \rangle} \langle \sigma_i \rangle_{tr} \langle \sigma_j \rangle_{tr} \right) = -J \sum_{\langle i,j \rangle} \left(\frac{\partial \langle \sigma_i \rangle_{tr}}{\partial h_k} \langle \sigma_j \rangle_{tr} + \langle \sigma_i \rangle_{tr} \frac{\partial \langle \sigma_j \rangle_{tr}}{\partial h_k} \right) \tag{B.4.54}$$

$$= -J \sum_j \left(\frac{\partial \langle \sigma_j \rangle_{tr}}{\partial h_k} \sum_{i=\text{n.n. of } j} \langle \sigma_i \rangle_{tr} \right) \tag{B.4.55}$$

and

$$-\frac{\partial}{\partial h_k} \left(\sum_j (h_j - b_j) \langle \sigma_j \rangle_{tr} \right) = -\langle \sigma_k \rangle_{tr} + \sum_j \frac{\partial b_j}{\partial h_k} \langle \sigma_j \rangle_{tr} - \sum_j (h_j - b_j) \frac{\partial \langle \sigma_j \rangle_{tr}}{\partial h_k} \tag{B.4.56}$$

Putting all the terms together, we see that the second term in Eq. B.4.56 cancels out Eq. B.4.53, and the third term in Eq. B.4.56 combined with Eq. B.4.55 satisfies Eq. B.4.47 for each j, and therefore becomes 0. That leaves us with:

$$\frac{\partial F_{var}}{\partial h_k} = -\langle \sigma_k \rangle_{tr} = -m_k. \tag{B.4.57}$$

For a more elegant solution, we can write:

$$\frac{\partial F_{var}}{\partial h_k} = \frac{\partial F_{var}}{\partial h_k} + \sum_j \frac{\partial F_{var}}{\partial b_j} \frac{\partial b_j}{\partial h_k}, \tag{B.4.58}$$

where the first term is for explicit $\{h\}$ dependence in F_{var}. Since F_{var} is minimized w.r.t. the parameters $\{b\}$, we know that $\partial F_{var}/\partial b_j = 0$. The only term with an explicit $\{h\}$ dependence in F_{var} is $\sum_j -h_j \langle \sigma_j \rangle_{tr}$, which leads to:

$$\frac{\partial F_{var}}{\partial h_k} = -\langle \sigma_k \rangle_{tr}. \tag{B.4.59}$$

Consequences of the mean-field solution with uniform h

Let us briefly examine the behavior of the system for an uniform applied field, *i.e.* $h_j = h$ for all j. The self-consistency equation Eq. B.4.47 then becomes:

$$b_j = h + J \sum_{i \,\text{n.n. of}\, j} \tanh(\beta b_i) \qquad (B.4.60)$$

3. Explain how we could motivate the solution of Eq. B.4.60 to be translationally invariant, i.e., $b_j = b$ for all j.

   Hint: What can you say about minimizing the ground-state $(T = 0)$ variational free energy?

   If we set the temperature to zero, i.e., $\beta \to \infty$, then Eq. B.4.60 gives (assuming that $b_j \neq 0$):

   $$\lim_{\beta \to \infty} b_j = h + zJ, \qquad (B.4.61)$$

   where z is the number of nearest neighbors. Then the variational parameters that optimize F_{var} are translationally invariant at $T = 0$. This motivates us to look for a set of solutions where each b_j is identical for the general temperature-dependent case.

   Once we have assumed the solution is translationally invariant, we will see that there are two phases: a disordered phase at $T > T_c$ in which $\langle \sigma_j \rangle \to 0$ as $h \to 0$ and an ordered phase at $T < T_c$ in which $\langle \sigma_j \rangle \to \text{sign}(h) m(T)$ with $m(T) > 0$ as $h \to 0$.

4. From Eq. B.4.60, find T_c. *Hint:* Make a sketch of the graphs of b and $\tanh(\beta b)$; what happens when you increase β? Set $h = 0$.

   Setting $h = 0$, Eq. B.4.60 becomes:

   $$b = zJ \tanh(\beta b). \qquad (B.4.62)$$

   We can see that there's always a solution at $b = 0$. Are there other solutions? Making a plot of the L.H.S. and R.H.S, we can see that as we increase β, two more solutions can occur. The critical temperature occurs when the derivatives w.r.t. b are equal at $b = 0$ for both sides of the equation:

   $$1 = zJ \left. \frac{\partial \tanh(\beta_c b)}{\partial b} \right|_{b=0}, \qquad (B.4.63)$$

   $$1 = zJ\beta_c, \qquad (B.4.64)$$

   $$T_c = zJ. \qquad (B.4.65)$$

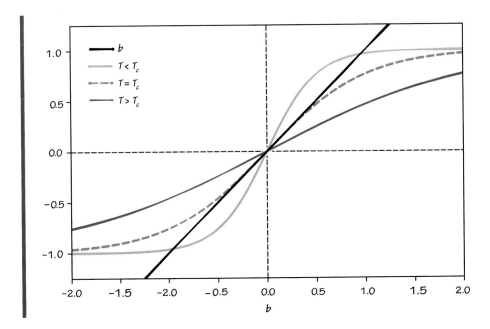

5. From Eq. B.4.60 show that $m(T) = \tanh[\beta(zJm(T) + h)]$. Remember that $m = \langle\sigma\rangle_{tr}$ when the variational parameters minimize F_{var}.

$$m(T) = \langle\sigma\rangle_{tr}\Big|_{\frac{\partial F_{var}}{\partial b} = 0} \qquad (B.4.66)$$

$$= \tanh(\beta b), \qquad (B.4.67)$$

$$\text{Eq. B.4.60} \implies b = h + zJm(T), \qquad (B.4.68)$$

$$m(T) = \tanh(\beta b) \qquad (B.4.69)$$

$$= \tanh[\beta(zJm(T) + h)]. \qquad (B.4.70)$$

6. We have argued above that for $T > T_c$, $m = 0$. For T near T_c (i.e., $T_c \gg (T_c - T) > 0$), m is very small. Find an expression for $m(T)$ near T_c. Again set $h = 0$.

Let's take $h = 0$ again. Near T_c, m is very small, so we can make the following expansion:

$$m = \tanh(\beta zJm) \qquad (B.4.71)$$

$$= \beta zJm - \frac{(\beta zJm)^3}{3} + \ldots. \qquad (B.4.72)$$

Solving for m, we get (from $T_c = zJ$ above):

$$m = \pm \frac{\sqrt{3}T}{T_c^{\frac{3}{2}}} (T_c - T)^{\frac{1}{2}}. \tag{B.4.73}$$

Notice the power-law dependence of $(T_c - T)^{\frac{1}{2}}$ around T_c. The value $\frac{1}{2}$ is a special critical exponent, which we will come back to later in the course.

B.4.5 Low-Temperature Analysis of Mean-Field Equations

Even though we can't solve the mean-field equations analytically, we can still get approximate expressions in certain limiting cases. This is known as asymptotic analysis, since the results will not be exact but "asymptotic." You'll get a chance to think carefully about what that means. For now, let's take a look at the low-temperature limit of the equation

$$\tilde{b} = \beta z J \tanh \tilde{b} \tag{B.4.74}$$

where $\tilde{b} \equiv \beta b$.

1. How does $\tilde{b}$ behave as $T \to 0$? *Hint:* Remember that $-1 \le \tanh x \le 1$ for all x.

   The $T \to 0$ limit corresponds to $\beta \to \infty$. Since the tanh on the right-hand side of Eq. B.4.74 is always finite, this means that the entire right-hand side goes to infinity as $T \to 0$. Therefore, $\tilde{b}$ gets very large as $T \to 0$.

2. With this in mind, write down the first few terms in a series approximation of Eq. B.4.74. Remember that $\tanh(x) = (1 - e^{-2x})/(1 + e^{-2x})$. *Hint:* What is the appropriate small parameter to expand in?

   Substituting in the expression for $\tanh(x)$, we have

   $$\tilde{b} = \beta z J \frac{1 - e^{-2\tilde{b}}}{1 + e^{-2\tilde{b}}}.$$

   Be careful here! Since $\tilde{b} \gg 1$, this means that $u = e^{-2\tilde{b}}$ is very small. This is the parameter that we should be expanding in! Using the standard expansion $1/(1 + u) = 1 - u + u^2 + \ldots$, we have

   $$\tilde{b} = \beta z J \left(1 - e^{-2\tilde{b}}\right) \frac{1}{1 + e^{-2\tilde{b}}}$$
   $$= \beta z J \left(1 - e^{-2\tilde{b}}\right) \left[1 - e^{-2\tilde{b}} + e^{-4\tilde{b}} + \ldots\right]$$
   $$= \beta z J \left[1 - 2e^{-2\tilde{b}} + 2e^{-4\tilde{b}} + \ldots\right].$$

3. Write down an expression for $\tilde{b}$ in terms of T and $T_c = zJ$. You only need the leading-order term of $\tilde{b}$ for this and the rest of this worksheet.

Keeping just the first term in the series gives $\tilde{b} = \beta zJ[1 + \ldots]$ or $\tilde{b} \simeq \beta zJ = T_c/T$.

4. Using your answer from (3), come up with a series expression for $m = \pm \tanh(\tilde{b})$. (The $\pm$ reflects the two possible states of broken symmetry.) Write down the dominant term and the leading-order correction. What is the size of the next-order correction?

Substituting in $\tilde{b} = T_c/T$ gives us $m = \tanh(T_c/T)$. We use the same expansion as previously:

$$m = \pm \tanh(T_c/T)$$

$$= \pm \left(1 - e^{-2T_c/T}\right) \frac{1}{1 + e^{-2T_c/T}}$$

$$= \pm \left(1 - e^{-2T_c/T}\right) \left(1 - e^{-2T_c/T} + e^{-4T_c/T}\right)$$

$$= \pm \left[1 - 2e^{-2T_c/T} + 2e^{-4T_c/T} - \ldots\right]$$

$$= \pm \left[1 - 2e^{-2T_c/T} + \mathcal{O}(e^{-4T_c/T})\right].$$

5. Which term(s) are important as $T \to 0$?

The first term is dominant, and the second term is subdominant because $e^{-2T_c/t}$ is small as $T \to 0$. The third term is even smaller.

6. What happens to the relative importance of the higher order corrections in your expression as $T \to 0$?

The relative importance of the corrections is the ratio between second and third terms, which is on the order of $e^{-4T_c/T}/e^{-2T_c/T} = e^{-2T_c/T}$. This goes to zero very quickly as $T \to 0$. Thus, even the corrections become unimportant at low enough T!

It is intriguing to relate your result with what we got in the low T expansions for m from the previous chapter:

7. At precisely $T = 0$, what is the ground state? Compare this to the dominant term in m.

The ground-state is $m = \pm 1$—all the spins pointing up, or all pointing down.

8. At low nonzero T, what are the lowest energy excitations? How much of an energy cost are they? What is the corresponding Boltzmann factor? Compare this to the 1st-order term in your expression for m.

The lowest energy excitations are the flips of a single spin. Since each spin has z neighbors, and since the energy of each bond goes from $-J$ to $+J$, this is a total energy change of $2zJ$. The corresponding Boltzmann factor is $e^{-2zJ/T} = e^{-2T_c/T}$, which is exactly the leading-order correction in our expression for m! Furthermore, the factor of 2 can be explained by noting that a spin-flip from $+1$ to -1 causes the total magnetization to change by $1 - (-1) = 2$.

It is a triumph for mean-field theory that it agrees with the m computed in a low-T expansion.

B.4.6   The Variational Free Energy F_{var} as a Function of the Order Parameter

We are now going to express the variational free energy (F_{var}) as a function of $\{m\}$ instead of $\{b\}$ by using the definition:

$$-\frac{\partial F_{tr}}{\partial b_j} = m_j = \tanh(\beta b_j). \tag{B.4.75}$$

This allows us to use the order parameters $\{m\}$ as the variational parameters. Now is a good time to do W4.3 if you have not yet done so.

1. Derive the following expression:

$$F = \sum_j V(m_j) + \frac{J}{2} \sum_{<i,j>} (m_i - m_j)^2, \tag{B.4.76}$$

where

$$V(m_j) = \frac{T}{2} \ln\left[\frac{1 - m_j^2}{4}\right] + m_j T \tanh^{-1}(m_j) - \frac{zJ}{2} m_j^2 - h_j m_j. \tag{B.4.77}$$

Hint: The following identity may come in handy:

$$\cosh^2 x - \sinh^2 x = 1 \implies 1 - \tanh^2 x = 1/\cosh^2 x. \tag{B.4.78}$$

If you can't get the answer going forward from F_{var}, try also working backward.

From before, we know that:

$$F_{var} = \sum_j -T \ln[2\cosh(\beta b_j)] - J \sum_{<i,j>} m_i m_j + \sum_j (b_j - h_j) m_j, \tag{B.4.79}$$

where

$$m_j = \tanh(\beta b_j) \tag{B.4.80}$$

$$\implies b_j = T\tanh^{-1}(m_j) \tag{B.4.81}$$

From the hyperbolic trig identity:

$$-T\ln[2\cosh(\beta b_j)] = T\ln\left(\frac{1}{2\cosh(\beta b_j)}\right) \tag{B.4.82}$$

$$= \frac{T}{2}\ln\left(\frac{1}{4\cosh^2(\beta b_j)}\right) \tag{B.4.83}$$

$$= \frac{T}{2}\ln\left(\frac{1-\tanh^2(\beta b_j)}{4}\right) \tag{B.4.84}$$

$$= \frac{T}{2}\ln\left(\frac{1-m_j^2}{4}\right). \tag{B.4.85}$$

For the second term in Eq. B.4.79, observe that:

$$\sum_{<i,j>}(m_i - m_j)^2 = -2\sum_{<i,j>}m_i m_j + z\sum_j m_j^2 \tag{B.4.86}$$

$$\implies -\sum_{<i,j>}m_i m_j = \frac{1}{2}\sum_{<i,j>}(m_i - m_j)^2 - \frac{z}{2}\sum_j m_j^2. \tag{B.4.87}$$

Putting this all together, we get:

$$F_{var} = \sum_j\left[\frac{T}{2}\ln\left(\frac{1-m_j^2}{4}\right) + m_j T\tanh^{-1}(m_j) - \frac{zJ}{2}m_j^2 - h_j m_j\right] \tag{B.4.88}$$

$$+ \frac{J}{2}\sum_{<i,j>}(m_i - m_j)^2. \tag{B.4.89}$$

2. As long as $|m_j|$ are all small, we can expand this in powers of m_j. Using the identity

$$\tanh^{-1}(x) = \frac{1}{2}\ln\left[\frac{1+x}{1-x}\right], \tag{B.4.90}$$

show that

$$V(m_j) = V_0 - h_j m_j + \frac{\alpha}{2}m_j^2 + \frac{u}{4}m_j^4 + \dots, \tag{B.4.91}$$

where

$$V_0 = -T\ln(2), \tag{B.4.92}$$

$$\alpha = T - zJ = T - T_c, \tag{B.4.93}$$

$$u = \frac{T}{3}. \tag{B.4.94}$$

The one expansion we'll need is

$$\ln(1 + x) = x - \frac{x^2}{2} + \frac{x^3}{3} + \mathcal{O}(x^4). \tag{B.4.95}$$

So, keeping terms up to m_j^4:

$$\frac{T}{2} \ln\left[\frac{1 - m_j^2}{4}\right] = \frac{T}{2}\left[\ln(1 - m_j^2) - 2\ln 2\right] \tag{B.4.96}$$

$$\approx -T\ln 2 - \frac{T}{2}\left(m_j^2 + \frac{m_j^4}{2}\right). \tag{B.4.97}$$

And also:

$$m_j T \tanh^{-1}(m_j) = \frac{m_j T}{2} \ln\left[\frac{1 + m_j}{1 - m_j}\right] \tag{B.4.98}$$

$$\approx \frac{m_j T}{2}\left[m_j - \frac{m_j^2}{2} + \frac{m_j^3}{3} - \left(-m_j - \frac{m_j^2}{2} - \frac{m_j^3}{3}\right)\right] \tag{B.4.99}$$

$$= m_j T\left(m_j + \frac{m_j^3}{3}\right) = T\left(m_j^2 + \frac{m_j^4}{3}\right). \tag{B.4.100}$$

Putting it all together:

$$V(m_j) = -T\ln 2 - h_j m_j + \frac{T - zJ}{2} m_j^2 + \frac{T}{12} m_j^4. \tag{B.4.101}$$

We can see a lot from Eqs. B.4.76 and B.4.77. Firstly, any spatially non-uniform solution of m_j will always have higher free energy than a uniform solution so long as $J > 0$. Secondly, for $T > T_c$ there is at least a local minimum of the free energy with all $m_j = 0$ (and is in fact a global minimum). For $T < T_c$, we can see that $m_j = 0$ is a local maximum of the free energy. If we use the expansion Eq. B.4.91, we can determine that there are two minimum free energy solutions by solving $\partial \tilde{V}/\partial m_j = 0$:

$$m_j = m = \pm\sqrt{|\alpha|/u} \sim (T_c - T)^{1/2}. \tag{B.4.102}$$

This is approximate, but it is valid as long as $|m|$ is small, which is to say $|\alpha|$ is small, which in turn is to say $(T_c - T) \ll u$. However, the symmetry of the solution is exact; since it contains only even powers of m, V is an even function of m, so if there is a minimum of the free energy with $m_j = m$, there is an equally good minimum of the free energy for $m_j = -m$.

B.4.7 The Mean-Field Critical Exponents β, γ, δ

From the previous part, we have:

$$F_{var} = V_0 - hm + \frac{\alpha}{2}m^2 + \frac{u}{4}m^4 + \ldots, \qquad (B.4.103)$$

where

$$V_0 = -T \ln 2,$$

$$\alpha = T - zJ$$

$$= T - T_c,$$

$$u = \frac{T}{3}.$$

Since we have now proven that the value of the order parameter m minimizes the free energy, the phase of the system is determined by the equation:

$$\frac{\partial F_{var}}{\partial m} = 0$$

$$\implies -h + \alpha m + um^3 = 0. \qquad (B.4.104)$$

We ignore higher order terms in m since the first terms necessarily dominates the behavior of the system.

Critical exponent β

The first critical exponent concerns the power-law behavior of the order parameter m near the critical temperature. We are looking for the form $m \propto (T_c - T)^\beta$. This case requires that $h = 0$, since any nonzero h leads to a symmetry-broken state at any given temperature. Solve Eq. B.4.104 and show that $\beta = 1/2$.

Eq. B.4.104 solves to:

$$m = \pm\sqrt{\frac{-\alpha}{u}}. \qquad (B.4.105)$$

Since $\alpha = T - T_c$, a real solution only exists for $T \leq T_c$, leading to $m \propto \sqrt{T_c - T}$ with $\beta = 1/2$.

Critical exponent γ

The next critical exponent concerns the susceptibility of the order parameter, evaluated at zero applied field:

$$\chi = \frac{\partial m}{\partial h}\bigg|_{h=0} \propto |T_c - T|^{-\gamma}. \qquad (B.4.106)$$

Solve for χ from Eq. B.4.104 and show that $\gamma = 1$. *Hint:* Remember implicit differentiation?

We can calculate the susceptibility from Eq. B.4.104 with implicit differentiation:

$$\frac{\partial}{\partial h}\left[-h + \alpha m + um^3\right] = 0,$$

$$-1 + \alpha\frac{\partial m}{\partial h} + 3um^2\frac{\partial m}{\partial h} = 0, \tag{B.4.107}$$

$$\frac{\partial m}{\partial h} = \frac{1}{\alpha + 3um^2},$$

Above T_c, $m = 0$ and thus $\chi \propto (T - T_c)^{-1}$. Below T_c, $m^2 = -\alpha/u$, thus $\chi = \frac{1}{2(T_c-T)}$. In either case, $\gamma = 1$.

Critical exponent δ

The third critical exponent we will study for now is the field dependence of the order parameter on the phase boundary, i.e., $T = T_c$. We are looking for the form $m \propto h^{1/\delta}$. Again, from Eq. B.4.104, show that $\delta = 3$.

In this case, $\alpha = 0$, thus Eq. (B.4.104) becomes:

$$-h + um^3 = 0 \implies m = \left(\frac{h}{u}\right)^{1/3}, \tag{B.4.108}$$

and thus $\delta = 3$.

B.5   Chapter 5 Worksheet Solutions

B.5.1   Liquid Crystals and Broken Symmetries

Liquid crystals are fascinating phases with properties (and symmetries!) between those of a liquid and a solid. Many liquid crystals are made of molecules with a rigid, rod-like structure. Because of the long shape, these molecules can have a tendency to point along the same axis in space. Here is an example. It is called N-(heptoxy-benzylidene)-p-n-pentylaniline.

All liquid crystal phases have orientational order. Some liquid crystal phases also exhibit positional order (regularities in the positions of molecules). This worksheet illustrates how orientational and positional orders are related to rotational and translational symmetries.

phase	symmetry			
	arbitrary rotations around the x-axis	arbitrary rotations around y- or z-axis	arbitrary translations along the x-axis	arbitrary translations along y- or z-axis
liquid				
solid				
nematic				
smectic A				
smectic C				

phase	symmetry			
	arbitrary rotations around the x-axis	arbitrary rotations around y- or z-axis	arbitrary translations along the x-axis	arbitrary translations along y- or z-axis
liquid	✓	✓	✓	✓
solid	×	×	×	×
nematic	✓	×	✓	✓
smectic A	✓	×	×	✓
smectic C	×	×	×	✓

1. In the liquid phase, the position and orientation of each molecule is *random*. Fill out the first row of the table with a ✓ if the symmetry is preserved, otherwise put a × if the symmetry is broken in any way.

   Since the molecules in a liquid are randomly positioned and randomly oriented, rotations will have no effect on any observable property. Similarly, the liquid phase is invariant to translations in any direction. Thus, all of the symmetries listed here are preserved: ✓.

2. In the (crystalline) solid phase, molecules are arranged with a defined orientation in a regular 3D lattice. Fill out the second row of the table.

For concreteness, assume a cubic lattice, with axes oriented along x, y, and z. Arbitrary rotations around the x-axis are no longer symmetries, because a rotation by $45°$ will transform the cube into a different position. The rotational symmetries around the y and z axes are also broken.

Since there is a regular, repeating structure along each of the x, y, and z directions, the translational symmetry is broken along each of these directions. (For instance, translation by half of a period leads to a different repeating pattern.) Therefore, the continuous translational symmetries are broken: ×.

(Notice that, however, the symmetries are not *completely* broken; there still remain certain rotations and translations which preserve the lattice structure of the crystal. For instance, if it were a cubic crystal, a rotation by 90 degrees or 180 degrees about one of the axes would still be a symmetry operation.)

3. In the nematic phase, the molecules are oriented so that the long axis tends to point in a shared direction (known as the director), but they are otherwise unordered. Their motion is unhindered, as in a liquid. Fill out the third row. (Define $\hat{x}$ as the orientation of the director.)

Now there is anisotropy because the molecules are preferentially oriented with the long axis in the $\hat{x}$ direction. Let us think about what this means from a symmetry perspective. An arbitrary rotation about the x-axis is still a valid symmetry, because the orientation of the director is maintained. (There is no ordering of any sort along the y or z direction.) However, an arbitrary rotation about the y- or z-axis is *not* a symmetry, because such a rotation would change the orientation of the director. The orientational order thus (partially) breaks the rotational symmetry.

The nematic phase has no positional order, so it maintains the translational symmetry in all directions.

A smectic liquid crystal phase maintains the orientational order of a nematic phase, but also exhibits positional ordering. Specifically, the molecules in a smectic phase tend to arrange themselves into regularly spaced layers of planes. It is like a liquid because separate smectic planes can freely slide over one another.

4. In the smectic A phase, the molecules within a smectic plane form a two-dimensional fluid, with random positions. The director is perpendicular to the smectic plane. Fill out the fourth row.

Now the picture is of long rod-shaped molecules oriented along the x-direction, and furthermore separated into $x − y$ planes in a regular, repeating spatial pattern. As with the nematic, this phase is symmetric to rotations about the x-axis, as these rotations do not affect the orientations of the director or of the smectic planes. However, the smectic A phase is not symmetric with respect to rotations around the y- or z-axis, as these rotations *do* have an effect on these order parameters.

The difference compared to the nematic is that a smectic phase has positional ordering—specifically, the center of mass positions of the molecules has a regular, repeating pattern in the x direction. This breaks the translational symmetry in the x direction. (For instance, a translation along the x direction by half of a period changes the locations of the smectic planes.) However, because the molecules are unordered *within* the $y - z$ plane, the smectic A phase is still symmetric with respect to arbitrary translations in the y and z directions.

(Note that the formation of smectic planes by itself is sufficient to break the rotational symmetry!)

5. The smectic C phase is identical to the smectic A phase, except that the director is tilted away from the smectic plane normal. Fill out the fifth row. (Define the smectic plane as the $y - z$ plane here; the director is *not* in the $\hat{x}$ direction.)

Now the smectic plane lies in the $y - z$ plane, but the molecules are oriented so that they are not perpendicular to the plane. (They do, however, tend to orient in the same direction!) Because of this, the rotational symmetry is further broken. Rotations about the x direction alter the orientation of the director: this is the difference compared to the smectic A phase. Rotations about the y or z direction alter the orientation of the smectic planes. Thus, the symmetry of arbitrary rotations in the x, y, and z directions are all broken. (Are there any rotational symmetries which remain?)

The positional ordering is the same as the that of smectic A phase; there is ordering in the x direction but not in the y or z direction.

6. Depending on the temperature, N-(heptoxy-benzylidene)-p-n-pentylaniline can exhibit a whole bunch of different phases, including two additional phases called smectic H and smectic B!

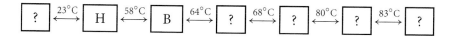

Five of the boxes are marked "?"—these correspond to the five phases in the table you just filled out. Based on the symmetries, can you figure out which phase belongs in which box?

Based on what we learned in this chapter, the high-temperature phase should respect all the symmetries. As the temperature is reduced, we expect more and more symmetries to be broken. Thus, the five phases in the table should be arranged in an order based on how many symmetries are broken, with the phases breaking the most symmetries at the lowest temperature. From highest to lowest temperature, these are: liquid, nematic, smectic A, smectic C, and solid. (An important point here is that these phases have a strict ordering: as the temperature is reduced, each subsequent phase breaks additional symmetries on top of the already broken symmetries.)

Hopefully this illustrates why symmetry is such a powerful concept in classifying the phases of matter!

$$\boxed{\text{solid}} \overset{23^\circ\text{C}}{\longleftrightarrow} \boxed{\text{smectic H}} \overset{58^\circ\text{C}}{\longleftrightarrow} \boxed{\text{smectic B}} \overset{64^\circ\text{C}}{\longleftrightarrow} \boxed{\text{smectic C}} \ldots$$

$$\ldots \overset{68^\circ\text{C}}{\longleftrightarrow} \boxed{\text{smectic A}} \overset{80^\circ\text{C}}{\longleftrightarrow} \boxed{\text{nematic}} \overset{83^\circ\text{C}}{\longleftrightarrow} \boxed{\text{liquid}}$$

B.5.2 Interpolating between the XY and Clock Models

As we have stressed in the text, the symmetry of a model is of supreme importance, often more so than the exact form of the Hamiltonian. In this worksheet we will think about the symmetries of the Hamiltonian

$$H = -J \sum_{\langle ij \rangle} \cos(\theta_i - \theta_j) - h \sum_j \cos(q\theta_j), \tag{B.5.1}$$

where q is an integer, and the spins on each site are an angle $0 \le \theta_j < 2\pi$.

1. For $h = 0$, what is this model equivalent to?

   For $h = 0$ the Hamiltonian is

   $$H = -J \sum_{\langle ij \rangle} \cos(\theta_i - \theta_j),$$

   which is the XY model (Eq. 5.12)

2. Consider a symmetry transformation of the form $\theta_j \to \theta_j + \Delta\theta \;\; \forall j$. If $h = 0$, what is the set of $\Delta\theta$ which leave the Hamiltonian invariant?

   Since the XY Hamltionian has a continuous rotational symmetry, all rotations between 0 and 2π are valid symmetry operations. We can see this by substituting in $\theta_j \to \theta_j + \Delta\theta$ and seeing that the Hamiltonian is the same:

   $$-J \sum_{\langle ij \rangle} \cos[(\theta_i + \Delta\theta) - (\theta_j + \Delta\theta)] = -J \sum_{\langle ij \rangle} \cos[\theta_i - \theta_j] = H.$$

Now let's turn on the "clock anisotropy" term with $h > 0$. (Isotropy means all spin directions are equivalent; anisotropy means they are not.)

3. In the case $q = 1$, what is the ground state of the model? What are the remaining symmetries (if any) of the Hamiltonian?

> The second term is like an external magnetic field; the energy is minimized by $\theta = 0$. Each spin experiences this potential identically and independently. Furthermore, the first term (the interaction term) is minimized when neighboring spins point in the same direction. Therefore, the ground-state of the model is the configuration where all of the spins point in the direction $\theta_j = 0 \, \forall j$.
>
> Any remaining symmetries of the form $\theta_j \to \theta_j + \Delta\theta$ must satisfy $\cos(\theta_j) = \cos(\theta_j + \Delta\theta)$. This is true only if $\Delta\theta$ is a factor of 2π. A rotation by a factor of 2π is the same as no rotation, i.e., the identity operation. Thus, there are no symmetries of the Hamiltonian (apart from the identity).

4. What about $q = 2$? Argue that this case is similar in spirit to an Ising model.

> In the case $q = 2$, the potential from the clock anisotropy term has two minima, at $\theta = 0$ and $\theta = \pi$, both with the same energy. This is as illustrated below. (As before, the first term is minimized when all spins point in the same direction.) So there are two ground-states: one where all the spins point in the $\theta = 0$ direction, and another where all spins are in the $\theta = \pi$ direction.
>
> This is similar to the Ising model because there are two equivalent ground-states.

Note that even though the spins are not Ising spins, the $q = 2$ case is still fundamentally an Ising model because of its symmetry.

5. More generally, for $q > 2$, how many ground-states are there, and what is the symmetry of the model?

> The answer can be inferred by extrapolation: there is one ground-state for $q = 1$ and two ground-states for $q = 2$, so it stands to reason that there are q ground-states for general q. To see this, let's look at the second term, $-h\cos(q\theta_j)$. This is an oscillating function which repeats q times every 2π radians. Since $-\cos x$ is minimized whenever x is a multiple of 2π, it follows that $-h\cos(q\theta)$ is minimized whenever
>
> $$q\theta = 2\pi n \text{ for some integer } n \implies \theta = 2\pi n/q.$$
>
> There are q such minima for $0 \le \theta < 2\pi$.
>
> The symmetry of the model is very similar. For a rotation by $\Delta\theta$ to be a symmetry, it must satisfy $-h\cos(q\theta_j) = -h\cos(q\theta_j + q\Delta\theta)$. This is only true if

$q\Delta\theta$ is a multiple of 2π, i.e.,

$$q\Delta\theta = 2\pi n \text{ for some integer } n \implies \Delta\theta = 2\pi n/q.$$

Since there are q such rotations, this is the symmetry of the q-state clock model.

6. Explain the following phrase to someone at your table: "the clock anisotropy term reduces the continuous rotational symmetry to a q-fold discrete rotational symmetry."

   In the $h = 0$ case, we saw that *any* angle $\Delta\theta$ between 0 and 2π was a symmetry in the Hamiltonian: the energy did not change if all the spins were rotated by $\Delta\theta$. This is what is known by the "continuous rotational symmetry." In contrast, if $h \neq 0$ then this is no longer the case. Specifically, the only values of $\Delta\theta$ which preserve the energy are multiples of $2\pi/q$. Since there are exactly q such rotation angles, this is known as a q-fold discrete rotational symmetry.

7. For each of the following situations, give a qualitative description of the phase of matter and sketch out the distribution over the angle θ_j of all spins on the lattice. Use a mean-field approximation.

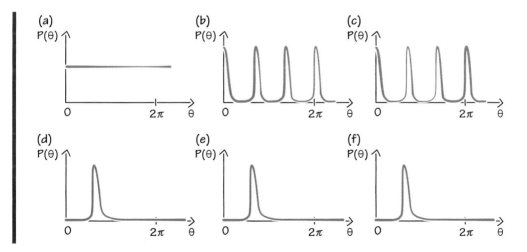

(a) Very high temperatures, $T \gg h$ and $T \gg J$

   Since the energy of thermal fluctuations exceeds any other energy scale in the problem (both the interaction strength and the anisotropy strength), the spins will be completely scrambled and every direction of θ is equally likely. There are no broken symmetries.

(b) $h \gg T$ and $J = 0$

Since the externally applied field is very strong compared to T, the periodic grooves in the potential are very deep. If the spin direction deviates even slightly from these minima, the energy gets large, so the probability of deviation from the minima is very low ($\sim e^{-h/T}$). Therefore, each spin will have a strong tendency to lock into one of the q minima at $2\pi n/q$.

Without an interaction term, there is no tendency for spins to align, and the direction of every spin is independent of the direction of the others. Each of the q preferred directions has the same energy and so is equally likely. Therefore, the spins are equally distributed among the q directions $\theta = 2\pi n/q$. The continuous rotational symmetry is broken but there remains a q-fold discrete rotational symmetry.

(c) $h \gg T$ and $0 < J \ll T$

If the interaction strength J is nonzero but small compared to the temperature, we can safely assume that the interactions have minimal effect. There is a very slight preference for nearby spins to point in the same direction, but not enough to cause global ordering. Therefore, the phase of matter is identical to that in the case in part (b).

(d) $h \gg T$ and $J \gg T$

Now that the interactions are large compared to the temperature, there will be a strong tendency for neighboring spins to align. Under the mean-field approximation, at low enough temperatures, this will lead to an ordered phase of broken symmetry, where one of the q ground-states is selected. This means that the vast majority of the spins will point in a particular one of the q directions.

(e) $J \gg T$ and $h = 0$

This is a pure XY model at low temperatures. Under a mean-field approximation, there is ordering and all the spins will tend to align in the same direction. The direction of ordering is spontaneously chosen. Since there is no clock anisotropy, any angle $0 \leq \theta < 2\pi$ is possible.

(f) $J \gg T$ and $0 < h \ll T$

As with part (e), since $J \gg T$, the state will be ordered. However, due to the clock anisotropy term, the direction of ordering will be along one of the q angles $\theta = 2\pi n/q$. Even the slightest anisotropy will cause this. This is because the effect of anisotropy is amplified by the macroscopic number of aligned spins. The difference in energy of ordering along a random direction, compared to one of the q clock directions, is on the order of hN—so in the limit $N \to \infty$ it becomes infinitely more favorable to align along one of the q clock directions.

It is possible to extrapolate between the $h \gg J$ and the $h \ll J$ cases. We will revisit this idea of approximate symmetry in more detail in future Section 7.5.2.

B.5.3 Determining the Symmetry and Order Parameter

Each of the scenarios below describes a phase transition where a discrete symmetry is broken. For each one, please (a) state the number of distinct broken-symmetry phases, (b) give an appropriate paradigmatic model which could be used to describe the system, and (c) describe a suitable order parameter, ϕ.

Hint 1: Remember that your order parameter should be zero in the symmetric state, but nonzero in the broken-symmetry state—and should be different for each of the possible states of broken symmetry!

Hint 2: For the 3-state clock model, the three states of broken symmetry have unit vectors of $\mathbf{S_1} = (1, 0)$, $\mathbf{S_2} = (-1/2, \sqrt{3}/2)$, and $\mathbf{S_3} = (-1/2, -\sqrt{3}/2)$. You can leave your expression for the order parameter in terms of $\mathbf{S_1}, \mathbf{S_2}$, and $\mathbf{S_3}$.

1. A ferromagnetic transition in a tetragonal crystal, where the spontaneous magnetization arises in any of the $-a$, $+a$, $-b$, and $+b$ directions.

   There are four discrete broken-symmetry phases, and this situation is represented by the **4-state clock model**, rather explicitly. The direction of magnetization, $\vec{m}$, can be taken to be the order parameter, as it is zero in the disordered state and nonzero in the ordered state. Specifically, in the $a - b$ plane, the order parameter will spontaneously point in the 12:00, 3:00, 6:00, or 9:00 direction, as in a 4-state clock model.

2. A structural transition from a cubic to a tetragonal crystal. (Remember that in a tetragonal unit cell, all three axes are perpendicular and two have the same length.)

   Suppose that the axes of the unit cell are oriented along the x-, y-, and z-axes. In the cubic unit cell, the lattice constants are identical for all three axes; however, in the transition to the tetragonal unit cell, one of the three axes spontaneously lengthens (or shortens) compared to the other two. As such, there are three equivalent broken-symmetry states. This can be modeled by a **3-state clock model**.

   Defining an appropriate order parameter is a little tricky. We wish to define a two-dimensional vector which points in the 12:00, 4:00, or 8:00 direction in the tetragonal state, but is zero in the cubic phase. It would make sense if each

one of the three clock directions corresponds to one of the possible tetragonal states. One possibility is to multiply each clock vector by one of the lattice
constants, and sum them up:

$$\phi = \ell_x \mathbf{S_1} + \ell_y \mathbf{S_2} + \ell_z \mathbf{S_3},$$

where ℓ_i is the length of the unit cell in the i axis. Since $\mathbf{S_1} + \mathbf{S_2} + \mathbf{S_3} = 0$ and
$\ell_x = \ell_y = \ell_z$ in the cubic phase, the order parameter is zero in the disordered
phase. You can also check that the order parameter points in the $\mathbf{S_1}$ direction
if the x-axis is lengthened in the tetragonal phase, $\mathbf{S_2}$ if the y-axis is lengthened,
and $\mathbf{S_3}$ if the z-axis is lengthened.

When a gas is placed in contact with a surface, the gas molecules can adsorb to
the surface if the interactions are favorable. This is how many important industrial chemical reactions are catalyzed. (Is adsorption energetically favorable or
unfavorable? How about entropically?)

Adsorption is energetically favorable, due to the formation of physical or chemical
bonds between the adatoms and the surface. However, it is entropically unfavorable because the adatoms are taken out of the gas phase and immobilized (effectively) onto a surface—the translational degrees of freedom are lost. Therefore,
adsorption is more thermodynamically favorable at lower temperatures!

3. Adsorption of atoms onto the sites of a 2D square lattice. The atoms are large
   enough that each adsorbed atom excludes all others from the adjacent sites, but
   not the diagonals (see figure). *Hint:* How many ground states are there if the
   lattice is full?

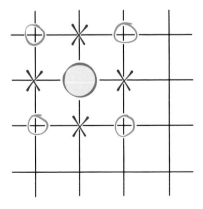

Since adsorption is energetically favorable, the ground-state is one where
adatoms are packed onto the lattice as densely as possible. This is achieved
when the adatoms are packed as in a checkerboard, occupying sites alternately such that exactly half of the sites are occupied. In the ground-state, one

sublattice is occupied and the other is empty. The sites corresponding to the white squares of the checkerboard form one sublattice, and the black squares are the other sublattice. Evidently, there are two such ground-states.

At nonzero T in a broken-symmetry phase, the difference between the sublattices is not black and white, but the concentrations of adatoms on the two sublattices still differ. There are two such states, so an Ising model would serve as a useful description of this process. An appropriate order parameter could be the difference in concentration between the two sublattices.

4. Same as part 3 except on a triangular lattice. Each adsorbed atom excludes exactly six directly adjacent sites.

If we try to pack adatoms onto the lattice as densely as possible without allowing any two to occupy adjacent sites, we end up with the following three configurations. Counting carefully, we see that exactly one-third of the sites of the triangular lattice are occupied. In fact, there are three identical, interpenetrating sublattices here, marked in red, blue, and green. In the ground-state, one of the three sublattices is completely occupied, and the other two are empty. There are three ground-states.

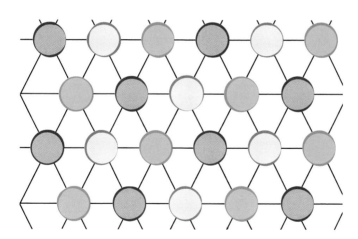

Following logic similar to that of part (3), the broken-symmetry state is a phase where one of the three sublattices has a greater concentration of adatoms than the other two. There are three equivalent broken-symmetry phases, so a useful model would be a **3-state clock model**. An order parameter can be defined, in analogy with (2), as

$$\phi = n_A S_1 + n_B S_2 + n_C S_3,$$

where n_A, n_B, and n_C are the concentrations of adatoms on the three sublattices, and S_i are the three equally spaced clock directions.

In the following chapter we will see why (at mean-field level) the $q = 3$ clock model has a discontinuous transition while the $q = 4$ clock model has a continuous transition. Knowing this, which of these transitions do you expect to be continuous and which discontinuous?

> Based on what we learned this chapter, many properties of a phase transition are determined by its symmetry. Therefore, we can figure out the properties by turning to model systems (i.e., Ising or clock model) with the same symmetry!
>
> 1. This has a 4-state clock symmetry, so we expect it to be continuous.
> 2. This has a 3-state clock symmetry, so we expect it to be discontinuous.
> 3. This has Ising symmetry, so we expect it to be continuous.
> 4. This has a 3-state clock symmetry, so we expect it to be discontinuous.

B.6 Chapter 6 Worksheet Solutions

B.6.1 Deducing the Landau Free Energy from Symmetry Considerations

Whereas we built up our intuition for the Ising model by considering microscopic interactions, universality implies a powerful general approach to describe the behavior near a critical point in terms of a free energy F expressed as an expansion in orders of ϕ, where the allowed powers of ϕ are determined by the symmetry (breaking) associated with the phase transition. This is known as Landau theory.

1. Let $F_L(T, h; \phi)$ be the free energy of an arbitrary(!) Ising system with volume V, which is a function of the temperature T and symmetry-breaking field h, as well as an order parameter ϕ (which we assume is spatially uniform for now). Why are we justified to claim

$$F_L(T, h; \phi) = \int_V d\vec{r} \; [\mathcal{F}_L(T; \phi) - h\phi]?$$

What is the interpretation of the two terms inside the brackets?

> We are writing the free energy as a volume integral of a free energy density over the system. Since we will end up taking the limit of an infinitely large system, it is advantageous to use the *density* because it stays finite and well defined.
>
> Of the two terms, one is the "intrinsic" free energy density in the absence of a symmetry-breaking field. The other one expresses how exactly the symmetry-breaking field breaks the energy—it couples linearly to the order parameter. (This is the generalization of how an applied magnetic field is linearly related to the energy of a single spin.)

2. The most general possible expansion in ϕ up to fourth order is

$$\mathcal{F}(T; \phi) = \mathcal{F}_0(T) + b(T)\phi + \frac{\alpha(T)}{2}\phi^2 + \frac{s(T)}{3}\phi^3 + \frac{u(T)}{4}\phi^4.$$

In what regime of the phase diagram would you expect this expansion to be accurate, and why?

As with any Taylor expansion, we expect this to be most accurate if ϕ is small. This is the case near a (continuous) phase transition, where ϕ starts off at zero and continuously grows larger as the temperature decreases past the transition temperature.

3. Because this is an Ising system, the free energy must be invariant under the symmetry $\phi \to -\phi$ if there is no symmetry-breaking field. Based on this, which of the terms in $\mathcal{F}$ are allowed? Which must be zero?

The invariance of the free energy may be expressed as $F_L(T, h = 0, \phi) = F_L(T, h = 0, -\phi)$. This would imply that

$$\mathcal{F}(T; \phi) = \mathcal{F}(T; -\phi).$$

This means that only even powers of ϕ are allowed, and the odd terms are not:

$$\mathcal{F}(T; \phi) = \mathcal{F}_0(T) + \frac{a(T)}{2}\phi^2 + \frac{u(T)}{4}\phi^4.$$

4. What is the restriction on u to obtain a physically realistic free energy? (Remember that the equilibrium values of ϕ are the minimums of the free energy.)

It is helpful to sketch out the shape of $\mathcal{F}(\phi) = \frac{\alpha}{2}\phi^2 + \frac{u}{4}\phi^4$ for varying values of the coefficients α and u. (We will leave out the T-dependence for simplicity, and we drop the constant term because it doesn't affect the value of ϕ which minimizes $\mathcal{F}$.)

Consider the large-ϕ behavior. In this regime the quartic term dominates. If u is positive, then as $\phi \to \infty$ the free energy will grow to be very large and positive—whereas if u is negative, then the free energy (as written) goes to negative infinity as $\phi \to \infty$. But this wouldn't make much sense, since, as the equilibrium value of ϕ is supposed to be the minimum of the free energy this leads to unbounded value of the order parameter. The free energy has to be bounded below for it to be physically meaningful. Therefore, we need to have $u > 0$.

5. At what value of α will the phase transition occur?

For $\alpha > 0$, the global minimum is at $\phi = 0$, but for $\alpha < 0$, the point $\phi = 0$ is a local maximum and there are two equivalent local minima at $\bar{\phi} = \pm\sqrt{|\alpha|/u}$. Therefore, as we found with the mean-field Ising solution, the phase transition occurs at $\alpha = 0$.

6. Remember that $\mathcal{F}_0$, α, and u each depend on T. Since we are near the critical point, let's expand in powers of a dimensionless "distance" from the critical point,

$$t \equiv \frac{T - T_c}{T_c},$$

which is sometimes known as a "reduced temperature." If a function $g(T)$ is analytic and amenable to Taylor expansion around $T = T_c$, we can say

$$g(T) = g_0 + g_1 t + \frac{g_2}{2}t^2 + O(t^3). \tag{B.6.1}$$

How are the coefficients g_0, g_1, and g_2 related to (derivatives of) $g(T)$?

Let's remind ourselves the formula for a Taylor expansion around $T = T_c$:

$$g(T) = g(T_c) + g'(T_c)(T - T_c) + \frac{1}{2}g''(T_c)(T - T_c)^2 + \mathcal{O}\left((T - T_c)^2\right). \tag{B.6.2}$$

This can be rewritten using the definition of the reduced temperature:

$$g = g(T_c) + g'(T_c)T_c \frac{T - T_c}{T_c} + \frac{1}{2}g''(T_c)T_c^2 \frac{T - T_c^2}{T_c^2} + \mathcal{O}\left((T - T_c)^2\right). \tag{B.6.3}$$

Therefore,

$$g_0 = g(T_c), \quad g_1 = g'(T_c)T_c, \quad g_2 = g''(T_c)T_c^2. \tag{B.6.4}$$

7. Near the critical point, we need to keep only the leading-order term of the expansion in t. What are the leading-order terms for $\mathcal{F}_0$, α, and u?

Generally, our Landau coefficients have expansions:

$$\mathcal{F}_0(T) = \mathcal{F}_0(T_c) + \mathcal{F}_1 t + \ldots, \tag{B.6.5}$$

$$\alpha(T) = \alpha_0 + \alpha_1 t + \frac{\alpha_2}{2}t^2 + \ldots, \tag{B.6.6}$$

$$u(T) = u_0 + u_1(T) + \ldots. \tag{B.6.7}$$

However, since $\alpha(T_c) = 0$, the constant term for $\alpha(T)$ is zero, meaning that its leading-order term is linear in t! For the other coefficients, the first nonzero term is constant—meaning that to leading order, there is no t dependence.

8. Compare your generic expression for $\mathcal{F}(T, \phi)$ to the variational free energy density you got from the mean-field Ising model,

$$\frac{F_{var}}{N} = f_{var} = -T \log 2 + \frac{T - zJ}{2} m^2 + \frac{T}{12} m^4 + \mathcal{O}(m^6).$$

The two agree if we make the identifications

$$\phi = m; \quad \mathcal{F}_0 = -T \log 2; \quad \alpha = T - zJ; \quad u = T/3.$$

If we wish to further break down the T-dependence, we can say $T = T_c$ and

$$\mathcal{F}(T_c) = -T_c \log 2, \qquad\qquad \mathcal{F}_1 = -T_c \log 2,$$

$$\alpha_1 = 1/T_c, \qquad\qquad \alpha_2 = 0,$$

$$u_0 = \frac{T_c}{3}, \qquad\qquad u_1 = \frac{T_c}{3}.$$

Notice that there are no odd powers of ϕ anywhere, as required by symmetry. Also, in this case α is exactly linear in $T - T_c$, and there are no higher powers of $T - T_c$.

What about the critical exponents? Remember that β, γ, and δ are defined by:

$$m \propto (T_c - T)^\beta \qquad \left.\frac{\partial m}{\partial h}\right|_{h=0} \propto |T_c - T|^{-\gamma} \qquad m|_{T=T_c} \propto h^{1/\delta}. \qquad (B.6.8)$$

Turns out, since the Landau free energy has the exact same form as the mean-field Ising model, all of the critical exponents are the same! So $\beta = 1/2$, $\gamma = 1$, and $\delta = 3$. This is not a coincidence, as we will explore much later in the book.

B.6.2   Understanding the Discontinuous Transition with a ϕ^3 Term

Depending on the symmetry of the system, the Landau free energy can have various permitted and forbidden terms which can affect the properties of the phase transition!

1. As a warm-up, suppose that we were studying a system where the graph of the free energy looks something like Figure B.6.1. What is the equilibrium value of ϕ? Are there any metastable solutions (local minima)?

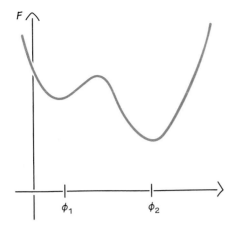

FIGURE B.6.1. A potential with two unequal minima.

The equilibrium value of ϕ is ϕ_2 because that is the value which minimizes $\mathcal{F}_L$. There is a metastable, local minimum at ϕ_1, but it is not the equilibrium solution since $\mathcal{F}_L(\phi_2) < \mathcal{F}_L(\phi_1)$.

For a 3-clock model, the Landau free energy permits a ϕ^3 term:[11]

$$\mathcal{F}_L = \frac{\alpha}{2}\phi^2 - \frac{s}{3}\phi^3 + \frac{u}{4}\phi^4 + \dots. \tag{B.6.9}$$

Before we do any calculations, let's take a look at the graph of $\mathcal{F}_L(\phi)$ for various values of the parameter α. Each curve is a different value of α.

2. There is one value of ϕ which is a (local) minimum of $\mathcal{F}_L$, regardless of α here. What is it?

$\phi = 0$ is always a local minimum (for $\alpha > 0$).

3. For some values of α this is the only minimum. For some values of α there is an *another* minimum at a different value of ϕ—let's call it $\bar{\phi}$. Separating these cases is a special value of $\alpha = \alpha^*$. Trace over this special curve with your pencil and label it α^*.

---

11. For the purposes of this worksheet, we'll take $s > 0$. The symbol ϕ represents the magnitude of the order parameter—which, since we're talking about a 3-clock model, can either point in the $0°$, $120°$, or $240°$ direction on a clock face.

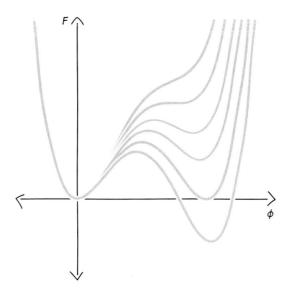

FIGURE B.6.2. A plot of Eq. B.6.9 for varying values of α.

α^* is the point at which the second minimum is born. The curve at exactly $\alpha = \alpha_c$ has an inflection point, and then as α decreases past α_c, the inflection point splits ("bifurcates") into a local min and a local max.

4. For $\alpha < \alpha^*$, is $\bar{\phi}$ necessarily the equilibrium state? Why or why not?

$\bar{\phi}$ is *not* necessarily the equilibrium state—just because it is a local minimum doesn't necessarily mean it's a global minimum!

5. There is another special value of $\alpha = \alpha_c$ which separates the case where $\bar{\phi}$ is the equilibrium state and where it isn't. Trace over this curve and label it α_c (with a different color if you have colored markers!)

Suppose we start at $\alpha = \alpha^*$ and decrease α. As we do this, the secondary minimum at $\bar{\phi}$ decreases in free energy. The phase transition occurs when $\mathcal{F}_L(\bar{\phi})$ decreases to a point where it is lower in free energy than the other minimum. Therefore, $\alpha = \alpha_c$ is the curve where the two minima are equal in free energy.

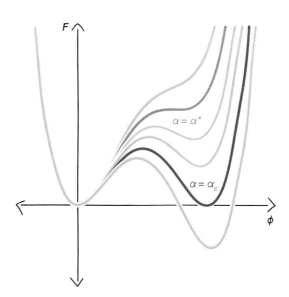

6. Sketch a graph of how the equilibrium value of ϕ depends on α, making sure to label α^* and α_c on your x-axis. Is this transition continuous or discontinuous?

On the basis of all the curves plotted in Figure B.6.2, the minimum value of ϕ looks something like this. For large α, the $\phi = 0$ is the only minimum and hence is the equilibrium state. Once α decreases past α^*, the second minimum at $\bar{\phi}$ appears, but initially it is higher in free energy, so the equilibrium state is still $\phi = 0$. Only until $\alpha < \alpha_c$ does $\bar{\phi}$ become the global free energy minimum.

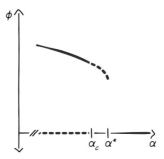

This is a discontinuous phase transition because for α immediately above α_c, ϕ is zero, whereas for α even slightly below α_c, it jumps discontinuously to a nonzero value of $\phi = \bar{\phi}$.

Now that we understand the overall behavior of this model, let's calculate the values of α^* and α_c.

7. By setting $d\mathcal{F}_L/d\phi = 0$, solve for the values of ϕ which extremize $\mathcal{F}_L$. Should you take the $+$ or the $-$ sign for $\bar{\phi}$? What does the other sign correspond to?

$$\frac{d\mathcal{F}_L}{d\phi} = \alpha\phi - s\phi^2 + u\phi^3 = \phi(u\phi^2 - s\phi + \alpha) = 0 \qquad \text{(B.6.10)}$$

$$\Longrightarrow \phi = 0 \ \text{ or } \ u\phi^2 - s\phi + \alpha = 0. \qquad \text{(B.6.11)}$$

As observed earlier, there is always a $\phi = 0$. Using the quadratic equation, we have

$$\phi = \frac{s \pm \sqrt{s^2 - 4\alpha u}}{2u}. \qquad \text{(B.6.12)}$$

There are two solutions to the quadratic equation if $s^2 - 4\alpha u$ is positive; should we take the $+$ or the $-$ root? Well, looking at the graph for any $\alpha < \alpha^*$, we see that from left to right, there is the minimum at $\phi = 0$, then a local max, then a local min. The $-$ root corresponds to the local max (on the left) and the $+$ root to the local min (on the right). So we want the $+$ root in $\bar{\phi}$.

8. For what values of α is your expression for $\bar{\phi}$ valid? Use this fact to determine α^*.

This isn't totally necessary, but if we rewrite $\bar{\phi}$ slightly we can better see how it depends on α:

$$\bar{\phi} = \frac{1}{2u}\left(s + \sqrt{s^2 - 4\alpha u}\right) \qquad \text{(B.6.13)}$$

$$= \frac{1}{2u}\left(s + \sqrt{s^2 - 4s^2\alpha u/s^2}\right) \qquad \text{(B.6.14)}$$

$$= \frac{1}{2u}\left(s + \sqrt{s^2(1 - 4\alpha u/s^2)}\right) \qquad \text{(B.6.15)}$$

$$= \frac{1}{2u}\left(s + s\sqrt{1 - 4\alpha u/s^2}\right) \qquad \text{(B.6.16)}$$

$$= \frac{s}{2u}\left(1 + \sqrt{1 - \frac{4u}{s^2}\alpha}\right). \qquad \text{(B.6.17)}$$

This expression is valid only if the quantity inside the square root is positive:

$$1 - \frac{4u}{s^2}\alpha > 0 \quad \Longrightarrow \quad 1 > \frac{4u}{s^2}\alpha \quad \Longrightarrow \quad \frac{s^2}{4u} > \alpha. \qquad \text{(B.6.18)}$$

Therefore, $\boxed{\alpha^* = s^2/4u}$.

9. Give an equation in terms of $\bar{\phi}$ which you can use to solve for α_c.

α_c is the point at which the $\phi = \bar{\phi}$ minimum is equal in free energy to the $\phi = 0$ minimum:

$$\mathcal{F}_L(\bar{\phi}) = \mathcal{F}_L(0) = 0. \tag{B.6.19}$$

10. It's possible but a little tedious to solve for α_c directly with this, so instead let's use another way. If you look back to the curve you labeled α_c, you may notice an interesting feature of quartic functions: the local maximum is exactly halfway between $\phi = 0$ and $\phi = \bar{\phi}$. This is the only value of α where this is the case! Use this fact to determine α_c.

As we determined in part (7), the local max has the $-$ sign in the quadratic formula, whereas $\bar{\phi}$ has the $+$ sign. Setting the local max to half of $\bar{\phi}$ results in

$$\frac{s}{2u}\left(1 - \sqrt{1 - \frac{4u}{s^2}\alpha}\right) = \frac{1}{2}\frac{s}{2u}\left(1 + \sqrt{1 - \frac{4u}{s^2}\alpha}\right) \tag{B.6.20}$$

$$\implies 1 - \sqrt{1 - \frac{4u}{s^2}\alpha} = \frac{1}{2} + \frac{1}{2}\sqrt{1 - \frac{4u}{s^2}\alpha} \tag{B.6.21}$$

$$\implies \frac{1}{2} = \frac{3}{2}\sqrt{1 - \frac{4u}{s^2}\alpha} \tag{B.6.22}$$

$$\implies \frac{1}{9} = 1 - \frac{4u}{s^2}\alpha \tag{B.6.23}$$

$$\implies \frac{8}{9} = \frac{4u}{s^2}\alpha \tag{B.6.24}$$

$$\implies \boxed{\alpha_c = \frac{2s^2}{9u}}. \tag{B.6.25}$$

B.7 Chapter 7 Worksheet Solutions

B.7.1 The Meaning of the Correlation Length

As we begin to study how order parameters can vary in space, the concept of the correlation length will repeatedly appear. Let's spend some time to think about this.

1. As a warm-up, let's think back to the exact solution of the 1D Ising model. We had found an expression for the correlation function:

$$\langle \sigma_i \sigma_j \rangle = e^{-|i-j|/\xi}, \quad \xi = |\log(\tanh \beta J)|^{-1}. \tag{3.34}$$

(a) What is the physical interpretation of $\langle \sigma_i \sigma_j \rangle$?

The correlation function describes whether the fluctuations between the spins i and j are correlated, i.e., if spin i is up, how much more likely than random chance it is that spin j is also up.

(b) Sketch a graph of $\langle \sigma_i \sigma_j \rangle$ as a function of $|i - j|$, and give the intuition behind the shape of the graph.

The correlation function is an exponentially decaying function of the distance between the spins; it is larger if the spins are closer and smaller if they are further.

(c) Interpret ξ.

ξ sets the lengthscale of the decay of correlation. If two spins are separated by distances far greater than ξ, they are essentially uncorrelated. This means that a local patch of aligned spins roughly has a size of ξ.

2. We will soon be studying a few different situations involving spatial variation in the magnetization, m_j. As a starting point we will use the Ising Hamiltonian

$$H = -J \sum_{\langle ij \rangle} \sigma_i \sigma_j - \sum_j h_j \sigma_j,$$

and for concreteness, you can have in your mind the picture of a 2D square lattice.

(a) In this Hamiltonian, each site has its own h_j. What might this represent physically?

The h_j is a magnetic field which affects only the spin on site j. It describes some perturbation to the system centered at site j. For instance, it could represent a localized impurity or defect in an otherwise regular crystal— or more generally serve to model some sort of heterogeneity in a material. As another example, near the surface of a material there could be a thin, oxidized layer, and maybe this affects the physics in a way that can be modeled by a nonzero h_j on the surface.

(b) Suppose that we are at $T > T_c$, and all the h_j are equal to zero except for $h_i > 0$ on a specific site i. On which sites j would the magnetization m_j be most affected?

Note that since we are above T_c, the m_j would be zero everywhere if there were no perturbing fields. If $h_i > 0$, the magnetization on site i would evidently be affected so that $m_i > 0$. Since i is coupled to its neighbors, its neighbors would have positive average spin as well, but to a lesser extent because of thermal fluctuations. Similarly, the neighbors of neighbors would be affected, but even lesser so. By this reasoning, there would be a region of spins near i where the magnetization $m_j > 0$. The spins closer to i would have a larger magnetization, and as you got out further from i, the magnitude of the response would eventually die away.

(c) Give an interpretation of what is described by the nonlocal susceptibility,

$$\chi_{ji} \equiv \frac{dm_j}{dh_i}.$$

The nonlocal susceptibility describes how much a localized perturbation at site i would affect the average spin at site j.

(d) Under what circumstances might χ_{ji} be large? Small? Can you guess how χ_{ji} could depend on i and j? How might the correlation length come into play?

As with the regular susceptibility, we would expect the nonlocal susceptibility to get very large near a critical point. In addition, however, there is now an additional dependence on the distance between sites i and j: as argued above, we would expect χ_{ji} to be larger for i and j closer together, and smaller if they are further apart. In all likelihood, the correlation length would define the rough size of the region around a perturbation where there is a substantial effect on the magnetization.

3. The correlation length also plays an important role in justifying the thermodynamic limit.

(a) Let's say we wanted to study a system of size L. Instead of treating it as a finite-size system, let us approximate it as an infinite-size system. Do you expect our approximation to be better if $\xi \ll L$ or if $\xi \gg L$? Why? (There are a couple of reasons here!)

We need $\xi \ll L$ for our approximation to be good. There are a few ways to see this.

One key difference between a finite system and an infinite system is that the finite system has a boundary, and the boundary conditions will affect various observables. We can think of the boundary conditions as some sort of nonzero h_i; it then stands to reason that only spins within a distance ξ of the boundary are significantly affected by it. If $\xi \ll L$, then only a very small portion of the system is "near" the surface, and so boundary effects are negligible. If, however, $\xi \gg L$, then the entire system is influenced by the boundary, and it would not be fair to ignore the boundary.

Another way to see this is by dividing the system into smaller, independent subsystems. The existence of a correlation length means that two points separated by a distance greater than $\sim \xi$ are essentially uncorrelated with each other. This implies that if we divided our system into chunks of volume $\sim \xi^d$ (for a d-dimensional system), then each of the chunks would be statistically independent. The number of these subsystems is $(L/\xi)^d$.

If $\xi \ll L$, then there are a large number of subsystems. If we are thinking about a property of the system which is an average over all the subsystems, such as the average magnetization, then by the central limit theorem we would be taking the average of a very large number of variables—so the error would decrease in proportion to $(L/\xi)^{-d/2}$. This is the justification for the thermodynamic limit for many observables. In contrast, if $L \ll \xi$ then even two points on opposite ends of the system would be substantially correlated and the thermodynamic limit would not make much sense.

(b) Let's reconsider our expression from part (1). Show that ξ diverges in the limit $T \to 0$ for the 1D Ising model.

Let's take a look at the expression again,

$$\xi = |\log(\tanh \beta J)|^{-1},$$

and build it from the inside out. $T \to 0$ corresponds to $\beta \to \infty$, which means that the argument of the tanh approaches infinity. This means that the tanh itself approaches 1. The log of something close to 1 is very close to zero. Finally, we have to take this to the -1 power, which means that the overall quantity ξ approaches infinity.

(c) What does this mean if we wanted to study our finite-size system at very low T?

Since ξ gets larger and larger at low T, eventually at low enough T, the correlation length ξ will approach the size of the system, L. Below this temperature, the thermodynamic limit will fail to capture some aspects of our finite-size system. In fact, no matter the size of the system, there will be eventually a point past which the correlation length is longer than the system size, and the thermodynamic limit will have some errors!

We'll see later in this chapter how ξ also diverges when you get close to a critical point. By the same logic, you can see how the thermodynamic limit might suffer similar issues once you get sufficiently close to a critical point such that ξ is comparable to L.

B.7.2 Correlation Function of the Ising Model with Spatial Variation

To capture situations where the magnetization is not spatially uniform, let us consider the Ising Hamiltonian where there is a separate, local magnetic field for each site j more carefully:

$$H = -J \sum_{\langle ij \rangle} \sigma_i \sigma_j - \sum_j h_j \sigma_j.$$

As before, we'll use a trial Hamiltonian $H_{tr} = - \sum_j b_j \sigma_j$.

1. Let's begin with a short warm-up. In previous chapters, we had considered a spatially uniform field ($h_j = h$ for all j) and we derived the following mean-field equations:

    $$b_j = J \sum_{i=\text{n.n. } j} m_i + h \quad \text{for each site } j,$$

    where $m_i = \langle \sigma_i \rangle_{tr}$. What is the interpretation of each term in this equation? If we had a spatially varying h_j, what would the equation be?

    b_j is the average effective local magnetic field felt by the spin on site j. It has two contributions: one from the average spins of its neighbors, and another from the externally applied magnetic field. Therefore, in the case with site-specific h_j, the second term will become h_j:

    $$b_j = J \sum_{i=\text{n.n. } j} m_i + h_j. \tag{B.7.1}$$

2. If $T > T_c$ and the h_j are zero on all the sites, then what is m_i?

    For $T > T_c$ we are in the disordered phase. Without any externally applied magnetic fields there will be no magnetization, so $m_i = 0$ for every single site.

    Let's find out how m_i is affected if h_j is not zero everywhere. To make our lives easier, let's say that h_j is small and that the resulting m_i are also small, so that we can use a linear approximation.

3. Use $m_i = \tanh(b_i/T)$ to eliminate $\{b\}$ from the mean-field equations, and use the expansion $\operatorname{arctanh}(x) = x + x^3/3 + \ldots$ to linearize the equations.

    Inverting the expression, we have

    $$b_i = T \operatorname{arctanh} m_i = T(m_i + m_i^3/3 + \ldots) \approx T m_i,$$

    where we want to keep only the leading-order term since m_i is small. This gives us

    $$T m_j = J \sum_{i=\text{n.n. } j} m_i + h_j \tag{B.7.2}$$

    to lowest order in m_i.

4. Take a derivative with respect to h_k to find a set of linear equations for the nonlocal susceptibility, $\chi_{jk} = dm_j/dh_k$.

    Taking the derivative, we get:

    $$T \frac{dm_j}{dh_k} = J \sum_{i=\text{n.n. } j} \frac{dm_i}{dh_k} + \frac{dh_j}{dh_k}.$$

    (To keep the indices straight: there is the free index j labeling each mean-field equation, a dummy index i for summing over neighbors, and another free index, k, for the site where we want to turn on a small local perturbation, h_k.) Since $dh_a/dh_b = \delta_{ab}$ (the Kronecker delta), this simplifies to

$$T\chi_{jk} = J \sum_{i=\text{n.n.}\,j} \chi_{ik} + \delta_{jk}.$$

5. For later convenience, we'll want to rewrite these equations as

$$\delta_{jk} = (T - T_c)\chi_{jk} + J \sum_i \Delta_{ji}\chi_{ik}$$

so that the sum runs over *all* the lattice sites, not just the nearest neighbors.[12] What should be the definition of Δ_{ji}? (Remember $T_c = zJ$.)

Let's rearrange things a little to start resembling our target:

$$\delta_{jk} = T\chi_{jk} - J \sum_{i=\text{n.n.}\,j} \chi_{ik}.$$

There is a $-T_c\chi_{jk}$ missing on the right-hand side, so let's add and subtract this to get

$$\delta_{jk} = (T - T_c)\chi_{jk} + T_c\chi_{jk} - J \sum_{i=\text{n.n.}\,j} \chi_{ik}.$$

Now if we compare the equations, we're pretty close; we just need these two things to be equal:

$$J \sum_i \Delta_{ji}\chi_{ik} = T_c\chi_{jk} - J \sum_{i=\text{n.n.}\,j} \chi_{ik}$$

or, using $T_c = zJ$,

$$J \sum_i \Delta_{ji}\chi_{ik} = zJ\chi_{jk} - J \sum_{i=\text{n.n.}\,j} \chi_{ik}.$$

The left side has one term for every single site i on the lattice, but the right side has one term for j and one for each the neighbors of j. This tells us that Δ_{ji} is something for $i = j$, something else if i is the nearest neighbor of j, and zero otherwise:

$$\Delta_{ji} = \begin{cases} z & \text{if} & i = j, \\ -1 & \text{if} & i \text{ and } j \text{ are nearest neighbors}, \\ 0 & & \text{otherwise}. \end{cases} \qquad (B.7.3)$$

6. Let $\vec{R}_a$ be the position coordinate of lattice site a. Argue why χ_{ij} depends only on $\vec{R}_i - \vec{R}_j$. Show that the same is true of Δ_{ji}.

χ_{ij} tells us how much the average spin on site i is affected by a local perturbing field on site j. Since the Hamiltonian is translationally invariant, the absolute position of i or j shouldn't matter—χ_{ij} should only depend on the relative

---

12. We also pulled out a $-T_c$ so that the coefficient of the other term vanishes as $T \to T_c$

displacement between i and j, or $\vec{R}_i - \vec{R}_j$. The same is true of Δ_{ij}, because as it is defined, it depends only on the relative positions of i and j.

7. Let $\chi(\vec{r})$ be a function of position where $\chi(\vec{R}_i - \vec{R}_j) = \chi_{ij}$. Define $\Delta(\vec{r})$ similarly. What is the interpretation of $\chi(\vec{r})$? For what values of $\vec{r}$ is $\chi(\vec{r})$ defined?

$\chi(\vec{r})$ describes the extent to which a small, local perturbation at a site will affect the magnetization at another site separated by $\vec{r}$ from the perturbation. If $\vec{r}_0$ is the site where the perturbation is applied, then $\chi(\vec{r})$ is defined only where $\vec{r}_0 + \vec{r}$ falls on a lattice site. The set of such $\vec{r}$ is the lattice vectors.

8. Letting $\vec{R}_k = \vec{0}$, $\vec{R}_j = \vec{r}$, and $\vec{R}_i = \vec{r}\,'$, rewrite your mean-field equations in terms of $\chi(\vec{r})$ and $\Delta(\vec{r} - \vec{r}\,')$.

This is a straightforward substitution:

$$\delta_{\vec{r},0} = (T - T_c)\chi(\vec{r} - \vec{0}) + J\sum_{\vec{r}\,'}\Delta(\vec{r} - \vec{r}\,')\chi(\vec{r} - \vec{0}),$$

$$\delta_{\vec{r},0} = (T - T_c)\chi(\vec{r}) + J\sum_{\vec{r}\,'}\Delta(\vec{r} - \vec{r}\,')\chi(\vec{r}\,').$$

Note that it is tricky to solve this equation, because $\chi(\vec{r})$ depends on $\chi(\vec{r}\,')$ for other sites $\vec{r}\,'$—all the equations are coupled together!

To proceed further, we'll need some more machinery, the Fourier Transform. Suppose that $\chi(\vec{r})$ can be expressed in the form

$$\chi(\vec{r}) = \int_{\vec{k}} e^{i\vec{k}\cdot\vec{r}}\tilde{\chi}(\vec{k}),$$

where $\int_{\vec{k}}$ means $\frac{1}{(2\pi)^d}\int d^d k$. For now, don't worry about the range of $\vec{k}$ we are integrating, or about the normalization of the integral—we will treat this as an ansatz to help us solve for $\chi(\vec{r})$ in the mean-field equations.

9. Substitute this ansatz for $\chi(\vec{r})$, and massage the equation until you get something which looks like $\int_{\vec{k}} e^{i\vec{k}\cdot\vec{r}}[\ldots\ldots] = 0$. Write the resulting expression in terms of $g(\vec{k}) = \sum_{\vec{r}} e^{-i\vec{k}\cdot\vec{r}}\Delta(\vec{r})$.
   Hint 1: $\delta_{\vec{r},0} = \int_{\vec{k}} e^{i\vec{k}\cdot\vec{r}} \cdot 1$.
   Hint 2: You may have to insert $e^{i\vec{k}\cdot\vec{r}}e^{-i\vec{k}\cdot\vec{r}} = 1$ into one of your terms.

Substituting the ansatz yields

$$\delta_{\vec{r},0} = (T - T_c)\left(\int_{\vec{k}} e^{i\vec{k}\cdot\vec{r}}\tilde{\chi}(\vec{k})\right) + J\sum_{\vec{r}\,'}\Delta(\vec{r} - \vec{r}\,')\left(\int_{\vec{k}} e^{i\vec{k}\cdot\vec{r}\,'}\tilde{\chi}(\vec{k})\right),$$

and applying the first hint gives

$$\int_{\vec{k}} e^{i\vec{k}\cdot\vec{r}} \cdot 1 = (T - T_c)\left(\int_{\vec{k}} e^{i\vec{k}\cdot\vec{r}}\tilde{\chi}(\vec{k})\right) + J\sum_{\vec{r}'} \Delta(\vec{r} - \vec{r}')\left(\int_{\vec{k}} e^{i\vec{k}\cdot\vec{r}'}\tilde{\chi}(\vec{k})\right).$$

Comparing this expression to our final target expression, we see that each of the terms has the correct form (integral $\int_{\vec{k}} e^{i\vec{k}\cdot\vec{r}}$ of something), except for the second term on the right, which has $\vec{r}'$ instead of $\vec{r}$. To convert the $\vec{r}'$ into $\vec{r}$, we use the second hint:

$$e^{i\vec{k}\cdot\vec{r}'} = e^{i\vec{k}\cdot\vec{r}'} e^{i\vec{k}\cdot\vec{r}} e^{-i\vec{k}\cdot\vec{r}} = e^{i\vec{k}\cdot\vec{r}} e^{-i\vec{k}\cdot(\vec{r}-\vec{r}')}.$$

Now, we can pull out the $\int_{\vec{k}} e^{i\vec{k}\cdot\vec{r}}$ from each term to give

$$\int_{\vec{k}} e^{i\vec{k}\cdot\vec{r}} \cdot 1 = \int_{\vec{k}} e^{i\vec{k}\cdot\vec{r}}\left[(T - T_c)\tilde{\chi}(\vec{k}) + J\sum_{\vec{r}'} \Delta(\vec{r} - \vec{r}')e^{-i\vec{k}\cdot(\vec{r}-\vec{r}')}\tilde{\chi}(\vec{k})\right],$$

and subtracting $\int_{\vec{k}} e^{i\vec{k}\cdot\vec{r}} \cdot 1$ from both sides gives

$$0 = \int_{\vec{k}} e^{i\vec{k}\cdot\vec{r}}\left[(T - T_c)\tilde{\chi}(\vec{k}) + \underbrace{J\sum_{\vec{r}'} \Delta(\vec{r} - \vec{r}')e^{-i\vec{k}\cdot(\vec{r}-\vec{r}')}}_{g(\vec{k})}\tilde{\chi}(\vec{k}) - 1\right].$$

As a final step, notice that the braced portion bears a strong resemblance to $g(\vec{k})$ as we've defined it in the chapter; the only difference is that the sum should be over $\vec{r} - \vec{r}'$. But this is equivalent to a sum over $\vec{r}'$ because these are just different ways of saying "sum over all the lattice sites" due to translational symmetry. So, we can simplify this to

$$0 = \int_{\vec{k}} e^{i\vec{k}\cdot\vec{r}}\left[(T - T_c)\tilde{\chi}(\vec{k}) + Jg(\vec{k})\tilde{\chi}(\vec{k}) - 1\right].$$

10. Since your expression is zero for any $\vec{r}$, it means that the integrand $[\ldots\ldots]$ vanishes. Use this to solve for $\tilde{\chi}(\vec{k})$.

$$0 = (T - T_c)\tilde{\chi}(\vec{k}) + Jg(\vec{k})\tilde{\chi}(\vec{k}) - 1 \tag{B.7.4}$$

$$\implies 1 = \left[(T - T_c) + Jg(\vec{k})\right]\tilde{\chi}(\vec{k}) \tag{B.7.5}$$

$$\implies \tilde{\chi}(\vec{k}) = \frac{1}{(T - T_c) + Jg(\vec{k})}. \tag{B.7.6}$$

11. Finally, substitute this in your ansatz to arrive at an integral expression for χ_{ij}. Success!

This gives

$$\chi(\vec{r}) = \int_{\vec{k}} \frac{e^{i\vec{k}\cdot\vec{r}}}{(T - T_c) + Jg(\vec{k})}. \tag{B.7.7}$$

B.7.3 Estimating a Correlation Length from the Lattice Correlation Function

We have derived the following expression for the nonlocal susceptibility of an Ising model on a lattice:

$$\chi(\vec{R}_i - \vec{R}_j) = \int_{\vec{k}} \frac{e^{i\vec{k}\cdot(\vec{R}_i - \vec{R}_j)}}{(T - T_c) + Jg(\vec{k})}, \tag{B.7.8}$$

where $\vec{R}_i$ is the position of lattice site i, $g(\vec{k}) = \sum_{\vec{r}} e^{-i\vec{k}\cdot\vec{r}} \Delta(\vec{r})$, and

$$\Delta(\vec{R}_i - \vec{R}_j) = \begin{cases} z & \text{if} & i = j, \\ -1 & \text{if} & i \text{ and } j \text{ are nearest neighbors,} \\ 0 & & \text{otherwise.} \end{cases} \tag{B.7.9}$$

The integral cannot be evaluated numerically, but you can argue how it depends on the distance between the lattice sites i and j. For concreteness, let's take a 2D square lattice of spacing a, where

$$\int_{\vec{k}} = \frac{1}{4\pi^2} \int_{-\pi/a}^{\pi/a} \int_{-\pi/a}^{\pi/a} dk_x \, dk_y.$$

1. What is $g(\vec{k})$ for small $|\vec{k}|$?

   On a square lattice, there are $z = 4$ nearest neighbors, and the nearest-neighbor vectors are $\pm a\hat{x}$ and $\pm a\hat{y}$. Therefore,

   $$g(\vec{k}) = \sum_{\vec{r}} e^{-i\vec{k}\cdot\vec{r}} \Delta(\vec{r})$$

   $$= z e^{-i\vec{k}\cdot 0}{}^{\,1} - \sum_{\vec{r} = \pm a\hat{x}, \pm a\hat{y}} e^{-i\vec{k}\cdot\vec{r}}$$

   $$= 4 - \left(e^{-ik_x a} + e^{+ik_x a} + e^{-ik_y a} + e^{+ik_y a} \right)$$

   $$= 4 - 2\cos(k_x a) - 2\cos(k_y a).$$

   For small k_x and k_y, we can use the expansion $\cos x = 1 - x^2/2 + \dots$ to say

   $$g(\vec{k}) = 4 - 2[1 - (k_x a)^2/2] - 2[1 - (k_y a)^2/2]$$

   $$= k_x^2 a^2 + k_y^2 a^2$$

   $$= |\vec{k}|^2 a^2.$$

2. Sketch the integrand without $e^{i\vec{k}\cdot\vec{r}}$ as a function of $|\vec{k}| = k$. (You can use the small $|\vec{k}|$ approximation.) What is the approximate width of the function?

This part of the integrand is

$$\frac{1}{(T - T_c) + Jk^2 a^2},$$

which is a bell-shaped curve centered around $k = 0$. It is positive everywhere and goes to zero as $k \to \infty$.

The width of the hump can be estimated by looking at where it attains half of its maximum value $1/(T - T_c)$,

$$\frac{1}{2(T - T_c)} = \frac{1}{(T - T_c) + Jk^2 a^2}$$

$$\implies T - T_c = Jk^2 a^2$$

$$\implies k \sim a^{-1}\sqrt{(T - T_c)/J}.$$

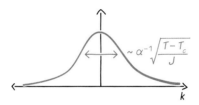

3. Sketch the real part of the numerator. How does this graph depend on $|\vec{r}| = r$?

The real part of the numerator is $\mathrm{Re}[e^{ikr}] = \cos(kr)$, which is a function of k which oscillates between positive and negative values. If r is larger, then the oscillations are more frequent.

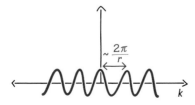

4. Multiply the two of these together. Sketch a graph of the real part of resulting function vs. k for the cases where r is large and where r is small. In which case do you expect the integral to be large and in which case small, and why? What does this result imply about how a system responds to local perturbations?

When you multiply the two together, you get an oscillating function inside of a bell-shaped envelope. The parts of the curve which lie below the x-axis will cancel out the parts which lie above the x-axis. If r is large, then there will be very rapid oscillations, and the positive bits above the x-axis will basically cancel out all the negative bits, and you will end up with an overall integral close to zero. In contrast, if r is small, then the period of oscillation will be long and most of the integral will lie under one sinusoidal period, so you will have no cancellation effect, and the integral will be large.

This tells us something which is perhaps intuitively clear: if we apply a small local perturbation, then the nearby sites will be substantially affected whereas the faraway sites will not be.

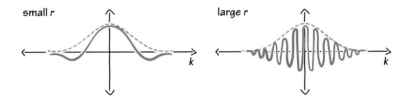

5. How big does r have to be for the integral to be small?

If the oscillations are frequent enough that there are many periods inside the hump of the bell curve, then there will be lots of cancellation and the integral will be close to zero. Therefore, the criterion for large r is that the period of oscillation in k space be smaller than the rough width of the bell curve:

$$1/r \lesssim \sqrt{\frac{T - T_c}{Ja^2}}$$

or $r \gtrsim a\sqrt{J/(T - T_c)} \sim \xi$. This is our first estimate of the correlation length! We'll return to this later in the chapter, and it's okay if you didn't manage to derive it this time around since it's a little tricky.

So for $r \lesssim \xi$, the integral will be substantial, whereas for $r \gtrsim \xi$, the integral will be close to zero.

B.7.4   Deriving the Ginzburg-Landau Functional from Symmetry Arguments

As discussed in the chapter, Ginzburg-Landau theory is an extension of Landau theory where spatial variations are allowed. Instead of considering just the symmetry of the order parameter, we now also have to take into account the spatial symmetry of the system. In this exercise we will demonstrate how to construct the free energy from symmetry arguments alone, using the Ising model as our example.

As before, the free energy of the system is

$$F = \int d\vec{r}\, \mathcal{F}, \tag{B.7.10}$$

such that the free energy of the system is the volume integral of the free energy density $\mathcal{F}$, which is a local property. When we constructed the free energy density in the previous chapter, we considered only the uniform state. Here, to make the argument as general as possible, we write the free energy density as:

$$\mathcal{F} = \mathcal{F}\left(\vec{r}, \phi(\vec{r}), \partial_i \phi, \partial_{ij} \phi, \ldots\right). \tag{B.7.11}$$

The arguments of the free energy density now reflect the fact that spatial variations are allowed, with explicit dependence on the position $\vec{r}$, and spatially varying order parameter $\phi(\vec{r})$. More importantly (and interestingly), the free energy density can now depend on *spatial derivatives* of the order parameter, denoted by the subscripted partial derivatives.

As before, we assume we are in the vicinity of a phase transition and therefore ϕ is a small quantity. Furthermore, we assume that the spatial variations are slow, such that the spatial derivatives are good quantities to expand on.

1. Given the form of Eq. B.7.11, show that the Taylor expansion in terms of the spatial derivatives to second order can be written as

$$\mathcal{F}\left(\vec{r}, \phi(\vec{r}), \partial_i \phi, \partial_{ij} \phi, \ldots\right) = V(\vec{r}, \phi(\vec{r}))$$

$$+ \sum_i (\partial_i \phi) \nu_i(\vec{r}, \phi(\vec{r}))$$

$$+ \sum_{ij} (\partial_i \partial_j \phi) \frac{\eta_{ij}(\vec{r}, \phi(\vec{r}))}{2}$$

$$+ \sum_{ij} (\partial_i \phi)(\partial_j \phi) \frac{\kappa_{ij}(\vec{r}, \phi(\vec{r}))}{2} + \ldots, \tag{B.7.12}$$

where

$$\mathcal{F}(\vec{r}, \phi(\vec{r}), 0, 0, \ldots) = V(\vec{r}, \phi(\vec{r}))$$

and $\nu_i, \eta_{ij}, \kappa_{ij} \ldots$ are general coefficients of expansion (the factors of $1/2$ are included by convention). This is sometimes called a gradient expansion.

A gradient expansion works exactly like a regular Taylor expansion. Recall how we define a series expansion for a function of one variable x about $x = 0$:

$$f(x) = f(0) + f_1 x + f_2 x^2 + \ldots. \tag{B.7.13}$$

We then use the properties of derivatives to find what each f_i is. For functions of more than one variable, the procedure is similar. Let's say we have

$f(a, x, y, z)$, $a \neq 0$, and we are expanding about $x = y = z = 0$:

$$f(a, x, y, z) = f(a, 0, 0, 0) + f_x(a)x + f_y(a)y + f_z(a)z \tag{B.7.14}$$

$$+ f_{xx}(a)x^2 + f_{xy}(a)xy + f_{xz}(a)xz + \dots \tag{B.7.15}$$

We can still use the properties of partial derivatives to find each f_i, f_{ij}, etc., but each coefficient will be function of a.

Now, we have our function $\mathcal{F}\left(\vec{r}, \phi(\vec{r}), \partial_i\phi, \partial_{ij}\phi, \dots\right)$. $\vec{r}$ and $\phi(\vec{r})$ are like the variable a above, while the derivatives of ϕ are like the coordinates x, y, z above. So, to expand about all orders of the spatial derivative equal to 0, we have:

$$\mathcal{F}\left(\vec{r}, \phi(\vec{r}), \partial_i\phi, \partial_{ij}\phi, \dots\right) = \mathcal{F}(\vec{r}, \phi(\vec{r}), 0, 0, \dots) \tag{B.7.16}$$

$$+ (\partial_x\phi)\nu_x + (\partial_y\phi)\nu_y + (\partial_z\phi)\nu_z$$

$$+ (\partial_x\partial_x\phi)\frac{\eta_{xx}}{2} + (\partial_x\partial_y\phi)\frac{\eta_{xy}}{2} + (\partial_x\partial_z\phi)\frac{\eta_{xz}}{2} + \dots$$

$$+ (\partial_x\phi)(\partial_x\phi)\frac{\kappa_{xx}}{2} + (\partial_x\phi)(\partial_y\phi)\frac{\kappa_{xy}}{2} + (\partial_x\phi)(\partial_z\phi)\frac{\kappa_{xz}}{2} + \dots$$

$$+ \dots,$$

where ν_i, η_{ij}, and κ_{ij} can each be functions of $\vec{r}$ and $\phi(\vec{r})$. We've included here the convention to attach the factor $1/2$ on terms with η_{ij} and κ_{ij}.

Note that each appearance of the differential operator increases the order of the term by 1. For example, all of these terms are of the 3rd order in derivatives: $\partial_i\partial_j\partial_k\phi$, $(\partial_i\partial_j\phi)(\partial_k\phi)$, and $(\partial_i\phi)(\partial_j\phi)(\partial_k\phi)$. So when asked to keep terms to a certain order in derivatives, make sure not to miss terms.

We relabel the first term as $V(\vec{r}, \phi(\vec{r}))$, and write everything down using summations to get:

$$\mathcal{F} = V(\vec{r}, \phi(\vec{r}))$$

$$+ \sum_i (\partial_i\phi)\nu_i(\vec{r}, \phi(\vec{r}))$$

$$+ \sum_{ij} (\partial_i\partial_j)\eta_{ij}(\vec{r}, \phi(\vec{r}))$$

$$+ \sum_{ij} (\partial_i\phi)(\partial_j\phi)\frac{\kappa_{ij}(\vec{r}, \phi(\vec{r}))}{2} + \dots,$$

which is the form of Eq. B.7.12.

2. Given that the system has translational symmetry,

$$\phi(\vec{r}) = \phi(\vec{r} + \vec{R}) \tag{B.7.17}$$

for any $\vec{R}$, prove that there is no explicit spatial dependence on $\vec{r}$ in any of the terms, i.e.

$$\mathcal{F}\left(\vec{r}, \phi(\vec{r}), \partial_i \phi, \partial_{ij} \phi, \dots\right) = \mathcal{F}\left(\phi(\vec{r}), \partial_i \phi, \partial_{ij} \phi, \dots\right) \tag{B.7.18}$$

or, more explicitly,

$$V(\vec{r}, \phi(\vec{r})) \equiv V(\phi(\vec{r})), \qquad \nu_i(\vec{r}, \phi(\vec{r})) \equiv \nu_i(\phi(\vec{r})),$$

$$\eta_{ij}(\vec{r}, \phi(\vec{r})) \equiv \eta_{ij}(\phi(\vec{r})), \qquad \kappa_{ij}(\vec{r}, \phi(\vec{r})) \equiv \kappa_{ij}(\phi(\vec{r})). \tag{B.7.19}$$

Let us incorporate (B.7.17) into (B.7.10):

$$F[\phi(\vec{r})] = F[\phi(\vec{r} + \vec{R})] \tag{B.7.20}$$

$$= \int d^d \vec{r} \Big\{ V(\vec{r}, \phi(\vec{r} + \vec{R})) + (\partial_i \phi) \nu_i(\vec{r}, \phi(\vec{r} + \vec{R}))$$

$$+ (\partial_i \partial_j \phi) \frac{\eta_{ij}(\vec{r}, \phi(\vec{r} + \vec{R}))}{2} + (\partial_i \phi)(\partial_j \phi) \frac{\kappa_{ij}(\vec{r}, \phi(\vec{r} + \vec{R}))}{2} + \dots \Big\}.$$

Now, let's make a variable substitution, $\vec{r}' = \vec{r} + \vec{R}$, so $\vec{r} = \vec{r}' - \vec{R}$, and $d^d \vec{r} = d^d \vec{r}'$:

$$\dots = \int d^d \vec{r}' \Big\{ V(\vec{r}' - \vec{R}, \phi(\vec{r}')) + (\partial_i \phi) \nu_i(\vec{r}' - \vec{R}, \phi(\vec{r}')) \tag{B.7.21}$$

$$+ (\partial_i \partial_j \phi) \frac{\eta_{ij}(\vec{r}' - \vec{R}, \phi(\vec{r}'))}{2} + (\partial_i \phi)(\partial_j \phi) \frac{\kappa_{ij}(\vec{r}' - \vec{R}, \phi(\vec{r}'))}{2} + \dots \Big\}$$

Remember that $\vec{r}'$ is just a dummy integration variable; it could be relabeled as $\vec{r}' \to \vec{r}$. Let us do just that, and compare Eq. B.7.21 to Eq. B.7.20, and we find that to satisfy $F[\phi(\vec{r})] = F[\phi(\vec{r} + \vec{R})]$ for any $\phi(\vec{r})$ and $\vec{R}$,

$$V(\vec{r}, \phi(\vec{r})) = V(\vec{r} - \vec{R}, \phi(\vec{r})), \qquad \nu_i(\vec{r}, \phi(\vec{r})) = \nu_i(\vec{r} - \vec{R}, \phi(\vec{r})),$$

$$\eta_{ij}(\vec{r}, \phi(\vec{r})) = \eta_{ij}(\vec{r} - \vec{R}, \phi(\vec{r})), \qquad \kappa_{ij}(\vec{r}, \phi(\vec{r})) = \kappa_{ij}(\vec{r} - \vec{R}, \phi(\vec{r}))$$

must all be true for any $\vec{R}$. Thus, we can conclude that none of these functions have any explicit position dependence on $\vec{r}$.

3. Assume that the system has hypercubic spatial symmetry (e.g., a square lattice in 2D or a cubic lattice in 3D), meaning that rotation by $\pi/2$ along any primary axis does not change the free energy of the system. Eliminate all unnecessary terms of spatial derivative in your previous expression to show that:

$$F = \int d^d \vec{r} \left\{ V(\phi(\vec{r})) + |\nabla \phi|^2 \frac{\kappa(\phi(\vec{r}))}{2} + \dots \right\}. \tag{B.7.22}$$

Hint: This part can be intuitively quick but formally tedious. Don't be afraid to peek at the solutions for the formal justification!

Hint: You might need the vector identity $\nabla \cdot (f\nabla g) = f\nabla^2 g + \nabla f \cdot \nabla g$.

A hypercubic rotational symmetry is a generalization of cubic rotational symmetry to higher dimensions. Let's first do an example in 3D to see how this symmetry works; then we'll generalize it to higher dimensions.

In 3D, with $\vec{r} = (x, y, z)$, a $\pi/2$ rotation about one of the axes will leave that coordinate the same, and apply to the other two a rotation matrix of the form $\begin{bmatrix} 0 & 1 \\ -1 & 0 \end{bmatrix}$. For example, a $\pi/2$ rotation about the z-axis will have the rotation matrix:

$$R^z_{\pi/2} = \begin{bmatrix} 0 & 1 & 0 \\ -1 & 0 & 0 \\ 0 & 0 & 1 \end{bmatrix}. \tag{B.7.23}$$

So:

$$\vec{r}' = R^z_{\pi/2}\vec{r} \tag{B.7.24}$$

$$= \begin{bmatrix} 0 & 1 & 0 \\ -1 & 0 & 0 \\ 0 & 0 & 1 \end{bmatrix}\begin{bmatrix} x \\ y \\ z \end{bmatrix} = \begin{bmatrix} y \\ -x \\ z \end{bmatrix}. \tag{B.7.25}$$

Let's take this example, and see what happens in 3D when we rotate the coordinates of $\phi(\vec{r}) \rightarrow \phi(R^z_{\pi/2}\vec{r})$. Note that we've removed the explicit position dependence from all the terms per results of the previous section:

$$F[\phi(\vec{r}')] = \int d^d\vec{r}\Big\{ V(\phi(y,-x,z)) + (\partial_i\phi)\nu_i(\phi(y,-x,z))$$

$$+ (\partial_i\partial_j\phi)\frac{\eta_{ij}(\phi(y,-x,z))}{2} + (\partial_i\phi)(\partial_j\phi)\frac{\kappa_{ij}(\phi(y,-x,z))}{2} + \dots \Big\},$$

where i and j are summed over $\{x, y, z\}$ in each term.

Now, let's make a variable substitution, which we are free to do:

$$\tilde{\vec{r}} = (\tilde{x}, \tilde{y}, \tilde{z}) = (y, -x, z). \tag{B.7.26}$$

This substitution has some interesting effects. Analogously to a rotation, the Jacobian of this variable substitution is 1, so $d^d\vec{r} = d^d\vec{r}' = d^d\tilde{\vec{r}}$. Let's also write out an example term to see what it does:

$$\partial_x\partial_y\phi(y,-x,z)\eta_{xy}(\phi(y,-x,z)) = (-\partial_{\tilde{y}})\partial_{\tilde{x}}\phi(\tilde{x},\tilde{y},\tilde{z})\eta_{xy}(\phi(\tilde{x},\tilde{y},\tilde{z})). \tag{B.7.27}$$

Note that η_{xy} doesn't change, because the name is just a label, not actual coordinates. Once again, we can feel free to relabel $\tilde{\vec{r}} \rightarrow \vec{r}$. This term in the context of the integral becomes:

$$\int d^d \vec{r} \left\{ \ldots - \partial_y \partial_x \phi(x,y,z) \eta_{xy}(\phi(x,y,z)) + \ldots \right\}. \tag{B.7.28}$$

What we have shown here is that when we rotate the coordinates of ϕ, it's equivalent to keeping the orientation of ϕ and rotating the associated differential operators in the opposite direction instead!

Now let's see what happens when we require $F[\phi(\vec{r})] = F[\phi(\vec{r}')]$ for any $\phi(\vec{r})$ and any $\pi/2$ rotated coordinate $\vec{r}'$. There are no differential operators associated with V, so no additional constraints will apply. The ν_i terms gives us:

$$(\partial_x \phi)\nu_x + (\partial_y \phi)\nu_y = -(\partial_y \phi)\nu_x + (\partial_x \phi)\nu_y. \tag{B.7.29}$$

By matching partial derivatives, we see that $\nu_x = \nu_y = 0$.

The terms η_{ij} and κ_{ij}, which we can think of as matrices, must be symmetric due to the commutative property of partial derivatives. Since $\partial_i \partial_j \phi = \partial_j \partial_i \phi$ and $(\partial_i \phi)(\partial_j \phi) = (\partial_j \phi)(\partial_i \phi)$, exchanging i with j must be invariant.

For now, let us think only about rotations that exchange x and y, i.e., $\pi/2$ rotations along z. For the η_{ij} term, $F[\phi(\vec{r})] = F[\phi(\vec{r}')]$ leads to:

$$(\partial_x \partial_x \phi)\eta_{xx} + (\partial_x \partial_y \phi)\eta_{xy} + (\partial_y \partial_x \phi)\eta_{yx} + (\partial_y \partial_y \phi)\eta_{yy}$$
$$= (\partial_y \partial_y \phi)\eta_{xx} - (\partial_y \partial_x \phi)\eta_{xy} - (\partial_x \partial_y \phi)\eta_{yx} + (\partial_x \partial_x \phi)\eta_{yy}. \tag{B.7.30}$$

Matching partial derivatives, we see that $\eta_{xx} = \eta_{yy}$ and $\eta_{xy} = -\eta_{yx}$. Knowing that η_{ij} must be a symmetric matrix, we get $\eta_{xy} = \eta_{yx} = 0$.

For the κ_{ij} term, the same argument as above leads to $\kappa_{xx} = \kappa_{yy}$ and $\kappa_{xy} = \kappa_{yx} = 0$.

If we apply this kind of argument to all allowed symmetry rotations, we can prove similar relations to pairwise coordinate labels. In the end we get that $\nu_i = 0$, $\eta_{ij} = \delta_{ij}\eta$, and $\kappa_{ij} = \delta_{ij}\kappa$. This leads to the term $\kappa \sum_i (\partial_i \phi)^2 / 2 = \kappa |\nabla \phi|^2 / 2$.

Although we did this analysis in 3D so we don't blow our minds thinking about rotations in higher dimensions, the same analysis extends to higher dimensions as well. We are used to considering rotations about some axis, but it's also equivalent (and more generally applicable) if we think about rotations as defined on a plane. In higher dimensions, a $\pi/2$ rotation in a plane defined by two primary axes still leaves all other coordinates the same, and applies a rotation matrix proportional to $\begin{bmatrix} 0 & 1 \\ -1 & 0 \end{bmatrix}$ on the coordinates involved in the rotation. As a result, all the previous analysis of pairwise coordinates still holds.

Having proved that $\eta_{ij}(\phi) = \delta_{ij}\eta(\phi)$, let's see how to eliminate it completely. We can rewrite the terms involved as:

$$(\partial_i\partial_j\phi)\eta_{ij}(\phi) = (\partial_i\partial_j\phi)\delta_{ij}\eta(\phi)$$

$$= (\partial_i\partial_i\phi)\eta(\phi)$$

$$= \eta(\phi)(\nabla^2\phi)$$

$$= \nabla\cdot(\eta(\phi)\nabla\phi) - (\nabla\eta(\phi)\cdot\nabla\phi), \qquad \text{(B.7.31)}$$

where the last step is a vector identity. Let's put this term back into the integral, and use the divergence theorem:

$$\int d^d\vec{r}\,[\nabla\cdot(\eta(\phi)\nabla\phi) - (\nabla\eta(\phi)\cdot\nabla\phi)] \qquad \text{(B.7.32)}$$

$$= \int_{\partial V} dS\,\hat{n}\cdot[\eta(\phi)\nabla\phi] - \int d^d\vec{r}\,(\nabla\eta(\phi)\cdot\nabla\phi). \qquad \text{(B.7.33)}$$

The first term becomes a surface integral. For a thermodynamic system, we either take the volume of our system to very large, or we consider periodic boundary conditions. In the first case, the surface contribution becomes negligible as the volume of the system becomes very large. In the second case, periodic boundary conditions by definition give 0 for the surface integral regardless of the size of the system. In either case, we can ignore that term.

Then for the second term, using the chain rule (remember that $\phi = \phi(\vec{r})$):

$$\ldots = -\int d^d\vec{r}\,(\nabla\eta(\phi)\cdot\nabla\phi) \qquad \text{(B.7.34)}$$

$$= -\int d^d\vec{r}\left(\frac{\partial\eta(\phi)}{\partial\phi}\nabla\phi\cdot\nabla\phi\right) \qquad \text{(B.7.35)}$$

$$= -\int d^d\vec{r}\,\frac{\partial\eta(\phi)}{\partial\phi}|\nabla\phi|^2. \qquad \text{(B.7.36)}$$

In the end, we see that we get a term in the integral proportional to $|\nabla\phi|^2$! Since κ itself is an undetermined function, we can absorb Eq. B.7.36 into it and rename the new coefficient $\kappa(\phi)$. In the end, the η_{ij} term does not add new dependencies on the derivatives, and we arrive at Eq. B.7.22.

4. Assume that ϕ is small, and that our system obeys the symmetry

$$F[\phi] = F[-\phi]. \qquad \text{(B.7.37)}$$

Expand the first term in the expression B.7.22 to obtain the final Ginzburg-Landau free energy:

$$F[\phi] = \int d^d\vec{r}\left\{V_0 + \frac{\alpha}{2}\phi(\vec{r})^2 + \frac{u}{4}\phi(\vec{r})^4 + \frac{\kappa}{2}[\nabla\phi(\vec{r})]^2 + \ldots\right\}. \qquad \text{(B.7.38)}$$

Expanding $V(\phi)$ about $\phi = 0$ is just a Taylor series in ϕ:

$$V(\phi) = V_0 + V_1\phi + \frac{V_2}{2}\phi^2 + \frac{V_3}{3}\phi^3 + \frac{V_4}{4}\phi^4 + \ldots. \tag{B.7.39}$$

However, since we are imposing the condition,

$$F[\phi] = F[-\phi] \implies V[\phi] = V[-\phi],$$

this implies that:

$$V_1 = -V_1, \qquad V_3 = -V_3, \qquad \ldots \tag{B.7.40}$$

Thus all odd terms in the expansion must be zero. Renaming the coefficients, we get:

$$V(\phi) = V_0 + \frac{\alpha}{2}\phi^2 + \frac{u}{4}\phi^4 + \ldots. \tag{B.7.41}$$

B.7.5   *Solving for the Continuum Correlation Length

Let us add to the Landau-Ginzburg free energy we have just derived a term that represents an external perturbation $-h(\vec{r})\phi(r)$. For example, this can be an externally applied magnetic field h coupled to the magnetization so that the free energy respects a global symmetry $F[\phi; h] = F[-\phi; -h]$. The total free energy is now:

$$
\begin{aligned}
F[\phi; h] &= \int d^d\vec{r}\left\{ V_0 - h(\vec{r})\phi(\vec{r}) + \frac{\alpha}{2}\phi(\vec{r})^2 + \frac{u}{4}\phi(\vec{r})^4 + \frac{\kappa}{2}\left[\nabla\phi(\vec{r})\right]^2 + \ldots\right\} \\
&= \int d^d\vec{r}\,\mathcal{F}\left[\phi(\vec{r}), \nabla\phi(\vec{r}); h(\vec{r})\right]. \tag{B.7.42}
\end{aligned}
$$

1. We are trying to minimize the free energy F with respect to the order parameter ϕ. Since $F[\phi; h]$ is a functional of $\phi(\vec{r})$, the minimization entails finding the functional form of ϕ, which calls for the use of Euler-Lagrange equation

$$\frac{\partial\mathcal{F}}{\partial\phi} - \nabla\cdot\frac{\partial\mathcal{F}}{\partial(\nabla\phi)} = 0, \tag{B.7.43}$$

where $\partial/\partial\nabla\phi$ means the partial derivative w.r.t. each component of the spatial derivative $\nabla\phi = (\partial_x\phi, \partial_y\phi, \ldots)$, and is therefore a vector. If you need a recap on the calculus of variations, see Appendix A.4.

Show that the differential equation for ϕ that minimizes F is:

$$\alpha\phi + u\phi^3 - \kappa\nabla^2\phi = h(\vec{r}). \tag{B.7.44}$$

The free energy density (minus the constant part) is

$$\mathcal{F} = -h(\vec{r})\phi + \frac{\alpha}{2}\phi^2 + \frac{u}{4}\phi^4 + \frac{\kappa}{2}\nabla\phi\cdot\nabla\phi,$$

390 APPENDIX B. WORKSHEET SOLUTIONS

where the last term is indeed

$$\frac{\kappa}{2}\nabla\phi\cdot\nabla\phi = \frac{\kappa}{2}\sum_\gamma (\partial_\gamma\phi)^2.$$

Thus, the second term in the Euler-Lagrange equation is:

$$-\nabla\cdot\frac{\partial\mathcal{F}}{\partial(\nabla\phi)} = -\frac{\kappa}{2}\nabla\cdot\sum_\gamma\frac{\partial(\partial_\gamma\phi)^2}{\partial(\partial_\gamma\phi)}\hat{x}_\gamma$$

$$= -\frac{\kappa}{2}\nabla\cdot\sum_\gamma 2(\partial_\gamma\phi)\hat{x}_\gamma$$

$$= -\kappa\nabla^2\phi,$$

where ∇^2 is the Laplacian operator. Combined with the first term in the Euler-Lagrange equation, we get:

$$-h + \alpha\phi + u\phi^3 - \kappa\nabla^2\phi = 0.$$

Correlation length measures the spatial extent of the response to a local perturbation, so we cannot assume a spatially uniform ϕ anymore. We can model a small local perturbation at $\vec{r}_0$ of strength h_0 with a field $h = h_0\delta(\vec{r}-\vec{r}_0)$ where h_0 is small.

2. With out loss of generality, assume that $\vec{r}_0 = 0$ such that the perturbation is at the origin. Use Eq. (B.7.44) to derive a form of ϕ for both $T > T_c$ and $T < T_c$.

   Hint 1: Assume that the order parameter takes the form

   $$\phi(\vec{r}) = \phi_0 + \tilde{\phi}(\vec{r}), \tag{B.7.45}$$

   where the spatial variation is a small perturbation from the uniform solution ϕ_0 (i.e., when $h = 0$).

   Hint 2: The differential equation involving a Laplacian can be most easily solved by Fourier transforms. See Appendix A.6 for some more help.

Substituting $h = h_0\delta(\vec{r})$ into Eq. B.7.44, we have:

$$\alpha\phi + u\phi^3 - \kappa\nabla^2\phi = h_0\delta(\vec{r}). \tag{B.7.46}$$

Now since we said h_0 is a small perturbation, we expect that $\phi(\vec{r})$ will have the form of a small correction $\tilde{\phi}(\vec{r})$ on the order of h_0 from its uniform solution both above and below T_c:

$$\phi(\vec{r}) = \phi_0 + \tilde{\phi}(\vec{r}) + \mathcal{O}\left(h_0^2\right), \tag{B.7.47}$$

where

$$\phi_0 = \begin{cases} 0 & \text{for } T > T_c \\ \sqrt{\dfrac{|\alpha|}{u}} & \text{for } T < T_c. \end{cases} \tag{B.7.48}$$

Using this form of $\phi(\vec{r})$ in Eq. B.7.46 and keeping to first order in $\tilde{\phi}$, for $T > T_c$ we get:

$$\alpha\tilde{\phi}(\vec{r}) - \kappa\nabla^2\tilde{\phi}(\vec{r}) = h_0\delta(\vec{r}). \tag{B.7.49}$$

And for $T < T_c$:

$$2\alpha\tilde{\phi}(\vec{r}) - \kappa\nabla^2\tilde{\phi}(\vec{r}) = h_0\delta(\vec{r}). \tag{B.7.50}$$

Since the forms of the equations are exactly the same, let's just focus on the $T > T_c$ case for now. We can solve this differential equation with Fourier transforms. Recall that the (non-unitary) Fourier transform and its inverse are defined as

$$\mathbb{F}\{g(\vec{r})\} \equiv G(\vec{k}) = \int d^d\vec{r}\, e^{-i\vec{k}\cdot\vec{r}} g(\vec{r}), \tag{B.7.51}$$

$$\mathbb{F}^{-1}\{G(\vec{k})\} \equiv g(\vec{r}) = \frac{1}{(2\pi)^d} \int d^d\vec{k}\, e^{i\vec{k}\cdot\vec{r}} G(\vec{k}), \tag{B.7.52}$$

from which we can prove (with integration by parts) that:

$$\mathbb{F}\{\nabla^2 g(\vec{r})\} = -|\vec{k}|^2 G(\vec{k}). \tag{B.7.53}$$

We can also see the Fourier transform of the delta function $\delta(\vec{r})$ is:

$$\mathbb{F}\{\delta(\vec{r})\} = 1. \tag{B.7.54}$$

Let's then define the Fourier transform

$$\mathbb{F}\{\tilde{\phi}(\vec{r})\} = \tilde{\Phi}(\vec{k}), \tag{B.7.55}$$

and take the Fourier transform of the entirety of Eq. B.7.49:

$$\alpha\tilde{\Phi}(\vec{k}) + \kappa|\vec{k}|^2\tilde{\Phi}(\vec{k}) = h_0, \tag{B.7.56}$$

$$\tilde{\Phi}(\vec{k}) = \frac{h_0}{\alpha + \kappa|\vec{k}|^2}. \tag{B.7.57}$$

Now we just have to do the inverse transform to get $\tilde{\phi}(\vec{r})$:

$$\tilde{\phi}(\vec{r}) = \mathbb{F}^{-1}\{\tilde{\Phi}(\vec{k})\} \tag{B.7.58}$$

$$= \frac{h_0}{(2\pi)^d} \int d^d\vec{k}\, \frac{e^{i\vec{k}\cdot\vec{r}}}{\alpha + \kappa|\vec{k}|^2}. \tag{B.7.59}$$

The integral can be done for $d = 1$ and $d = 3$ using contour integration, as well as in general for arbitrary d with another method. We'll write down the results here, but the derivation is included in Appendix A.6. Note, we'll define $q = \sqrt{\alpha/\kappa}$.

In $d = 1$:

$$\tilde{\phi}(x) = \frac{h_0}{2\sqrt{\alpha\kappa}} e^{-q|x|}. \tag{B.7.60}$$

In $d = 3$:

$$\tilde{\phi}(\vec{r}) = \frac{h_0}{4\pi\kappa} \frac{e^{-q|\vec{r}|}}{|\vec{r}|}. \tag{B.7.61}$$

And in general for any d, for large $q|\vec{r}|$:

$$\tilde{\phi}(\vec{r}) \approx \frac{h_0}{2\kappa} \frac{q^{(d-3)/2}}{\left[2\pi|\vec{r}|\right]^{(d-1)/2}} e^{-q|\vec{r}|}. \tag{B.7.62}$$

This derivation is for $T > T_c$. For $T < T_c$, the analysis is completely identical, with the exception that the parameter $\alpha \to 2|\alpha|$ everywhere it occurs in q.

3. Give a definition of the correlation length from your answer (or the solution!) to the first part, and show that the correlation length ξ follows:

$$\xi \propto |T - T_c|^{-\nu}, \tag{B.7.63}$$

where $\nu = 1/2$ for both $T > T_c$ and $T < T_c$.

From the above derivation, we see that there's always an exponential decaying term in the response $\tilde{\phi}(\vec{r})$. We define the correlation length ξ as the lengthscale of the decay:

$$\xi = q^{-1}. \tag{B.7.64}$$

For $T > T_c$, the form is:

$$\xi = \sqrt{\frac{\kappa}{\alpha}} \propto |T - T_c|^{-1/2}. \tag{B.7.65}$$

For $T < T_c$, the only difference is that $\alpha \to 2|\alpha|$ everywhere, so the correlation length becomes:

$$\xi = \sqrt{\frac{\kappa}{2|\alpha|}} \propto |T - T_c|^{-1/2}. \tag{B.7.66}$$

4. From your answer to the previous part, show that there is a critical ratio:

$$\frac{\xi(T \to T_c^+)}{\xi(T \to T_c^-)} = \sqrt{2}. \tag{B.7.67}$$

From above using ξ for above and below T_c:

$$\frac{\xi(T \to T_c^+)}{\xi(T \to T_c^-)} = \lim_{\alpha \to 0} \frac{\sqrt{\kappa/\alpha}}{\sqrt{\kappa/2|\alpha|}} = \sqrt{2} \qquad \text{(B.7.68)}$$

5. A more subtle critical exponent governs the nature of the thermodynamic fluctuations at criticality $(T = T_c)$. Away from criticality, as we've just seen, the correlations fall exponentially with distance. A key feature of a system *at* criticality is that it has no characteristic lengthscale—it is fractal. As a consequence, correlation functions can only fall as a power of distance:

$$\langle \phi(\vec{r})\phi(\vec{r}') \rangle \propto |\vec{r} - \vec{r}'|^{-(d-2)+\eta} \text{ at } T = T_c \qquad \text{(B.7.69)}$$

where η is yet another critical exponent, which in the present case takes the value $\eta = 0$. Show that this is the case by evaluating this for $d = 3$.

We showed in the last section of Appendix A.6 that for $\kappa = 0$, the response has the form

$$\tilde{\phi}(\vec{r}) = \frac{h_0}{4\kappa\pi^{d/2}|\vec{r}|^{d-2}}\Gamma\left(\frac{d}{2} - 1\right) \implies \eta = 0, \qquad \text{(B.7.70)}$$

which also holds explicitly for $d = 3$, as mentioned in the appendix.

B.7.6 Susceptibility of the XY Model for $T < T_c$

Suppose we have a system of XY spins. There is an intruiging piece of physics about the way in which the spins respond to an externally applied field: below T_c, the susceptibility depends on the direction in which the field is applied! This worksheet gives you the chance to see why this is the case.

1. Let's start by considering the simpler case of an Ising order parameter, ϕ. Assume that everything is spatially uniform for now.

   (a) Suppose that $T > T_c$. What is ϕ if there are no applied fields? If we apply a small field of magnitude h, how does ϕ respond, in terms of χ?

   $\phi = 0$ in the absence of externally applied fields. If we apply a small field of magnitude h, then $\phi \simeq h\chi$.

   (b) Now consider $T < T_c$. The equilibrium value of $\phi = \bar{\phi}$ is nonzero. What is the new equilibrium value, in terms of χ, if we apply a small field h?

   Now $\phi \simeq \bar{\phi} + h\chi$, because the field changes ϕ by an amount proportional to χ.

2. Let's see what is different for the case of a two-component XY order parameter, $\phi = \phi_1\hat{\mathbf{e}}_1 + \phi_2\hat{\mathbf{e}}_2$. Remember that the spin can point in any direction in an abstract two-dimensional plane.

(a) Write down the expression for the energy in the presence of an externally applied field, $\mathbf{h} = h_1\hat{\mathbf{e}}_1 + h_2\hat{\mathbf{e}}_2$.

The energy is the dot product of the field and the spin,
$-\mathbf{h} \cdot \phi = -h_1\phi_1 - h_2\phi_2$.

(b) If $T > T_c$, do you expect the susceptibility to depend on the direction of the applied field? Write an expression for ϕ in the presence of a small field, $\mathbf{h}$, in terms of χ.

The susceptibility shouldn't depend on the direction of the applied field—after all, the XY model has a symmetry where all the orientations are equivalent, and this symmetry is not broken for $T > T_c$. So, $\phi \simeq \chi\mathbf{h} = h_1\chi\hat{\mathbf{e}}_1 + h_2\chi\hat{\mathbf{e}}_2$ here still.

(c) Now consider $T < T_c$. What is the equilibrium order parameter in terms of a magnitude $|\phi| = \bar{\phi}$ and an arbitrary angle $0 \le \theta < 2\pi$?

Since the symmetry is broken below T_c, the magnitude is nonzero, but the angle is arbitrary and spontaneously chosen. The order parameter is a vector of magnitude $\bar{\phi}$ in a direction of θ, so $\phi = \bar{\phi}\cos(\theta)\hat{\mathbf{e}}_1 + \bar{\phi}\sin(\theta)\hat{\mathbf{e}}_2$.

(d) Without loss of generality, assume that $\theta = 0$ so that ϕ points in the $\hat{\mathbf{e}}_1$ direction. Let's apply a weak magnetic field—we can apply it in either the $\hat{\mathbf{e}}_1$ or the $\hat{\mathbf{e}}_2$ direction. Do you expect ϕ to respond the same in both cases? Why or why not? *Hint:* Are the $\hat{\mathbf{e}}_1$ and $\hat{\mathbf{e}}_2$ directions equivalent?

In contrast to the case for $T > T_c$, the response is probably different depending on whether we apply a field in the $\hat{\mathbf{e}}_1$ or the $\hat{\mathbf{e}}_2$ direction. This is because the two directions are not equivalent any longer—one of them is parallel to the direction of the order parameter, whereas one of them is perpendicular to it.

Let $\hat{\mathbf{e}}_\parallel = \cos(\theta)\hat{\mathbf{e}}_1 + \sin(\theta)\hat{\mathbf{e}}_2$ be a unit vector in the direction of the order parameter, $\phi = \bar{\phi}\hat{\mathbf{e}}_\parallel$. Anything *parallel* to $\hat{\mathbf{e}}_\parallel$ is called longitudinal, and anything *perpendicular* to it is transverse. Let $\chi_\parallel$ and $\chi_\perp$ denote the longitudinal and transverse susceptibility, respectively.

3. We can think of our magnetic field as having a longitudinal and a transverse component, $\mathbf{h} = h_\parallel\hat{\mathbf{e}}_\parallel + h_\perp\hat{\mathbf{e}}_\perp$ (where $\hat{\mathbf{e}}_\perp$ is an in-plane unit vector perpendicular to $\hat{\mathbf{e}}_\parallel$, i.e. $\vec{\mathbf{e}}_\perp = -\sin(\theta)\vec{\mathbf{e}}_1 + \cos(\theta)\vec{\mathbf{e}}_2$).

(a) Write down the expression for ϕ in the presence of small $\mathbf{h}$, in terms of $\chi_\parallel$ and $\chi_\perp$.

The resulting order parameter will be the original equilibrium order parameter, plus the response due to the magnetic field. The longitudinal response will be $\chi_\parallel h_\parallel\hat{\mathbf{e}}_\parallel$, and the transverse response will be $\chi_\perp h_\perp\hat{\mathbf{e}}_\perp$, so overall,

$$\phi = \bar{\phi}\,\hat{\mathbf{e}}_\parallel + \chi_\parallel h_\parallel\hat{\mathbf{e}}_\parallel + \chi_\perp h_\perp\hat{\mathbf{e}}_\perp.$$

(b) What do you think is larger, $\chi_\parallel$ or $\chi_\perp$? *Hint:* How is the magnitude/direction of the order parameter affected by a longitudinal versus a transverse perturbation?

A longitudinal perturbation affects only the magnitude of ϕ. In comparison, a transverse perturbation rotates the direction θ of the vector and has a slight effect on the magnitude as well. Since the free energy depends only on the magnitude of ϕ, and not the angle, rotations in ϕ do not encounter any resistance. Therefore, the system will be more susceptible to magnetic fields in the transverse direction, $\chi_\perp > \chi_\parallel$.

Now that we understand what longitudinal and transverse susceptibility mean, let's explicitly calculate $\chi_\perp(\vec{r})$ and $\chi_\parallel(\vec{r})$ as functions of position. For T slightly below T_c we can use a Ginzburg-Landau form of the free energy,

$$\mathcal{F}_{var}[\phi] = \int d\vec{r} \left[V_0 + \frac{\alpha}{2}\phi^2 + \frac{u}{4}\phi^4 + \frac{\kappa}{2}|\nabla\phi|^2 - \mathbf{h}\cdot\phi \right],$$

where $\phi^2 = \phi_1^2 + \phi_2^2$, $\phi^4 = (\phi^2)^2$, and $|\nabla\phi|^2 = \sum_\alpha |\nabla\phi_\alpha|^2$. (Remember $\alpha < 0$ since we're below T_c). If $\mathbf{h}(\vec{r}) = 0$, then $\phi(\vec{r})$ is spatially uniform with magnitude $\bar{\phi} = \sqrt{|\alpha|/u}$. Without loss of generality, let's assume the ordering is in the $\theta = 0$ direction, so $\bar{\phi} = \bar{\phi}\hat{\mathbf{e}}_1$.

4. If we apply a weak magnetic field, then the order parameter shouldn't change much, so we can use an ansatz,

$$\phi(\vec{r}) = \bar{\phi} + \delta\phi(\vec{r}),$$

where $\delta\phi = \delta\phi_1\hat{\mathbf{e}}_1 + \delta\phi_2\hat{\mathbf{e}}_2$ is the response to the perturbation. Which component describes the longitudinal response? The transverse response?

Since we assumed the ordering was in the $\hat{\mathbf{e}}_1$ direction, the $\delta\phi_1$ component describes the longitudinal response, and the $\delta\phi_2$ component describes the transverse response.

5. Write down expressions for ϕ^2, ϕ^4, and $|\nabla\phi|^2$, keeping terms up to quadratic order in $\delta\phi_1$ and $\delta\phi_2$.

We have $\phi = (\bar{\phi} + \delta\phi_1)\hat{\mathbf{e}}_1 + \delta\phi_2\hat{\mathbf{e}}_2$, so

$$\phi^2 = (\bar{\phi} + \delta\phi_1)^2 + (\delta\phi_2)^2$$
$$= \bar{\phi}^2 + 2\bar{\phi}\delta\phi_1 + (\delta\phi_1)^2 + (\delta\phi_2)^2,$$
$$\phi^4 = \left(\bar{\phi}^2 + 2\bar{\phi}\delta\phi_1 + (\delta\phi_1)^2 + (\delta\phi_2)^2 \right)^2$$
$$= \bar{\phi}^4 + 4\bar{\phi}^3\delta\phi_1 + (4\bar{\phi}^2 + 2\bar{\phi}^2)(\delta\phi_1)^2 + 2\bar{\phi}^2(\delta\phi_2)^2 + \ldots$$
$$= \bar{\phi}^4 + 4\bar{\phi}^3\delta\phi_1 + 6\bar{\phi}^2(\delta\phi_1)^2 + 2\bar{\phi}^2(\delta\phi_2)^2 + \ldots,$$

$$|\nabla\phi|^2 = |\nabla\phi_1|^2 + |\nabla\phi_2|^2$$

$$|\nabla\phi_1|^2 = |\nabla\bar{\phi} + \nabla\delta\phi_1|^2 = |\nabla\delta\phi_1|^2$$

$$|\nabla\phi_2|^2 = |\nabla\delta\phi_2|^2$$

6. Substitute these expressions into the free energy and show that

$$\mathcal{F}_{var}[\phi] = \mathcal{F}_{var}[\bar{\phi}] + \int d\vec{r}\left[|\alpha|(\delta\phi_1)^2 + \frac{\kappa}{2}|\nabla\delta\phi_1|^2 - h_1\delta\phi_1\right]$$

$$+ \int d\vec{r}\left[+\frac{\kappa}{2}|\nabla\delta\phi_2|^2 - h_2\delta\phi_2\right].$$

Plugging in each of these, we get

$$\mathcal{F}_{var}[\phi] = \int d\vec{r}\Big[V_0 + \frac{\alpha}{2}\left(\bar{\phi}^2 + 2\bar{\phi}\delta\phi_1 + (\delta\phi_1)^2 + (\delta\phi_2)^2\right)$$

$$+ \frac{u}{4}\left(\bar{\phi}^4 + 4\bar{\phi}^3\delta\phi_1 + 6\bar{\phi}^2(\delta\phi_1)^2 + 2\bar{\phi}^2(\delta\phi_2)^2\right)$$

$$+ \frac{\kappa}{2}\left(|\nabla\delta\phi_1|^2 + |\nabla\delta\phi_2|^2\right)$$

$$- h_1\delta\phi_1 - h_2\delta\phi_2\Big].$$

To simplify this, let's group together the terms which are zero order in $\delta\phi$, first order, and second order:

$$\mathcal{F}_{var}[\phi] = \int d\vec{r}\Big[\underbrace{V_0 + \frac{\alpha}{2}\bar{\phi}^2 + \frac{u}{4}\bar{\phi}^4}_{\mathcal{F}_{var}[\bar{\phi}]}$$

$$+ (\alpha\bar{\phi} + u\bar{\phi}^3)\delta\phi_1$$

$$+ \left(\frac{\alpha}{2} + \frac{u}{4}6\bar{\phi}^2\right)\delta\phi_1^2 + \left(\frac{\alpha}{2} + \frac{u}{4}2\bar{\phi}^2\right)\delta\phi_2^2$$

$$+ \frac{\kappa}{2}\left(|\nabla\delta\phi_1|^2 + |\nabla\delta\phi_2|^2\right) - h_1\delta\phi_1 - h_2\delta\phi_2\Big].$$

We recognize that the zero-order term is simply the free energy evaluated at $\phi = \bar{\phi}$. The coefficient of the 1st-order term vanishes because

$$\alpha\sqrt{\frac{-\alpha}{u}} + u\sqrt{\frac{-\alpha^3}{u^3}} = \sqrt{\frac{-\alpha}{u}}\left(\alpha + u\frac{-\alpha}{u}\right) = 0.$$

The coefficient of the 2nd-order terms evaluate to:

$$\frac{\alpha}{2} + \frac{u}{4}6\bar{\phi}^2 = \frac{\alpha}{2} + \frac{u}{4}\frac{-6\alpha}{u} = \frac{\alpha}{2} - \frac{3\alpha}{2} = -\alpha$$

and

$$\frac{\alpha}{2} + \frac{u}{4}2\bar{\phi}^2 = \frac{\alpha}{2} + \frac{u}{4}\frac{-2\alpha}{u} = \frac{\alpha}{2} - \frac{\alpha}{2} = 0.$$

Interestingly, the coefficient of the $(\delta\phi_2)^2$ term vanishes! So, we are left with

$$\mathcal{F}_{var}[\phi] = \mathcal{F}_{var}[\bar{\phi}] + \int d\vec{r}\left[|\alpha|(\delta\phi_1)^2 + \frac{\kappa}{2}\left(|\nabla\delta\phi_1|^2 + |\nabla\delta\phi_2|^2\right) - h_1\delta\phi_1 - h_2\delta\phi_2\right].$$

7. What does the lack of a $(\delta\phi_2)^2$ term imply about the nature of the transverse susceptibility? Does this agree with your qualitative assessment earlier?

The coefficient of the $(\delta\phi_{1/2})^2$ term can be thought of as a spring constant which describes the restoring force, or how much $\delta\phi_{1/2}$ "wants" to go back to zero. Since it's equal to zero for $\delta\phi_2$, this implies that a transverse perturbation encounters no restoring force, in contrast to the longitudinal case. This agrees with our qualitative assessment from earlier that the transverse susceptibility is larger.

In fact, you may remember that the coefficient of the $(\delta\phi_{1/2})^2$ was proportional to $T - T_c$ in the case of the Ising model—and since this coefficient is zero, the transverse susceptibility behaves as if we are exactly at the critical point. In this case, there is no correlation length, and the correlations decay as a power law of the distance.

One other way to think of this is that to leading order, there is no penalty for transverse deviations. These just lead to a rotation of the angle of the order parameter, not to a change in magnitude. There is, however, still the κ term, which discourages nearby sites from having $\delta\phi_2$ that varies too rapidly in space.

B.7.7 A Domain Wall in an Ising System

So far, in a phase with broken Ising symmetry, we have assumed that one of the two states is chosen across the entire extent of the system. However, there can be situations where different regions of the system occupy different broken-symmetry states. These regions are known as domains, and the boundaries between them are known as domain walls.

1. Consider an Ising system for $T < T_c$ in a continuum limit where the state of the system is described by an order parameter field, ϕ. Suppose that
   - far to the left there is a slight negative magnetic field,
   - far to the right there is a slight positive magnetic field, and
   - around the origin there are no magnetic fields.

   Do you expect ϕ to be positive or negative for $x \ll 0$? How about for $x \gg 0$? Sketch a graph of what you expect $\phi(x)$ to look like.

I expect $\phi < 0$ for $x \ll 0$ and $\phi > 0$ for $x \gg 0$ because of the effect of the weak magnetic fields. Therefore, there will be the $\phi < 0$ equilibrium solution far to the left, the $\phi > 0$ equilibrium far to the right, and some characteristic width W in the middle where the two will interpolate.

2. If T is not too far below T_c, then the free energy of the system has a Landau-Ginzberg form,

$$\mathcal{F}[\phi] = \int d\vec{r} \left[\frac{\alpha}{2} \phi^2 + \frac{u}{4} \phi^4 + \frac{\kappa}{2} \left(\frac{\partial \phi}{\partial x} \right)^2 \right] \tag{B.7.71}$$

where $\alpha = T - T_c < 0$. Let $\bar{\phi} = \sqrt{-\alpha/u}$. What is $\lim_{x \to \infty} \phi(x)$? How about $\lim_{x \to -\infty} \phi(x)$?

These are $+\bar{\phi}$ and $-\bar{\phi}$, respectively.

3. Suppose that our domain wall has a width W.

   (a) Label W in your sketch from part (1).

   (b) Give an argument for why W cannot be too small.

   If W is too small, then the slope of the graph will be very large. (In the lattice picture, this corresponds to neighboring sites having very different values of ϕ.) This is unfavorable because neighboring sites have a preference to have similar values of ϕ, owing to the κ term (or the interaction term).

   (c) Give an argument for why W cannot be too large.

   If W is too large, then there is a large range of x where the order parameter is far from either of the equilibrium values, $\bar{\phi}$ or $-\bar{\phi}$. (In the lattice picture, this corresponds to a large number of sites.) This is unfavorable because the single-site part of the free energy is minimized by ϕ close to $\bar{\phi}$ or $-\bar{\phi}$.

4. Let's come up with a rough estimate for W. In this "calculation" you are free to drop any constant factors; we just want to find out how W depends on the parameters α, u, and κ.

   (a) Sketch $\frac{\alpha}{2} \phi^2 + \frac{u}{4} \phi^4$. What is the height of the "free energy barrier" to cross from one minimum to the other?

   This is a two-well potential with minima at $\phi = \pm\bar{\phi}$. The depth of the wells is

   $$\mathcal{F}(\bar{\phi}) = -\frac{\alpha^2}{2u} + \frac{u\alpha^2}{4u^2} = -\frac{\alpha^2}{4u}, \tag{B.7.72}$$

   so the barrier height is $\sim \alpha^2/u$.

(b) The free energy cost of a domain wall is the difference in free energy between a state without a domain wall and a state with a domain wall. What is the cost that arises from the first two terms in Eq. B.7.71, $\frac{\alpha}{2}\phi^2 + \frac{u}{4}\phi^4$?

   Hint: To estimate an integral, multiply a typical value of the integrand by the range of x where it is nonzero.

   In a state without a domain wall, the free energy is simply $-\alpha^2/4u$, integrated over the length of the system. A state *with* a domain wall has the same free energy density, except for the region of the space where there is a domain wall. Therefore, the difference between these two involves just the region of space with the domain wall:

$$\int_{-W/2}^{W/2}\left(\frac{\alpha}{2}\phi(x)^2 + \frac{u}{4}\phi(x)^4\right)dr - \left[\frac{-\alpha^2}{4u}W\right].$$

   The exact value of the integral will depend on the exact functional form of $\phi(x)$; however, to estimate it we just need to know a typical value of the integrand. A reasonable guess is that the integrand is typically something like half of the barrier height, so we may estimate the integral as

$$\int_{-W/2}^{W/2}\frac{1}{2}\frac{-\alpha^2}{4u}dx - \left[\frac{-\alpha^2}{4u}W\right] = \frac{\alpha^2}{8u}W.$$

   We shouldn't take the factor of $1/8$ here too literally because this is just an estimate.

(c) Now let's look at the last term. What is $\frac{\partial\phi}{\partial x}$ roughly?

   This is the slope of the graph. Over the range of x of width W, $\phi(x)$ changes from $-\bar\phi$ to $\bar\phi$, so here the slope is on average $2\bar\phi/W$. Otherwise the slope is zero.

(d) Determine, up to a constant, the free energy cost of the domain wall arising from the last term in the free energy.

   In the absence of the domain wall, the last term is zero because there is no spatial variation. With a domain wall, there is a simple integral

$$\int dx\,\frac{\kappa}{2}\left(\frac{\partial\phi}{\partial x}\right)^2.$$

   We can estimate this integral, as before, by multiplying the average value of the integrand with the width of x where it is nonzero. We have just determined the average of $\frac{\partial\phi}{\partial x}$ is $2\bar\phi/W$ over the range of x of width W. Therefore, the integral is roughly

$$\int dx \, \frac{\kappa}{2} \left(\frac{\partial \phi}{\partial x} \right)^2 \simeq W \times \frac{\kappa}{2} \left(\frac{2\bar{\phi}}{W} \right)^2$$

$$= W \frac{\kappa}{2} \frac{4|\alpha|}{uW^2}$$

$$= \frac{2\kappa|\alpha|}{u} \frac{1}{W}.$$

Note that $|\alpha| = -\alpha$ since $\alpha < 0$.

(e) Estimate W by minimizing the sum of parts (b) and (d).

The overall free energy cost of the domain wall is about

$$\frac{\alpha^2}{8u} W + \frac{2\kappa|\alpha|}{u} \frac{1}{W}.$$

Note that the first term grows with W, whereas the second term shrinks with W, so somewhere in the middle there must be a W which minimizes the free energy cost. This is given by taking the derivative with respect to W and setting it to zero:

$$0 = \frac{\alpha^2}{8u} - \frac{2\kappa|\alpha|}{u} \frac{1}{W^2}.$$

$$\implies \frac{\alpha^2}{8u} \simeq \frac{2\kappa|\alpha|}{u} \frac{1}{W^2}$$

$$\implies W \simeq \sqrt{\frac{2\kappa|\alpha|}{u} \frac{8u}{\alpha^2}} = \boxed{4\sqrt{\kappa/|\alpha|}}.$$

Again, note that because of our gross approximations, we don't expect the factor of 4 out front to really mean anything. However, we do expect the general scaling of the result to be valid; that is, we do think that W will be proportional to $\sqrt{\kappa/|\alpha|}$.

5. You can actually arrive at this same result from just dimensional analysis. Let E denote the dimension of free energy, M the dimension of ϕ, and L the dimension of length.

(a) What are the dimensions of α, u, and κ?

Matching dimensions on the left and right sides of the equation, we see that:

$$[\alpha] = \frac{E}{M^2 L}, \quad [u] = \frac{E}{M^4 L}, \quad [\kappa] = \frac{EL}{M^2},$$

(b) What is the only combination that gives dimension of length?

Indeed the only combination that gives dimensions of length is $\sqrt{\kappa/\alpha}$.

B.8   Chapter 8 Worksheet Solutions

B.8.1   Gaussian Integrals and Wick's Theorem (Isserlis' Theorem)

Here we will prove a few properties of Gaussian integrals, as well as Wick's (probability) theorem.

Review on Gaussian integrals

First we will review some basic facts about Gaussian integrals of the form:

$$\int_{-\infty}^{\infty} dx\, x^{2n} e^{-x^2/2\lambda}. \tag{B.8.1}$$

If you are already very familiar with this, you can skip to the next section.

1. A general (nonnormalized, zero-centered) Gaussian distribution is:

$$G(x; \lambda) = e^{-x^2/2\lambda}. \tag{B.8.2}$$

Show that the definite integral

$$I = \int_{-\infty}^{\infty} dx\, G(x; \lambda) = \sqrt{2\pi\lambda}. \tag{B.8.3}$$

Hint: Use cylindrical coordinates $x^2 + y^2 = \rho^2$ to evaluate the quantity

$$I^2 = \int_{-\infty}^{\infty} \int_{-\infty}^{\infty} dx\, dy\, G(x; \lambda) G(y; \lambda). \tag{B.8.4}$$

By evaluating the I^2, we can make use of the coordinate transformation:

$$\int_{-\infty}^{\infty} \int_{-\infty}^{\infty} dx\, dy \to 2\pi \int_{0}^{\infty} \rho\, d\rho. \tag{B.8.5}$$

The factor of 2π is due to the integrand having no angular dependence. This gives us:

$$I^2 = 2\pi \int_{0}^{\infty} \rho\, d\rho\, e^{-\rho^2/2\lambda}. \tag{B.8.6}$$

You can do this integral by parts, or just a little pattern recognition:

$$I^2 = 2\pi\lambda \int_{0}^{\infty} \frac{\rho}{\lambda}\, d\rho\, e^{-\rho^2/2\lambda} \tag{B.8.7}$$

$$= 2\pi\lambda \left(-e^{-\rho^2/2\lambda}\right)\Big|_{0}^{\infty} = 2\pi\lambda. \tag{B.8.8}$$

So the original integral is $I = \sqrt{2\pi\lambda}$.

2. Since $G(x; \lambda)$ is even, all integrals of the form

$$\int_{-\infty}^{\infty} dx\, x^k\, G(x; \lambda) \tag{B.8.9}$$

evaluate to zero for odd powers k. Thus, only integrals of the form

$$\int_{-\infty}^{\infty} dx\, x^{2n} e^{-x^2/2\lambda} \tag{B.8.10}$$

are nonzero for integer $n \geq 0$. Show that:

$$\int_{-\infty}^{\infty} dx\, x^{2n} e^{-x^2/2\lambda} = \lambda(2n-1) \int_{-\infty}^{\infty} dx\, x^{2n-2} e^{-x^2/2\lambda}. \tag{B.8.11}$$

Hint: One way to do this is by integration by parts.

Let us use integration by parts with the identification (with a small label change $n \to m$ for now)

$$u' = x^{2m},$$

$$v = e^{-x^2/2\lambda},$$

such that after integration by parts ($\int dx\, u'v = uv - \int dx\, uv'$), we have:

$$\int_{-\infty}^{\infty} dx\, x^{2m} e^{-x^2/2\lambda} \tag{B.8.12}$$

$$= \frac{x^{2n+1}}{2n+1} e^{-x^2/2\lambda}\Big|_{-\infty}^{\infty} \overset{0}{\searrow} - \int_{-\infty}^{\infty} dx\, \frac{-x^{2m+2}}{\lambda(2m+1)} e^{-x^2/2\lambda}. \tag{B.8.13}$$

Letting $m = n - 1$, we find, as advertised:

$$\int_{-\infty}^{\infty} dx\, x^{2n} e^{-x^2/2\lambda} = \lambda(2n-1) \int_{-\infty}^{\infty} dx\, x^{2n-2} e^{-x^2/2\lambda}. \tag{B.8.14}$$

3. For our last integration exercise, prove the following integrals. *Hint:* Complete the square for the exponents, i.e., find c for $-x^2/2\lambda + bx = -(x/\sqrt{2\lambda} - c)^2 + c^2$.

Let us complete the square first, which will allow us to do the integral comfortably with a simple change of variable. Matching coefficients for

$$\frac{-x^2}{2\lambda} + bx = -\left(\frac{x}{\sqrt{2\lambda}} - c\right)^2 + c^2$$

$$= -\frac{x^2}{2\lambda} + \frac{2cx}{\sqrt{2\lambda}}$$

$$\implies c = b\sqrt{\frac{\lambda}{2}}.$$

(a)
$$Z = \int_{-\infty}^{\infty} dx \, e^{-x^2/2\lambda + bx} = e^{b^2\lambda/2}\sqrt{2\pi\lambda}.$$

Using the completed square of the exponent, we can make a variable substitution:

$$\frac{y}{\sqrt{2\lambda}} = \frac{x}{\sqrt{2\lambda}} - c. \tag{B.8.15}$$

Since y is just an offset from the original variable x, $dy = dx$, and the bounds of integration $\pm\infty$ are not affected. The integral then becomes:

$$\int_{-\infty}^{\infty} dy \, e^{-y^2/2\lambda + b^2\lambda/2} = e^{b^2\lambda/2}\sqrt{2\pi\lambda}. \tag{B.8.16}$$

(b)
$$\int_{-\infty}^{\infty} dx \, x \, e^{-x^2/2\lambda + bx} = b\lambda Z.$$

With the same variable substitution as above,

$$y + c\sqrt{2\lambda} = x \implies x = y + b\lambda,$$

the integral becomes:

$$\int_{-\infty}^{\infty} dy \, (y + b\lambda) \, e^{-y^2/2\lambda + b^2\lambda/2} = b\lambda e^{b^2\lambda/2}\sqrt{2\pi\lambda} = b\lambda Z.$$

Note that the odd part of the integral drops out.

(c)
$$\int_{-\infty}^{\infty} dx \, x^2 e^{-x^2/2\lambda + bx} = (\lambda + b^2\lambda^2)Z.$$

Finally, using all the tricks from before, we get

$$\int_{-\infty}^{\infty} dy \, (y + b\lambda)^2 \, e^{-y^2/2\lambda + b^2\lambda/2}$$

$$= e^{b^2\lambda/2} \int_{-\infty}^{\infty} dy \, (y^2 + 2yb\lambda + b^2\lambda^2) \, e^{-y^2/2\lambda}$$

$$= e^{b^2\lambda/2}\sqrt{2\pi\lambda}(\lambda + b^2\lambda^2) = (\lambda + b^2\lambda^2)Z,$$

Where we used the results from previous parts. Note that the odd integrand term once again drops out.

Since you now have all the tricks in the bag to do Gaussian integrals, we will give you the following expressions to save you from the tedium (but feel free to prove them!):

$$\int_{-\infty}^{\infty} dx \, x^3 e^{-x^2/2\lambda + bx} = (3b\lambda^2 + b^3\lambda^3)Z,$$

$$\int_{-\infty}^{\infty} dx \, x^4 e^{-x^2/2\lambda + bx} = (3\lambda^2 + 6b^2\lambda^3 + b^4\lambda^4)Z.$$

Wick's (probability) theorem

A multivariate (multidimensional) Gaussian distribution of N variables can be written as:

$$G(\{x\};\{\lambda\}) = \prod_{j}^{N} \exp(-x_j^2/2\lambda_j),$$

where each x_j is an independent random variable. We can define a "partition function" or normalization factor as

$$Z = \int \prod_{j}^{N} \left[dx_j \, \exp(-x_j^2/2\lambda_j) \right].$$

Since the x_js are independent, this integral factorizes into:

$$Z = \prod_{j} \int dx_j \, \exp(-x_j^2/2\lambda_j) = \prod_{j} \sqrt{2\pi\lambda_j} = \prod_{j} Z_j. \tag{B.8.17}$$

With this in place, we can define the expectation values of various quantities, in general of the form:

$$\langle A(\{x\}) \rangle = Z^{-1} \int A(\{x\}) \prod_{j} \left[dx_j \, \exp(-x_j^2/2\lambda_j) \right]. \tag{B.8.18}$$

Let us now calculate some expectation values. You may want to consult results from the previous sections.

As a freebie, $\langle x_j \rangle = 0$ for any j, since the integrand is odd for that j.

(a) Show that $\left\langle \prod_{j}(x_j)^{n_j} \right\rangle = \prod_{j} \left\langle (x_j)^{n_j} \right\rangle$ for any nonnegative integer n_j.

By definition:

$$\left\langle \prod_{j}(x_j)^{n_j} \right\rangle = Z^{-1} \int \prod_{j} \left[dx_j \, (x_j)^{n_j} \exp(-x_j^2/2\lambda_j) \right]$$

$$= \prod_{j} \frac{\int dx_j \, (x_j)^{n_j} \exp(-x_j^2/2\lambda_j)}{Z_j}.$$

Since each x_j is independent, we can factorize it as such. Notice that each factor in this product (the boxed part) is in fact equal to:

$$\langle (x_k)^{n_k} \rangle = Z^{-1} \int (x_k)^{n_k} \prod_{j} \left[dx_j \, \exp(-x_j^2/2\lambda_j) \right]$$

$$= \boxed{\frac{\int dx_k \, (x_k)^{n_k} \exp(-x_k^2/2\lambda_k)}{Z_k}} \prod_{j \neq k} \frac{\int dx_j \, \exp(-x_j^2/2\lambda_j)}{Z_j}^{\; 1}.$$

Thus

$$\left\langle \prod_j (x_j)^{n_j} \right\rangle = \prod_j \left\langle (x_j)^{n_j} \right\rangle.$$

(b) Show that $\langle x_j x_k \rangle = \delta_{jk} \lambda_j$, and thus $\langle (x_j)^2 \rangle = \lambda_j$.

If $j \neq k$, then using the previous result, we have $\langle x_j x_k \rangle = \langle x_j \rangle \langle x_k \rangle = 0$.

If $j = k$, then $\langle x_j x_k \rangle = \langle (x_j)^2 \rangle = \langle (x_k)^2 \rangle = \lambda_j$, invoking the result from previous sections.

Combining these two statements, we get $\langle x_j x_k \rangle = \delta_{jk} \lambda_j$, and $\langle (x_j)^2 \rangle = \lambda_j$ follows as well.

(c) Show that, specifically, $\langle (x_k)^4 \rangle = 3 \langle (x_k)^2 \rangle^2$.

We invoke the general result of Gaussian integrals from previous sections:

$$\langle (x_k)^4 \rangle = Z^{-1} \int (x_k)^4 \prod_j \left[dx_j \exp(-x_j^2/2\lambda_j) \right]$$

$$= \frac{\int (x_k)^4 dx_k \exp(-x_k^2/2\lambda_k)}{Z_k} \prod_{j \neq k} \frac{\int dx_j \exp(-x_j^2/2\lambda_j)}{Z_j}^{\,1}$$

$$= \frac{\lambda_k^2 (4-1)(2-1) Z_k}{Z_k} = 3 \langle (x_k)^2 \rangle^2.$$

(d) Argue that

$$\langle x_p x_q x_r x_s \rangle = \langle x_p x_q \rangle \langle x_r x_s \rangle + \langle x_p x_r \rangle \langle x_q x_s \rangle + \langle x_p x_s \rangle \langle x_q x_r \rangle \qquad \text{(B.8.19)}$$

is true.

Using results from part (b), we want to show that

$$\langle x_p x_q x_r x_s \rangle = \delta_{pq} \delta_{rs} \lambda_p \lambda_r + \delta_{pr} \delta_{qs} \lambda_p \lambda_q + \delta_{ps} \delta_{qr} \lambda_p \lambda_q. \qquad \text{(B.8.20)}$$

First, if any of these indices occurs only once, for instance, if p is not equal to any of q, r, s, then the overall expression is zero, since $\langle x_p \rangle = 0$. Thus, nonzero terms must contain pairs of identical indices, for example, $p = q$, giving a factor of $\langle (x_p)^2 \rangle = \lambda_p$. This selection rule is represented by the Kronecker deltas, selecting all the possible combinations of paired indices.

In the case where all four indices are the same, we get $\langle (x_p)^4 \rangle = 3 \langle (x_p)^2 \rangle^2 = 3\lambda_p^2$, which the expression correctly recovers.

Thus we have demonstrated Eq. B.8.19 to be true.

406 APPENDIX B. WORKSHEET SOLUTIONS

Here have proved Wick's probability theorem up to the 4th-order expectation values, but this can be further generalized to higher (even) powers, which can be proven with combinatorics.[13]

Importantly, for the expectation value of a single variable to the power of $2n$:

$$\langle x^{2n} \rangle = (2n-1)!! \langle x^2 \rangle^n, \tag{B.8.21}$$

where !! is the double factorial

$$(2n-1)!! = (2n-1)(2n-3)\cdots 3 \cdot 1. \tag{B.8.22}$$

As an example of the higher order general expression, the sixth-order one is:

$$\begin{aligned}
&\langle x_1 x_2 x_3 x_4 x_5 x_6 \rangle \\
&= \langle x_1 x_2 \rangle \langle x_3 x_4 \rangle \langle x_5 x_6 \rangle + \langle x_1 x_2 \rangle \langle x_3 x_5 \rangle \langle x_4 x_6 \rangle + \langle x_1 x_2 \rangle \langle x_3 x_6 \rangle \langle x_4 x_5 \rangle \\
&+ \langle x_1 x_3 \rangle \langle x_2 x_4 \rangle \langle x_5 x_6 \rangle + \langle x_1 x_3 \rangle \langle x_2 x_5 \rangle \langle x_4 x_6 \rangle + \langle x_1 x_3 \rangle \langle x_2 x_6 \rangle \langle x_4 x_5 \rangle \\
&+ \langle x_1 x_4 \rangle \langle x_2 x_3 \rangle \langle x_5 x_6 \rangle + \langle x_1 x_4 \rangle \langle x_2 x_5 \rangle \langle x_3 x_6 \rangle + \langle x_1 x_4 \rangle \langle x_2 x_6 \rangle \langle x_3 x_5 \rangle \\
&+ \langle x_1 x_5 \rangle \langle x_2 x_3 \rangle \langle x_4 x_6 \rangle + \langle x_1 x_5 \rangle \langle x_2 x_4 \rangle \langle x_3 x_6 \rangle + \langle x_1 x_5 \rangle \langle x_2 x_6 \rangle \langle x_3 x_4 \rangle \\
&+ \langle x_1 x_6 \rangle \langle x_2 x_3 \rangle \langle x_4 x_5 \rangle + \langle x_1 x_6 \rangle \langle x_2 x_4 \rangle \langle x_3 x_5 \rangle + \langle x_1 x_6 \rangle \langle x_2 x_5 \rangle \langle x_3 x_4 \rangle .
\end{aligned} \tag{B.8.23}$$

Wick's theorem for linear functions

We can generalize the above to a generalized linear function. Consider the same multivariate Gaussian distribution of N variables,

$$G(\{x\}; \{\lambda\}) = \prod_j^N \exp(-x_j^2/2\lambda_j),$$

and a general linear function of these variables,

$$F_a = \sum_j f_j^a x_j,$$

where f_j^a are the coefficients for each of the terms linear in x_j.

We will prove a few identities of the expectation value of linear functions using Wick's probability theorem. Again, as a freebie, $\langle F_a \rangle = \sum_j f_j^a \langle x_j \rangle = 0$.

1. Show that:

$$\langle F_a F_b \rangle = \sum_j f_j^a f_j^b \lambda_j.$$

---

13. See L. Isserlis, *Biometrika*, **12**, 134–139 (1918).

$$F_a F_b = \sum_{j,k} (f_j^a x_j)(f_k^b x_k),$$

$$\langle F_a F_b \rangle = \sum_{j,k} f_j^a f_k^b \langle x_j x_k \rangle$$

$$= \sum_{j,k} f_j^a f_k^b \delta_{jk} \lambda_j$$

$$= \sum_j f_j^a f_j^b \lambda_j,$$

where in the last step we sum over the index k.

2. Show that:

$$\langle F_a F_b F_c \rangle = 0.$$

$$F_a F_b F_c = \sum_{i,j,k} (f_i^a x_i)(f_j^b x_j)(f_k^c x_k),$$

$$\langle F_a F_b F_c \rangle = \sum_{i,j,k} f_i^a f_j^b f_k^c \langle x_i x_j x_k \rangle .$$

However, since there are three terms in $\langle x_i x_j x_k \rangle$, there must be at least one index occurring an odd number of times, thus giving $\langle x_i x_j x_k \rangle = 0$. Therefore:

$$\langle F_a F_b F_c \rangle = 0.$$

3. Show that:

$$\langle F_a F_b F_c F_d \rangle = \langle F_a F_b \rangle \langle F_c F_d \rangle + \langle F_a F_c \rangle \langle F_b F_d \rangle + \langle F_a F_d \rangle \langle F_b F_c \rangle .$$

$$F_a F_b F_c F_d = \sum_{p,q,r,s} (f_p^a x_p)(f_q^b x_q)(f_r^c x_r)(f_s^d x_s),$$

$$\langle F_a F_b F_c F_d \rangle = \sum_{p,q,r,s} f_p^a f_q^b f_r^c f_s^d \langle x_p x_q x_r x_s \rangle .$$

Now we invoke Wick's theorem, and sum over appropriate indices:

$$\langle F_a F_b F_c F_d \rangle$$

$$= \sum_{p,q,r,s} f_p^a f_q^b f_r^c f_s^d (\delta_{pq}\delta_{rs}\lambda_p\lambda_r + \delta_{pr}\delta_{qs}\lambda_p\lambda_q + \delta_{ps}\delta_{qr}\lambda_p\lambda_q)$$

$$= \sum_{p,r} (f_p^a f_p^b \lambda_p)(f_r^c f_r^d \lambda_r) + \sum_{p,q} (f_p^a f_p^c \lambda_p)(f_q^b f_q^d \lambda_q) + \sum_{p,q} (f_p^a f_p^d \lambda_p)(f_q^b f_q^c \lambda_q)$$

$$= \langle F_a F_b \rangle \langle F_c F_d \rangle + \langle F_a F_c \rangle \langle F_b F_d \rangle + \langle F_a F_d \rangle \langle F_b F_c \rangle .$$

Indeed, Wick's theorem for general linear functions generalizes the same way as the basic Wick's theorem.

Expectation of linear functions for Gaussian random variables

1. Consider a linear function $f = f_0 + f_k x_k$ for some specific value of k (no summations here). By completing the square a previously, show that $\langle e^f \rangle = \exp\left\{ \langle f \rangle + (1/2)[\langle f^2 \rangle - \langle f \rangle^2] \right\}$. *Hint:* Calculate $\langle f \rangle$ and $\langle f^2 \rangle$ along the way.

Just as how we completed the square the first time:

$$\langle e^f \rangle = Z^{-1} \int e^{f_0} e^{f_k x_k} \prod_j \left[dx_j \exp(-x_j^2/2\lambda_j) \right]$$

$$= \frac{e^{f_0} \int dx_k \exp(-x_k^2/2\lambda_k + f_k x_k)}{Z_k} \prod_{j \neq k} \underbrace{\frac{\int dx_j \exp(-x_j^2/2\lambda_j)}{Z_j}}_{1}$$

$$= \frac{e^{f_0} \exp(f_k^2 \lambda_k/2) Z_k}{Z_k} = \exp(f_0 + f_k^2 \lambda_k/2).$$

This can be rewritten in terms of expectation values of f. First we have $\langle f \rangle = f_0$, since x_k is an odd power. We also have:

$$\langle f^2 \rangle = \langle f_0^2 + 2 f_0 f_k x_k + f_k^2 x_k^2 \rangle$$

$$= f_0^2 + f_k^2 \lambda_k.$$

Thus:

$$f^2 \lambda_k/2 = \frac{\langle f^2 \rangle - \langle f \rangle^2}{2}$$

$$\implies \langle e^f \rangle = \exp\left\{ \langle f \rangle + (1/2)[\langle f^2 \rangle - \langle f \rangle^2] \right\}.$$

2. Now let us consider a generalized linear function $f = f_0 + \sum_j f_j x_j$. Show that the expression $\langle e^f \rangle = \exp\left\{ \langle f \rangle + (1/2)[\langle f^2 \rangle - \langle f \rangle^2] \right\}$ is still true for the generalized function.

Following the previous logic:

$$\langle e^f \rangle = Z^{-1} \int e^{f_0} \prod_j \left[dx_j e^{f_j x_j} \exp(-x_j^2/2\lambda_j) \right]$$

$$= e^{f_0} \prod_j \frac{\int dx_j \exp(-x_j^2/2\lambda_j + f_j x_j)}{Z_j}$$

$$= e^{f_0} \prod_j \frac{\exp(f_j^2 \lambda_j/2) Z_j}{Z_j} = \exp\left(f_0 + \sum_j f_j^2 \lambda_j/2\right).$$

As before, $\langle f \rangle = f_0$. On the other hand, $\langle f^2 \rangle$ is a bit more interesting. If we expand out the square, we get:

$$f^2 = f_0^2 + 2f_0 \sum_j f_j x_j + \sum_{j,k} f_j f_k x_j x_k. \tag{B.8.24}$$

The expectation value is therefore:

$$\langle f^2 \rangle = f_0^2 + \sum_{j,k} f_j f_k \delta_{jk} \lambda_j$$

$$= f_0 + \sum_j f_j^2 \lambda_j.$$

Thus:

$$\sum_j f_j^2 \lambda_j/2 = \frac{\langle f^2 \rangle - \langle f \rangle^2}{2}$$

$$\implies \langle e^f \rangle = \exp\left\{\langle f \rangle + (1/2)[\langle f^2 \rangle - \langle f \rangle^2]\right\}.$$

B.8.2 Mean-Field Approximations to the 0D ϕ^4 Theory

Let us consider the ϕ^4 theory, but with only a single site. This is similar to considering the statistical properties of the single site Ising model. Let us consider the Hamiltonian

$$H(\phi) = \frac{\tilde{\alpha}}{2}\phi^2 + \frac{u}{4}\phi^4, \tag{B.8.25}$$

so we can consider that ϕ is the order parameter. For this exercise, let us work in units where the temperature is always 1, such that $T = \beta = 1$. The partition function is then defined as

$$Z = \int_{-\infty}^{\infty} d\phi\, e^{-H(\phi)}, \tag{B.8.26}$$

and the free energy as

$$F = -\ln Z.$$

Since the Hamiltonian is even in ϕ, only expectation values of even powers of ϕ will be nonzero, defined as:

$$\langle \phi^{2n} \rangle = Z^{-1} \int_{-\infty}^{\infty} d\phi\, \phi^{2n} e^{-H(\phi)}. \tag{B.8.27}$$

1. We will begin with some rescaling of parameters. We will assume $u > 0$ to ensure the energy is bounded below. By using a variable substitution of the form $\theta = \phi u^{1/4}$, and an appropriate definition of α, show that:

(a)
$$F(\tilde{\alpha}, u) = \frac{\ln u}{4} + \mathcal{F}(\alpha),$$

where

$$\mathcal{F}(\alpha) = F(\tilde{\alpha} = \alpha, u = 1). \qquad \text{(B.8.28)}$$

Since $\phi = u^{-1/4}\theta$, $d\phi = u^{-1/4}d\theta$, and the partition function can be rewritten as follows:

$$Z = \int d\phi \, \exp\left(-\frac{\tilde{\alpha}}{2}\phi^2 - \frac{u}{4}\phi^4\right)$$

$$= u^{-1/4} \int d\theta \, \exp\left(-\frac{\tilde{\alpha}}{2\sqrt{u}}\theta^2 - \frac{\theta^4}{4}\right).$$

Defining $\alpha = \tilde{\alpha}/\sqrt{u}$, we have:

$$Z = u^{-1/4} \int d\theta \, \exp\left(-\frac{\alpha}{2}\theta^2 - \frac{\theta^4}{4}\right). \qquad \text{(B.8.29)}$$

The free energy is then

$$F = -\ln Z = \frac{\ln u}{4} - \ln\left[\int d\theta \, \exp\left(-\frac{\alpha}{2}\theta^2 - \frac{\theta^4}{4}\right)\right],$$

in which we can recognize the second term as

$$-\ln\left[Z(\tilde{\alpha} = \alpha, u = 1)\right] = F(\tilde{\alpha} = \alpha, u = 1).$$

(b)
$$\langle\phi^{2n}\rangle = u^{-n/2}G_{2n}(\alpha),$$

where

$$G_{2n}(\alpha) = \langle\phi^{2n}\rangle_{\tilde{\alpha}=\alpha, u=1}. \qquad \text{(B.8.30)}$$

Using the same definition of α as above, the expectation value becomes:

$$\langle\phi^{2n}\rangle = Z^{-1} \int d\phi \, \phi^{2n} \exp\left(-\frac{\tilde{\alpha}}{2}\phi^2 - \frac{u}{4}\phi^4\right)$$

$$= \frac{1}{Zu^{1/4}} \int d\theta \, \frac{\theta^{2n}}{u^{n/2}} \exp\left(-\frac{\alpha}{2}\theta^2 - \frac{\theta^4}{4}\right).$$

We can now substitute in the expression for Z we found in the previous part,

$$\left\langle \phi^{2n} \right\rangle = \frac{u^{1/4}}{u^{n/2}u^{1/4}} \frac{\int d\theta \, \theta^{2n} \exp\left(-\frac{\alpha}{2}\theta^2 - \frac{\theta^4}{4}\right)}{\int d\theta \, \exp\left(-\frac{\alpha}{2}\theta^2 - \frac{\theta^4}{4}\right)}$$

$$= u^{-n/2} G_{2n}(\alpha),$$

where we can see that the quotient of the two integrals, after a change of variable $\theta \to \phi$, is equivalent to $\left\langle \phi^{2n} \right\rangle_{\tilde{\alpha}=\alpha, u=1} = G_{2n}(\alpha)$.

The point of this exercise is show that the physics is in fact entirely contained in the (rescaled) coefficient α, and therefore in the functions $\mathcal{F}(\alpha)$ and $G_{2n}(\alpha)$, which only depend on α.

2. While the functions $\mathcal{F}(\alpha)$ and $G_{2n}(\alpha)$ *can* be evaluated exactly in terms of special functions, this is not particularly illuminating. Here let us use a variational approximation and see how well we can do compared to the exact solution. Since we have shown that α is the physically important parameter, we will work with $\tilde{\alpha} = \alpha$ and $u = 1$ such that $F = \mathcal{F}$ and $G_{2n} = \left\langle \phi^{2n} \right\rangle$.

Consider a trial Hamiltonian:

$$H_{tr} = \frac{\mu}{2}\phi^2.$$

(a) Evaluate the variational free energy $F_{var} = F_{tr} + \left\langle H - H_{tr} \right\rangle_{tr}$.

Let us start with the trial partition function:

$$Z_{tr} = \int d\phi \, \exp\left(-\frac{\mu\phi^2}{2}\right)$$

$$= \sqrt{\frac{2\pi}{\mu}}.$$

The trial free energy is then:

$$F = -\ln Z_{tr} = \frac{1}{2}\ln\left(\frac{\mu}{2\pi}\right).$$

The second term in the variation free energy is:

$$\left\langle H - H_{tr} \right\rangle_{tr} = \left\langle \frac{\alpha - \mu}{2}\phi^2 + \frac{\phi^4}{4} \right\rangle_{tr}$$

$$= \frac{\alpha - \mu}{2}\left\langle \phi^2 \right\rangle_{tr} + \frac{1}{4}\left\langle \phi^4 \right\rangle_{tr}$$

$$= \frac{\alpha - \mu}{2\mu} + \frac{3}{4\mu^2}.$$

The last step makes use of previous Gaussian integrals, since the trial ensemble is Gaussian distributed.

Putting it together, we get:

$$F_{var} = \frac{1}{2}\ln\left(\frac{\mu}{2\pi}\right) + \frac{\alpha - \mu}{2\mu} + \frac{3}{4\mu^2}.$$

(b) Minimize F_{var} to find value of the variational parameter μ that best approximates the system. Keeping in mind that $\mu > 0$ due to the $\ln\mu$ term in the free energy, show that

$$\mu = \frac{\alpha + \sqrt{\alpha^2 + 12}}{2}.$$

Solving for $\frac{dF_{var}}{d\mu} = 0$, assuming $\mu > 0$, we get

$$0 = \frac{dF_{var}}{d\mu} = \frac{1}{2\mu} - \frac{\alpha}{2\mu^2} - \frac{6}{4\mu^3},$$

$$0 = \mu^2 - \alpha\mu - 3,$$

$$\mu = \frac{\alpha + \sqrt{\alpha^2 + 12}}{2}$$

by taking the positive root of the solution to ensure $\mu > 0$.

(c) Evaluate $G_2(\alpha)$ in the trial ensemble. Substitute in the optimal μ from the last step and expand $G_2(\alpha)$ to three leading orders, for both small and large α. *Hint:* To expand about $\alpha \to \infty$, substitute in $\gamma = \alpha^{-1}$, expand about $\gamma = 0$, then reverse the substitution.

In the trial ensemble,

$$G_{2n}|_{tr} = \langle \phi^{2n} \rangle_{tr},$$

which we can evaluate quickly with our knowledge of Gaussian integrals. Thus for $n = 1$:

$$G_2(\alpha) = \langle \phi^2 \rangle_{tr} = \mu^{-1}. \tag{B.8.31}$$

Knowing that $\mu^{-1} = \frac{2}{\alpha + \sqrt{\alpha^2 + 12}}$, we can perform the prescribed expansions and get:

$$G_2(\alpha) = \begin{cases} \dfrac{1}{\sqrt{3}} - \dfrac{\alpha}{6} + \dfrac{\alpha^2}{24\sqrt{3}} + \cdots & \text{about } \alpha = 0 \\[2ex] \dfrac{1}{\alpha} - \dfrac{3}{\alpha^3} + \dfrac{18}{\alpha^5} + \cdots & \text{about } \alpha \to \infty. \end{cases}$$

3. So far we have used a trial Hamiltonian that respected the symmetry of the system. However, in general we can have spontaneously broken symmetries in the trial Hamiltonian as well. Consider now the trial Hamiltonian:

$$H_{tr} = \frac{\mu}{2}\phi^2 - h\phi.$$

(a) Evaluate the variational free energy $F_{var} = F_{tr} - \langle H - H_{tr} \rangle_{tr}$.

The distribution is now an offset Gaussian, which is a bit more tedious. However, since we have done much work in Chapter 8 worksheets, we can now use the results. Start again with the trial partition function,

$$Z_{tr} = \int d\phi \, \exp\left(-\frac{\mu\phi^2}{2} + h\phi\right)$$

$$= e^{h^2/2\mu}\sqrt{\frac{2\pi}{\mu}},$$

which gives us the trial free energy:

$$F_{tr} = -\ln Z_{tr} = \frac{1}{2}\ln\left(\frac{\mu}{2\pi}\right) - \frac{h^2}{2\mu}.$$

The second term in the variational free energy is more tedious:

$$\langle H - H_{tr} \rangle_{tr} = \frac{\alpha - \mu}{2}\langle\phi^2\rangle_{tr} + \frac{1}{4}\langle\phi^4\rangle_{tr} + h\langle\phi\rangle_{tr}.$$

Since the trial ensemble is not a zero-centered Gaussian, even the odd terms have nonzero expectation value. We can invoke the various integrals we performed before, and get:

$$\langle H - H_{tr} \rangle_{tr} = \frac{\alpha - \mu}{2}\overbrace{\left(\frac{1}{\mu} + \frac{h^2}{\mu^2}\right)}^{\langle\phi^2\rangle_{tr}=} + \frac{1}{4}\overbrace{\left(\frac{3}{\mu^2} + \frac{6h^2}{\mu^3} + \frac{h^4}{\mu^4}\right)}^{\langle\phi^4\rangle_{tr}=} + h\overbrace{\left(\frac{h}{\mu}\right)}^{\langle\phi\rangle_{tr}=}$$

$$= -\frac{1}{2} + \frac{h^2 + \alpha}{2\mu} + \frac{3 + h^2\alpha}{4\mu^2} + \frac{3h^2}{2\mu^3} + \frac{h^4}{4\mu^4}.$$

Putting it all together, we get:

$$F_{var} = \frac{1}{2}\ln\left(\frac{\mu}{2\pi}\right) - \frac{1}{2} + \frac{\alpha}{2\mu} + \frac{3 + h^2\alpha}{4\mu^2} + \frac{3h^2}{2\mu^3} + \frac{h^4}{4\mu^4}. \tag{B.8.32}$$

(b) Minimize F_{var} to find values of the parameters μ and h that best approximate the system. Show that there are two solutions (remember $\mu > 0$ by necessity):

$$h = 0, \ \mu = \frac{\alpha + \sqrt{\alpha^2 + 12}}{2}$$

and

$$h = \sqrt{-\alpha\mu^2 - 3\mu}, \ \mu = -\alpha + \sqrt{\alpha^2 - 6}.$$

Taking derivatives of F_{var}, we get the system of equations:

$$\frac{\partial F_{var}}{\partial h} = 0 \implies \qquad h\left(\frac{h^2}{\mu} + \alpha\mu + 3\right) = 0,$$

$$\frac{\partial F_{var}}{\partial \mu} = 0 \implies \qquad \frac{1}{2} - \frac{\alpha}{2\mu} - \frac{2\alpha h^2 + 3}{\mu^2} - \frac{9h^2}{2\mu^3} - \frac{h^4}{\mu^4} = 0.$$

Quartic equations are hard to solve, so let us solve for h first and get two solutions:

$$h = \begin{cases} 0 \\ \sqrt{-\alpha\mu^2 - 3\mu}. \end{cases}$$

Here we took only the positive root for h, because F_{var} depends only on even powers of h and therefore the two roots are equivalent.

Substituting $h = 0$ into $\partial F_{var}/\partial \mu = 0$ and solving for μ, we get

$$\mu^2 - \alpha\mu - 3 = 0 \implies \mu = \frac{\alpha + \sqrt{\alpha^2 + 12}}{2},$$

which unsurprisingly is the same solution from the previous part when $h = 0$ by assumption.

For the second solution, we can substitute in the nonzero solution of h in $\partial F_{var}/\partial \mu = 0$ and solve for μ:

$$\left.\frac{\partial F_{var}}{\partial \mu}\right|_{h=\sqrt{-\alpha\mu^2 - 3\mu}} = 0$$

$$2\alpha + \mu + \frac{6}{\mu} = 0$$

$$\implies \mu = -\alpha + \sqrt{\alpha^2 - 6}.$$

(c) Interpret these solutions. According to the mean-field approximation, is there a phase transition for some critical value of $\alpha = \alpha_c$? Should there be a phase transition?

As part of our calculations for the previous part, we found that the expectation value of the order parameter is:

$$\langle \phi \rangle_{tr} = \frac{h}{\mu}. \tag{B.8.33}$$

Thus, the value of h determines whether there is a finite (symmetry-broken) order parameter and therefore a phase transition.

The first set of solutions,

$$h = 0, \ \mu = \frac{\alpha + \sqrt{\alpha^2 + 12}}{2},$$

Is valid for all values of α, and describes the high-symmetry state. The second set of solutions, however, describes a different state:

$$h = \sqrt{-\alpha\mu^2 - 3\mu}, \ \mu = -\alpha + \sqrt{\alpha^2 - 6}.$$

For h to be real, we must have:

$$-\alpha\mu^2 - 3\mu \geq 0,$$

$$\alpha\mu \leq -3.$$

Since $\mu > 0$ is essential, it means that α must be negative. For μ to be real and positive, we need:

$$\alpha^2 - 6 \geq 0 \implies \alpha \leq -\sqrt{6}.$$

We choose the negative root because $\mu > 0$ is mandatory, and this satisfies the condition $\alpha\mu \leq -3$. This implies that there may be a phase transition at a critical value of α_c for $\alpha \leq \sqrt{6}$.

Hint: For your reference, here are the various free energies plotted as a function of α.

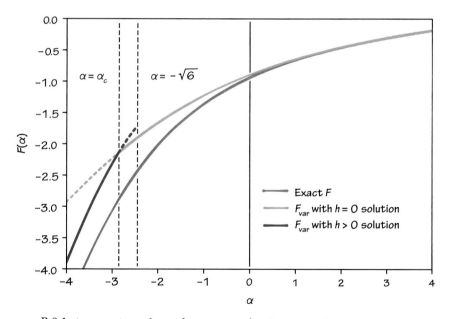

FIGURE B.8.1. A comparison of exact free energy and various approximations to it.

As we can see here, the mean-field approximation very closely follows the exact free energy. However, the kink in the approximation (at the intersection of the blue and red lines) indicates a discontinuity in the derivative of the free energy, which suggests a phase transition. The critical value α_c is in fact less than the upper bound we have calculated, $\alpha_c < -\sqrt{6}$.

However, since this is a single site system, we know there cannot be a phase transition, which is reflected by the smooth exact free energy curve.

B.10   Chapter 10 Worksheet Solutions

B.10.1   Renormalization Group of the 1D Ising Chain

As a concrete and tractable example, let us calculate the critical properties of the 1D Ising model using the method of renormalization group.

Recall that the Hamiltonian of the 1D Ising model is:

$$H = -J \sum_j \sigma_j \sigma_{j+1} - h \sum_j \sigma_j + \sum_j \varepsilon_0.$$

Here we have included an additive constant term in the energy of ε_0 per site, which does not modify the physics. We will see why we added this term shortly. H can be rewritten in terms of a function $K(\sigma_j, \sigma_{j+1})$:

$$H = \sum_j^N K(\sigma_j, \sigma_{j+1})$$

$$K(\sigma_a, \sigma_b) = -J\sigma_a\sigma_b - \frac{h}{2}(\sigma_a + \sigma_b) + \varepsilon_0.$$

We will work in units where $T = \beta = 1$, so the partition function can then be written as the trace of a product over transfer matrices:

$$Z = \mathrm{Tr}\left[t^N\right],$$

$$t = e^{-\varepsilon_0}\begin{bmatrix} e^{J+h} & e^{-J} \\ e^{-J} & e^{J-h} \end{bmatrix}.$$

The idea is to coarse-grain every two sites into one, renormalize the coupling constants while doubling the number of sites, and repeat the procedure to see which parameters are relevant and irrelevant.

1. We start by decimating the original chain into blocks of two spins each. Show that this amounts to calculating $Z = \mathrm{Tr}\left[(t^2)^{N/2}\right]$, and calculate $\tilde{t} = t^2$. This corresponds to the transfer matrix between block spins.

The partition function written out explicitly as a summation over indices is

$$Z = \sum_{\sigma_1=\pm1}\sum_{\sigma_2=\pm1}\cdots\sum_{\sigma_N=\pm1} t_{\sigma_1\sigma_2}t_{\sigma_2\sigma_3}\cdots t_{\sigma_N\sigma_1},$$

where the σ_js index the transfer matrices. We want to rewrite it such that only every other summation remains, meaning we can just sum over (for example) the even indices. Assuming N is even, this leaves us with

$$Z = \sum_{\sigma_1=\pm 1} \cdots \sum_{\sigma_{N-1}=\pm 1} \left(\sum_{\sigma_2=\pm 1} t_{\sigma_1\sigma_2} t_{\sigma_2\sigma_3} \right) \cdots \left(\sum_{\sigma_N=\pm 1} t_{\sigma_{N-1}\sigma_N} t_{\sigma_N\sigma_1} \right)$$

$$= \sum_{\sigma_1=\pm 1} \sum_{\sigma_3=\pm 1} \cdots \sum_{\sigma_{N-1}=\pm 1} t^2_{\sigma_1\sigma_3} t^2_{\sigma_3\sigma_5} \cdots t^2_{\sigma_{N-1}\sigma_1}$$

$$= \text{Tr}\left[(t^2)^{N/2} \right],$$

where we now have a new transfer matrix $\tilde{t} = t^2$, which we can find with a matrix multiplication:

$$\tilde{t} = e^{-2\varepsilon_0} \begin{bmatrix} e^{-2J} + e^{2J+2h} & e^h + e^{-h} \\ e^h + e^{-h} & e^{-2J} + e^{2J-2h} \end{bmatrix}.$$

2. We now want to map a new chain of N block spins onto the original chain, thus giving us a new partition function $\tilde{Z} = \text{Tr}\left[\tilde{t}^N \right]$ where each transfer matrix has the form:

$$\tilde{t} = e^{-\tilde{\varepsilon}_0} \begin{bmatrix} e^{\tilde{J}+\tilde{h}} & e^{-\tilde{J}} \\ e^{-\tilde{J}} & e^{\tilde{J}-\tilde{h}} \end{bmatrix}.$$

Since $\tilde{t} = t^2$, show that:

$$\tilde{J} = J + \frac{1}{4}\ln\left[\frac{\left(e^{-4J} + e^{-2h}\right)\left(e^{-4J} + e^{2h}\right)}{4\cosh^2(h)} \right], \tag{B.10.1}$$

$$\tilde{h} = h + \frac{1}{2}\ln\left[\frac{\cosh(h+2J)}{\cosh(h-2J)} \right], \tag{B.10.2}$$

$$\tilde{\varepsilon}_0 = 2\varepsilon_0 - \frac{1}{4}\ln\left[8\cosh^2(h)\left(\cosh(2h) + \cosh(4J)\right) \right]. \tag{B.10.3}$$

Writing out $\tilde{t} = t^2$:

$$e^{-\tilde{\varepsilon}_0} \begin{bmatrix} e^{\tilde{J}+\tilde{h}} & e^{-\tilde{J}} \\ e^{-\tilde{J}} & e^{\tilde{J}-\tilde{h}} \end{bmatrix} = e^{-2\varepsilon_0} \begin{bmatrix} e^{-2J} + e^{2J+2h} & e^h + e^{-h} \\ e^h + e^{-h} & e^{-2J} + e^{2J-2h} \end{bmatrix}.$$

This gives us three equations:

$$-\tilde{\varepsilon}_0 + \tilde{J} + \tilde{h} = -2\varepsilon_0 + \ln\left(e^{-2J} + e^{2J+2h} \right), \tag{B.10.4}$$

$$-\tilde{\varepsilon}_0 + \tilde{J} - \tilde{h} = -2\varepsilon_0 + \ln\left(e^{-2J} + e^{2J-2h} \right), \tag{B.10.5}$$

$$-\tilde{\varepsilon}_0 - \tilde{J} = -2\varepsilon_0 + \ln\left(e^h + e^{-h} \right). \tag{B.10.6}$$

Eqs. B.10.4 and B.10.5 give us:

$$\tilde{h} = h + \frac{1}{2} \ln \left[\frac{\cosh(h + 2J)}{\cosh(h - 2J)} \right]. \tag{B.10.7}$$

Eqs. B.10.4 and B.10.5 also give us:

$$-\tilde{\varepsilon}_0 + \tilde{J} = -2\varepsilon_0 + \frac{1}{2} \ln \left[e^{4J} + e^{-4J} + e^{2h} + e^{-2h} \right]. \tag{B.10.8}$$

Now we can solve for $\tilde{J}$ and get

$$\tilde{J} = J + \frac{1}{4} \ln \left[\frac{\left(e^{-4J} + e^{-2h} \right) \left(e^{-4J} + e^{2h} \right)}{4 \cosh^2(h)} \right] \tag{B.10.9}$$

as well as $\tilde{\epsilon}_0$:

$$\tilde{\varepsilon}_0 = 2\varepsilon_0 - \frac{1}{4} \ln \left[8 \cosh^2(h) \left(\cosh(2h) + \cosh(4J) \right) \right]. \tag{B.10.10}$$

Of course, a healthy amount of algebra has been omitted, but nothing particularly tricky occurs!

3. There are of course many ways to simplify the equations, but the reason we wrote out Eqs. B.10.1 to B.10.3 the way we did is so that now we can express them as beta functions:

$$\tilde{J} - J = \beta_J(J, h),$$

$$\tilde{h} - h = \beta_h(J, h),$$

$$\tilde{\varepsilon}_0 - 2\varepsilon_0 = f(J, h).$$

We note here that the ε_0 is not a physically significant parameter, but naturally arises as a part of the RG iteration process, so we use the label $f(J, h)$ instead of calling it a beta function.

A fixed point occurs when beta functions are equal to zero. Show that the fixed point in $J - h$ parameter space is $(J, h) = (0, 0)$, and determine the "relevance" of J and h.

The two relevant equations we want to solve are:

$$\beta_J(J, h) = 0 \implies \frac{1}{4} \ln \left[\frac{\left(e^{-4J} + e^{-2h} \right) \left(e^{-4J} + e^{2h} \right)}{4 \cosh^2(h)} \right] = 0,$$

$$\beta_h(J, h) = 0 \implies \frac{1}{2} \ln \left[\frac{\cosh(h + 2J)}{\cosh(h - 2J)} \right] = 0.$$

$\beta_h(J, h) = 0$ is satisfied only when $h = 0$. For all other values of $h > 0$, $\beta_h > 0$, meaning that h increases exponentially without bound as the RG iteration occurs. This means that h is always relevant.

For $h = 0$,

$$\beta_J(J, h = 0) = \frac{1}{2} \ln\left(\frac{1 + e^{-4J}}{2}\right),$$

$$\frac{1}{2} \ln\left(\frac{1 + e^{-4J}}{2}\right) = 0 \implies J = 0,$$

which gives us the fixed point $(J, h) = (0, 0)$. For $J > 0$, $\beta(J, h = 0) < 0$, which means that J decreases exponentially toward the $J = 0$ fixed points, and is therefore irrelevant.

We have been doing our calculations with $T = \beta = 1$, which means that the parameter J is indeed actually J/T. The $J = 0$ fixed point, interpreted as temperature, means that $T \to \infty$ is really the fixed point, and therefore no finite temperature phase transition occurs.

There is in fact another fixed point at $J \to \infty$ (check that there is!) we neglected to analyze. This corresponds to the "ordered phase" of the 1D Ising model at $T = 0$, but as we have discovered that J flows toward 0 for any finite J, the $T = 0$ fixed point is unstable and does not indicate a true phase transition.

4. Finally, let us briefly take a look at the correlation length ξ. From the above analysis, we can see that, if we start with some coupling constant J_0, then we can write down a recurrence relation for the coupling constant after n RG iterations:

$$J_n = J_{n-1} + \beta_J(J_{n-1}, h = 0) = J_{n-1} + \frac{1}{2} \ln\left[\frac{1 + e^{-4J_{n-1}}}{2}\right].$$

(a) Assume that we are far away from the fixed point, so J_0 is large and therefore e^{-4J} is small. Show that:

$$J_n = J_0 - \frac{n}{2} \ln(2) + \mathcal{O}\left(e^{-4J_{n-1}}\right). \tag{B.10.11}$$

Give an approximate limit on n for this approximation to hold.

If we take the large J expansion of $\beta_J(J, h = 0)$, we get:

$$\frac{1}{2} \ln\left[\frac{1 + e^{-4J}}{2}\right] = -\frac{1}{2} \ln(2) + \frac{e^{-4J}}{2} + \mathcal{O}\left(e^{-8J}\right).$$

For large J, the first term dominates, and thus the RG iteration can be approximated by

$$J_n = J_0 - \frac{n}{2} \ln(2) + \mathcal{O}\left(e^{-4J_{n-1}}\right),$$

as every iteration reduces the value of J by $\ln(2)/2$.

Of course, this only makes sense if J_n remains large after each iteration, which sets a limit on the number of iterations n such that:

$$J_0 \gg \frac{n}{2} \ln(2),$$

$$n \ll \frac{2J_0}{\ln(2)}.$$

(b) We are now going to push the limit of this approximation, that is, use a value of n as large as possible such that Eq. B.10.11 will break down for any larger n. Let us call this value n', and the corresponding value $J_{n'} = a \sim \mathcal{O}(1)$, indicating that it is on the order of unity. Justify this limit on the approximation.

The reason that Eq. B.10.11 holds up to a certain value of n is that the next-order term $\sim e^{-4J_{n-1}}$ is small. The approximation breaks down once this term is approximately on the same order as the leading term:

$$\frac{\ln(2)}{2} = e^{-4J},$$

$$J = \frac{\ln(2) - \ln\left[\ln(2)\right]}{4} \approx 0.265.$$

Of course this is not exactly when the approximation breaks down, but we see that it is roughly when $J \sim \mathcal{O}(1)$.

(c) The correlation length ξ is defined through the two-site correlation function:

$$\langle \sigma_0 \sigma_L \rangle_J = e^{-L/\xi(J)},$$

where L is the distance between the two sites. Physically, the correlation length should depend on the coupling strength J, and for $J \sim \mathcal{O}(1)$, let $\xi(J) \equiv \xi_1$.

Consider now the separation distance $L = l2^{n'}$ (the same n' from part 4b), where l is the distance between each site. Since each RG iteration coarse-grains each two adjacent spins into one, the correlation of two sites at distance L with coupling J_0 corresponds to the correlation of two sites at distance l after n' RG iterations with coupling $J_{n'}$. In equation, this means:

$$\langle \sigma_0 \sigma_L \rangle_{J_0} = \langle \sigma_0 \sigma_l \rangle_{J_{n'}}. \tag{B.10.12}$$

Using the approximations:

$$J_{n'} = a \sim \mathcal{O}(1),$$

$$\xi(a) = \xi_1 \sim l,$$

show that $\xi(J_0) = \left[\xi_1 e^{-2a}\right] e^{2J_0}$.

First, let us rewrite the expression Eq. B.10.12:

$$\langle\sigma_0\sigma_L\rangle_{J_0} = \langle\sigma_0\sigma_l\rangle_{J_{n'}},$$

$$e^{-L/\xi(J_0)} = e^{-l/\xi(J_{n'})},$$

$$\frac{l2^{n'}}{\xi(J_0)} = \frac{l}{\xi(J_{n'})},$$

$$\xi(J_0) = \xi(J_{n'})2^{n'}.$$

Now, let us invoke that $J_{n'} = a$, and use the expression from 4b,

$$a = J_{n'} = J_0 - \frac{n'}{2}\ln(2), \tag{B.10.13}$$

where we (with prior justification) ignore the higher order terms. This gives us:

$$2^{n'} = e^{2J_0 - 2a}.$$

Finally, since $J_{n'} = a \sim \mathcal{O}(1)$, $\xi(J_{n'}) = \xi_1$. Putting it all together, we get:

$$\xi(J_0) = \left[\xi_1 e^{-2a}\right]e^{2J_0}. \tag{B.10.14}$$

(d) Recall that the exact solution for the 1D Ising model gives us the correlation length:

$$\xi_{exact} = \left|\ln\left[\tanh(\beta J)\right]\right|^{-1}. \tag{B.10.15}$$

Assuming that $\beta J \gg 1$, expand ξ_{exact} to leading order in $e^{\beta J}$. Does it match up with our RG calculation?

Assuming that $\beta J \gg 1$, since $\tanh(\beta J) \le 1$, we can drop the absolute value and rewrite ξ_{exact} as:

$$\xi_{exact} = \left[-\ln\left(\frac{e^{\beta J} - e^{-\beta J}}{e^{\beta J} + e^{-\beta J}}\right)\right]^{-1}. \tag{B.10.16}$$

Since we want to expand around $e^{\beta J} \to \infty$, let us make the substitution $x = e^{-\beta J}$ and expand about $x = 0$:

$$\left[-\ln\left(\frac{1/x - x}{1/x + x}\right)\right]^{-1} = \frac{1}{2x^2} - \frac{x^2}{6} + \cdots. \tag{B.10.17}$$

Plugging back in the substitution, we get

$$\xi_{exact} = \frac{1}{2}e^{2\beta J} + \cdots, \tag{B.10.18}$$

which matches our RG approximation, making the identification $J_0 = \beta J$, since we have been working with $T = \beta = 1$ units.

With knowledge of only the leading term in the beta function, we have successfully computed the correct J_0 dependence of the correlation length. There is a number of order 1 in the prefactor—$[\xi_a e^{-2a}]$—that we cannot compute from simple RG analysis. This requires a more complete solution. But the important (exponential) temperature dependence of the correlation length is seen directly from knowledge of the leading-order behavior of the beta function. (Recall, with the temperature included, J_0 really means J_0/T.)

B.10.2   Perturbative Renormalization Group of the ϕ^4 Theory in d Dimensions

In this exercise, we will carry out the renormalization group calculation using perturbation expansion of the ϕ^4 Hamiltonian described in the chapter. Before any calculations, let us detail the setup of the problem first.

Setting up the system

Consider a Hamiltonian density for a translationally invariant system with a global $\theta \to -\theta$ symmetry:

$$\mathcal{H}[\theta(\vec{r})] = \mathcal{H}_0 + \mathcal{H}_1,$$

$$\mathcal{H}_0 = \frac{\kappa}{2} |\nabla \theta|^2,$$

$$\mathcal{H}_1 = V(\theta)$$

$$= V_0 + \frac{V_2}{2} \theta^2 + \frac{V_4}{4} \theta^4 + \frac{V_6}{6} \theta^6 + \cdots,$$

where $\theta = \theta(\vec{r})$ is a position-dependent real scalar field in d dimensions. The two terms of the Hamiltonian density are $\mathcal{H}_0$, which is the noninteracting part, and $\mathcal{H}_1 = V$, which we call the interacting part. If $V = 0$, then this is a noninteracting system corresponding to the Gaussian model, which we have solved exactly before. Here we will see what happens when we turn on a small V, which we will write as an expansion in terms even in θ (due to the global symmetry), and develop the RG equations perturbatively.

The (total) Hamiltonian is:

$$H = \int_\Omega d^d \vec{r} \, (\mathcal{H}_0 + \mathcal{H}_1) = H_0 + H_1. \tag{B.10.19}$$

We will consider the system in d-dimensional space with periodic boundary conditions over length L, such that the domain[14] of the volume integral is $\Omega = L^d$.

We will define a quantity S (sometimes called the "action") as:

$$S = \frac{1}{T}(H_0 + H_1) = S_0 + S_1. \tag{B.10.20}$$

The Boltzmann factor is then defined in terms of the action:

$$e^{-S} = e^{-H_0/T} e^{-H_1/T}. \tag{B.10.21}$$

The partition function is consequently:

$$Z = \int D\theta\, e^{-S}$$
$$= \int D\theta\, \exp\left[-\frac{1}{T}\int_\Omega d^d\vec{r}\,\left(\frac{\kappa}{2}|\nabla\theta|^2 + V(\theta)\right)\right], \tag{B.10.22}$$

where $\int D\theta$ indicates the functional integral over all configurations of θ over Ω.

As the last (but critical) step of our setup, let us consider some smallest lengthscale l below which $\theta(\vec{r})$ cannot meaningfully vary (so l is akin to the lattice constant of a microscopic model). This allows us to consider the field $\theta(\vec{r})$ as a sum of its Fourier components $\theta_{\vec{k}}$:

$$\theta(\vec{r}) = L^{-d/2} \sum_{|\vec{k}|<\Lambda} e^{i\vec{k}\cdot\vec{r}}\theta_{\vec{k}}, \tag{B.10.23}$$

where $\Lambda \propto l^{-1}$ is the (spatial) frequency cutoff imposed due to the microscopic lengthscale, and is for the time being an arbitrary value. This allows us to define the functional integral as:[15]

$$\int D\theta = \prod_{|\vec{k}|<\Lambda} \int d\theta_{\vec{k}} \tag{B.10.24}$$

The notation here really means, written out for the components $\vec{k} = k_j \hat{k}$,

$$\prod_{|\vec{k}|<\Lambda} = \prod_{0\le k_1} \prod_{0\le k_2} \cdots \prod_{0\le k_d}^{|\vec{k}|<\Lambda}, \tag{B.10.25}$$

where $k_j = \frac{2\pi n_j}{L}$ for integer n_j, such that all the $\vec{k}$ points are in the first "hyperoctant."

---

14. We will play a bit fast and loose here, with Ω indicating both the domain and its volume.

15. The logic is that the functional integral, after defining a cutoff lengthscale l, is

$$\int D\theta = \prod_{l\le|\vec{r}_n|}^{\Omega} \int d\theta(\vec{r}_n)$$

for all "lattice points" $\vec{r}_n$ separated by l inside Ω. Since the Jacobian of the Fourier transform is 1, we can rewrite this integral in Fourier space. This is performed more explicitly in the chapter on the self-consistent Gaussian approximation.

Since in general Fourier decomposition gives complex coefficients, the components $\theta_{\vec{k}}$ are really

$$\theta_{\vec{k}} = \theta_{\vec{k}}^{Re} + i\theta_{\vec{k}}^{Im}, \tag{B.10.26}$$

and the integration means:

$$\int d\theta_{\vec{k}} = \int d\theta_{\vec{k}}^{Re} \int d\theta_{\vec{k}}^{Im}. \tag{B.10.27}$$

Overview of RG procedure

In the following diagram, we will take a bird's-eye view of a single RG step we will carry out:

$$
\begin{array}{ccc}
F(\kappa, V_0, V_2, V_4, \cdots; \Lambda) & \xrightarrow[\text{Averaging over } \tilde{\Lambda} \le |\vec{k}| < \Lambda]{\text{Coarse-graining}} & \tilde{F}(\kappa, \tilde{V}_0, \tilde{V}_2, \tilde{V}_4, \cdots; \tilde{\Lambda}) \\
\text{Rescaling} \downarrow {\scriptstyle \kappa \to 1, \Lambda \to 1} & & \text{Rescaling} \downarrow {\scriptstyle \kappa \to 1, \tilde{\Lambda} \to 1} \\
F'(\kappa = 1, E_0, \alpha, u, \cdots; \Lambda = 1) & \xrightarrow[\alpha \to \tilde{\alpha},\ u \to \tilde{u}, \cdots]{\text{RG flow}} & \tilde{F}'(\kappa = 1, \tilde{E}_0, \tilde{\alpha}, \tilde{u}, \cdots; \tilde{\Lambda} = 1)
\end{array}
$$

The initial physical system is represented by the free energy F, with parameters κ, V_0, ..., and a lengthscale cutoff of Λ^{-1}. There are two operations we can perform on this system.

- We can nondimensionalize the parameters, such that the physics is rewritten in terms of dimensionless quantities in units of κ and Λ. This is represented by the free energy F' and dimensionless parameters E_0, α, ... The lengthscale Λ^{-1} and Gaussian coupling strength κ are normalized to 1. In some sense, this procedure yields the fundamental description of the system for given physical parameters.
- We can perform a coarse-graining procedure, where we average away a subset of short lengthscale physics. The resulting system, represented by $\tilde{F}$ and parameters $\tilde{V}_0$, ..., will reflect the remaining longer lengthscale physics. The cutoff lengthscale will be updated as well to a new value $\tilde{\Lambda}^{-1}$ to reflect the coarse-graining.

To quantify how coarse-graining affected the physics, we will need to nondimensionalize the new system $\tilde{F}$ as well and rewrite it in terms of dimensionless quantities in units of κ and $\tilde{\Lambda}$. This is represented by the free energy $\tilde{F}'$ and new dimensionless parameters $\tilde{E}_0$, $\tilde{\alpha}$, ... Once again the new lengthscale $\tilde{\Lambda}^{-1}$ and Gaussian coupling κ are normalized to 1.

Now that parameters in F' and $\tilde{F}'$ are on equal footing, we can examine how coarse-graining affected the physics, and devise a description for the "RG flow" of the physical parameters. Since now we can consider the coarse-grained system $\tilde{F}$ as the new initial system, this is an iterative process. If coarse-graining is performed with infinitesimal steps (which it is), then the flow of parameters can be formulated as differential equations.

Nondimensionalization via rescaling

RG calculation involves iteratively coarse-graining and rescaling parameters, and we will start with the rescaling step. We will first rescale lengths using the natural lengthscale of the system, Λ^{-1} (remember Λ has units of inverse length),

$$\vec{x} = \Lambda \vec{r} \quad \text{and} \quad \vec{q} = \frac{\vec{k}}{\Lambda}, \tag{B.10.28}$$

which preserves the dot product $\vec{k} \cdot \vec{r} = \vec{q} \cdot \vec{x}$. This rescales the volume as

$$\Omega' = \Omega \Lambda^d$$

and (very importantly) the limits of the Fourier frequencies

$$|\vec{k}| < \Lambda \to |\vec{q}| < 1.$$

1. We will also need to rescale the field θ. First, the energy of the system is measured in units of temperature T, and we would like to express all coupling constants in units of T. Since we are considering $\mathcal{H}_1 = V$ as a perturbation to $\mathcal{H}_0 = \kappa |\nabla \theta|^2 / 2$, we will rescale parameters such that $\kappa/T \to 1$. We will define a new field,

$$\phi(\vec{x}) = z\theta(\vec{r}), \tag{B.10.29}$$

with z being the scaling factor for the field, such that after rescaling the noninteracting action is

$$S_0[\theta] = \frac{1}{T} \int_\Omega d^d\vec{r} \, \frac{\kappa}{2} |\nabla_{\vec{r}} \theta(\vec{r})|^2 \stackrel{\text{def.}}{\equiv} S_0'[\phi] = \frac{1}{2} \int_{\Omega'} d^d\vec{x} \, |\nabla_{\vec{x}} \phi(\vec{x})|^2, \tag{B.10.30}$$

so that the action is a dimensionless quantity. Show that

$$z = \sqrt{\frac{\kappa}{T}} \, \Lambda^{(2-d)/2}. \tag{B.10.31}$$

We begin with substituting in the coordinate transformation $\vec{x} = \Lambda \vec{r}$ into the original $S_0[\theta]$:

$$\vec{x} = \Lambda \vec{r} \implies \begin{cases} dr_j = \dfrac{dx_j}{\Lambda} \\[2ex] \displaystyle\int_\Omega d^d\vec{r} = \int_{\Omega'} \dfrac{d^d\vec{x}}{\Lambda^d} \\[2ex] \hat{r}_j = \hat{x}_j, \end{cases}$$

$$\nabla_{\vec{r}}\,\theta(\vec{r}) = \hat{r}_j \frac{\partial}{\partial r_j} \frac{\phi(\vec{x})}{z}$$

$$= \frac{\Lambda \hat{x}_j}{z} \frac{\partial}{\partial x_j} \phi(\vec{x})$$

$$= \frac{\Lambda}{z} \nabla_{\vec{x}}\,\phi(\vec{x}).$$

Plugging it into $S_0[\theta] \stackrel{\text{def.}}{=} S_0'[\phi]$:

$$S_0[\theta] = \frac{1}{T} \int_\Omega d^d\vec{r}\,\frac{\kappa}{2}\,|\nabla_{\vec{r}}\,\theta(\vec{r})|^2$$

$$= \frac{\kappa}{2T}\frac{\Lambda^2}{z^2\Lambda^d} \int_{\Omega'} d^d\vec{x}\,|\nabla_{\vec{x}}\,\phi(\vec{x})|^2$$

$$\stackrel{\text{def.}}{=} \frac{1}{2} \int_{\Omega'} d^d\vec{x}\,|\nabla_{\vec{x}}\,\phi(\vec{x})|^2 = S_0'[\phi].$$

We can solve for z:

$$\frac{\kappa\Lambda^{2-d}}{z^2 T} = 1$$

$$z = \sqrt{\frac{\kappa}{T}}\,\Lambda^{(2-d)/2}.$$

2. Now let us turn our attention to the interaction term $\mathcal{H}_1 = V(\theta)$. We want to rescale V such that it is dimensionless. Considering that it originally has units of energy per volume, we would like to define a new interaction potential,

$$V'[\phi(\vec{x})] = \frac{V[\theta(\vec{r})]}{T\Lambda^d}, \tag{B.10.32}$$

such that the new potential V' is dimensionless and a function of the new field ϕ. We will assign it an expansion in terms even in ϕ,

$$V'(\phi) = E_0 + \frac{\alpha}{2}\phi^2 + \frac{u}{4}\phi^4 + \frac{u_6}{6}\phi^6 + \cdots, \tag{B.10.33}$$

which satisfies

$$S_1[\theta] = \frac{1}{T} \int_\Omega d^d\vec{r}\,V[\theta(\vec{r})] \stackrel{\text{def.}}{=} S_1'[\phi] = \int_{\Omega'} d^d\vec{x}\,V'[\phi(\vec{x})], \tag{B.10.34}$$

such that the action is again a dimensionless quantity. Show that:

$$E_0 = \frac{V_0}{T\Lambda^d}, \qquad \alpha = \frac{V_2}{\kappa\Lambda^2}, \qquad u = \frac{TV_4}{\kappa^2}\Lambda^{d-4}, \qquad u_6 = \frac{T^2V_6}{\kappa^3}\Lambda^{2d-6}.$$

Let us first write out the expansion for S'_1,

$$S'_1[\phi] = \int_{\Omega'} d^d \vec{x} \, V'[\phi(\vec{x})]$$

$$= \int_{\Omega'} d^d \vec{x} \left(E_0 + \frac{\alpha}{2} \phi^2 + \frac{u}{4} \phi^4 + \frac{u_6}{6} \phi^6 + \cdots \right),$$

and compare it with the expansion for S_1:

$$S_1[\theta] = \frac{1}{T} \int_{\Omega} d^d \vec{r} \left(V_0 + \frac{V_2}{2} \theta^2 + \frac{V_4}{4} \theta^4 + \frac{V_6}{6} \theta^6 + \cdots \right)$$

$$= \frac{1}{T\Lambda^d} \int_{\Omega'} d^d \vec{x} \left(V_0 + \frac{V_2}{2z^2} \phi^2 + \frac{V_4}{4z^4} \phi^4 + \frac{V_6}{6z^6} \phi^6 + \cdots \right).$$

Comparing terms in power of ϕ, we find:

$$E_0 = \frac{V_0}{T\Lambda^d}, \qquad \alpha = \frac{V_2}{\kappa\Lambda^2}, \qquad u = \frac{TV_4}{\kappa^2} \Lambda^{d-4}, \qquad u_6 = \frac{T^2 V_6}{\kappa^3} \Lambda^{2d-6}.$$

By nondimensionalizing the parameters and rescaling the frequency cutoff, we can now compare (apples to apples!) the parameters before and after the coarse-graining step, which we will now work on.

Coarse-graining with momentum shell RG

The method we will use to coarse-grain the system is called "momentum shell RG." In terms of the Fourier components $\theta_{\vec{k}}$, larger $\vec{k}$ corresponds to higher spatial frequency, or shorter lengthscale, features. The idea is to average out the physics over an "outer shell" in $\vec{k}$ space, and see how the rescaled parameters (E_0, α, u, u_6, ...) change.

Let us introduce a frequency $\tilde{\Lambda}$ that serves as the cutoff for "fast modes," such that:

$$\theta(\vec{r}) = \theta_s(\vec{r}) + \theta_f(\vec{r}), \tag{B.10.35}$$

where

$$\theta_s(\vec{r}) = L^{-d/2} \sum_{|\vec{k}| < \tilde{\Lambda}} e^{i\vec{k}\cdot\vec{r}} \theta_{\vec{k}},$$

$$\theta_f(\vec{r}) = L^{-d/2} \sum_{\tilde{\Lambda} \le |\vec{k}| < \Lambda} e^{i\vec{k}\cdot\vec{r}} \theta_{\vec{k}},$$

so θ_f contains the configurations of θ with short-range "fast" features, and θ_s contains the remaining long wavelength "slow" features. Our goal in this section is to integrate away the fast features to obtain a new effective interaction potential:

$$V(\theta) \xrightarrow{\text{RG}} \tilde{V}(\theta_s).$$

1. Let us first examine the Boltzmann factor of the noninteracting Hamiltonian, e^{-S_0}, in a bit more detail. Show that:

$$S_0[\theta] = \frac{\kappa}{2T} \sum_{|\vec{k}|<\Lambda} |\vec{k}|^2 \left| \theta_{\vec{k}} \right|^2. \tag{B.10.36}$$

Hint 1: Remember that $\nabla e^{i\vec{k}\cdot\vec{r}} = i\vec{k}e^{i\vec{k}\cdot\vec{r}}$.
Hint 2: Remember that:

$$\int_L dx \, \exp\left(i\frac{2\pi n}{L}x\right) = \begin{cases} L & \text{for } n = 0 \\ 0 & \text{for integer } n \neq 0. \end{cases} \tag{B.10.37}$$

Hint 3: Since θ is a real-valued function, by definition of Fourier transforms, complex conjugation gives you:

$$\theta_{\vec{k}}^* = \theta_{-\vec{k}}.$$

Starting with the definition of $S_0[\theta]$:

$$S_0 = \frac{\kappa}{2T} \int_\Omega d^d\vec{r} \, |\nabla\theta|^2.$$

The gradient term is:

$$\nabla\theta = \nabla\left(L^{-d/2} \sum_{|\vec{k}|<\Lambda} e^{i\vec{k}\cdot\vec{r}}\theta_{\vec{k}}\right)$$

$$= L^{-d/2} \sum_{|\vec{k}|<\Lambda} i\vec{k}e^{i\vec{k}\cdot\vec{r}}\theta_{\vec{k}}.$$

Since this is a vector, we can write the magnitude squared as (with $\vec{k}$ and $\vec{k}'$ being two dummy variables):

$$|\nabla\theta|^2 = \nabla\theta \cdot \nabla\theta$$

$$= L^{-d}\left(\sum_{|\vec{k}|<\Lambda} i\vec{k}e^{i\vec{k}\cdot\vec{r}}\theta_{\vec{k}}\right) \cdot \left(\sum_{|\vec{k}'|<\Lambda} i\vec{k}'e^{i\vec{k}'\cdot\vec{r}}\theta_{\vec{k}'}\right)$$

$$= -L^{-d} \sum_{|\vec{k}|<\Lambda} \sum_{|\vec{k}'|<\Lambda} (\vec{k}\cdot\vec{k}') \, e^{i(\vec{k}+\vec{k}')\cdot\vec{r}}\theta_{\vec{k}}\theta_{\vec{k}'}.$$

Let us now perform the volume integral:

$$\int_\Omega d^d\vec{r}\, |\nabla\theta|^2 = -L^{-d} \sum_{|\vec{k}|<\Lambda} \sum_{|\vec{k}'|<\Lambda} (\vec{k}\cdot\vec{k}')\, \theta_{\vec{k}}\theta_{\vec{k}'} \int_\Omega d^d\vec{r}\, e^{i(\vec{k}+\vec{k}')\cdot\vec{r}}.$$

Since the values of $\vec{k}$ and $\vec{k}'$ are in fact of the form $\vec{k} = 2\pi\vec{n}/L$, where $\vec{n}$ is a vector with integer components, the volume integral gives us a Kronecker delta between the indices (per hint 2):

$$\int_\Omega d^d\vec{r}\, e^{i(\vec{k}+\vec{k}')\cdot\vec{r}} = L^d\delta_{\vec{k},-\vec{k}'}. \tag{B.10.38}$$

Plugging this into the previous expression and summing over $\vec{k}'$, we get:

$$\int_\Omega d^d\vec{r}\, |\nabla\theta|^2 = -L^{-d} \sum_{|\vec{k}|<\Lambda} \sum_{|\vec{k}'|<\Lambda} (\vec{k}\cdot\vec{k}')\, \theta_{\vec{k}}\theta_{\vec{k}'} L^d\delta_{\vec{k},-\vec{k}'}$$

$$= -\sum_{|\vec{k}|<\Lambda} (-\vec{k}\cdot\vec{k})\, \theta_{\vec{k}}\theta_{\vec{k}}^* = \sum_{|\vec{k}|<\Lambda} |\vec{k}|^2 \left|\theta_{\vec{k}}\right|^2.$$

Putting the multiplicative factors, we get:

$$S_0[\theta] = \frac{\kappa}{2T} \sum_{|\vec{k}|<\Lambda} |\vec{k}|^2 \left|\theta_{\vec{k}}\right|^2.$$

2. The previous exercise shows that S_0 can be written as a sum of terms each with independent wave vector $\vec{k}$, so we can indeed separate it into two terms just like we did $\theta = \theta_s + \theta_f$, as

$$S_0[\theta] = S_0[\theta_s] + S_0[\theta_f], \tag{B.10.39}$$

where

$$S_0[\theta_s] = \frac{\kappa}{2T} \sum_{|\vec{k}|<\tilde{\Lambda}} |\vec{k}|^2 \left|\theta_{\vec{k}}\right|^2,$$

$$S_0[\theta_f] = \frac{\kappa}{2T} \sum_{\tilde{\Lambda}\le|\vec{k}|<\Lambda} |\vec{k}|^2 \left|\theta_{\vec{k}}\right|^2,$$

such that the functional integral can be rewritten as

$$\int D\theta = \int D\theta_s D\theta_f. \tag{B.10.40}$$

Now, let us formally define what it means to integrate away the short wavelength features. The statistical properties of the original system are determined by the partition function

$$Z = \int D\theta\, \exp\left[-S_0[\theta_s] - S_0[\theta_f] - \frac{1}{T}\int_\Omega d^d\vec{r}\, V(\theta)\right]. \tag{B.10.41}$$

If we want to average over just the fast components, we need to define an appropriate partition function,

$$Z_f = \int D\theta_f \, e^{-S_0[\theta_f]},$$

such that thermal average of any operator over the short wavelength physics is defined as

$$\left\langle \hat{A}[\theta_s, \theta_f] \right\rangle_f = Z_f^{-1} \int D\theta_f \, \hat{A}[\theta_s, \theta_f] e^{-S_0[\theta_f]}. \tag{B.10.42}$$

With this, we would like to define a new effective action $\tilde{S}[\theta_s]$ depending only on the slow components of the field, such that the new Boltzmann factor is defined by:

$$e^{-\tilde{S}[\theta_s]} \stackrel{\text{def.}}{\equiv} \frac{e^{-S_0[\theta_s]}}{Z_f} \int D\theta_f \, \exp\left[-S_0[\theta_f] - \frac{1}{T}\int_\Omega d^d\vec{r}\, V(\theta_s + \theta_f)\right] \tag{B.10.43}$$

$$= e^{-S_0[\theta_s]} \left\langle \exp\left[-\frac{1}{T}\int_\Omega d^d\vec{r}\, V(\theta_s + \theta_f)\right]\right\rangle_f. \tag{B.10.44}$$

Show that this definition lets us naturally calculate the expectation value of operators depending only on θ_s as

$$\left\langle \hat{O}[\theta_s]\right\rangle = Z^{-1} \int D\theta \, \hat{O}[\theta_s]e^{-S} = \tilde{Z}^{-1} \int D\theta_s \, \hat{O}[\theta_s]e^{-\tilde{S}[\theta_s]}, \tag{B.10.45}$$

where:

$$\tilde{Z} = \int D\theta_s \, e^{-\tilde{S}[\theta_s]},$$

And thus describes the coarse-grained system with high frequency components averaged away.

Let us explicitly write out Eq. B.10.45 in full with the slow and fast components identified:

$$\left\langle \hat{O}[\theta_s]\right\rangle = Z^{-1} \int D\theta_s D\theta_f \, \hat{O}[\theta_s] \exp\left[-S_0[\theta_s] - S_0[\theta_f] - \frac{1}{T}\int_\Omega d^d\vec{r}\, V(\theta_s + \theta_f)\right],$$

$$Z = \int D\theta_s D\theta_f \, \exp\left[-S_0[\theta_s] - S_0[\theta_f] - \frac{1}{T}\int_\Omega d^d\vec{r}\, V(\theta_s + \theta_f)\right].$$

We can rearrange to do the integral over θ_f first and get:

$$\left\langle \hat{O}[\theta_s]\right\rangle = Z^{-1} \int D\theta_s \, \hat{O}[\theta_s] \exp\left[-S_0[\theta_s]\right]$$

$$\int D\theta_f \, \exp\left[-S_0[\theta_f] - \frac{1}{T}\int_\Omega d^d\vec{r}\, V(\theta_s + \theta_f)\right]$$

$$(\text{B.10.43}) \Rightarrow = \frac{Z_f}{Z} \int D\theta_s \, \hat{O}[\theta_s] \exp\left[-\tilde{S}[\theta_s]\right],$$

$$Z = \int D\theta_s \, \exp\left[-S_0[\theta_s]\right]$$

$$\int D\theta_f \, \exp\left[-S_0[\theta_f] - \frac{1}{T} \int_\Omega d^d\vec{r} \, V(\theta_s + \theta_f)\right]$$

$$= Z_f \int D\theta_s \, \exp\left[-\tilde{S}[\theta_s]\right] = Z_f \tilde{Z}.$$

Combining this together, we get:

$$\langle \hat{O}[\theta_s]\rangle = \tilde{Z}^{-1} \int D\theta_s \, \hat{O}[\theta_s] e^{-\tilde{S}[\theta_s]}.$$

3. The formal manipulations from the last part shows us that the physics of the long wavelength components can indeed be described by the new action $\tilde{S}[\theta_s]$ defined by:

$$e^{-\tilde{S}[\theta_s]} = e^{-S_0[\theta_s]} \left\langle \exp\left[-\frac{1}{T} \int_\Omega d^d\vec{r} \, V(\theta_s + \theta_f)\right]\right\rangle_f. \tag{B.10.46}$$

So far, this is an exact expression that we cannot evaluate exactly in general. We would like to approximate it perturbatively in powers of V. Notice that by taking the logarithm we get:

$$\tilde{S}[\theta_s] = S_0[\theta_s] - \ln\left\{\left\langle \exp\left[-\frac{1}{T} \int_\Omega d^d\vec{r} \, V(\theta_s + \theta_f)\right]\right\rangle_f\right\}. \tag{B.10.47}$$

An important identity can help us evaluate the second term:

$$\ln\left[\langle e^{\lambda x}\rangle\right] = \lambda \langle x\rangle + \frac{\lambda^2}{2}\left[\langle x^2\rangle - \langle x\rangle^2\right]$$

$$+ \frac{\lambda^3}{6}\left[\langle x^3\rangle - 3\langle x^2\rangle\langle x\rangle + 2\langle x\rangle^3\right] + \cdots. \tag{B.10.48}$$

This is called the cumulant expansion, which works for the random variable x with any distribution.[16] To first order in V, the logarithm terms gives:

---

16. This expansion can be proved by noting

$$\langle e^{\lambda x}\rangle = \left\langle 1 + \lambda x + \frac{\lambda^2}{2}x^2 + \frac{\lambda^3}{6}x^3 + \cdots\right\rangle$$

$$= 1 + \lambda\langle x\rangle + \frac{\lambda^2}{2}\langle x^2\rangle + \frac{\lambda^3}{6}\langle x^3\rangle + \cdots$$

and that

$$\ln(1+y) = y - \frac{y^2}{2} + \frac{y^3}{3} + \cdots.$$

We recover the cumulant expansion by collecting terms in powers of λ.

$$\ln\left\{\left\langle\exp\left[-\frac{1}{T}\int_\Omega d^d\vec{r}\, V(\theta_s+\theta_f)\right]\right\rangle_f\right\} = -\frac{1}{T}\int_\Omega d^d\vec{r}\,\left\langle V(\theta_s+\theta_f)\right\rangle_f + \cdots.$$

(B.10.49)

Show that:

$$\left\langle V(\theta_s+\theta_f)\right\rangle_f = V_0 + V_s(\theta_s) + \left\langle V_f(\theta_f)\right\rangle_f$$

(B.10.50)

$$+ \frac{3V_4}{2}\theta_s^2\left\langle\theta_f^2\right\rangle_f + \frac{V_6}{2}\left[5\theta_s^4\left\langle\theta_f^2\right\rangle_f + 15\theta_s^2\left\langle\theta_f^2\right\rangle_f^2\right] + \cdots,$$

where:

$$V_s(\theta_s) = \frac{V_2}{2}\theta_s^2 + \frac{V_4}{4}\theta_s^4 + \frac{V_6}{6}\theta_s^6 + \cdots,$$

$$V_f(\theta_f) = \frac{V_2}{2}\theta_f^2 + \frac{V_4}{4}\theta_f^4 + \frac{V_6}{6}\theta_f^6 + \cdots.$$

Hint 1: Look at the definitions of $\left\langle\hat{O}\right\rangle_f$ and $S_0[\theta_f]$ again. Is there anything familiar (perhaps from the Gaussian integrals we considered in Chapter 8 worksheets)?

Hint 2: You might recall that the expectation value of linear functions F in Gaussian distributions is:

$$\left\langle F^{2n}\right\rangle = (2n-1)(2n-3)\cdots(1)\left\langle F^2\right\rangle.$$

From the definition of $\left\langle\hat{O}\right\rangle_f$ and $S_0[\theta_f]$, we can see that the expectation value is in fact evaluated with Gaussian distributions:

$$S_0[\theta_f] = \frac{\kappa}{2T}\sum_{\tilde{\Lambda}\leq|\vec{k}|<\Lambda}|\vec{k}|^2\left|\theta_{\vec{k}}\right|^2.$$

$$\left\langle\hat{O}\right\rangle_f = Z_f^{-1}\int D\theta_f\,\hat{O}e^{-S_0[\theta_f]}$$

$$= Z_f^{-1}\left(\prod_{\tilde{\Lambda}\leq|\vec{k}|<\Lambda}\int d\theta_{\vec{k}}^{\mathrm{Re}}d\theta_{\vec{k}}^{\mathrm{Im}}\right)\hat{O}\exp\left[-\frac{\kappa}{2T}\sum_{\tilde{\Lambda}\leq|\vec{k}|<\Lambda}|\vec{k}|^2\left|\theta_{\vec{k}}\right|^2\right]$$

$$= Z_f^{-1}\left(\prod_{\tilde{\Lambda}\leq|\vec{k}|<\Lambda}\int d\theta_{\vec{k}}^{\mathrm{Re}}d\theta_{\vec{k}}^{\mathrm{Im}}\,\hat{O}\exp\left[-\frac{\kappa}{2T}|\vec{k}|^2\left(\left(\theta_{\vec{k}}^{\mathrm{Re}}\right)^2+\left(\theta_{\vec{k}}^{\mathrm{Im}}\right)^2\right)\right]\right).$$

If we expand out $V(\theta_s+\theta_f)$, we get:

$$V(\theta) = V_0 + \frac{V_2}{2}\left(\theta_s^2 + 2\theta_s\theta_f + \theta_f^2\right)$$

(B.10.51)

$$+ \frac{V_4}{4}\left(\theta_s^4 + 4\theta_s^3\theta_f + 6\theta_s^2\theta_f^2 + 4\theta_s\theta_f^3 + \theta_f^4\right) + \cdots.$$

Recall from the Gaussian integral worksheet from Chapter 8 on Wick's theorem (WS.B.8.1) that we can calculate the expectation value of general linear functions easily. The Fourier transform of θ_f is such a function. Thus, the expectation value of any odd power of θ_f is automatically zero. Factors of θ_s is unaffected by the expectation value operation, since we are only averaging over the fast degrees of freedom. Thus, counting up all the nonzero terms, we have:

$$\langle V(\theta_s + \theta_f) \rangle = V_0 + \frac{V_2}{2}\theta_s^2 + \frac{V_4}{4}\theta_s^4 + \frac{V_6}{6}\theta_s^6 + \cdots$$

$$+ \frac{V_2}{2}\langle \theta_f^2 \rangle_f + \frac{V_4}{4}\langle \theta_f^4 \rangle_f + \frac{V_6}{6}\langle \theta_f^6 \rangle_f + \cdots$$

$$+ \frac{6V_4}{4}\theta_s^2 \langle \theta_f^2 \rangle_f + \frac{V_6}{6}\left[15\theta_s^4 \langle \theta_f^2 \rangle_f + 15\theta_s^2 \langle \theta_f^4 \rangle_f \right] + \cdots.$$

Finally, using the properties of Gaussian integrals (or equivalently Wick's theorem):

$$\langle \theta_f^{2n} \rangle_f = (2n-1)(2n-3)\cdots(1)\langle \theta_f^2 \rangle_f.$$

We can simplify the expression to:

$$\langle V(\theta_s + \theta_f) \rangle_f = V_0 + V_s(\theta_s) + \langle V_f(\theta_f) \rangle_f$$

$$+ \frac{3V_4}{2}\theta_s^2 \langle \theta_f^2 \rangle_f + \frac{V_6}{2}\left[5\theta_s^4 \langle \theta_f^2 \rangle_f + 15\theta_s^2 \langle \theta_f^2 \rangle_f^2 \right] + \cdots.$$

4. The above exercise tells us that we only have to calculate $\langle \theta_f^2 \rangle_f$, and all other terms can be obtained from it. Let us calculate this quantity.

(a) Since θ_f is real, $\theta_f = \theta_f^*$:

$$\theta_f^2 = \theta_f \theta_f^* = L^{-d} \sum_{\tilde{\Lambda} \leq |\vec{k}| < \Lambda} \sum_{\tilde{\Lambda} \leq |\vec{k}'| < \Lambda} e^{i(\vec{k} - \vec{k}') \cdot \vec{r}} \theta_{\vec{k}} \theta_{\vec{k}'}^*.$$

Use this expression to show that

$$\langle \theta_f^2 \rangle_f = L^{-d} \sum_{\tilde{\Lambda} \leq |\vec{k}| < \Lambda}^{\text{Shell}} \frac{T}{\kappa |\vec{k}|^2}, \qquad (B.10.52)$$

where now the summation is over all hyperoctants (i.e., components of $\vec{k}$ can now be negative), and is thus a "shell" in d-dimensional $\vec{k}$ space.

Let us first perform the calculation in $d = 1$ and generalize it to d dimensions. In 1D, $\vec{k} = k$, $\vec{r} = r$, and

$$\sum_{\tilde{\Lambda} \leq |\vec{k}| < \Lambda} \rightarrow \sum_{\tilde{\Lambda} \leq k < \Lambda},$$

where $k > 0$ since $\tilde{\Lambda} > 0$. By definition of the complex conjugate, $\theta_k^* = \theta_k^{Re} - i\theta_k^{Im}$, so

$$\theta_f^2 = L^{-1} \sum_{\tilde{\Lambda} \le k < \Lambda} \sum_{\tilde{\Lambda} \le k' < \Lambda} e^{i(k-k')r} \left(\theta_k^{Re} + i\theta_k^{Im}\right) \left(\theta_{k'}^{Re} - i\theta_{k'}^{Im}\right). \qquad (B.10.53)$$

Let us take a single term in the summation and see how it behaves as an expectation value. In 1D, labelling the integration variables with k_j,

$$\langle \hat{O} \rangle_f = Z_f^{-1} \left(\prod_{\tilde{\Lambda} \le k_j < \Lambda} \int d\theta_{k_j}^{Re} d\theta_{k_j}^{Im} \, \hat{O} \exp\left[-\frac{\kappa}{2T} k_j^2 \left(\left(\theta_{k_j}^{Re}\right)^2 + \left(\theta_{k_j}^{Im}\right)^2 \right) \right] \right),$$

and letting $\hat{O} = \left(\theta_k^{Re} + i\theta_k^{Im}\right)\left(\theta_{k'}^{Re} - i\theta_{k'}^{Im}\right)$, we find that

$$\langle \hat{O} \rangle_f = \langle \theta_k^{Re} \theta_{k'}^{Re} \rangle + i \overset{0}{\langle \theta_k^{Im} \theta_{k'}^{Re} - \theta_k^{Re} \theta_{k'}^{Im} \rangle} + \langle \theta_k^{Im} \theta_{k'}^{Im} \rangle$$

$$= \lambda_k \delta_{k,k'} + \lambda_k \delta_{k,k'},$$

where $\lambda_k = T/\kappa k^2$ from the Gaussian weights. The cross terms average to zero, since there is no overlap of integration variables and only odd powers exist. Plugging this back into Eq. B.10.53, we have:

$$\langle \theta_f^2 \rangle_f = L^{-1} \sum_{\tilde{\Lambda} \le k < \Lambda} \sum_{\tilde{\Lambda} \le k' < \Lambda} e^{i(k-k')r} 2 \frac{T}{\kappa k^2} \delta_{k,k'}$$

$$= L^{-1} \sum_{\tilde{\Lambda} \le k < \Lambda} 2 \frac{T}{\kappa k^2}.$$

Since the summand is even in k, we can break up the factor of 2 and rewrite the summation symmetrically as:

$$\langle \theta_f^2 \rangle_f = L^{-1} \left(\sum_{\tilde{\Lambda} \le k < \Lambda} \frac{T}{\kappa |k|^2} + \sum_{-\Lambda \le k < -\tilde{\Lambda}} \frac{T}{\kappa |k|^2} \right) = L^{-1} \sum_{\tilde{\Lambda} \le |k| < \Lambda}^{\text{Shell}} \frac{T}{\kappa |k|^2}.$$

This procedure is in fact completely generalizable to d dimensions, so now with a simple change of variables we have:

$$\langle \theta_f^2 \rangle_f = L^{-d} \sum_{\tilde{\Lambda} \le |\vec{k}| < \Lambda}^{\text{Shell}} \frac{T}{\kappa |\vec{k}|^2}.$$

(b) Use the continuum approximation

$$\left(\frac{2\pi}{L}\right)^d \sum_{\vec{k}} \to \int d^d \vec{k},$$

to show that

$$\langle \theta_f^2 \rangle_f = \frac{A_d}{(2\pi)^d} \frac{T}{\kappa} \frac{\Lambda^{d-2}}{(d-2)} \left[1 - \left(\frac{\tilde{\Lambda}}{\Lambda} \right)^{d-2} \right],$$ (B.10.54)

where A_d is the surface area of a d-dimensional hypersphere.[17]

Picking up from the previous part, the continuum approximation changes summation over discrete $\vec{k}$ points in a "momentum shell" to an integral with d-dimensional spherical symmetry, so:

$$L^{-d} \sum_{\tilde{\Lambda} \leq |\vec{k}| < \Lambda}^{\text{Shell}} \frac{T}{\kappa |\vec{k}|^2} \to \int_{k=\tilde{\Lambda}}^{k=\Lambda} \frac{d^d\vec{k}}{(2\pi)^d} \frac{T}{\kappa k^2}$$

$$= \int_{\tilde{\Lambda}}^{\Lambda} \frac{A_d k^{d-1} dk}{(2\pi)^d} \frac{T}{\kappa k^2}$$

$$= \frac{A_d}{(2\pi)^2} \frac{T}{\kappa} \int_{\tilde{\Lambda}}^{\Lambda} k^{d-3} dk$$

$$= \frac{A_d}{(2\pi)^d} \frac{T}{\kappa} \frac{\Lambda^{d-2}}{(d-2)} \left[1 - \left(\frac{\tilde{\Lambda}}{\Lambda} \right)^{d-2} \right].$$

(c) We have now performed the bulk of the coarse-graining step, which is to integrate over $\tilde{\Lambda} \to \Lambda$ in $\vec{k}$ space. Since eventually we want to consider this as a continuous process in some RG time t, let us choose:

$$\tilde{\Lambda} = e^{-\delta t} \Lambda,$$

where δt is a small increment in RG time t. Show that:

$$\langle \theta_f^2 \rangle_f = \frac{X_d \delta t}{z^2} + \mathcal{O}(\delta t^2),$$ (B.10.56)

where $X_d = A_d / (2\pi)^d$.

---

17. This is the factor that comes in when doing volume integrals with spherical symmetry. In 2D, $A_2 = 2\pi$ since

$$\int d^2\vec{r} = \int 2\pi\rho\, d\rho,$$

And in 3D, $A_3 = 4\pi$ since:

$$\int d^3\vec{r} = \int 4\pi r^2\, dr$$ (B.10.55)

In general,

$$A_d = \frac{2\pi^{d/2}}{\Gamma(d/2)},$$

and it is a finite small positive number for all d.

Recall from the definition of the scaling factor z that:

$$z^2 = \frac{\kappa}{T\Lambda^{d-2}}. \tag{B.10.57}$$

Then by definition of $\tilde{\Lambda}$ and X_d:

$$\begin{aligned}
\langle \theta_f^2 \rangle_f &= \frac{X_d}{z^2(d-2)} \left[1 - e^{-\delta t(d-2)} \right] \\
&= \frac{X_d}{z^2(d-2)} \left[1 - \left(1 - \delta t(d-2) + \mathcal{O}(\delta t^2) \right) \right] \\
&= \frac{X_d \,\delta t}{z^2} + \mathcal{O}(\delta t^2).
\end{aligned}$$

Constructing the RG differential equations

We now have all the ingredients ready, or mise en place if you will, to assemble the differential equations that describe the RG flow. The original physical system is described by the action:

$$S[\theta] = T^{-1} \int_\Omega d^d\vec{r} \left[\frac{\kappa}{2} |\nabla \theta|^2 + V_0 + \frac{V_2}{2}\theta^2 + \frac{V_4}{4}\theta^4 + \frac{V_6}{6}\theta^6 + \cdots \right]. \tag{B.10.58}$$

After coarse-graining away the fast components θ_f, we are left with an effective action as a functional of the slow components θ_s (see Eqs. B.10.47–B.10.50), which we would like to write in the form:

$$\tilde{S}[\theta_s] = T^{-1} \int_\Omega d^d\vec{r} \left[\frac{\kappa}{2} |\nabla \theta_s|^2 + \tilde{V}(\theta_s) \right], \tag{B.10.59}$$

where

$$\tilde{V}(\theta_s) = \tilde{V}_0 + \frac{\tilde{V}_2}{2}\theta_s^2 + \frac{\tilde{V}_4}{4}\theta_s^4 + \frac{\tilde{V}_6}{6}\theta_s^6 + \cdots. \tag{B.10.60}$$

We will first find the new coefficients $\tilde{V}_{2n}$ in terms of the original coefficients and parameters, then see how their rescaled versions ($\tilde{E}_0, \tilde{\alpha}, \tilde{u}, \cdots$) changed compared to the original system and derive the flow equations.

1. From Eq. B.10.50, show that:

$$\begin{aligned}
\tilde{V}_0 &= V_0 + \frac{V_2}{2} \langle \theta_f^2 \rangle_f + \frac{3V_4}{4} \langle \theta_f^2 \rangle_f^2 + \frac{5V_6}{2} \langle \theta_f^2 \rangle_f^3 + \cdots, \\
\tilde{V}_2 &= V_2 + 3V_4 \langle \theta_f^2 \rangle_f + 15V_6 \langle \theta_f^2 \rangle_f^2 + \cdots, \\
\tilde{V}_4 &= V_4 + 10V_6 \langle \theta_f^2 \rangle_f + \cdots, \\
\tilde{V}_6 &= V_6 + \cdots.
\end{aligned}$$

Writing out all the terms in Eq. B.10.50 to combined sixth power in θ_f and θ_s, we get

$$\langle V(\theta_s + \theta_f)\rangle_f = V_0 + \frac{V_2}{2}\theta_s^2 + \frac{V_4}{4}\theta_s^4 + \frac{V_6}{6}\theta_s^6 + \cdots$$

$$+ \frac{V_2}{2}\langle\theta_f^2\rangle_f + \frac{3V_4}{2}\langle\theta_f^2\rangle_f \theta_s^2 + \frac{5V_6}{2}\langle\theta_f^2\rangle_f \theta_s^4 + \cdots$$

$$+ \frac{V_4}{4}3\langle\theta_f^2\rangle^2 + \frac{15V_6}{2}\langle\theta_f^2\rangle^2 \theta_s^2 + \cdots$$

$$+ \frac{V_6}{6}15\langle\theta_f^2\rangle^3 + \cdots$$

$$= \tilde{V}_0 + \frac{\tilde{V}_2}{2}\theta_s^2 + \frac{\tilde{V}_4}{4}\theta_s^4 + \frac{\tilde{V}_6}{6}\theta_s^6 + \cdots,$$

where we used Wick's theorem for higher order terms in $\langle V_f(\theta_f)\rangle_f$. Collecting terms in orders of θ_s and comparing terms, we find

$$\tilde{V}_0 = V_0 + \frac{V_2}{2}\langle\theta_f^2\rangle_f + \frac{3V_4}{4}\langle\theta_f^2\rangle_f^2 + \frac{5V_6}{2}\langle\theta_f^2\rangle_f^3 + \cdots,$$

$$\tilde{V}_2 = V_2 + 3V_4\langle\theta_f^2\rangle_f + 15V_6\langle\theta_f^2\rangle_f^2 + \cdots,$$

$$\tilde{V}_4 = V_4 + 10V_6\langle\theta_f^2\rangle_f + \cdots,$$

$$\tilde{V}_6 = V_6 + \cdots.$$

2. Since we have integrated away all features for $\tilde{\Lambda} \le |\vec{k}| < \Lambda$, the cutoff is now $\tilde{\Lambda}$, which will be the new lengthscale for rescaling the parameters. Following the exact same procedure as previously, the whole suite of rescaled parameters is:

$$\tilde{\vec{x}} = \tilde{\Lambda}\vec{r}, \qquad \tilde{z} = \sqrt{\frac{\kappa}{T}}\tilde{\Lambda}^{(2-d)/2}, \qquad \tilde{\phi}(\tilde{x}) = \tilde{z}\theta_s(\vec{r}), \qquad \tilde{V}'(\tilde{\phi}) = \frac{\tilde{V}(\theta_s)}{T\tilde{\Lambda}^d},$$

$$\tilde{V}'(\tilde{\phi}) = \tilde{E}_0 + \frac{\tilde{\alpha}}{2}\tilde{\phi}^2 + \frac{\tilde{u}}{4}\tilde{\phi}^4 + \frac{\tilde{u}_6}{6}\tilde{\phi}^6 + \cdots,$$

$$\tilde{E}_0 = \frac{\tilde{V}_0}{T\tilde{\Lambda}^d}, \qquad \tilde{\alpha} = \frac{\tilde{V}_2}{\kappa\tilde{\Lambda}^2}, \qquad \tilde{u} = \frac{T\tilde{V}_4}{\kappa^2}\tilde{\Lambda}^{d-4}, \qquad \tilde{u}_6 = \frac{T^2\tilde{V}_6}{\kappa^3}\tilde{\Lambda}^{2d-6}.$$

Recall that to view coarse-graining as a continuous process in some RG time t, we have defined $\tilde{\Lambda} = e^{-\delta t}\Lambda$. We can then consider the rescaled parameters to be continuous functions of t in the following sense (take $\tilde{u}$ and u, for example):

$$u_\Lambda = u(t),$$

$$\tilde{u}_{\tilde{\Lambda}} = u(t + \delta t)$$

$$= u(t) + \frac{du}{dt}\bigg|_t \delta t + \mathcal{O}(\delta t^2).$$

On the other hand, $\tilde{u}$ is a function of δt and can be written as:

$$\tilde{u}(\delta t) = \tilde{u}(0) + \frac{d\tilde{u}}{d(\delta t)}\bigg|_{\delta t \to 0} \delta t + \mathcal{O}(\delta t^2).$$

Since $\tilde{u}(0) = u_\Lambda = u(t)$, if we consider an infinitesimal δt, we can construct a differential equation for $u(t)$ by

$$\frac{du}{dt} = \frac{d\tilde{u}}{d(\delta t)}\bigg|_{\delta t \to 0}, \tag{B.10.61}$$

and the construction in Eq. B.10.61 applies for all of the rescaled parameters. Use this to show that:

$$\frac{dE_0}{dt} = E_0 d + \frac{X_d}{2}\alpha,$$

$$\frac{d\alpha}{dt} = 2\alpha + 3X_d u,$$

$$\frac{du}{dt} = (d-4)u + 10X_d u_6,$$

$$\frac{du_6}{dt} = (2d-6)u_6 + \mathcal{O}(u_8),$$

where u_8 is the coefficient for 8th-order term in the expansion for V'.

Hint 1: Since this amounts to finding the 1st-order expansion in δt, $e^{\lambda\delta t} = 1 + \lambda\delta t$ is a perfectly fine expansion to use everywhere. You also only need to keep terms to order δt elsewhere.

Hint 2: Look back at expressions for $E_0, \alpha, \dots$, etc. to simplify your results.

Recall that $\langle \theta_f^2 \rangle_f = \delta t(X_d T \Lambda^{d-2}/\kappa)$ to first order in δt. Writing out the coarse-grained rescaled parameters to first order in δt, and thus obtaining the differential equations, we have

$$\tilde{E}_0 = \frac{\tilde{V}_0}{T\Lambda^d} e^{d\delta t}$$

$$= \frac{1}{T\Lambda^d}(1 + d\delta t)\left(V_0 + \frac{V_2}{2}\langle \theta_f^2 \rangle_f\right) + \mathcal{O}(\delta t^2)$$

$$= E_0 + \left(\frac{V_0 d}{T\Lambda^d} + \frac{V_2 X_d}{2\Lambda^2}\right)\delta t + \mathcal{O}(\delta t^2)$$

$$\implies \frac{dE_0}{dt} = E_0 d + \frac{X_d}{2}\alpha,$$

$$\tilde{\alpha} = \frac{\tilde{V}_2}{\kappa\Lambda^2}e^{2\delta t}$$

$$= \frac{1}{\kappa\Lambda^2}(1 + 2\delta t)\left(V_2 + 3V_4\left\langle\theta_f^2\right\rangle_f\right) + \mathcal{O}(\delta t^2)$$

$$= \alpha + \left(\frac{2V_2}{\kappa\Lambda^2} + \frac{3X_d TV_4}{\kappa^2}\Lambda^{d-4}\right)\delta t + \mathcal{O}(\delta t^2)$$

$$\implies \frac{d\alpha}{dt} = 2\alpha + 3X_d u,$$

$$\tilde{u} = \frac{T\Lambda^{d-4}\tilde{V}_4}{\kappa^2}e^{(4-d)\delta t}$$

$$= \frac{T\Lambda^{d-4}}{\kappa^2}[1 + (4-d)\delta t]\left(V_4 + 10V_6\left\langle\theta_f^2\right\rangle_2\right) + \mathcal{O}(\delta t^2)$$

$$= u + \left((d-4)\frac{TV_4\Lambda^{d-4}}{\kappa^2} + 10X_d\frac{T^2}{\kappa^3}\Lambda^{2d-6}\right)\delta t + \mathcal{O}(\delta t^2)$$

$$\implies \frac{du}{dt} = (d-4)u + 10X_d u_6,$$

$$\tilde{u}_6 = \frac{T^2\Lambda^{2d-6}\tilde{V}_6}{\kappa^3}e^{(6-2d)\delta t}$$

$$= \frac{T^2\Lambda^{2d-6}}{\kappa^3}[1 + (6-2d)\delta t]\left[V_6 + \mathcal{O}(V_8)\right]\delta t + \mathcal{O}(\delta t^2)$$

$$= u_6 + (6-2d)\frac{T^2 V_6}{\kappa^3}\Lambda^{2d-6}\delta t + \mathcal{O}(\delta t^2)$$

$$\implies \frac{du_6}{dt} = (2d-6)u_6 + \mathcal{O}(u_8),$$

where we have used previous expressions to simplify the results:

$$E_0 = \frac{V_0}{T\Lambda^d}, \qquad \alpha = \frac{V_2}{\kappa\Lambda^2}, \qquad u = \frac{TV_4}{\kappa^2}\Lambda^{d-4}, \qquad u_6 = \frac{T^2 V_6}{\kappa^3}\Lambda^{2d-6}.$$

INDEX

$1/N$ expansion, 127, 234–35. See also Gaussian model in the $N \to \infty$ limit

ϕ^4 theory: dependence on dimension, 193, 195–96; Ginzburg-Landau-like functional, 191–96; Landau-like free energy, 134–37; mean-field treatment in 0D, 202–4, 409–16; in the renormalization group, 235–40, 246–56, 422–39; variational treatment with the self-consistent Gaussian approximation, 191–94

$O(N)$ model
  definition, 127
  in the ordered phase, 196–99
  with random field disorder, 212–14
  in the self-consistent Gaussian approximation, 194–99
  for values of N: $N < 0$, 195; $N = 1$ (see Ising model); $N = 2$ (see XY model); $N = 3$ (see Heisenberg model); in the $N \to \infty$ limit, 195

abstraction versus reality, 57, 218
alloy, 60, 111–12
almost continuous broken symmetry, 119n15, 130, 138, 169–70
amplitude ratio. See critical amplitude ratio
analyticity, 3, 84–85
anisotropic treatment of the square-lattice Ising model, 88–90
anomalous dimension, 14, 220, 229
antiferromagnets, 40–41, 59–60, 111–12
approximation techniques. See mean-field theory, low-temperature expansion, high-temperature expansion
asymptotic analysis, 81; of the Ising model near $T = 0$, 102, 347–49; of the Ornstein-Zernicke law, 270–71

basin of attraction, 231
beta functions, 229
bicritical point, 144–46
binomial theorem, 41, 66, 310–11
bipartite lattice, 111–12
Boltzmann constant, xv, 20
Bose condensation, 120
bottom-up versus top-down approach, 132–33, 151
bound current, 118
brass, 111–12
Bravais lattice, 113–14
Brillouin zone, 155, 159, 185; definition of, 263–64
broken symmetry
  definition of, 42–43, 106–9
  examples of, 111–21
  of specific materials: crystals, 7–8, 114; ferromagnets, 116–19; Ising model, 83; liquid crystals, 114–16 (see also liquid crystals); order-disorder transitions, 111–13; superconductors, 120–21, 241; superfluids, 119–20
  spontaneous nature of, 7–9, 42, 83

calculus of variations, 264–65
canonical ensemble, xv, 20, 34
central limit theorem, 273–75; the $O(N)$ model in the $N \to \infty$ limit, 195; from a renormalization group perspective, 232–34; self-averaging, 207; universality, 16, 242; the validity of mean-field theory, 87
clock model, 124–25; 2-state, 138, 167 (see also Ising model); 3-state, 92–93, 138–40, 148–50, 367–72; 4-state, 119, 140–41, 167; interpolation to XY model, 129–30, 357–61; Landau theory of, 137–41

441

Helmholtz free energy. See free energy, Helmholtz
high-temperature expansion, 57–58
holographic model, 29
homotopy, 170

imaginary time, 276–77
impurities. See quenched disorder
Imry-Ma Argument, 211–12
ineluctable, 164
infrared convergence versus divergence, 190
intensive quantity, 22–23
internal energy, 21
internal symmetry, 120. See also rotational symmetry in spin space
invariance, 32, 121–22, 136
irrelevant perturbations, 223–25, 230
Ising model
    anisotropic square-lattice, 88–90
    critical temperature of, 54–55, 88
    definition of, 39–40
    ground state of, 40–41
    Hamiltonian of, 62, 301–2
    with long-range interactions, 285–87
    low-energy excitations of, 44, 70, 324
    low-energy temperature expansion of (see low-temperature expansion)
    mean-field approximation of, 77–80, 100–101, 339–47
    physical realizations of, 60, 111–12, 124
    qualitative behavior of, 42–43, 79–81
    with random field disorder, 211–12
    relation to ferromagnetism, 40, 116–19
    solution in 0D, 49
    solution in 1D: with combinatorics, 50–51, 66, 310–11; lack of a phase transition, 46, 48, 53; with transfer matrices, 51–53, 66–72, 312–28
    solution in 2D, 53–54
    with spatially varying fields, 152–55
    symmetry of, 106
    variations on, 54–55

Jensen's inequality, 75, 95, 331–33

K-L divergence, 91–92
kinks. See domain walls
Kubo formula, 31

Landau free energy, 132
Landau theory, 132–50; derivation from symmetry considerations, 147–48, 364–67; of the q-state clock model, 136–37; of the XY model, 136–37
Larkin-Ovchinikov-Lee-Rice Argument, 212–14
latent heat, 4, 140
lattice: bipartite, 111–12; Bravais, 113–14; dual, 56; hexagonal, 56, 61; hypercubic, 41; structure, effects of, 41, 54–55, 155; structure, encoded in $g(k)$, 188; triangular, 41, 56, 61, 92, 130; vector, 8
lengthscales. See correlation length, coherence length
linear response function, 31, 159
liquid crystals, 27n8, 114–16, 127–29, 353–57
local field, 73, 77, 86–87
long-range order, 2, 42, 116n9, 211–14; formal definition of, 109
longitudinal and transverse fluctuations: in the $O(N)$ model, 197; in the XY model, 163–64, 181–82, 393–97
loops, 56–58
Lorentzian, 188, 242, 267–70, 275n4
low-temperature expansion, 46–48, 55–57; detailed calculation, 69–72, 323–28; formal presentation, 259–60
lower critical dimension. See critical dimension

macroscopic wavefunction, 120–21, 126, 288–89
magnetic moment, 41
magnetization: of classical systems, 118, 127; definition of, 41–42; notation for, xv
marginal interactions, 230
mean-field equations. See self-consistency relations
mean-field theory
    of 3-state clock model, 92–93 (see also clock model, 3-state)
    calculation of critical exponents, 83–85, 104–5, 352–53
    intuition for, 73
    of Ising model, 77–80, 100–101, 339–47; low-temperature asymptotic analysis, 81, 102, 347–49
    justification via variational principle, 95–100